PRACTICAL MATLAB® APPLICATIONS FOR ENGINEERS

Handbook of Practical MATLAB® for Engineers

Practical MATLAB® Basics for Engineers

Practical MATLAB® Applications for Engineers

PRACTICAL MATLAB® FOR ENGINEERS

PRACTICAL MATLAB®
APPLICATIONS FOR
ENGINEERS

Misza Kalechman

Professor of Electrical and Telecommunication Engineering Technology

New York City College of Technology

City University of New York (CUNY)

CRC Press
Taylor & Francis Group
Boca Raton   London   New York

CRC Press is an imprint of the
Taylor & Francis Group, an **informa** business

This book was previously published by Pearson Education, Inc.

CRC Press
Taylor & Francis Group
6000 Broken Sound Parkway NW, Suite 300
Boca Raton, FL 33487-2742

© 2009 by Taylor & Francis Group, LLC
CRC Press is an imprint of Taylor & Francis Group, an Informa business

No claim to original U.S. Government works
Printed in the United States of America on acid-free paper
10 9 8 7 6 5 4 3 2 1

International Standard Book Number-13: 978-1-4200-4776-9 (Softcover)

Library of Congress Cataloging-in-Publication Data

Kalechman, Misza.
    Practical MATLAB applications for engineers / Misza Kalechman.
        p. cm.
    Includes bibliographical references and index.
    ISBN 978-1-4200-4776-9 (alk. paper)
    1. Engineering mathematics--Data processing. 2. MATLAB. I. Title.

TK153.K179 2007
620.001'51--dc22
                                                             2008000269

Visit the Taylor & Francis Web site at
http://www.taylorandfrancis.com

and the CRC Press Web site at
http://www.crcpress.com

Contents

Preface vii

Author ix

1 Time Domain Representation of Continuous and Discrete Signals 1
    1.1 Introduction 1
    1.2 Objectives 4
    1.3 Background 4
    1.4 Examples 58
    1.5 Application Problems 93

2 Direct Current and Transient Analysis 101
    2.1 Introduction 101
    2.2 Objectives 103
    2.3 Background 104
    2.4 Examples 138
    2.5 Application Problems 208

3 Alternating Current Analysis 223
    3.1 Introduction 223
    3.2 Objectives 224
    3.3 Background 226
    3.4 Examples 267
    3.5 Application Problems 310

4 Fourier and Laplace 319
    4.1 Introduction 319
    4.2 Objectives 321
    4.3 Background 322
    4.4 Examples 376
    4.5 Application Problems 447

5 DTFT, DFT, ZT, and FFT 457
    5.1 Introduction 457
    5.2 Objectives 458
    5.3 Background 459
    5.4 Examples 505
    5.5 Application Problems 556

6 Analog and Digital Filters 561
    6.1 Introduction 561
    6.2 Objectives 562
    6.3 Background 563
    6.4 Examples 599
    6.5 Application Problems 660

Bibliography 667

Index 671

Preface

Practical MATLAB® Applications for Engineers introduces the reader to the concepts of MATLAB® tools used in the solution of advanced engineering course work followed by engineering and technology students. Every chapter of this book discusses the course material used to illustrate the direct connection between the theory and real-world applications encountered in the typical engineering and technology programs at most colleges. Every chapter has a section, titled Background, in which the basic concepts are introduced and a section in which those concepts are tested, with the objective of exploring a number of worked-out examples that demonstrate and illustrate various classes of real-world problems and its solutions.

The topics include

- Continuous and discrete signals
- Sampling
- Communication signals
- DC (direct current) analysis
- Transient analysis
- AC (alternating current) analysis
- Fourier series
- Fourier transform
- Spectra analysis
- Frequency response
- Discrete Fourier transform
- Z-transform
- Standard filters
- IRR (infinite impulse response) and FIR (finite impulse response) filters

For product information, please contact
The MathWorks, Inc.
3 Apple Hill Drive
Natick, MA 01760-2098 USA
Tel: 508 647 7000
Fax: 508-647-7001
E-mail: info@mathworks.com
Web: www.mathworks.com

Author

Misza Kalechman is a professor of electrical and telecommunication engineering technology at New York City College of Technology, part of the City University of New York.

Mr. Kalechman graduated from the Academy of Aeronautics (New York), Polytechnic University (BSEE), Columbia University (MSEE), and Universidad Central de Venezuela (UCV; electrical engineering).

Mr. Kalechman was associated with a number of South American universities where he taught undergraduate and graduate courses in electrical, industrial, telecommunication, and computer engineering; and was involved with applied research projects, design of laboratories for diverse systems, and installations of equipment.

He is one of the founders of the Polytechnic of Caracas (Ministry of Higher Education, Venezuela), where he taught and served as its first chair of the Department of System Engineering. He also taught at New York Institute of Technology (NYIT); Escofa (officers telecommunication school of the Venezuelan armed forces); and at the following South American universities: Universidad Central de Venezuela, Universidad Metropolitana, Universidad Catolica Andres Bello, Universidad the Los Andes, and Colegio Universitario de Cabimas.

He has also worked as a full-time senior project engineer (telecom/computers) at the research oil laboratories at Petroleos de Venezuela (PDVSA) Intevep and various refineries for many years, where he was involved in major projects. He also served as a consultant and project engineer for a number of private industries and government agencies.

Mr. Kalechman is a licensed professional engineer of the State of New York and has written *Practical MATLAB for Beginners* (Pearson), *Laboratorio de Ingenieria Electrica* (Alpi-Rad-Tronics), and a number of other publications.

1

Time Domain Representation of Continuous and Discrete Signals

> This time, like all time, is a very good one, if we know what to do with it. Time is the most valuable and the most perishable of all possessions.
>
> **Ralph Waldo Emerson**

1.1 Introduction

Signals are physical variables that carry information about a particular process or event of interest. Signals are defined mathematically over a range and domain of interest, and constitute different things to different people.

To an electrical engineer, it may be

- A current
- A voltage
- Power
- Energy

To a mechanical engineer, it may be

- A force
- A torque
- A velocity
- A displacement

To an economist, it may be

- Growth (GNP)
- Employment rate
- Prime interest rate
- Inflation rate
- The stock market variations

To a meteorologist, it may be

- Atmospheric temperature
- Atmospheric humidity
- Atmospheric pressures or depressions
- Wind speed

To a geophysicist, it may be

- Seismic waves
- Tsunamis
- Volcanic activity

To a physician, it may be

- An electrocardiogram (EKG)
- An electroencephalogram (EEG)
- A sonogram

For a telecommunication engineer, it may be

- Audio sound wave (human voice or music)
- Video (TV, HDTV, teleconference, etc.)
- Computer data
- Modulated-waves (amplitude modulation [AM], frequency modulation [FM], phase modulation [PM], quadrature amplitude modulation [QAM], etc.)
- Multiplexed waves (time division multiplexing [TDM], statistical time division multiplexing [STDM], frequency division multiplexing [FDM], etc.)

From a block box diagram point of view, signals constitute inputs to a system, and their responses referred to as outputs. Since many of the measuring, recording, tracking, and processing instruments of signal activities are electrical or electronic devices, scientists and engineers usually convert any type of physical variations into an electrical signal.

Electrical signals can be classified using a variety of criteria. Some of the signal's classification criteria are

a. Signals may be functions of one or more than one independent variable generated by a single source or multiple sources.

b. Signals may be single or multidimensional.

c. Signals may be orthogonal or nonorthogonal, periodic or nonperiodic, even, odd, or present a particular symmetry.

d. Signals may be deterministic or nondeterministic (probabilistic).

e. Signals may be analog or discrete.

f. Signals may be narrow or wide band.

g. Signals may be power or energy signals.

In any case, signals are produced as a result of a process defined by a mathematical relation usually in the form of an equation, an algorithm, a model, a table, a plot, or a given rule.

A one-dimensional (1-D) signal is given by a mathematical expression consisting of one independent variable, for example, audio. A 2-D signal is a function of two independent variables, for example, a black and white picture. A full motion black and white video can be viewed as a 3-D signal, consisting of pictures (2-D) that are transmitted or processed at a particular rate. The dimension of a video signal can be increased by adding color (red, green, and blue), luminance, etc.

Deterministic and probabilistic signals is another broad way to classify signals. Deterministic signals are those signals where each value is unique, while nondeterministic signals are those whose values are not specified. They may be random or defined by statistical values such as noise. In this book, the majority of the signals are restricted to 1-D and 2-D, limited to one independent variable usually either time (t) or frequency (f or w), and

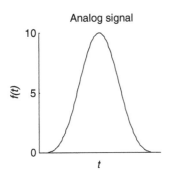
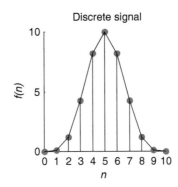

FIGURE 1.1
Analog and discrete signal representation.

deterministic such as current, voltage, power, or energy represented as vectors or matrices by MATLAB®.

In this book, following the widely accepted industrial standards, signals are classified in two broad categories

- Analog
- Discrete

Analog signals are signals capable of changing at any time. This type of signals is also referred as continuous time signals, meaning that continuous amplitude imply that the amplitude of the signal can take any value.

Discrete time signals, however, are signals defined at some instances of time, over a time interval $t \in [t_0, t_1]$. Therefore discrete signals are given as a sequence of points, also called samples over time such as $t = nT$, for $n = 0, \pm 1, \pm 2, \ldots, \pm N$, whereas all other points are undefined.

An analog or continuous signal is denoted by $f(t)$, whereas a discrete signal is represented by $f(nT)$ or in short without any loss of generality by $f(n)$, as indicated in Figure 1.1 by dots.

An analog signal $f(t)$ can be converted into a discrete signal $f(nT)$ by sampling $f(t)$ with a constant sampling rate T (a time also referred as T_s), where n is an integer over the range $-\infty < n < +\infty$ large but finite. Therefore a large, but finite number of samples also referred to as a sequence can be generated. Since the sampling rate is constant (T), a discrete signal can simply be represented by $f(nT)$ or $f(n)$, without any loss of information (just a scaling factor of T).

Continuous time systems or signals usually model physical systems and are best described by a set of differential equations. The analogous model for discrete models is described by a set of difference equations.

Signals that occur in nature are usually analog, but if a signal is processed by a computer or any digital device the continuous signal must be converted to a discrete sequence (using an analog to digital converter, denoted by A/D), or mathematically by a finite sequence of numbers that represent its amplitude at the sampling instances.

Discrete signals take the value of the continuous signals at equally spaced time intervals (nT). Those values can be considered an ordered sequence, meaning that the discrete signal represents mathematically the sequence: $f(0), f(1), f(2), f(3), \ldots, f(n)$.

The spacing T between consecutive samples of $f(t)$ is called the sampling interval or the sampling period (also referred to as T_s).

1.2 Objectives

After completing this chapter the reader should be able to

- Mathematically define the most important analog and discrete signals used in practical systems
- Understand the sampling process
- Understand the concept of orthogonal signal
- Define the most widely used orthogonal signal families
- Understand the concepts of symmetric and asymmetric signals
- Understand the concept of time and amplitude scaling
- Understand the concepts of time shifting, reversal, compression, and expansion
- Understand the reconstruction process involved in transforming a discrete signal into an analog signal
- Compute the average value, power, and energy associated with a given signal
- Understand the concepts of down-, up-, and resampling
- Define the concept of modulation, a process used extensively in communications
- Define the multiplexing process, a process used extensively in communications
- Relate mathematically the input and output of a system (analog or digital)
- Define the concept and purpose of a window
- Define when and where a window function should be used
- Define the most important window functions used in system analysis
- Use the window concept to limit or truncate a signal
- Model and generate different continuous as well as discrete time signals, using the power of MATLAB

1.3 Background

R.1.1   The sampling or Nyquist–Shannon theorem states that if a continuous signal $f(t)$ is band-limited* to f_m Hertz, then by sampling the signal $f(t)$ with a constant period $T \leq [1/(2.f_m)]$, or at least with a sampling rate of twice the highest frequency of $f(t)$, the original signal $f(t)$ can be recovered from the equally spaced samples $f(0)$, $f(T)$, $f(2T)$, $f(3T)$, ..., $f(nT)$, and a perfect reconstruction is then possible (with no distortion).

The spacing T (or T_s) between two consecutive samples is called the sampling period or the sampling interval, and the sampling frequency F_s is defined then as $F_s = 1/T$.

R.1.2   By passing the sampling sequence $f(nT)$ through a low-pass filter* with cutoff frequency f_m, the original continuous time function $f(t)$ can be reconstructed (see Chapter 6 for a discussion about filters).

---

* The concepts of band-limit and filtering are discussed in Chapters 4 and 6. At this point, it is sufficient for the reader to know that by sampling an analog function using the Nyquist rate, a discrete function is created from the analog function, and in theory the analog signal can be reconstructed, error free, from its samples.

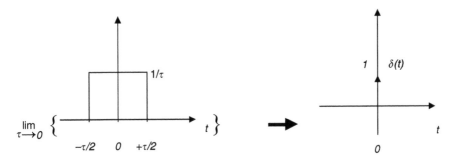

FIGURE 1.2
The impulse function $\delta(t)$.

R.1.3   Analytically, the sampling process is accomplished by multiplying $f(t)$ by a sequence of impulses. The concept of the unit impulse $\delta(t)$, also known as the Dirac function, is introduced and discussed next.

R.1.4   The unit impulse, denoted by $\delta(t)$, also known as the Dirac or the Delta function, is defined by the following relation:

$$\int_{-\infty}^{+\infty} \delta(t)dt = 1$$

meaning that the area under $[\delta(t)] = 1$,

where

$$\delta(t) = 0, \quad \text{for } t \neq 0$$

$$\delta(t) = 1, \quad \text{for } t = 0$$

$\delta(t)$ is an even function, that is, $\delta(t) = \delta(-t)$.

The impulse function $\delta(t)$ is not a true function in the traditional mathematical sense. However, it can be defined by the following limiting process:

by taking the limit of a rectangular function with an amplitude $1/\tau$ and width τ, when τ approaches zero, as illustrated in Figure 1.2.

The impulse function $\delta(t)$, as defined, has been accepted and widely used by engineers and scientists, and rigorously justified by an extensive literature referred as the generalized functions, which was first proposed by Kirchhoff as far back as 1882. A more modern approach is found in the work of K.O. Friedrichs published in 1939. The present form, widely accepted by engineers and used in this chapter is attributed to the works of S.L. Sobolov and L. Swartz who labeled those functions with the generic name of distribution functions.

Teams of scientists developed the general theory of generalized (or distribution) functions apparently independent from each other in the 1940s and 1950s, respectively.

R.1.5   Observe that the impulse function $\delta(t)$ as defined in R.1.4 has zero duration, undefined amplitude at $t = 0$, and a constant area of one. Obviously, this type of function presents some interesting properties when analyzed at one point in time, that is, at $t = 0$.

R.1.6   Since $\delta(t)$ is not a conventional signal, it is not possible to generate a function that has exactly the same properties as $\delta(t)$. However, the *Dirak* function as well as its derivatives $(d\delta(t)/dt)$ can be approximated by different mathematical models.

        Some of the approximations are listed as follows (Lathi, 1998):

$$\delta(t) = \text{limit}_{a \to 0} \left\{ \frac{1}{a} \left[\frac{\sin(\pi t/a)}{\pi t/a} \right] \right\} \quad \text{(using } \sin c\text{'s)}$$

$$\delta(t) = \text{limit}_{a \to 0} \left[\frac{e^{jt/a}}{\pi jt} \right] \quad \text{(using exponentials)}$$

$$\delta(t) = \text{limit}_{a \to 0} \left\{ \frac{e^{-t^2/4a^2}}{a\sqrt{\pi}} \right\} \quad \text{(using Gaussian)}$$

R.1.7   Multiplying a unit impulse $\delta(t)$ by a constant A changes the area of the impulse to A, or the amplitude of the impulse becomes A.

R.1.8   The impulse function $\delta(t)$ when multiplied by an arbitrary function $f(t)$ results in an impulse with the magnitude of the function evaluated at $t = 0$, indicated by

$$\delta(t)f(t) = f(0)\,\delta(t)$$

Observe that $f(0)\,\delta(t)$ can be defined as

$$f(0)\delta(t) = \begin{cases} 0 & \text{for } t \neq 0 \\ f(0) & \text{for } t = 0 \end{cases}$$

R.1.9   A shifted impulse $\delta(t - t_1)$ is illustrated in Figure 1.3. When the shifted impulse $\delta(t - t_1)$ is multiplied by an arbitrary function $f(t)$, the result is given by

$$\delta(t - t_1) \cdot f(t) = f(t_1) \cdot \delta(t - t_1)$$

R.1.10 The derivative of the unit *Dirak* $\delta(t)$ is called the unit doublet, denoted by $\dfrac{d[\delta(t)]}{dt} = \delta'(t)$, is illustrated in Figure 1.4.

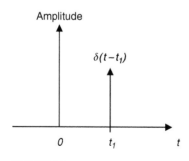

FIGURE 1.3
Plot of $\delta(t - t_1)$.

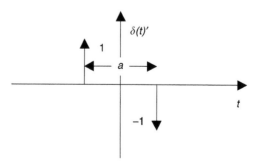

FIGURE 1.4
Plot of $\delta(t)'$ as a approaches zero.

R.1.11 Figure 1.4 indicates that the unit doublet cannot be represented as a conventional function since there is no single value, finite or infinite, that can be assigned to $\delta(t)'$ at $t = 0$.

R.1.12 Additional useful properties of the impulse function $\delta(t)$ that can be easily proven are stated as follows:

a. $\int_{-\infty}^{+\infty} f(t)\delta(t - t_o)dt = f(t_o)$

b. $\int_{-\infty}^{+\infty} f(t - t_o)\delta(t)dt = f(-t_o)$

c. $\int_{-\infty}^{+\infty} f(t - t_o)\delta(t - t_1)dt = f(t_1 - t_0)$

R.1.13 The unit impulse $\delta(t)$, the unit doublet $\delta(t)'$, and the higher derivatives of $\delta(t)$ are often referred as the impulse family. These functions vanish at $t = 0$, and they all have the origin as the sole support. At $t = 0$, all the impulse functions suffer discontinuities of increasing complexity, consisting of a series of sharp pulses going positive and negative depending on the order of the derivative.

As was stated $\delta(t)$ is an even function of t, and so are all its even derivatives, but all the odd derivatives of $\delta(t)$ return odd functions of t.

The preceding statement is summarized as follows:

$$\delta(t) = \delta(-t), \quad \delta'(t) = -\delta(-t)$$

or in general

$$\delta^{(2n)}(t) = \delta^{2n}(-t) \quad \text{(even case)}$$

$$\delta^{2n-1}(t) = -\delta^{2n-1}(-t) \quad \text{(odd case)}$$

R.1.14 A train of impulses denoted by the function $Impl(t)_T]$ defines a sequence consisting of an infinite number of impulses occurring at the following instants of time nT, ..., $-T$, T, $2T$, $3T$, ..., nT, as n approaches ∞. This sequence can be expressed analytically by

$$Impl[(t)_T] = \sum_{n=-\infty}^{\infty} \delta(t - nT)$$

illustrated in Figure 1.5.

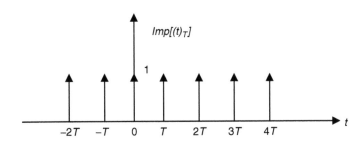

FIGURE 1.5
Plot of $Impl(t)_T]$.

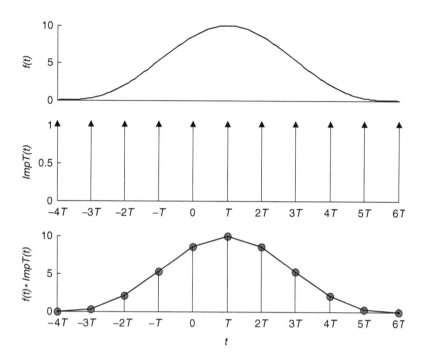

FIGURE 1.6
Plots illustrating the sampling process.

The expansion of the function $Imp[(t)_T]$ results in

$$Imp[(t)_T] = \sum_{n=-\infty}^{\infty} \delta(t - nT)$$

$$= \delta(t + nT) \cdots \delta(t + T) + \delta(t) + \delta(t - T) + \cdots + \delta(t - nT)$$

R.1.15 The sampling process is modeled mathematically by multiplying an arbitrary analog signal $f(t)$ by the train of impulses defined by $Imp[(t)_T]$. This process is illustrated graphically in Figure 1.6.

Analytically,

$$f(t) \cdot \sum_{n=-\infty}^{\infty} \delta(t - nT) = \sum_{n=-\infty}^{\infty} f(t)\delta(t - nT)$$

and the expanded discrete version of $f(t)$ is given by

$$f(n) = \sum_{k=-\infty}^{\infty} f(k)\delta(t - k) = \cdots + f(-2)\delta(t + 2) + f(-1)\delta(t + 1) + f(0)\delta(t)$$

$$+ f(1)\delta(t - 1) + f(2)\delta(t - 2) + \cdots + f(n)\delta(t - n)$$

assuming that $T = 1$, without any loss of generality.

In general, the set of samples given by $f(-n), f(-n+1), ..., f(0), f(1), ..., f(n-1), f(n)$, can be real or complex. $f(n)$ is called a real sequence if all its samples are real and a complex sequence if at least one sample is complex.

Observe that any (discrete) sequence $f(n)$ can be expressed by the equation

$$f(n) = \sum_{k=-\infty}^{\infty} f(k)\delta(n-k)$$

Examples of analog signals are often encountered in nature such as sound, temperatures, pressure, growth, and precipitations waves.

Discrete time signals or events are usually man-made functions such as weekly pay, monthly payment of a loan or mortgage, or the (U.S.) presidential election every 4 years.

Discrete signals are often confused with digital signals and binary signals. A digital signal $f(nT)$ or in short $f(n)$ is a discrete time signal whose values are one of a predefined finite set of values.

A binary signal is a discrete signal whose values consist of either zeros or ones.

An analog or continuous time function or signal can be transformed into a digital signal using an A/D. Conversely, a digital signal can be converted into an analog signal by means of a digital to analog converter (D/A).

Digital signals are frequently encoded using binary codes such as ASCII* into strings of ones and zeros because in this format they can be stored and processed by digital devices such as computers, and are in general more immune to noise and interference.

R.1.16 The discrete impulse sequence $\delta(n)$ also called the *Kronecker delta* sequence (named after the German mathematician Leopold Kronecker [1823–1891]) is defined analytically as follows and illustrated in Figure 1.7.

$$\delta(n) = \begin{cases} 1 & \text{for } n = 0 \\ 0 & \text{for } n \neq 0 \end{cases}$$

Note that the discrete impulse is similar to the analog version $\delta(t)$.

R.1.17 A discrete shifted impulse $\delta(n-m)$ is illustrated in Figure 1.8.

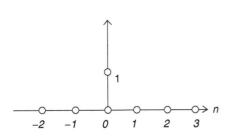

FIGURE 1.7
Plot of the discrete impulse $\delta(n)$.

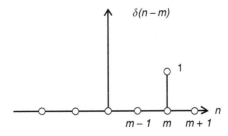

FIGURE 1.8
Plot of $\delta(n-m)$.

* The ASCII code is defined in Chapter 3 of *Practical MATLAB® Basics for Engineers*.

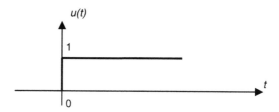

FIGURE 1.9
Plot of the step function $u(t)$.

The discrete shifted impulse function $\delta(n - m)m - 1 \cdots m \cdots m + 1n$ is defined as

$$\delta(n - m) = \begin{cases} 1 & \text{for } n = m \\ 0 & \text{for } n \neq m \end{cases}$$

R.1.18 Let us go back to the analog world. The analog unit step function denoted by $u(t)$ is illustrated in Figure 1.9.
Analytically, the analog unit step is defined by

$$u(t) = \begin{cases} 1 & \text{for } t \geq 0 \\ 0 & \text{for } t < 0 \end{cases}$$

The step is related to the impulse by the following relations:

$$\frac{du(t)}{dt} = \delta(t)$$

or in general

$$\frac{d}{dt}[u(t - t_o)] = \delta(t - t_o)$$

$$\int_{-\infty}^{\infty} \delta(t)dt = u(t)$$

or in general

$$u(t - t_o) = \int_{-\infty}^{t} \delta(\tau - t_o)d\tau = \begin{cases} 1 & \text{for } t > t_0 \\ 0 & \text{for } t < t_0 \end{cases}$$

The derivative of the unit step constitutes a break with the traditional differential and integral calculus. This new approach to the class of functions called singular functions is referred to as generalized or distributional calculus (mentioned in R.1.4).

R.1.19 The analog unit step $u(t)$ can be implemented by a switch connected to a voltage source of 1 V that closes instantaneously at $t = 0$, illustrated in Figure 1.10.

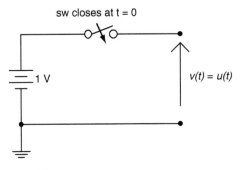

FIGURE 1.10
Circuit implementation of *u(t)*.

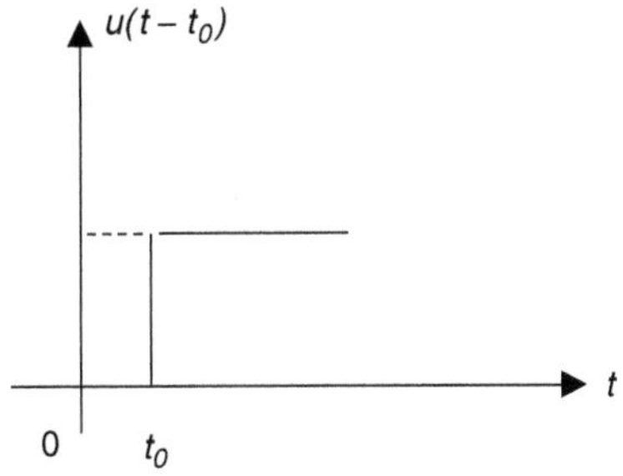

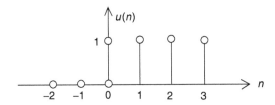

FIGURE 1.11
Plot of $u(t - t_0)$.

FIGURE 1.12
Plot of *u(n)*.

R.1.20 A right-shifted unit step, by t_0 units, denoted by $u(t - t_0)$ is illustrated in Figure 1.11. The shifted step function $u(t - t_0)$ is defined analytically by

$$u(t - t_0) = \begin{cases} 1 & \text{for } t \geq t_0 \\ 0 & \text{for } t < t_0 \end{cases}$$

R.1.21 A unit step sequence or the discrete unit step *u(n)* is illustrated in Figure 1.12. The unit discrete step *u(n)* is defined analytically by

$$u(n) = \begin{cases} 1 & \text{for } n \geq 0 \\ 0 & \text{for } n < 0 \end{cases}$$

R.1.22 A unit discrete step sequence *u(n)* can be constructed by a sequence of impulses indicated as follows:

$$u(n) = \sum_{k=0}^{\infty} \delta(n - k)$$

Observe that $\delta(n) = u(n) - u(n - 1)$.

R.1.23 A shifted and amplitude-scaled step sequence, $A\,u(n - m)$ is illustrated in Figure 1.13. The sequence $A\,u(n - m)$ is defined analytically by

$$A u(n - m) = \begin{cases} A & \text{for } n \geq m \\ 0 & \text{for } n < m \end{cases}$$

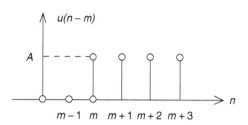

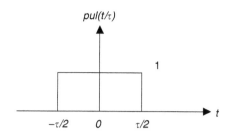

FIGURE 1.13
Plot of $u(n - m)$.

FIGURE 1.14
Plot of $pul(t/\tau)$.

R.1.24 The analog pulse function $pul(t/\tau)$ is illustrated graphically in Figure 1.14. The function $pul(t/\tau)$ is defined analytically by

$$pul(t/\tau) = \begin{cases} 1 & \text{for } -\tau/2 \leq t \leq \tau/2 \\ 0 & \text{for } -\tau/2 > t \quad \text{and} \quad t > \tau/2 \end{cases}$$

R.1.25 The analog pulse $pul(t/\tau)$ is related to the analog step function $u(t)$ by the following relation:

$$pul(t/\tau) = u(t + \tau/2) - u(t - \tau/2)$$

R.1.26 The discrete pulse sequence denoted by $pul(n/N)$ is given by

$$pul(n/N) = \begin{cases} 1 & \text{for } -N/2 \leq n \leq N/2 \\ 0 & \text{for } -N/2 > n \quad \text{and} \quad n > N/2 \end{cases}$$

For example, for $N = 11$ (odd), the discrete sequence is given by

$$pul(n/11) = \begin{cases} 1 & \text{for } -5 \leq n \leq 5 \\ 0 & \text{for } -5 > n \quad \text{and } n > 5 \end{cases}$$

The preceding function $pul(n/11)$ is illustrated in Figure 1.15.
Observe that the pulse function $pul(n/11)$ can be represented by the superposition of two discrete step sequences as

$$pul(n/11) = u(n + 5) - u(n - 6)$$

R.1.27 The analog unit ramp function denoted by $r(t) = t\,u(t)$ is illustrated in Figure 1.16. The unit ramp is defined analytically by

$$r(t) = \begin{cases} t & \text{for } t \geq 0 \\ 0 & \text{for } t < 0 \end{cases}$$

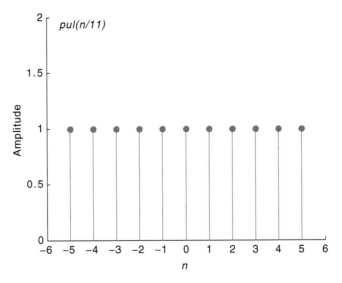

FIGURE 1.15
Plot of the function *pul(n/11)*.

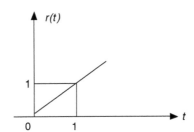

FIGURE 1.16
Plot of the analog unit ramp function $r(t) = t\,u(t)$.

R.1.28 The more general analog ramp function $r(t) = t\,u(t)$ (with time and amplitude scaled) is defined by

$$Ar(t - t_0) = \begin{cases} At & \text{for } t \geq t_0 \\ 0 & \text{for } t < t_0 \end{cases}$$

where A represents the ramp's slope and t_0 is the time shift with respect to the origin.

R.1.29 The discrete ramp sequence denoted by $r(n - m)u(n - m)$ is defined analytically by

$$r(n - m)u(n - m) = \begin{cases} n & \text{for } n \geq m \\ 0 & \text{for } n < m \end{cases}$$

R.1.30 The analog unit parabolic function $p_K(t)\,u(t)$ is defined by

$$p_K(t)u(t) = \begin{cases} \dfrac{1}{K!}t^K & \text{for } t \geq 0 \quad \text{and} \quad K = 2, 3, \ldots \\ 0 & \text{for } t < 0 \end{cases}$$

R.1.31 The discrete unit parabolic function $p_K(n)\,u(n)$ is defined by

$$p_K(n)u(n) = \begin{cases} \dfrac{1}{K!}n^K & \text{for } n \leq 0 \quad \text{and} \quad K = 2, 3, \ldots \\ 0 & \text{for } n < 0 \end{cases}$$

R.1.32 Observe that
   a. The unit ramp presents a sharp 45° corner at $t = 0$.
   b. The unit parabolic function presents a smooth behavior at $t = 0$.
   c. The unit step presents a discontinuity at $t = 0$.

R.1.33 The step, ramp, and parabolic functions are related by derivatives as follows:
   a. $(d/dt)[r(t)] = u(t)$
   b. $\dfrac{d}{dt}[p_2(t)] = r(t)u(t)$
   c. $(d/dt)[p_a(t)] = p_{a-1}(t)$

   Observe that the first relation makes sense for all $t \neq 0$, since at $t = 0$ a discontinuity occurs, whereas the second and third relations hold for all t.

R.1.34 Note that, in general, the product $f(t)$ times $u(t)\,[f(t)u(t)]$ defines the composite function given by

$$f(t)u(t) = \begin{cases} f(t) & \text{for } t \geq 0 \\ 0 & \text{for } t < 0 \end{cases}$$

R.1.35 A wide class of engineering systems employ sinusoidal and exponential* signals as inputs. A real exponential analog signal is in general given by

$$f(t) = Ae^{bt}$$

where $e = 2.7183$ (Neperian constant) and A and b are in most cases real constants. Observe that for $f(t) = Ae^{bt}$,

a. $f(t)$ is a decaying exponential function for $b < 0$.

b. $f(t)$ is a growing exponential function for $b > 0$.

   The coefficient b as exponent is referred to as the damping coefficient or constant. In electric circuit theory, the damping constant is frequently given by $b = 1/\tau$, where τ is referred as the time constant of the network (see Chapter 2).

   Note that the exponential function $f(t) = Ae^{bt}$ repeats itself when differentiated or integrated with respect to time, and constitutes the homogeneous solution of

---

* Recall that sinusoids are complex exponentials (Euler), see Chapter 4 of *Practical MATLAB® Basics for Engineers*.

the system differential equation.* Note also that when *b* is complex, then by Euler's equalities *f(t)* presents oscillations.

Finally, the more common equation $f(t) = Ae^{-bt}u(t)$ is defined analytically by

$$f(t) = \begin{cases} Ae^{-bt} & \text{for } t \geq 0 \\ 0 & \text{for } t < 0 \end{cases}$$

R.1.36 An exponential sequence can be defined by $f(n) = Aa^n$, for $-\infty \leq n \leq \infty$, where *a* can be a real or complex number.

R.1.37 The following example illustrates the form of an exponential function for various values for *A* and *b*.

Let us explore the behavior of the exponential function, by creating the script file *exponentials* that returns the following plots:

a. $f_1(t) = 4e^{(-t/2)}$

b. $f_2(t) = 4e^{(t/2)}$

c. $f_3(t) = 4e^{(-t/2)}u(t)$

d. $f_4(t) = -4e^{(t/2)}u(t)$

for $A = \pm 4$ and $b = \pm 1/2$, over the range $-3 \leq t \leq 3$, using 61 elements.

MATLAB Solution
```
% Script file: exponentials
t = -3:.1:3;
ft1 = 4*exp(-t./2);
ft2 = 4*exp(t./2);
ut _ 1 = [zeros(1,30) ones(1,31)];          % step with 61 elements
ft3 = ft1.*ut _ 1;
subplot(2,2,1);
plot(t, ft1);
axis([-3 3 -.5 18]);xlabel('t (time)')
title('f1(t)=4*exp(-t/2) vs. t');
ylabel('Amplitude [f1(t)]')
subplot(2,2,2);
plot(t, ft2); xlabel('t (time)')
axis([-3 3 0 20]);
title('f2(t) = 4*exp(t/2) vs. t');
ylabel('Amplitude [f2(t)]');
subplot(2,2,3);
plot(t, ft3);
axis([-3 3 -.5 5]);
title('f3(t) =4*exp(-t/2)*u(t) vs. t');
xlabel('t (time)')
ylabel('Amplitude[f3(t)]');
subplot(2,2,4);
ft4 =-1.*ft2.*ut _ 1;
plot(t, ft4);
axis([-3 3 -20 1]);
title('f4(t) = -4*exp(-t/2)*u(t) vs. t')
xlabel('t (time)'); ylabel('Amplitude[f4(t)]');
```

The script file *exponentials* is executed and the results are shown in Figure 1.17.

---

* See Chapter 7, *Practical MATLAB® Basics for Engineers*.

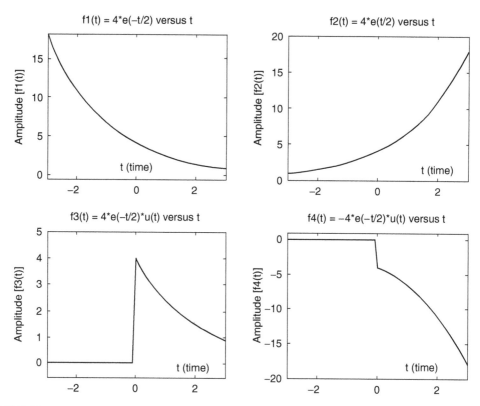

FIGURE 1.17
Plots of $f_1(t)$, $f_2(t)$, $f_3(t)$, and $f_4(t)$ of R.1.37.

R.1.38 A real exponential discrete sequence is defined by the equation of the form

$$f(n) = ac^n$$

where a and c are real constants.

Observe that the sequence given by $f(n)$ can converge or diverge depending on the value of c (less than or greater than one).

R.1.39 Recall that the general sinusoidal (analog) function is given by

$$f(t) = A\cos(\omega t + \alpha)$$

where A represents its amplitude (real value); ω is referred as the angular frequency and is given in radian/second; $\omega = 2\pi f$, where f is its frequency in hertz, or cycles per second ($f = 1/T$); and α is referred as the phase shift in radians or degrees (2π rad $= 360°$) (see Chapter 4 of the book titled *Practical MATLAB® Basics for Engineers* for additional details).

R.1.40 Recall that sinusoidal and exponential functions are related by Euler's identities; introduced and discussed in Chapter 4 of the book titled *Practical MATLAB® Basics for Engineers*, and repeated as follows:

$$e^{jwt} = \cos(wt) + j\sin(wt)$$

R.1.41 A sinusoidal discrete sequence is defined by the following equation:

$$f(n) = A\cos(2\pi n/N + \alpha)$$

where A is a real number and represents its amplitude, N the period given by an integer, α the phase angle in radians or degrees, and $2\pi/N$ its angular frequency in radians.

R.1.42 Clearly, a discrete time sequence may or may not be periodic. A discrete sequence is periodic if $f(n) = f(n + N)$, for any integer n, or if $2\pi/N$ can be expressed as $r\pi$, where r is a rational number.

R.1.43 For example, $cos(3n)$ is not a periodic sequence since $3 = r\pi$, and clearly r cannot be a rational number. On the other hand, consider the sequence $cos(0.2\pi n)$, that is periodic since $0.2\pi = r\pi$ or $r = 0.2 = 2/10$, where r is clearly a rational number, then the period is given by $N = 2\pi/0.2\pi$ or $N = 10$.

R.1.44 Observe that for the case of a continuous time sinusoidal function of the form $f(t) = A\,cos(w_o t)$, $f(t)$ is always periodic, with period $T = 2\pi/w_o$, for any w_o.

R.1.45 The most important signal, among the standard signals used in circuit analysis, electrical networks, and linear systems, in general, is the sinusoidal wave, in either of the following forms:

$$f(t) = sin(wt)$$

$$f(t) = cos(wt)$$

or most effective as a complex wave

$$f(t) = e^{jwt} = cos(wt) + j\,sin(wt) \quad \text{(Euler's identity)}$$

R.1.46 Let $f_n(t)$ be the family of exponential signals of the form

$$f_n(t) = e^{jwnt}$$

where

$$wn = nw_0, \quad \text{for } n = 0, \pm 1, \pm 2, \ldots, \pm\infty$$

where w_0 is called the fundamental frequency, wn's are called its harmonic frequencies (see Chapter 4, where $w_0 = 2\pi/T$). This family possesses the property called orthogonal, which means that the following integral over a period shown for the products of any two members of the family is either zero or a constant given by $2\pi/w_0$

$$\int_{-\pi/w_o}^{\pi/w_o} f_n(t) \cdot f_m{}^*(t)\,dt = \begin{cases} 2\pi/w_o & \text{for } n = m \\ 0 & \text{for } n \neq m \end{cases}$$

where $f_m(t)^*$ denotes the complex conjugate of $f_m(t)$. For example, if $f_m(t) = e^{jwnt}$, then $f_m(t)^* = e^{-jwnt}$. For the special case in which the orthogonal constant is one, the family is called orthonormal.

R.1.47 There are a number of orthonormal families. Some of the most frequently used orthonormal families in system analysis are

a. *Hermite*

b. *Laguerre*

c. *sinc* (where $sinc_n(t) = sin(t - n\pi)/[\pi(t - n\pi)]$)

R.1.48 The *Hermitian* orthonormal family of signals are generated starting from the *Gaussian* signal

$$Her_0 = e^{-[t^2/4]}$$

and all other members are generated by successive differentiations with respect to t.

The first members of the *Hermitian* family are indicated as follows:

$$Her_1 = te^{-t^{\wedge}2/4}$$

$$Her_2 = (t^2 - 1)e^{-t^{\wedge}2/4}$$

$$Her_3 = (t^3 - 3t)e^{-t^{\wedge}2/4}$$

$$Her_4 = (t^4 - 6t^2 + 3)e^{-t^{\wedge}2/4}$$

R.1.49 The polynomial factors in the expressions defined by Her_n are referred as the *Hermitian* polynomials, and the orthogonal interval is over the range $-\infty \leq t \leq +\infty$. The script file *Hermite*, given as follows, returns the plots of the first five members of the Hermite's family, over the range $-5 \leq t \leq +5$, in Figure 1.18.

MATLAB Solution
```
% Script file: Hermite
t =-5:.1:5;
Her _ 0 = exp(-t.^2./4);
Her _ 1 = t.*exp(-t.^2./4);
Her _ 2 = (t.^2-1).*exp(-t.^2./4);
Her _ 3 = (t.^3-3.*t).*exp(-t.^2./4);
Her _ 4 = (t.^4-6*t.^2+3).*exp(-t.^2./4);
plot(t,Her _ 0,'*:',t,Her _ 1,'d-.',t,Her _ 2,'h--',t,Her _ 3,'s-',t,Her _ 4,'p:')
xlabel('time')
ylabel(' Amplitude')
title('First five members of the Hermite family')
legend('Her 0','Her 1','Her 2','Her 3','Her 4')
```

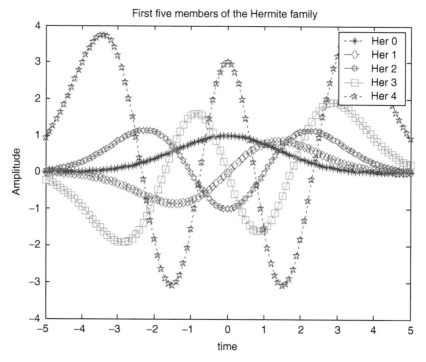

FIGURE 1.18
(See color insert following page 374.) Plots of the Hermite family of R.1.49.

R.1.50 The Laguerre orthonormal family of signals are generated starting from the function given by

$$Lag_0 = e^{-t/2} \quad \text{for } t > 0$$

and by successive differentiations with respect to t the other members of the family are generated, indicated as follows:

$$Lag_1 = (1 - t)e^{-t/2}$$

$$Lag_2 = (1 - 2t + 0.5t^2)e^{-t/2}$$

$$Lag_3 = (1 - 3t + 0.67t^2 - 0.166t^3)e^{-t/2}$$

R.1.51 The polynomial factors in the expressions shown in R.1.50 are referred as the Laguerre's polynomials, over the orthogonal interval given by $0 \le t \le +\infty$. The script file Laguerre returns the plots of the first four members of the Laguerre's family, over the range $0 \le t \le 5$, are shown in Figure 1.19.

```
% Script file: Laguerre
t = 0:.1:15;
Lag _ 0 = exp(-t./2);
Lag _ 1 = (1-t).*t.*exp(-t./2);
Lag _ 2 = (1-2.*t+.5.*t.^2).*exp(-t./2);
Lag _ 3 = (t.^3-3.*t).*exp(-t.^2./4);
plot(t,Lag _ 0,'*:',t,Lag _ 1,'d-.',t,Lag _ 2,'h--',t,Lag _ 3,'s-')
xlabel ('time')
ylabel ('Amplitude')
title('First four members of the Laguerre family')
legend ('Lag 0','Lag 1','Lag 2','Lag 3')
```

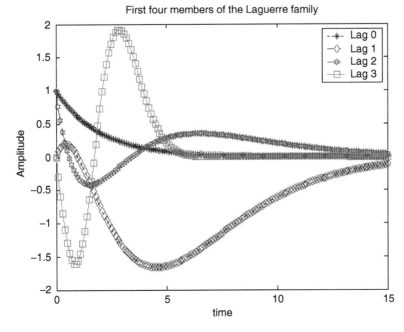

FIGURE 1.19
(See color insert following page 374.) Plots of the Laguerre family of R.1.51.

R.1.52 The family of time-shifted *sinc* functions are given by

$$sinc_n(t) = sin(t - n\pi)/[\pi(t - n\pi)]$$

for $n = 0, \pm1, \pm2, ..., \pm\infty$ forms and orthonormal family, over the range $-\infty \leq t \leq +\infty$, and are referred to as the *sinc* family.

R.1.53 The first five members of the *sinc* family are given as follows:

$$sinc_0(t) = sin(t)/[\pi t]$$

$$sinc_1(t) = sin(t - \pi)/[\pi(t - \pi)]$$

$$sinc_2(t) = sin(t - 2\pi)/[\pi(t - 2\pi)]$$

$$sinc_3(t) = sin(t - 3\pi)/[\pi(t - 3\pi)]$$

$$sinc_4(t) = sin(t - 4\pi)/[\pi(t - 4\pi)]$$

The script file *sinc_n*, shown as follows, returns the plot of the first four members of the *sinc* family, over the range $-5\pi \leq t \leq +5\pi$, indicated in Figure 1.20.

MATLAB Solution
```
% Script file: sinc _ n
t =-5*pi: 0.25:5*pi;
Sinc _ 0 = sin(t) ./ (pi*t);
Sinc _ 1 = sin(t- pi ) ./ (pi*(t-pi));
Sinc _ 2 = sin(t- 2*pi ) ./ (pi*(t-2*pi));
Sinc _ 3 = sin(t-3*pi ) ./ (pi*(t-3*pi));
plot(t,Sinc _ 0,'*:',t,Sinc _ 1,'d-.',t,Sinc _ 2,'h--',t,Sinc _ 3,'s-')
xlabel('time')
ylabel('Amplitude')
title('First four members of the sinc family')
legend('sinc 0','sinc 1','sinc 2','sinc 3')
```

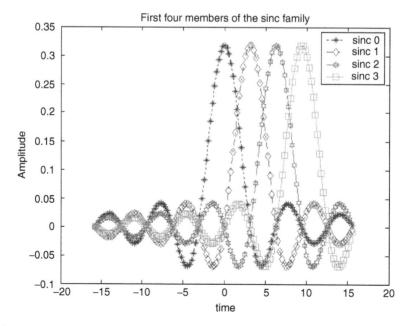

FIGURE 1.20
(See color insert following page 374.) Plots of the *sinc* family of R.1.53.

R.1.54 Another important class of signals are the signals used in the transmission and processing of information such as voice, data, and video, referred to as telecom (telecommunication) signals. Telecom signals are, broadly speaking, composed of

a. Modulated signals

b. Multiplex signals

R.1.55 An exponential analog-modulated sinusoidal signal is given by

$$f(t) = Ae^{at}\cos(\omega t + \alpha)$$

where the sinusoidal term is called the carrier, and the exponential Ae^{at} is called the envelope of the carrier that can represent a message or in general information.

R.1.56 An exponential discrete modulated sinusoid sequence is defined by

$$f(n) = ac^n\cos(2\pi n/N + \alpha)$$

Recall that a, c, N, and α were defined in R.1.38 and R.1.41.

R.1.57 Modulated signals are used extensively by electrical, telecommunication, computer, and information system engineers to deliver and process information.

The modulation process involves two signals referred as

a. The carrier (a high-frequency sinusoidal)

b. The information signal (i.e., the message that can be audio, voice, data, or video)

R.1.58 The modulation process is accomplished by varying one of the variables that defines the carrier (amplitude, frequency, or phase) in accordance with the instantaneous changes in the information signal. Information such as music or voice (audio) consists typically of low frequencies and is referred to as a base band signal. Base band signals cannot be transmitted in its maiden form because of physical limitations due to the distances involved in the transmission path, such as attenuation. Hence, to obtain an economically viable system that can, in addition, support a number of additional information channels (multiplexing), the information signal has to be boosted to higher frequencies through the modulation process.

R.1.59 AM is the process in which the amplitude of the high-frequency carrier is varied in accordance with the instantaneous variations of the information signal. This process is accomplished by multiplying the high-frequency carrier by the low-frequency component of the information signal.

AM signals present a constant frequency (which corresponds to the carrier's frequency) and phase variation.

A special type of AM signal used in the transmission of digital information is the amplitude shift keying (ASK) signals also known as on-off keying (OOK).

R.1.60 FM is a type of modulation in which the frequency of the carrier is varied in accordance with the instantaneous variations of the amplitude of the information signal. These types of signals present a constant magnitude and phase.

A special type of FM signal used in the transmission of digital information is the frequency shift keying (FSK) signal employed to modulate information that is originated from digital sources such as computers.

Modulators and demodulators used to transmit digital information are referred to as modems that stand for modulator–demodulator.

R.1.61 PM is a technique in which the phase of the carrier signal is varied in accordance with the instantaneous changes of the information signal.

Phase shift keying (PSK) is a special case of PM signals, in which the phase of the analog high-frequency carrier is varied in accordance with the information signal that is digital in nature. PM and FM are commonly referred to as angle modulation (for obvious reasons).

R.1.62 AM is also referred as linear modulation. It is a modulation technique that is bandwidth efficient. The bandwidth requirements vary between BW and 2 BW, where BW refers to the bandwidth of the information signal or message $m(t)$.* It is inefficient as far as power is concerned and its performance is poor in the presence of noise (compared with angle modulation FM or PM). AM is widely used in commercial broadcasting systems such as radio and TV, and in point-to-point communication systems.

R.1.63 Angle modulation (FM or PM) is commonly referred to as nonlinear modulation, and its most important characteristics are

- High BW requirements
- Good performance in the presence of noise
- High fidelity

Angle modulation is used in commercial broadcasting such as radio and TV with a superior quality of the reception of the information signal $m(t)$, compared with AM.

R.1.64 The time domain representation of the analog modulation signals is presented as follows:

a. *AM signal* $= A\,m(t)\cos(w_c t)$

b. *FM signal* $= A\cos\left[w_c t + 2\pi k_F \int_{-\infty}^{t} m(k)dk\right]$

c. *PM signal* $= A\,\cos[w_c t + 2\pi k_P m(t)]$

where w_c denotes the high-frequency carrier, $m(t)$ refers to the information or message signal, A represents the carrier amplitude, and k_P and k_F are constants that represent deviations.

R.1.65 Signals or sequences can be left- or right-sided.

a. A right-sided or causal sequence (or signal)[†] is defined by $f(n) = 0$, for $n < 0$.

b. A left-sided or noncausal sequence (or signal) is defined by $f(n) = 0$, for $n > 0$.

c. A two-sided sequence (or signal) is defined for all n $(-\infty < n < +\infty)$.

R.1.66 A symmetric or even function (or sequence) is defined by

$$f(t) = f(-t) \quad \text{(analog case)}$$

and

$$f(n) = f(-n) \quad \text{(discrete case)}$$

R.1.67 An asymmetric or odd function (or sequence) is defined by the following relations:

$$f(t) = -f(-t) \quad \text{(analog case)}$$

and

$$f(n) = -f(-n) \quad \text{(discrete case)}$$

---

* For a formal definition of BW see Chapter 4. At this point, it is sufficient for the reader to associate BW with the signal quality.

† For the case of continuous signals just replace n by t.

R.1.68   Any real function or sequence can be expressed as a sum of its even part (f_e) plus its odd part (f_o) as indicated by the following equation:

$$f(t) = f_e(t) + f_o(t)$$

where $f_e(t) = 1/2\,[f(t) + f(-t)]$ and $f_o(t) = 1/2\,[f(t) - f(-t)]$ for the analog case, and $f(n) = f_e(n) + f_o(n)$, where $f_e(n) = 1/2\,[f(n) + f(-n)]$ and $f_o(t) = 1/2\,[f(n) - f(-n)]$ for the discrete case, assuming the sequences are real.

R.1.69   The average value of the function $f(t)$ in the interval $-T/2 \le t \le T/2$ is given by

$$f_{ave} = \lim_{T\to\infty}\left\{\frac{1}{T}\int_{-T/2}^{T/2} f(t)dt\right\}$$

Observe that the average value f_{ave} is contained in the even portion of $f(t)\,\{f_e(t)\}$, since the contribution of the odd portion is always zero.

R.1.70   The general algebraic rules governing even and odd symmetric functions are summarized as follows:

   a. The sum of two even functions is also even.

   b. The product of two even functions is also even.

   c. The product of two odd functions is even.

   d. An even function squared becomes even.

   e. An odd function squared becomes even.

   f. The sum of two odd functions is also odd.

   g. The sum of an even plus an odd function is neither even nor odd.

   h. The product of an even by an odd function is odd.

R.1.71   Any analog signal $f(t)$, or discrete sequence $f(n)$, of the independent variables either t or n, can be transformed with respect to the independent variable (t or n) in the following ways:

   a. Time transformation

     i. Reversal or reflection returns $f(-t)$ or $f(-n)$.

     ii. Time scaling by a returns $f(at)$ or $f(an)$ {expansion ($a < 1$), or compression ($a > 1$)}.

     iii. Time shifting by t_o returns $f(t - t_o)$ or $f(n - n_o)$. If $t_o > 0$ then $f(t)$ is shifted to the right by t_o, and if $t_o < 0$ then $f(t)$ is shifted to the left by t_o.

   b. Amplitude transformation

     i. Inversion returns $-f(t)$ or $-f(n)$.

     ii. Amplification or attenuation by A returns $A\,f(t)$ or $A\,f(n)$. If $A > 1$, which means amplification and if $A < 1$, which means attenuation.

     iii. Direct current (DC) shifting by A returns $A + f(t)$ or $A + f(n)$. If $A > 0$, which means $f(t)$ moves up by A and if $A < 0$, which means $f(t)$ moves down by A.

R.1.72   Recall that given a continuous time signal $f(t)$, the signal $f(t - t_1)$ is the signal $f(t)$ shifted t_1 units to the right, and $f(t + t_2)$ represents the signal $f(t)$ shifted t_2 units to the left, where t_1 and t_2 are positive, real numbers.

R.1.73   For example, let $f(t)$ be the function shown in Figure 1.21.
     Sketch the functions $f(t - 1)$, $f(t - 2)$, $f(t + 1)$, and $f(t + 2)$.

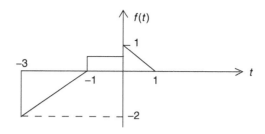

FIGURE 1.21
Plot of *f(t)* of R.1.73.

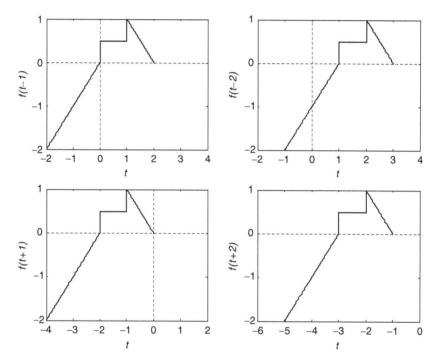

FIGURE 1.22
Plots of $f(t-1)$, $f(t-2)$, $f(t+1)$, and $f(t+2)$ of R.1.73.

ANALYTICAL Solution

The functions $f(t-1)$, $f(t-2)$, $f(t+1)$, and $f(t+2)$ are shown in Figure 1.22.

R.1.74 Given the continuous time signal $f(t)$, then by multiplying the independent variable t by -1, a reverse time function $f(-t)$ is created. The same can be said about the sequence $f(n)$ and its discrete reverse time sequence $f(-n)$.

R.1.75 For example, using the function defined in R.1.72, the reverse function $f(-t)$ is shown in Figure 1.23.

R.1.76 Given the function $f(t)$, then by multiplying the independent variable t by a real constant a, the function experiences the following changes:

a. If $a > 1$, then $f(t)$ is compressed in time by a factor of $1/a$.

b. If $a < 1$, then $f(t)$ is expanded in time by a factor of a.

R.1.77 For example, using the function defined in R.1.72, sketch the plots for $f(2t)$ and $f(t/2)$.

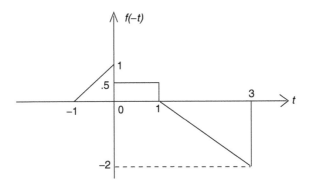

FIGURE 1.23
Plot of $f(-t)$ of the function defined in R.1.75.

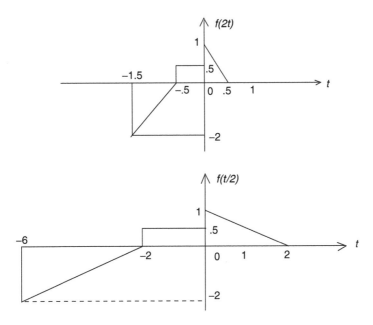

FIGURE 1.24
Plots of $f(2t)$ and $f(t/2)$ of R.1.73.

ANALYTICAL Solution

The functions for $f(2t)$ and $f(t/2)$ are shown in Figure 1.24.

Note that the concepts and definitions presented for the case of continuous time functions such as compression, expansion, time reversal, inversion, and time and amplitude shifting are equally applicable for the case of discrete sequences by changing the independent variable t to n.

R.1.78 Time signals (or sequences) encountered in real-world problems are in general real functions of t (or n). But sometimes it is useful to work with complex signals or sequences when performing systems analysis. Complex sequences can easily be expressed in terms of their real and imaginary parts of $f(t)$ or $f(n)$ as illustrated in the following expressions:

$$f(n) = real[\,f(n)] + j\,imag[\,f(n)] = a(n) + jb(n), \quad \text{for the discrete case}$$

or

$$f(t) = real[\,f(t)\,] + j\,imag[\,f(t)\,], \quad \text{for the analog case}$$

R.1.79 The complex conjugate sequence of $f(n)$ is denoted by $f^*(n)$, where

$$f^*(n) = real\,[f(n)] - j\,imag\,[f(n)] = a(n) - jb(n)$$

R.1.80 Let $f(n)$ be a complex sequence, where $f(n) = a(n) + jb(n)$. This sequence can further be decomposed into

$$f(n) = [a_e(n) + a_o(n)] + j[b_e(n) + b_o(n)]$$

where the subscripts e and o denote the even and odd parts, respectively, of the $a(n)$ and $b(n)$ of $f(n)$.

The same relation holds when n is replaced by t for the analog case.

R.1.81 A signal or sequence is periodic (with either period T or N) if the following relations hold

$$f(t) = f(t + kT) \quad \text{(analog)}$$

or

$$f(n) = f(n + kN) \quad \text{(discrete)}$$

for any $k = 0, \pm1, \pm2, \pm3, \ldots$.

When the signal or sequence does not satisfy the preceding relations, it is nonperiodic. A periodic signal is defined for all $t\ (-\infty, \infty)$. Periodic signals or sequences are basically ideal concepts. Most practical signals are basically nonperiodic.

R.1.82 The energy E of a signal $f(t)$ or sequence $f(n)$ is defined by

$$E = \int_{-\infty}^{+\infty} \left| f(t) \right|^2 dt < \infty$$

where $|f(t)|^2| = f(t) \cdot f(t)^*$, for the continuous case, and

$$E = \sum_{n=-\infty}^{+\infty} \left| f(n) \right|^2 \Delta n < \infty$$

for the discrete case.

R.1.83 A finite length sequence with finite magnitudes will always have finite energy or an infinite sequence with a finite number of samples may not have infinite energy.

R.1.84 A signal $f(t)$ defined over the range $t_o \le t \le t_1$, with a finite number of maxima and minima, is associated with a finite energy content E (in joules).

R.1.85 Let the energy of the signal $f(t)$ exist and be finite, then the signal $f(t)$ is referred to as an energy signal.

R.1.86 The average power of a finite discrete sequence $f(n)$ (or time-limited signal $f(t)$) is defined by the following equations:

$$P_{av} = \frac{1}{t - t_0} \int_{t_0}^{t_1} \left| f(t) \right|^2 dt \quad \text{(analog case)}$$

and by

$$P_{av} = \frac{1}{2N+1} \sum_{-N}^{+N} |f(n)|^2 \quad \text{(discrete case)}$$

R.1.87 Periodic signals are referred to as power signals, since they possess infinite energy.

R.1.88 An infinite energy signal with finite power is referred to as a power signal. A finite energy signal with infinite power is referred as an energy signal.

R.1.89 Recall that the MATLAB function *stem* returns the plot of a discrete sequence, whereas the *plot* command returns the plot of an analog (continuous) signal.

R.1.90 Recall that if *z* is complex then the MATLAB command *plot(z)* returns the continuous plot of *imag(z)* versus *real(z)*, whereas the command *stem(z)* returns the discrete plot of *real(z)* versus *n*.

R.1.91 The discrete unit impulse sequence $\delta(n)$ of length *N* can be obtained by using the MATLAB statement

$$Imp = [1\,zeros(1, N-1)]$$

Imp consists of an *N*-element row vector with one as the first element, followed by $N-1$ zeros, with the implicit assumption that the first element corresponds to $n = 0$, of the sequence $\delta(n)$.

R.1.92 The shifted unit impulse $\delta(n-k)$ of length *N* can be created by using the following MATLAB statement:

$$Impk = [zeros(1, k-1)\,1\,zeros(1, N-k)]$$

R.1.93 Another way to generate a unit impulse sequence of length $n = 2N+1$ with the unit impulse located at *k*, where *k* may be anywhere over the range $-N < n < N$, is by the following function file:

```
function [k,n] = Impfun(n1,n2,n3)
n = n2:1:n3;
k = [(n - n1)==0];
```

R.1.94 For example, the following script file *sequence_impulse* returns the discrete plot of the signal $x(n) = \delta(n-3)$, using a 21-element sequence over the range $-10 \le n \le 10$, and the function file *Impfun*:

MATLAB Solution
```
% Script file: sequence _ impulse
n = -10:1:10;
[k,n] = Impfun(3,-10,10);
stem(n,k)
title(' \delta(n - 3) vs. n')
xlabel('time index n')
ylabel('Amplitude')
a = min(n);
b = max(n);
c = min(k)-.5;
d = max(k)+.5;
axis([a b c d]);grid on;
```

See Figure 1.25.

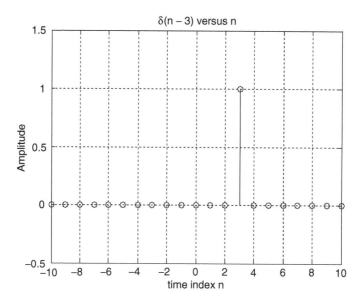

FIGURE 1.25
Plot of $x(n) = \delta(n - 3)$ of R.1.94.

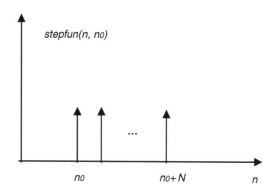

FIGURE 1.26
Plot of *stepfun(n, no)* of R.1.97.

R.1.95 A unit step sequence of length N can be generated using the following MATLAB command:

$$un = [ones(1, N)]$$

R.1.96 The shifted (or delayed) unit step sequence $u(n - k)$ can be created by executing the following MATLAB command:

$$unk = [zeros(1, k - 1)\ ones(1, N)]$$

Observe that the total number of elements of the sequence *unk* is $N + k - 1$.

R.1.97 The MATLAB function *stepfun(n, no)* returns the shifted step (by *no* units to the right) sequence shown in Figure 1.26. Recall that the *stepfun(n, no)* can be used with either analog or discrete arguments, defined as

$$stepfun(n, no) = u(n - no) = \begin{cases} 1 & \text{for } n \geq no \\ 0 & \text{for } n < no \end{cases}$$

The step function called *Heaviside* is indicated as follows:

```
function stepseq = Heaviside(x)
stepseq = (x>=0);
```

R.1.98 For example, write a program that returns $u(t)$ and $u(t - 2)$, using the function *Heaviside*, over the range $-10 \leq t \leq 10$.

MATLAB Solution
```
>> x = -10:0.1:10;
>> stepfun = Heaviside(x);
>> subplot(2, 1, 1);
>> plot(x, stepfun)
>> xlabel('t (time)')
>> title('u(t) vs. t')
>> ylabel('Amplitude.')
>> axis([-10 10 ,0.5 1.5])
>> subplot(2, 1, 2);
>> stepfun = Heaviside(x-2);
>> plot(x, stepfun)
>> axis([-10 10 ,0.5 1.5])
>> title('u(t-2) vs. t')
>> ylabel('Amplitude.');
>> xlabel('t (time)');
```

See Figure 1.27.

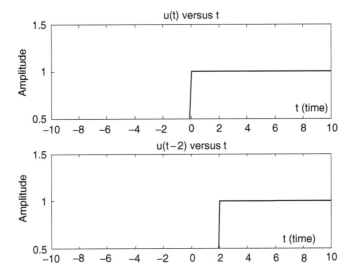

FIGURE 1.27
Plots of $u(t)$ and $u(t - 2)$ of R.1.98.

R.1.99 The MATLAB function *sign(t)* is defined as follows:

$$sign(t) = \begin{cases} 1 & t > 0 \\ 0 & t = 0 \\ -1 & t < 0 \end{cases}$$

The *sign(t)* function is illustrated in Figure 1.28.

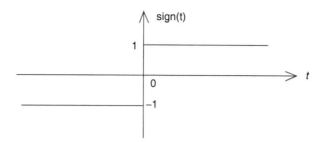

FIGURE 1.28
Plot of the function *sign(t)* of R.1.99.

Note that the function *sign(t)* can be created by using the step functions indicated as follows:

$$sign(t) = u(t) - u(-t) = -1 + 2u(t)$$

R.1.100 The MATLAB symbolic toolbox calls the impulse function $\delta(t)$ by using the name *Dirac(t)*.

R.1.101 The MATLAB symbolic toolbox calls the step function *u(t)* by using the name *Heaviside(t)*.

R.1.102 The MATLAB symbolic toolbox calls the function *sign(t)* by using the name *signum(t)*.

R.1.103 Let us gain some experience by using the MATLAB symbolic toolbox in evaluating the following expressions:

a. $\int_{-\infty}^{\infty} \delta(t)\,dt =$

b. $\int_{-2}^{3} u(t)\,dt =$

c. $\int_{-\infty}^{\infty} sign(t)\,dt =$

d. $u(t)\big|_{t=3} =$

e. $u(t)\big|_{t=-2} =$

f. $\int t\,u(t)\,dt =$

g. $\int_{1}^{2} t\,u(t)\,dt =$

h. $\int_{-1}^{2} t\,u(t)\,dt =$

i. $\int_{-1}^{2} t\,sign(t)\,dt =$

MATLAB Solution
```
>> syms t a
>> area _ impulse = int('Dirac(t)', -inf, inf)      % area of the impulse
                                                       δ(t)

        area _ impulse =
                    1

>> area _ step = int('Heaviside(t)', -2, 3)          % area of the step
                                                       from -2 to +3
```

```
        area _ step =
                      3

>> area _ sign = int('signum(t)', -2, 3)        % area of the sign from
                                                   -2 to +3

        area _ sign =
                      1

>> stept _ 3 = vpa('Heaviside(3)')              % evaluates u(t) at t =3

        stept _ 3 =
                      1

>> stepmin _ 2 = vpa('Heaviside(-2)')           % returns u(t) at t = -2

        stepmin _ 2 =
                      0

>> differstep = diff('Heaviside(t)')            % returns d(u(t))/dt

        differstep =
                      Dirac(t)

>> intramp = int('Heaviside(t)'*t)              % returns the integral
                                                   of t u(t) dt

        intramp =
                1/2*Heaviside(t)*t^2

>> area _ ramp12 = int('Heaviside(t)'*t,1,2)    % area of [t u(t)] from
                                                   t =1 to t =2

        area _ ramp12 =
                      3/2

>> area _ ut _ 12 = int('Heaviside(t)'*t,-1,2)  % area t u(t) from t = -1
                                                   to t =2

        area _ ut _ 12 =
                      2

>> area _ signt = int('signum(t)'*t,-1,2)       % area sign(t)*t from t=-1
                                                   to t=2

        area _ sign =
                      5/2
```

R.1.104 Create the script file *plot_ramp* that returns the plot of *t u(t)* versus *t*, over the range
$-1 \le t \le 3$, using *ezplot*.

MATLAB Solution
```
% Script file: plot _ ramp
ramp = ('Heaviside(t)'*t)                       % returns the ramp over
                                                   -1 ≤ t ≤3

ezplot(ramp, [-1 +3])                           % see plot Figure 1.29
title('heaviside(t)*t vs. t');
xlabel('t'); ylabel('t*u(t)')   ;
```

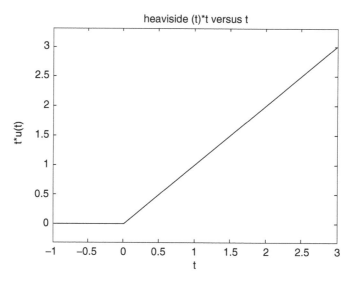

FIGURE 1.29
Plot of *t u(t)* of R.1.104.

R.1.105 The MATLAB command *square(t, a)* returns the periodic square wave with period
$T = 2 * \pi$, over the range defined by *t*, where *a* is a constant that indicates the percent of the period *T* for which the square wave is positive.

R.1.106 For example, create the script file *squares* that returns the plots over two cycles of
the square sequences, with period $T = 2\pi$, with the following specs:

 a. $f_1(t)$ versus *t*, with $mag[f_1(t)] = 1$, during 50% of the period *T* and $mag[f_1(t)] = -1$,
    during the remaining 50%

 b. $f_2(t)$ versus *t*, with $mag[f_2(t)] = 2$, during 25% of the period *T* and $mag[f_2(t)] = -2$,
    during the remaining 75%

 c. $f_3(t)$ versus *t*, with $mag[f_3(t)] = 3$, during 33% of the period *T* and $mag[f_3(t)] = -3$,
    during the remaining 67%

 d. $f_4(t)$ versus *t*, with $mag[f_4(t)] = 4$, during 75% of the period *T* and $mag[f_4(t)] = -4$,
    during the remaining 25%

MATLAB Solution
```
% Script file: squares
t = 0:.1*pi:4*pi;
f1 =square(t,50);
f2 =2*square(t,25);
f3 =3*square(t,33);
f4 =4*square(t,75);
subplot(2,2,1)
plot(t,f1)
ylabel('f1(t)')
axis([0 4*pi -1.5 1.5])
grid on;
title('Square(t,50) vs t')
subplot(2,2,2)
plot(t,f2)
```

```
ylabel( 'f2(t)')
axis([0 4*pi -2.5 2.5])
grid on;
title('2*Square(t,25) vs t')
subplot(2,2,3)
plot(t,f3)
axis([0 4*pi -4.5 4.5])
grid on;
ylabel('  f3(t) ');xlabel(' t (time)')
title('3*Square(t,33) vs t')
subplot(2,2,4)
plot(t,f4)
axis([0 4*pi -5.5 5.5])
grid on;
title('4*Square(t,75) vs t')
ylabel(' f4(t) ');xlabel(' t (time) ')
```

The script file *squares* is executed and the results are shown in Figure 1.30.

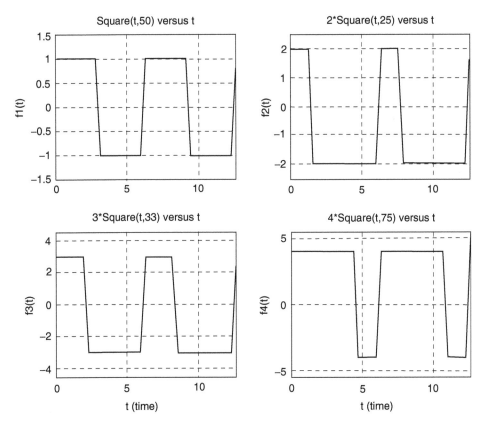

FIGURE 1.30
Plots of the function *square* of R.1.106.

R.1.107   The MATLAB command *sawtooth(t, b)* returns a triangular wave, with magnitudes between −1 and +1, and a period of $T = 2\pi$. The scalar b, between 0 and 1, indicates the percent of the period T with positive slope, where the maximum occurs at the end.

R.1.108 For example, create the script file *triangles* that returns the plots over two cycles of the triangular sequences, with period $T = 2\pi$, with the following specs:

    a. $f_1(t)$ versus t, with positive slope during 50% of the period T, and a swing from -1 to $+1$

    b. $f_2(t)$ versus t, with positive slope during 25% of the period T, and a swing from -2 to $+2$

    c. $f_3(t)$ versus t, with a positive slope during 33% of the period T, and a swing from -3 to $+3$

    d. $f_4(t)$ versus t, with a positive slope during 75% of the period T, and a swing from -4 to $+4$

MATLAB Solution
```
% Script file: triangles
t = 0:0.1*pi:4*pi;
f1 = sawtooth(t,.5);
f2 = 2*sawtooth(t,.25);
f3 = 3*sawtooth(t,.33);
f4 = 4*sawtooth(t,.75);
subplot(2,2,1)
plot(t,f1)
ylabel ('f1(t)')
axis ([0 4*pi -1.5 1.5])
grid on;
title ('Sawtooth(t,.50) vs t')
subplot(2,2,2)
plot(t,f2)
ylabel( 'f2(t)')
axis ([0 4*pi -2.5 2.5])
grid on;
title ('2*Sawtooth(t,.25) vs t')
subplot (2,2,3)
plot (t,f3)
axis ([0 4*pi -4.5 4.5])
grid on;
ylabel('  f3(t) ');xlabel(' t (time)')
title('3*Sawtooth(t,.33) vs t')
subplot(2,2,4)
plot(t,f4)
axis([0 4*pi -5.5 5.5])
grid on;
title('4*Sawtooth(t,.75) vs t')
ylabel(' f4(t) '); xlabel(' t (time) ')
```

The script file *triangles* is executed and the results are shown in Figure 1.31.

R.1.109 The MATLAB function *sinc(x)* evaluates the function defined by

$$sinc(x) = \frac{sin(\pi x)}{\pi x}$$

R.1.110 For example, the script file *sincs* returns the plot of the function *sinc(x)* over the range $5 \le x \le 5$, illustrated in Figure 1.32.

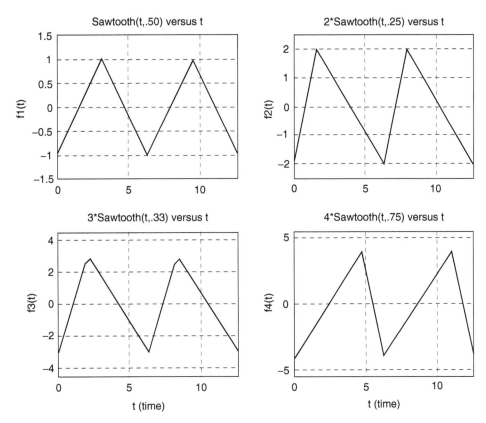

FIGURE 1.31
Plots of the function *sawtooth* of R.1.108.

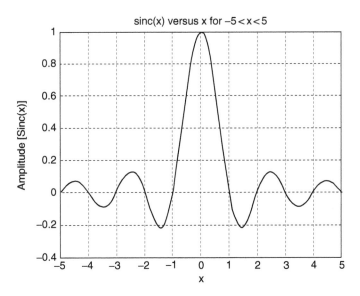

FIGURE 1.32
Plot of the function *sinc* of R.1.110.

```
MATLAB Solution
% Script file: sincs
x = -5:0.1:5;
y = sinc(x);
plot(x, y)
title('sinc(x) vs. x for -5<x<5')
xlabel('x');
ylabel('Amplitude [Sinc(x)]');
grid on;
```

R.1.111   The MATLAB function *tripuls(t, c)* returns a symmetric triangle with its base along the horizontal axis, with length c, centered at $t = 0$.

R.1.112   For example, create the script file *triang* that returns the plots of triangles with the following specs:

a. $f_1(t)$ versus t, with $peak[f_1(t)] = 1$ and a base length $= 3$

b. $f_2(t)$ versus t, with $peak[f_2(t)] = 2$ and a base length $= 5$

c. $f_3(t)$ versus t, with $peak[f_3(t)] = 3$ and a base length $= 10$

d. $f_4(t)$ versus t, with $peak[f_4(t)] = 4$ and a base length $= 12$

```
MATLAB Solution
% Script file: triang
t = -6:0.1:6;
f1 = tripuls(t,3);
f2 = 2*tripuls(t,5);
f3 = 3*tripuls(t,10);
f4 = 4*tripuls(t,12);
subplot(2,2,1)
plot(t,f1)
ylabel('Amplitude [f1(t)]');
xlabel(' t (time)')
axis([-6 6 -0.5 1.5])
title('tripuls(t,3) vs. t')
subplot(2,2,2)
plot(t,f2)
ylabel( 'Amplitude [f2(t)]')
xlabel(' t (time)')
axis([-6 6 -0.5 2.5])
title('2*tripuls(t,5) vs. t')
subplot(2,2,3)
plot(t,f3)
axis([-6 6 -0.5 4.5])
ylabel('  Amplitude [f3(t)] ');xlabel(' t (time)')
title('3*tripuls(t,10) vs. t')
subplot(2,2,4)
plot(t,f4)
axis([-6 6 -0.5 5.5])
title('4*tripuls(t,12) vs. t')
ylabel(' Amplitude [f4(t)] '); xlabel(' t (time) ')
```

The script file *triang* is executed and the results are shown in Figure 1.33.

R.1.113   The MATLAB function *rectpuls(t, d)* returns a symmetric rectangle with width d, centered at $t = 0$.

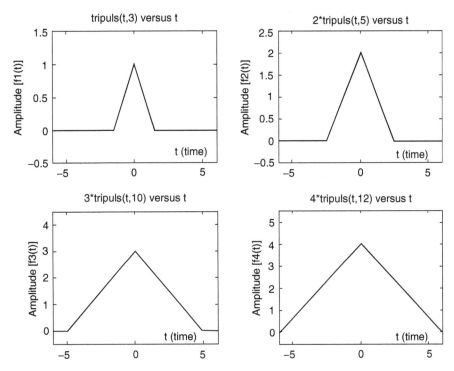

FIGURE 1.33
Plot of the function *tripuls* of R.1.112.

R.1.114 For example, create the script file *rect_pulses* that returns rectangle plots with the following specs:

a. $f_1(t)$ versus t, with $mag[f_1(t)] = 1$ and width $= 1$

b. $f_2(t)$ versus t, with $mag[f_2(t)] = 2$ and width $= 3$

c. $f_3(t)$ versus t, with $mag[f_3(t)] = 3$ and width $= 6$

d. $f_4(t)$ versus t, with $mag[f_4(t)] = 4$ and width $= 9$

MATLAB Solution
```
% Script file: rect_pulses
t = -6:.1:6;
f1 = rectpuls(t,1);
f2 = 2*rectpuls(t,3);
f3 = 3*rectpuls(t,6);
f4 = 4*rectpuls(t,9);
subplot(2,2,1)
plot (t,f1)
ylabel (' Amplitude [f1(t)]');xlabel('t (time)');
axis ([-6 6 -0.5 1.5])
title('Rectpuls(t,1) vs. t')
subplot(2,2,2)
plot(t,f2)
ylabel( 'Amplitude [f2(t)]'); xlabel('t (time)');
axis([-6 6 -0.5 2.5])
title('2*Rectpuls(t,3) vs. t')
subplot(2,2,3)
```

```
plot(t,f3)
axis([-6 6 -0.5 4.5])
ylabel(' Amplitude [f3(t)]');xlabel(' t (time)')
title('3*Rectpuls(t,6) vs. t')
subplot(2,2,4)
plot(t,f4)
axis([-6 6 -0.5 5.5])
title('4*Rectpuls(t,9) vs. t')
ylabel(' Amplitude [f4(t)] '); xlabel(' t (time) ')
```

The script file *rect_pulses* is executed and the results are shown in Figure 1.34.

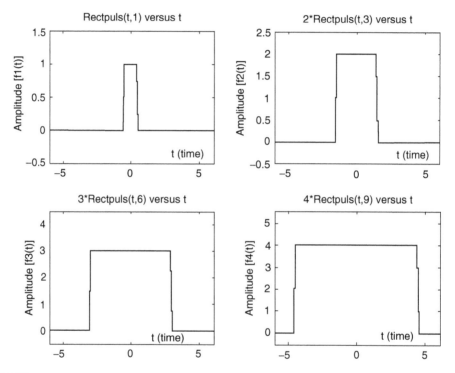

FIGURE 1.34
Plots of the function *rectpuls* of R.1.114.

R.1.115 The MATLAB function $y = pulstran(t, d, \text{'}f\text{'})$ returns a symmetric train of continuous or discrete functions 'f' with d periods over the range defined by t.

R.1.116 A more general MATLAB function is $[f, t] = gensig(\text{'}type\text{'}, T, range, Ts)$ that returns the periodic function f defined by *type (sin, square, or pulse)*, over the *range t*, with a sampling rate T_s.

R.1.117 For example, use the *gensig* command to create three cycles of a square periodic wave, with period $T = 3$ and a sampling rate $Ts = 0.1$ (Figure 1.35).

MATLAB Solution
```
>> [squar,t] = gensig('square',3,9,0.1);
>> plot(t,squar)
>> axis([0 9 0 1.2])
>> ylabel('Amplitude');xlabel('time')
>> title('Plot using [squar,t] = gensig(square,3,9,0.1)')
```

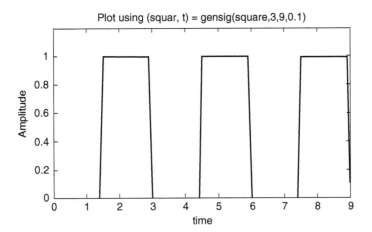

FIGURE 1.35
Plot of R.1.117.

R.1.118 As an additional example, create the script file *triang_pulses* that returns the plots of triangular waves with 3, 4, 5, and 2 cycles, respectively, and unit magnitude, over the range $0 \leq t \leq 1$.

MATLAB Solution
```
% Script file: triang _ pulses
% echo on
t =0:0.001:1;
subplot(2,2,1)
d1 = [0:.33:1];                        % 3 cycles
y1 = pulstran(t,d1,'tripuls',.25);
plot (t,y1)
title ('3 triangular cycles');
axis ([0 1 -0.5 1.5]);
ylabel('Amplitude') ;
subplot(2,2,2)
da = [0:.25:1];
y2 = pulstran(t,da,'tripuls',.25);     % 4 cycles
plot(t,y2);
title('4 triangular cycles');
axis([0 1 -0.5 1.5]);
ylabel('Amplitude') ;
subplot(2,2,3)
d3 = [0:.20:1];                        % 5 cycles
y3 = pulstran(t,d3,'tripuls',.1);
plot(t,y3);
title('5 triangular cycles');
axis ([0 1 -0.5 1.5]);
xlabel('time')
ylabel('Amplitude') ;
d4 = [0:.5:1];                         % 2 cycles
subplot(2,2,4)
y4 = pulstran(t,d4,'tripuls',.4);
plot (t,y4)
```

```
title ('2 triangular cycles');
axis ([0 1 -0.5 1.5]);
ylabel ('Amplitude') ;
xlabel ('time');
```

The script file *triang_pulses* is executed and the results are shown in Figure 1.36.

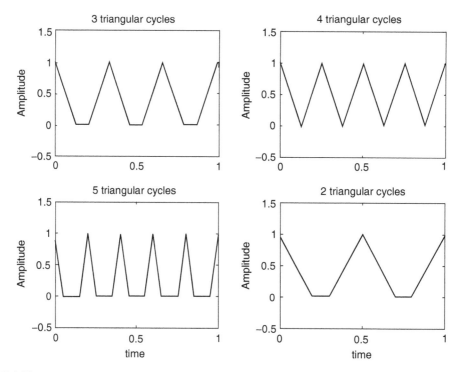

FIGURE 1.36
Plots of R.1.118.

R.1.119 Recall that system concepts such as transfer function, filter, and input and output signals were introduced in Chapter 7 of the book titled *Practical MATLAB® Basics for Engineers*.

The MATLAB command *[h, t] = impulse(P, Q, t)* returns as output *h*, when the input is a Dirac impulse applied to a filter or system whose transfer function is given by *P/Q*, where *P* and *Q* are polynomials expressed as row arrays whose coefficients are arranged in descending order of *s* (where $s = jw$ is the Laplace variable) and *t* is the time interval of interest (see Chapter 4 for more information).

R.1.120 The MATLAB command *[gs, t] = step(P, Q, t)* returns the step response of a linear system with a transfer function defined by $H = P/Q$, over a time interval *t*.

R.1.121 The MATLAB function *hn = dimpulse(P, Q, n)* returns the discrete impulse response consisting of *n* samples, applied to the discrete transfer function $H(z) = P(z)/Q(z)$, where $P(z)$ and $Q(z)$ represent the vector polynomial consisting of its coefficients arranged in decreasing powers of *z*, where $z = e^{jw}$.*

---

* See Chapter 5 for more information about *z*.

R.1.122 For example, let

$$H(Z) = \frac{2.24z^2 + 2.5z + 2.25}{z^2 + 0.5z + 0.68}$$

Create the script file *disc_imp* that returns the discrete response in the form of a table and a plot for the sequence of length $n = 10$. The results are shown in Figure 1.37.

MATLAB Solution
```
% Script file: disc _ imp
clc; clf;
n =10;
P = [2.24 2.5 2.25];
Q = [1 .5 .68];
hn = dimpulse(P,Q,n);
nn = 0:1:9;
results = [ nn' hn];
disp('************************************************************');
disp('The impulse response sequence h(n) for the first 10 samples is:');
disp('^^^^^^^^^^^^^^^^^^^^');
disp('    n        h(n)');
disp('^^^^^^^^^^^^^^^^^^^^');
disp(results);
disp('************************************************************');
yzero = zeros(1,10);
stem(nn,hn); hold on; plot(nn,yzero)
title('Discrete impulse response h(n) vs n')
xlabel('time index n');
ylabel('Amplitude [h(n)]');
```

The script file *disc_ imp* is executed and the results are shown as follows:

```
******************************************************************
The impulse response sequence h(n) for the first 10 samples is:
^^^^^^^^^^^^^^^^^^^^^^^^

    n         h(n)
^^^^^^^^^^^^^^^^^^^^^^^^

    0         2.2400
1.0000        1.3800
2.0000        0.0368
3.0000       -0.9568
4.0000        0.4534
5.0000        0.4239
6.0000       -0.5203
7.0000       -0.0281
8.0000        0.3679
9.0000       -0.1648
******************************************************************
```

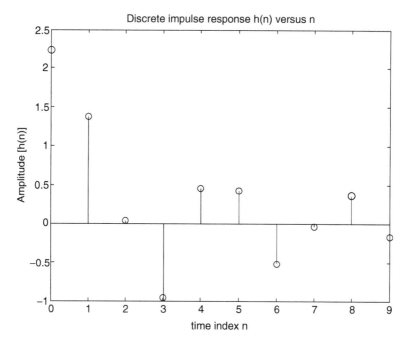

FIGURE 1.37
Discrete impulse response *h(n)* of R.1.122.

R.1.123 The MATLAB function *un = dstep(P, Q, n)* returns the discrete step response with length *n* of the system defined by the transfer function *H(z) = P(z)/Q(z)*.

R.1.124 For example, create the script file *disc_step* that returns the discrete response in the form of a table and a plot for a sequence of length *n = 10*, for the system defined in R.1.122.

MATLAB Solution
```
% Script file: disc _ step
clc; clf;
n = 10;
P = [2.24 2.5 2.25];
Q = [1 .5 .68];
un = dstep(P,Q,n);
nn = 0:1:9;
results = [nn' un];
disp('****************************************************************');
disp('The discrete step sequence u(n) for the first 10 samples is:');
disp('^^^^^^^^^^^^^^^^^^')
disp('    n         u(n)')
disp('^^^^^^^^^^^^^^^^^^')
disp(results);
disp('****************************************************************');
yzero = zeros(1,10);
stem (nn,un); hold on; plot (nn,yzero)
title ('Discrete step response u(n) vs n')
xlabel ('time index n');
ylabel ('Amplitude[u(n)]');
```

The script file *disc_step* is executed and the results are given as follows and in Figure 1.38.

```
***************************************************************
The discrete step sequence u(n) for the first 10 samples is:
^^^^^^^^^^^^^^^^^^^^^^^^^^
    n              u(n)
^^^^^^^^^^^^^^^^^^^^^^^^^^
    0              2.2400
    1.0000         3.6200
    2.0000         3.6568
    3.0000         2.7000
    4.0000         3.1534
    5.0000         3.5773
    6.0000         3.0570
    7.0000         3.0289
    8.0000         3.3968
    9.0000         3.2320
***************************************************************
```

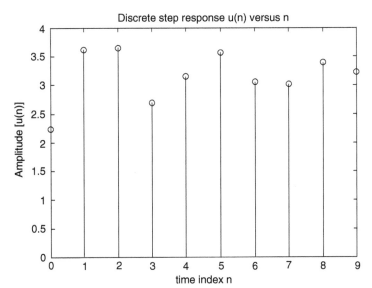

FIGURE 1.38
Discrete step response of R.1.124.

R.1.125 The MATLAB function $[y, n] = filter(b, a, x)$ defined in Chapter 7 of the book titled *Practical MATLAB® Basics for Engineers* returns the output sequence y, when the relation input (x) output (y) is given by the following system difference equation:

$$\sum_{k=0}^{N1} a_k\, y(n-k) = \sum_{l=0}^{N2} b_l\, x(n-l)$$

R.1.126 A discrete sequence may consist of either a finite or infinite number of samples. A finite length sequence is characterized by a finite number of nonzero samples. An infinite length sequence extends infinitely to the right or the left of the n-axis (discrete time). In either case, the infinite or finite sequences can be represented by a sum of weighted time-shifted impulse samples.

R.1.127 Reconstruction of an analog signal *f(t)* from the discrete sequence *f(n)* can be accomplished, if the analog signal *f(t)* is band-limited and converted into the discrete sequence *f(n)* by sampling it with a rate above the Nyquist/Shannon rate.

R.1.128 Recall that the sampling theorem (Nyquist/Shannon) states that an analog signal *f(t)*, band-limited to *fm* hertz, must be sampled at a rate greater than twice its highest frequency *fm*, to be able to reconstruct *f(t)* from its samples given by *f(n)* (R.1.1).

The reconstruction process is done by passing the sequence *f(n)* through an ideal analog low-pass filter with a cutoff frequency of *fm* (see Chapters 4 and 6 for information regarding frequency domain and filtering).

R.1.129 Since ideal sharp rectangular filters, with a cutoff frequency of *fm* are unrealizable, the practical sampling rate is often five or six times the frequency *fm*. At this point let us analyze the reconstruction process in the time domain, where the reconstructed signal *f(t)* can be approximated by the following equation:

$$f(t) = \sum_{n=-\infty}^{n=+\infty} f(n) \frac{\sin c(t - nT)}{T}$$

where

$$\sin c(t) = \frac{\sin(\pi t)}{\pi t}$$

R.1.130 If a discrete time signal *f(n)* was originally obtained by sampling a band-limited continuous time signal by using the Nyquist/Shannon sampling rate, then the process of up- or down-sampling the time continuous signal *f(t)* means that the original signal is sampled now by a higher or lower sampling rate.

R.1.131 The term up-sampling refers to increasing the sampling rate by an integer factor *L*. This process means that interpolation is required by placing additional samples between the original sampled function *f(n)*. This process is accomplished by a low-pass filter scaled to a cutoff frequency $w_c = \pi/L$.

R.1.132 The MATLAB function *interp* returns the sequence *fL* that consists of increasing the sequence *f(n)* defined by a vector *f* by an integer factor *L*.

The function *interp* can take any of the following forms:

a. *fL* = *interp(f, L)*

b. *fL* = *interp(f, LN, a)*

c. *[fL, h]* = *interp(f, L,N, a)*   (with default values of *L* = 4 and *a* = 0.5)

The resampled vector *fL* has a length given by *length(fL)* = *L* * *length(f)*.

The sample signal *f* is assumed to be band-limited scaled to $0 \le w \le a$, with $a \le 0.5$, where *h* represents the interpolation filter coefficient. Ideally $L \le 10$.

R.1.133 The term decimate is referred to the reduction or down-sampling the discrete signal *f* by an integer factor *M*, returning the sequence *fM*.

The MATLAB function *decimate* can take any of the following forms:

$$fM = decimate(f, M)$$

$$fM = decimate(f, M \cdot n)$$

$$fM = decimate(f, M \cdot 'fir')$$

$$fM = decimate(f, M, N, 'fir')$$

where

f is the input sampled vector $(f(n))$.

M the down-sampling integer rate factor, or the resampling rate is $1/M$ times the original rate. The resulting length is given by $length(fM) = [length\ (f)]/M$.

N the order of the (Chebyshev type-1) filter used to accomplish the resampling.

fir a 30-point low-pass filter forward direction only with cutoff frequency $w_c = \pi/M$, before resampling is done.

The decimation process of the first two options uses an *IIF* Chebyshev low-pass type-1 filter with forward and backward directions (see Chapter 6 for information regarding filters).

R.1.134 The MATLAB function *resample* returns the sequence $f\ LM$, consisting of f resampled with a rate that is the ratio of two integers given by L/M.

The syntax of the function *resample* with some options is indicated as follows:

$$fLM = resample(f, L, M)$$

$$fLM = resample(f, L, M, R)$$

$$fLM = resample(f, L, M, h)$$

$$[fLM, h] = resample(f, L, M)$$

where R is the input rate with a default value of 10. This function uses an *FIR*, and a Kaiser window with $\beta = 5$ (windows are presented later in this section).

R.1.135 The process referred to as multiplexing consists of merging two (or more) discrete sequences $f_1(n)$ and $f_2(n)$ into a single sequence by alternating the samples of $f_1(n)$ with $f_2(n)$. Then the resulting multiplexed, or in short *mux* sequence is given by

$$mux(f_1, f_2) = [f_1(1)\ f_2(1)\ f_1(2)\ f_2(2)\ f_1(3)\ f_2(3) \ldots f_1(n-1)\ f_2(n-1)\ f_1(n)\ f_2(n)]$$

The lengths of the sequences of f_1 and f_2 are assumed to be n (equal). Then the length of *mux* (f_1, f_2) sequence is $2n$. If the sequences are $f_1(n)$ and $f_2(m)$, where $n > m$, then by increasing the length of f_2 to n, by making the last $n - m$ samples zeros, the two sequences (lengths) became equal and the *mux command* can then be safely used.

R.1.136 Signals may also be classified according to the probability of predicting its behavior with certainty or with some sort of ambiguity into

a. Deterministic or nonrandom

b. Probabilistic, stochastic, or random

R.1.137 Deterministic signals can be expressed in terms of a well-defined process, table, rules, or by a mathematical relation (equation). These type of signals are fully predictable.

R.1.138 Random signals are not predictable; they are noise-like functions where particular values or samples are not important, but rather the statistical information over a large range of samples is, such as the expected value, the mean, and standard deviation. Most signals encountered in practical applications and in this text are real and deterministic. Observe that random signals cannot be reproduced, but may carry valuable information. The more unpredictable or random a signal is the more information it carries.

R.1.139 The process of limiting or truncating a function or sequence consisting of an infinite or a very large number of samples such as

$$f(n) = \sum_{l=-\infty}^{+\infty} F_l \delta(n - l)$$

is by approximating $f(n)$ by another $f_a(n)$, consisting of a finite number of samples $(2N + 1)$ given by

$$f_a(n) = \sum_{l=-N}^{+N} F_l \delta(n - l)$$

Mathematically, the truncation process is accomplished by multiplying the function $f(n)$ by another function $w(n)$ called rectangular window, where $w(n)$ is defined by

$$w(n) = \begin{cases} 1 & n < |N| \\ 0 & n > |N| \end{cases}$$

Observe that the lengths of $w(n)$ and $f_a(n)$ are $2N + 1$. The rectangular window is the simplest model used to truncate a function and all the weighing coefficients used for that purpose are one. Note that $w(n)$ is equivalent to the pulse function $pul(n/N)$.

R.1.140 The practical way to deal with a function $f(n)$, which has an infinite range, is by truncating $f(n)$. Therefore,

$$f_a(n) = f(n) * w(n)$$

An additional objective of $w(n)$ is to improve the smoothness of $f_a(n)$ by removing oscillations associated with a sharp truncation process.

R.1.141 Practical considerations of the truncation process and the use of different window models are better understood in the frequency domain with applications in filter design (see Chapters 4 and 6).

R.1.142 Many window models have been proposed by mathematicians and engineers over the last century. All the window models share similar properties such as

a. The sample located at $n = 0$ is multiplied (scaled) by 1 (unaffected).

b. The shape of $f(n)$ is relatively unaffected for $n < |N|$.

c. The shape of $f(n)$ is increasingly affected for the values of n in the vicinity of $|N|$.

d. The window coefficients range from 1 to 0, where 1 corresponds to the value at $n = 0$ and the smaller coefficients are for the larger values of n as n approaches N.

e. All window models are of finite length and the point of symmetry is located at the midpoint *[length (w(n))]/2*.

R.1.143 When an arbitrary function presents a discontinuity and is approximated by a large, but finite number of terms, a ripple is generated at the discontinuity with a magnitude of about 10% of the jump value. This behavior is referred to as the Gibb's phenomenon (see Chapter 4 for more details). The objective of the various window models is to reduce the Gibb's effect that translates into oscillations.

R.1.144 MATLAB offers the user a number of built-in window functions in the signal processing toolbox. Some of the window models most often used are known by the
following names:
a. Rectangular
b. Triangular
c. Hanning
d. Hamming
e. Kaiser
f. Chebyshev
g. Bartlett
h. Blackman

R.1.145 The MATLAB function *Hamming(N)* returns a vector with *N*-weighted points
referred to as the *Hamming* window.

R.1.146 Similarly, the MATLAB commands

Hanning(N)

Blackman(N)

return *N*-length vectors representing the *Hanning* or *Blackman* weighted-type
windows.

R.1.147 The three window models: Hamming, Hanning, and Blackman are based on
cosine functions. The mathematical equations used to generate the above window
sequences are defined as follows:
a. For the Hamming window,
$$w(n) = 0.54 + 0.46\cos(\pi * n/M)$$
b. For the Hanning (Van Hann) window,
$$w(n) = 0.54 + 0.46\cos((\pi * n)/(M + 1))$$
c. For the Blackman window,
$$w(n) = 0.5 + 0.46\cos((\pi * n)/M) + 0.08\cos((4 * \pi * n)/(2n + 1))$$
where $M = 1/2(L - 1)$ is an integer representing the midpoint and L is the
window's length.

R.1.148 The MATLAB command *Kaiser* (N, β) returns the N points of the Kaiser window,
where β is a constant, with the following range $1 \le \beta \le 10$, where β represents a
tradeoff between the side lobe height and its width. The Kaiser window is based
on the modified Bessel function $I_o(x)$ given as follows:

$$w(n) = \frac{I_o\beta\sqrt{1 - (n - M)^2/M}}{I_0(\beta)} \quad \text{for } n = 0, 1, 2, \ldots (L - 1)$$

where $M = 1/2(L - 1)$ is the midpoint and the weight β is over the range $0 \le \beta \le 10$.
This window is frequently used in the design of filters, where β controls the
stop- and pass-band ripples.

R.1.149 The MATLAB function *triang(N)* returns the N-point triangular window based on
the following equation:

$$w(n) = 1 - \frac{|n|}{M + 1}$$

R.1.150 The MATLAB function *boxcar(N)* returns the *N* points of a rectangular window defined by

$$w(n) = \begin{cases} 1 & |n| < M \\ 0 & |n| > M \end{cases}$$

R.1.151 The MATLAB function *Bartlett(N)* returns the *N* points of the Bartlett window based on the equation

$$w(n) = \frac{M - n}{M}$$

R.1.152 There are other popular window models such as Parabolic, Cauchy, and Gaussian defined below. The Parabolic or *Parzen* window is based on the equation given by

$$w(n) = 1 - \left[\frac{n - M}{M}\right]^2$$

R.1.153 The *Cauchy* window is defined by the equation given by

$$w(n) = \frac{M^2}{M^2 + \alpha^2(n - M)^2}$$

where α is an optional control character.

R.1.154 The *Gaussian* window is defined by the following equation:

$$w(n) = \exp\left[-\frac{1}{2}\alpha^2\left[\frac{n - M}{M}\right]^2\right]$$

where α is an optional control character.

R.1.155 All the proposed window models share the characteristic that the peak occurs in the middle (midpoint) of the window sequence, whereas at its edges (end points) the behavior tends to be smooth.

The behavior of some of the window models may present side lobes or undesirable shapes over some regions. In any case, an attempt is made to present only the most important and the widely accepted window models without exploring their behaviors.

R.1.156 For example, create the script file *windows* that returns the plots of the following windows:

a. *Hamming*

b. *Hanning*

c. *Blackman*

shown in Figure 1.39, using *N* = 31.

MATLAB Solution
```
>> n = -15:1:15;
>> WHAM = Hamming(31);
>> WHAN = Hanning(31);
>> WBLAC = Blackman(31);
```

```
>> subplot(3, 1, 1);
>> plot(n, WHAM);
>> title('Hamming(31) window');
>> ylabel('Amplitude');xlabel('n')
>> subplot(3, 1, 2);
>> plot(n, WHAN);
>> title('Hanning(31) window');
>> ylabel('Amplitude');xlabel('n')
>> subplot(3, 1, 3);
>> plot(n, WBLAC);
>> title('Blackman(31) window');
>> xlabel('n');
>> ylabel('Amplitude')
```

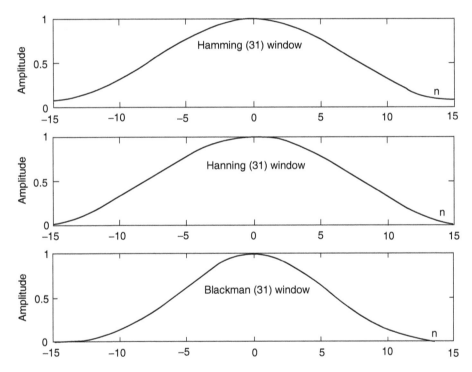

FIGURE 1.39
Plots of the *Hamming, Hanning,* and *Blackman* windows of R.1.156.

R.1.157 Observe that the *Hamming, Hanning,* and *Blackman* window models shown in Figure 1.39 present similar shapes. They are plotted on the same graph for comparison purposes in Figure 1.40, given by script file *compare_win*.

MATLAB Solution
```
% Script file:compare _ win
% This file returns the plots of the
% HAMMING, HANNING and BLACKMAN windows
% using 31 points.
%*****************************
clc;clf;
n = -15:1:15;
```

```
wham = hamming(31);
whan = hanning(31);
wblac = blackman(31);
plot(n,wham,'*',n,whan,'d',n,wblac,'o');
hold
plot(n,wham,n,whan,n,wblac);
title('HAMMING,HANNING and BLACKMAN windows ');
xlabel('points n');
ylabel('Amplitude')
legend('HAMM','HANN','BLAC');
```

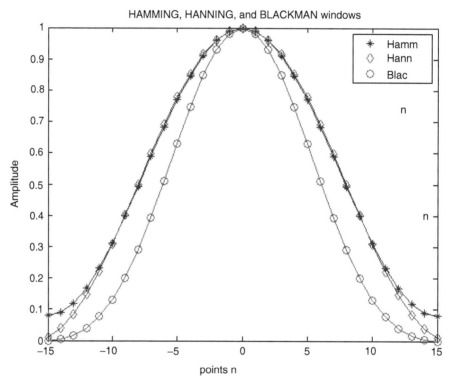

FIGURE 1.40
Plots of Hamm, Hann, and Blackman windows of R.1.157.

R.1.158 The Kaiser window is a controllable window (controlled by β), and is presented as follows, by the script file *Kaisers* for the following values of β: *3, 7,* and *10,* using an $N = 31$-point approximation.

MATLAB Solution
```
% Script file: Kaisers
% This file returns the plots
% KAISER window for betas: 3,7,10.
% using a 31 point approximation
%*******************************
clc;clf;
n = -15:1:15;
wk3 = kaiser(31,3);wk7=kaiser(31,7);
```

```
wk10 = kaiser(31,10);
plot(n,wk3,'*',n,wk7,'d',n,wk10,'o');
hold
plot(n,wk3,n,wk7',n,wk10);
title('KAISER windows for \beta=3,7, and 10 ');
xlabel('points n');
ylabel('Amplitude')
legend('\beta:3','\beta:7','\beta:10');
```

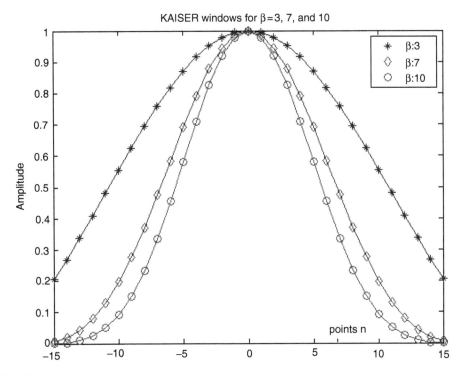

FIGURE 1.41
Plots of the Kaiser windows of R.1.158.

The script file *Kaisers* is executed and the resulting plots are shown in Figure 1.41. Observe from Figure 1.41 that the larger the β the sharper the shape.

R.1.159 The script file *win_tri_rect_bar* returns the plots of the triangular, rectangular, and Bartlett window plots using an $N = 31$ approximation. The resulting plots are shown in Figure 1.42.

MATLAB Solution
```
% Script file: win _ tri _ rect _ bar
% This file returns the plots of the
% TRIANGULAR, RECTANGULAR and
% BARTLETT windows
% using a 31 point approximation
%**********************************
clc; clf;
n = -15:1:15;
```

```
wtri = triang(31);
wrect = boxcar(31);
wbart = bartlett(31);
plot (n,wtri,'*',n,wrect,'d',n,wbart,'o');
hold
plot (n,wtri,n,wrect,n,wbart);
title ('TRIANGULAR,RECTANGULAR and BARTLETT windows ');
xlabel ('points n');
ylabel ('Amplitude')
legend ('TRIAN','RECT','BART');
axis([-18 18 0 1.3]);
```

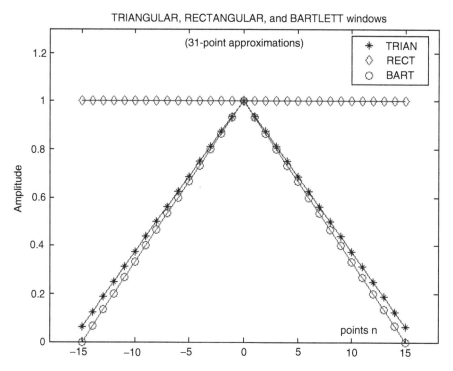

FIGURE 1.42
Plots of the triangular, rectangular, and Bartlett window of R.1.159.

R.1.160 For completeness, a random or noise-like signal is presented below.
    The MATLAB statement *sqrt(P)*randn(1, n)* returns a white Gaussian noise sequence with power *P*.

R.1.161 For example, create the script file *signal_noise* that returns the plots of the following signals:

a. Gaussian white noise

b. The signal—*signal = cos(2πn/64)*

c. The contaminated signal plus noise using *n = 128* elements. Let the white Gaussian signal power be *P = 2.7* watts

MATLAB Solution

```
% Script file: signal _ noise
n = 1:128; P =2.7;
figure(1)
subplot(2,1,1)
white _ noise = sqrt(P)*randn(1,128);
stem(n,white _ noise); hold on;plot(n,white _ noise);
xlabel('discrete time n');
ylabel('Amplitude'); title('Gaussian noise');
subplot(2,1,2)
signal = cos(2*pi*n/64);
stem(n,signal);hold on;plot(n,signal);
title('cosine wave with period=64')
xlabel('discrete time n');
ylabel('Amplitude');

figure(2)
signal _ noise = white _ noise+signal;
stem(n,signal _ noise);hold on;plot(n,signal _ noise);
title('cosine wave plus white Gaussian noise')
xlabel('discrete time n');
ylabel('Amplitude');
```

The script file *signal_noise* is executed as follows and the results are shown in Figures 1.43 and 1.44.

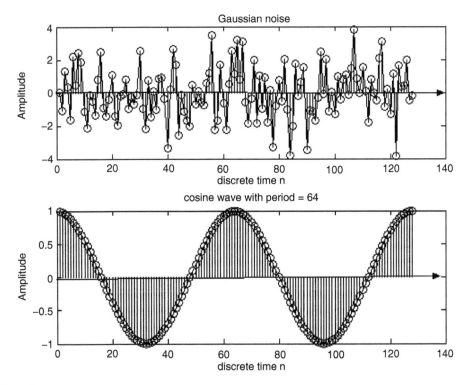

FIGURE 1.43
Plots of parts a and b of R.1.161.

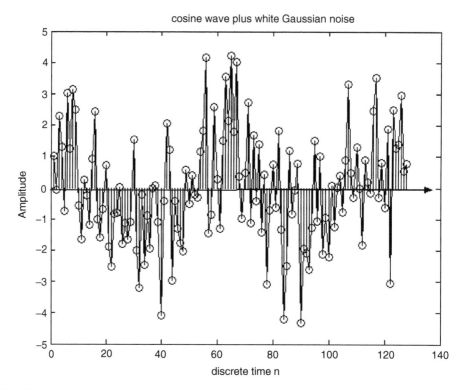

FIGURE 1.44
Plot of part c of R.1.161.

R.1.162 Sound waves can be represented by 1-D vectors while 2-D matrices can be used to represent images (black and white), whereas higher dimension matrices can be used to represent color images and video.

R.1.163 The MATLAB command *wavrecord(N, Fs)* takes the N audio elements sampled at a frequency of *Fs*, directly from an audio input device such as a microphone. The default value for *Fs* is *11,025* Hz. This function can only be used with Windows 95, 98, or NT machines.

R.1.164 The MATLAB command *wavplay(y, Fs)* sends the audio signal defined by the vector y, sampled at a frequency of *Fs* Hertz to an output audio device. Standard audio rates are *8000, 11,025, 22,050,* and *44,100* Hz. The MATLAB default rate is *11,025* Hz.

R.1.165 MATLAB provides with a sound file called *sound(y, Fs)*, and *soundsc(y, Fs)*, that sends the audio signal defined by y to an output audio device. y is assumed to have a magnitude range $-1.0 \leq y \leq 1.0$, and any values outside that range is clipped. The difference between *sound* and *soundsc* is that the latter is autoscale, and y is played as loud as possible. The MATLAB default sound rate is *8192* Hz.

  MATLAB also provides with a speech file named *mtlb.mat* that can be used for testing purposes, consists of 4001 elements, sampled at 1418 Hz.

R.1.166 The following example shows the script file *audio* that returns

  a. The plot of the sinusoidal audio signal is given by $y = cos(2\pi f_o t) + 3\,cos(0.5\pi f_o t)$; where $f_o = 1000$ Hz, and a sample frequency of $Fs = 8000$ Hz (implying that 8000 audio samples per second are processed).

  b. The speech file *mtlb* (Figure 1.45).

MATLAB Solution

```
% Script file: audio
subplot(2,1,1)
fo = 1000;
Fs = 8000;Ts =1/Fs;
t = 0:Ts:1;
y = cos(2*pi*fo.*t)+3*cos(0.5*pi*fo.*t);        % audio sequence
plot(t(1:100),y(1:100));
ylabel('Amplitude');
xlabel('time (sec)');
title('cos(2\pi fo t)+3 cos(0.5\pi fo t) vs.t ')

subplot(2,1,2)
load mtlb; Fs=1418; T=1/Fs;
x = 1:4001;xx=x.*T;                             % speech file
plot(xx,mtlb);
axis([0 4000*T -4 4]);
ylabel('Amplitude ');
title('[Speech file mtlb] vs. t');
xlabel('time (sec)')
```

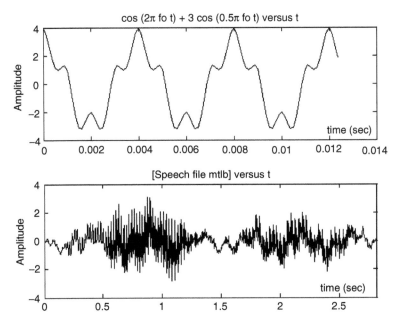

FIGURE 1.45
Plots of audio and speech files of R.1.166.

R.1.167 The MATLAB command *[y, Fs, nbits]* = *wavread('microsoft_file')* returns the vector *y* consisting of the audio wave samples from a *Microsoft file*, at a sampling frequency *Fs*, using *n* bits to encode each sample of *y*, whereas the amplitude of *y* is scaled over the range $-1 \le y \le +1$. Similarly, the command *wavwrite* returns a Microsoft wave file.

R.1.168 The MATLAB command *image(matrix_A)* returns *matrix_A* as an image, where each element of the *matrix_A* is defined by a color. The matrix dimensions may be 2-D ($m \times n$) or 3-D ($m \times n \times r$). In its simplest version, the elements of the 2-D ($n \times m$) matrix are used as indices in the current colormap to define its color.

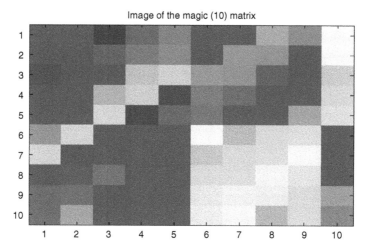

FIGURE 1.46
(See color insert following page 374.) Color plot of the magic matrix of R.1.169.

R.1.169 For example, the color image of the *magic(10)* matrix is processed and displayed as
follows, Figure 1.46:

MATLAB Solution
```
>> image(magic(10))
>> title('Image of the magic(10) matrix')
```

R.1.170 Images can be transformed and processed by the filtering functions presented in
Chapter 6. In general, the filtering process involves sophisticated mathematical
manipulations, that rely on the theory of complex variables and transform theory.

Precisely for these reasons MATLAB provides its users with simple filtering
commands, avoiding all the complicated mathematics used on image processing
defined for 2-D matrices such as *gradianr(matrix_A)* and *del1(matrix_A)*.

The command *gradianr(matrix_A)* returns the numerical gradient, whereas the
command *del1(matrix_A)* returns its derivative (the discrete Laplacian).

R.1.171 Let us illustrate some of the image processing techniques by performing the first
and second derivative using as an illustrative example, the matrix *y = magic(10)*.
The results are shown in Figures 1.47 and 1.48.

MATLAB Solution
```
y = magic(10);
>> ygrad = gradient(y);
>> figure(1);
>> image(ygrad)
>> title('gradient of magic(10)')
>> ydel2 = del2(y);
>> figure(2);
>> image(ydel2)
>> title('second derivative (del2) of magic(10)')
```

R.1.172 Animation and motion of video signals can be accomplished using MATLAB by
displaying pictures or figures one after the other, referred to as frames.

R.1.173 The MATLAB command *getframe* captures the content of the current figure window and creates with it a movie frame. The *getframe* command is placed usually

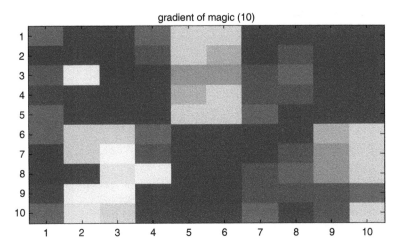

FIGURE 1.47
(See color insert following page 374.) First color derivative plot of the magic matrix of R.1.169.

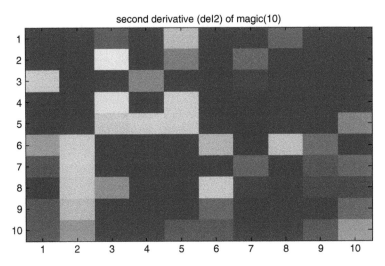

FIGURE 1.48
(See color insert following page 374.) Second color derivative plot of the magic matrix of R.1.169.

in a (*for-end*) loop, to assemble an array of movie frames, spaced over time by its looping index.

R.1.174   For example, the following MATLAB sequence can be used to assemble a movie, based on the contents of the current figure window over *n*.

MATLAB Solution
```
for frame=1:n
    M(frame) = getframe;   % captures the current figure as a frame
                                of M
end
```

R.1.175   The MATLAB command *movie(M)* displays the recorded *M* array containing all the frames of *M*, one frame after the other in sequential ascending order: *frame(1), frame(2), frame(3), …, frame(n − 1), frame(n).*

1.4   Examples

Example 1.1

Create the script file *disc_func* that returns the plot of the following discrete function (Figure 1.49):

$$f(n) = -5\delta(n + 4) + \delta(n) + 2\delta(n - 3)$$

MATLAB Solution
```
% Script file : disc _ func
n = -10:10;                                  % vector n from -10 to +10
fn = [zeros(1,6) -5 zeros(1,3) 1 zeros(1,2) 2 zeros(1,7)];
yzero = zeros(1,21);
% plot the function f(n)
stem(n,fn)                                   % plot the function f(n)
hold on;plot(n,yzero);
xlabel('time index, n')
ylabel('Amplitude')
axis([-10 10 -7 4])
title('f(n) = -5\delta(n+4) + \delta(n) +2\delta(n-3)')
```

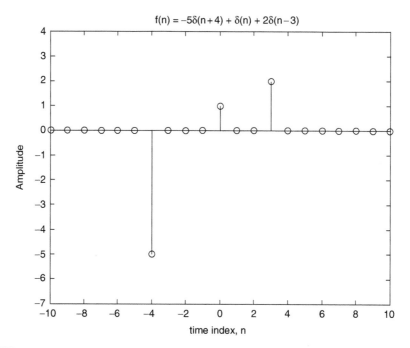

FIGURE 1.49
Plot of Example 1.1.

Example 1.2

Create the script file *analog_func* that returns the plot of the analog signal given by

$$f(t) = u(t) + u(t - 1) + u(t - 2) - 3u(t - 3)$$

over the range $-1 \le t \le 5$ (Figure 1.50).

MATLAB Solution

```
% Script file: analog _ func
t = -1:0.01:5;
t0 = 0;
ut = stepfun(t,t0);              % u(t)
t0 =1;
ut1 =stepfun(t,t0);             % u(t-1)
t0 = 2;
ut2 = stepfun(t,t0);                    % u(t-2)
t0 = 3;
ut3 = -3 * stepfun(t,t0);              % -3u(t-3)
fn = ut+ut1+ut2+ut3;
plot(t,fn);
axis([-1 5 -1 4])
title('f(t) = u(t) + u(t-1) + u(t-2) - 3u(t-3)   ')
xlabel('time t')
ylabel('Amplitude [f(t)]')
```

The script file *analog-func* is executed and the result is shown in Figure 1.50.

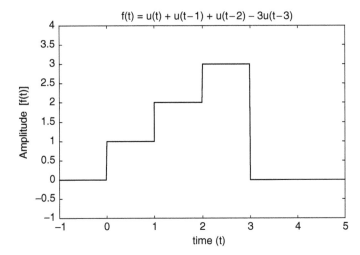

FIGURE 1.50
Plot of Example 1.2.

Example 1.3

Create the script file *sequences* that uses MATLAB to verify graphically that the discrete sequences $f_1(n)$ and $f_2(n)$, defined as follows, over the range $-10 \leq n \leq 10$ are equal:

$$f_1(n) = 3u(-n) * u(n + 5)$$

and

$$f_2(n) = 3[u(n + 5) - u(n + 1)]$$

MATLAB Solution

```
% Script file: sequences
clc; clf;
yzero = [zeros(1,21)]; n=-10:10;
n0 = 0;un = stepfun(n, n0);
```

```
urev = fliplr(un);n1 = -5;
un _ 5 = stepfun(n, n1); fn1=3*(urev .* un _ 5);
subplot(2,1,1)
stem(n,fn1);hold on; plot(n,yzero)
title('f1(n) vs. n');
ylabel('Amplitude [f1(n)]'); xlabel('Discrete time n')
un1=stepfun(n,1); fn2=3*(un _ 5 -un1);
axis([-10 10 -1 4])
subplot(2,1,2)
stem(n, fn2);hold on; plot(n,yzero);
title('f2(n) vs. n');
ylabel('Amplitude [f1(n)]'); xlabel(' Discrete time n ');
axis([-10 10 -1 4])
```

The script file *sequences* is executed and the results are shown in Figure 1.51.

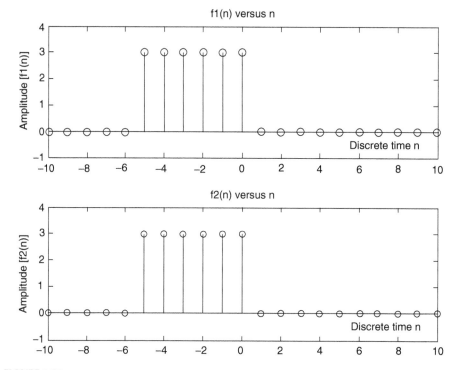

FIGURE 1.51
Plots of Example 1.3.

Example 1.4

Create the script file *graphs* that returns the plots of the following time functions:

$$f_1(t) = 4.5e^{-1.4t}\cos(8.3t + 1.25)u(t)$$

and

$$f_2(t) = f_1(-t)$$

over the range $-5 \le t \le 5$.

MATLAB Solution
```
% Script file: graphs
t = -5:.01:5; t0 = 0;
u = stepfun(t,t0);
f1 = 4.5*exp(-1.4.*t).*cos(8.3.*t+1.25).*u;
subplot(2,1,1);
plot(t,f1);axis([-5 5 -4 5]);
title('[f1(t)= 4.5 exp(-1.4.*t).*cos(8.3 *t + 1.25).*u(t)] vs. t');
ylabel('Amplitude [f1(t)]'); xlabel('t (time)');
f2 = fliplr(f1);
subplot(2,1,2);
plot(t,f2);axis([-5 5 -4 5]);
title('[f2(t)= f1(-t)] vs. t');
ylabel('Amplitude [f2(t)]'); xlabel('t (time)');
```
The script file *graphs* is executed and the results are shown in Figure 1.52.

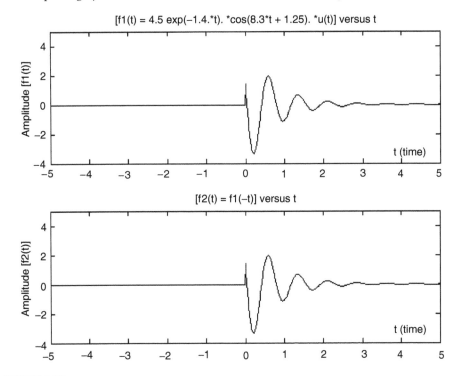

FIGURE 1.52
Plots of Example 1.4.

Example 1.5

Create the script file *disc_square_triang* that returns the plots of the following periodic discrete signals, with period $T = 2$, over the range $0 \le n \le 4$, using the square and sawtooth functions

$$f_1(n) = \begin{cases} 3 & 0 < n \\ -3 & 1 < n < 2 \end{cases}$$

$$f_2(n) = \begin{cases} 6n - 3 & 0 < n < 1 \\ -6n + 9 & 1 < n < 2 \end{cases}$$

MATLAB Solution

```
% Script file: disc _ square _ triang
n = 0:0.1:4; N = 2;
f1 = 3* square((2*pi*n/N),50);
f2 = 3* sawtooth((2*pi*n/N),0.5);
subplot(2,1,1);
stem(n,f1);
title(' f1(n) vs. n, using the square function')
xlabel('time index n');ylabel('Amplitude')
subplot(2,1,2)
stem(n,f2);title('f2(n) vs. n, using the sawtooth function')
xlabel('time index n');ylabel('Amplitude');
```

The script file *disc_square_triang* is executed and the results are shown in Figure 1.53.

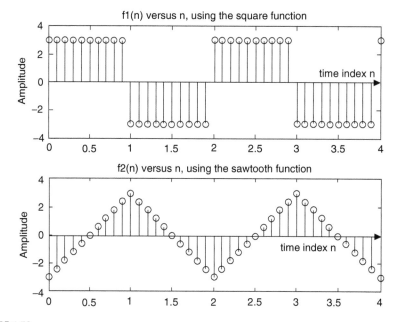

FIGURE 1.53
Plots of Example 1.5.

Example 1.6

Let $f(n) = 3e^{(-0.3+j(pi/4))n}$, over the range $0 \leq n \leq 15$. Create the script file *disc_plots* that returns the following plots:

a. Real $[f(n)]$ versus n

b. Imaginary $[f(n)]$ versus n

c. Magnitude of $[f(n)]$ versus n

d. Phase of $[f(n)]$ versus n

MATLAB Solution

```
% Script file: disc _ plots
n = 0:1:15;
a = -0.3+j*pi/4;yzero=zeros(1,length(n));
fn = 3*exp(a*n);
```

```
subplot(2,2,1);
stem (n, real(fn)); hold on; plot(n,yzero);
title ('real part of [f(n)] vs. n')
ylabel('real part of [f(n)]');
axis([0 15 -2 3])
subplot(2,2,2);
stem(n, imag(fn));
hold on;plot(n,yzero);
ylabel('imaginary part of [f(n)]')
title('imaginary part of [f(n)] vs. n')
ylabel('imaginary part of [f(n)]');
axis([0 15 -2 2])
subplot(2,2,3);
stem(n, abs(fn));
title('magnitude of [f(n)] vs. n')
xlabel('time index n')
ylabel('magnitude of [f(n)]')
subplot(2,2,4);
stem(n,(180/pi) * angle(fn));
hold on; plot(n,yzero);
title('phase of [f(n)] vs n')
xlabel('time index n')
ylabel('degrees')
```

The script file *disc_plots* is executed and the results are shown in Figure 1.54.

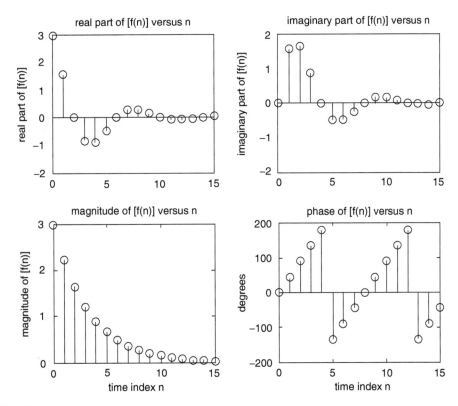

FIGURE 1.54
Plots of Example 1.6.

Example 1.7

Create the script file *analog_plots* that returns the plot of each of the real value exponential signals shown as follows over the range $-5 \le t \le 6$.

a. $f_1(t) = e^{-t}u(t)$
b. $f_2(t) = e^{-1}$
c. $f_3(t) = f_1(-t)$
d. $f_4(t) = -e^{-t}u(t)$
e. $f_5(t) = f_2(t) \cdot [(u(t-2) - u(t-3))]$
f. $f_6(t) = f_1(t) + f_1(-t)$
g. $f_7(t) = f_5(t-1)$
h. $f_8(t) = f_4(t) + [(u(t-2) - u(t-3)]$

MATLAB Solution
```
% Script file: analog _ plots
t = -6:0.01:6;
t0 = 0;
u0 = stepfun(t, t0);
f1 = exp(-t).*u0;
f2 = exp(-t);
f3 = fliplr(f1);
tt = fliplr(t);
f4 = -1.*f1;

figure(1)
subplot(2,2,1);
plot(t, f1);
title('f1(t) vs t,(Example 1.7)');
ylabel('f1(t)')
axis([-6 6 -.5 1.2]);grid on
subplot(2,2,2);
plot(t, f2);
title('f2(t) vs t,(Example 1.7)');
ylabel('f2(t)')
axis([-6 6 -10 200]);grid on
subplot(2,2,3);
plot(t, f3);
title('f3(t) vs t,(Example 1.7)');
ylabel('f3(t)')
axis([-6 6 -.5 1.2]);
grid on;
xlabel('t')
subplot(2,2,4);
plot(t, f4);grid on;
title('f4(t) vs t,(Example 1.7)');
ylabel('f4(t)')
axis([-6 6 -1.2 .5])
xlabel('t');
```

```
figure(2);
ut35 = stepfun(t, 2) - stepfun(t, 3);
f5 = f2.*ut35;
t _ 1 = t+1;
f8 = f4 + ut35;
subplot(2,2,1);
plot(t, f5);
axis([1 4 -.1 .2]);
title('f5(t) vs t,(Example 1.7)');
ylabel('f5(t)'); grid
subplot(2,2,2);
plot(t, f1+f3); axis([-6 6 -.5 1.5]);
title('f6(t) vs t,(Example 1.7)');
ylabel('f5(t)');grid
subplot(2,2,3);
plot(t _ 1, f5); axis([2 4 -.1 .25]);
title('f7(t) vs t,(Example 1.7)');
ylabel('f5(t)'); xlabel('t');;grid
subplot(2,2,4);
plot(t, f8); axis([-6 6 -1.2 1]);
title('f8(t) vs t,(Example 1.7)');
ylabel('f5(t)');xlabel('t'); grid
```

The resulting plots are shown in Figures 1.55 and 1.56.

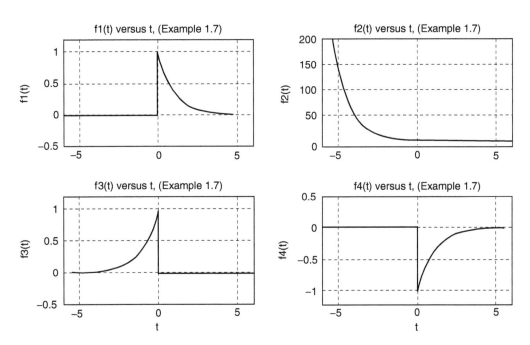

FIGURE 1.55
Plots of $f_1(t), f_2(t), f_3(t)$, and $f_4(t)$ of Example 1.7.

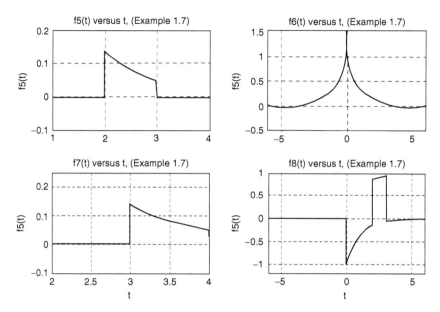

FIGURE 1.56
Plots of $f_5(t), f_6(t), f_7(t),$ and $f_8(t)$ of Example 1.7.

Example 1.8

Let $f(n)$ be given by

$$f(n) = \begin{cases} 0 & n < 0 \text{ and } n > 2 \\ 1 & 0 < n \leq 1 \\ n & 1 < n \leq 2 \end{cases}$$

Create the script file *f_n* that returns the following:

a. Plot of $[f(n)]$ versus n
b. Plot of $[f(n)]^2$ versus n
c. Plot of $[-f(n)]$ versus n
d. Plot of $[f(n-2)]$ versus n
e. The energy of $[f(n)]$
f. The power of $[f(n)]$

MATLAB Solution
```
% Script file: f _ n
fn = [zeros(1,100) ones(1,10) [1.1:0.1:2] zeros(1,81)];
n = -10:0.1:10;
subplot(2,2,1);
plot(n, fn);
title('f(n) vs n');ylabel('Amplitude');
axis([-1 5 -.2 2.3]);
subplot(2,2,2);
fnsquare = fn.^2;
plot(n, fnsquare);ylabel('Amplitude');
title('[f(n)]^2 vs. n');
```

```
axis([-1 5 -.2 4.3]);
subplot(2,2,3);
f3=-1.*fn;          % creates -f(n)
plot(n,f3);title('-f(n) vs. n');
axis([-1 5 -2.3 .2]);
ylabel('Amplitude');xlabel('time index n');
subplot(2,2,4)
shiftn = n+2;plot(shiftn, fn)
title('f(n-2) vs. n');
ylabel('Amplitude');xlabel('time index n');
axis([0 5 -.5 2.5]);
disp('********** RESULTS ARE :  ************************')
Energy _ fn = sum(fnsquare)
disp('    ')
Power _ fn = Energyfn/length(n)          % asuming, f (n) is periodic T=20
disp('(in joules and watts)')
disp('*************************************************')
```

The script file *f_n* is executed and the results are shown as follows and in Figure 1.57.

```
*************** RESULTS ARE:*******************
Energy _ fn =
            34.8500
Power _ fn =
            0.1734
(in joules and watts)
*************************************************
```

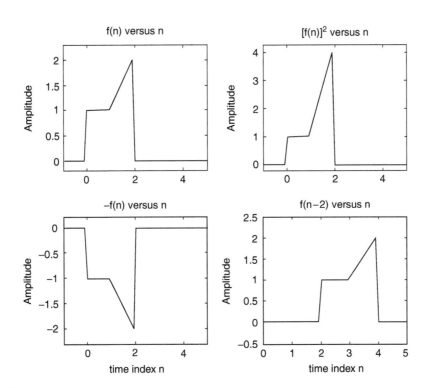

FIGURE 1.57
Plots of *f(n)*, *f(n)²*, *–f(n)*, and *f(n – 2)* of Example 1.8.

Example 1.9

Given the following analog transfer function:

$$H(s) = \frac{s^3 + 3s^2 + 5s - 3}{9s^5 - 6s^4 + 2s^3 - 4s^2 + s + 5}$$

Create the script file *responses* that returns the plots of the impulse and step responses over the range given by $30 \leq t \leq 40$.

MATLAB Solution
```
% Script file: responses
P= [1 3 5 -3];
Q= [9 -6 2 -4 1 5]; t = 0:0.5:50;
subplot(2,1,1);
Y1 = impulse(P,Q,t);
plot(t, Y1);
axis([30 40 -1e10 500e10])
title('analog impulse response');
xlabel('time');ylabel('Amplitude ')
subplot(2,1,2);
Y2 = step(P,Q,t);
plot(t, Y2);ylabel('Amplitude ');
axis([30 40 -16e9 1e13]);
title('analog step response'); xlabel ('time')
```

The script file *responses* is executed and the results are shown in Figure 1.58.

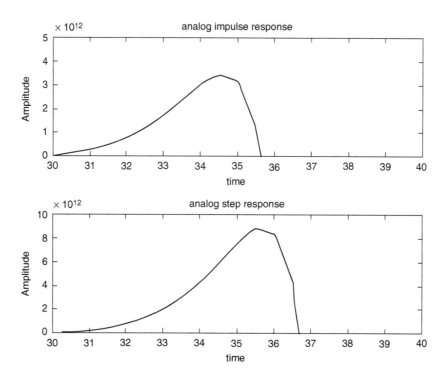

FIGURE 1.58
Plots of system responses of Example 1.9.

Example 1.10

Create the script file *period_seq* that returns the plots of the following periodic sequences defined as follows over one period:

a. $f_1(n) = \sum_{k=0}^{25}(-1)^k\, \delta(n-k)$, for $0 \le k \le 25$.

b. $f_2(n) = \begin{cases} \sin(n\pi) & \text{for } 0 \le n \le 1 \\ 0 & \text{for } 1 < n \le 2' \end{cases}$
over $0 \le n \le 6$.

c. $f_3(n) = n$, with period $-4 \le n \le 4$, over $-5 \le n \le 30$.

d. Random periodic function $f_4(n)$, with a period of 1, over $-1 \le n \le 3$.

MATLAB Solution
```
n = 1:25;
f1n = [(-1).^n.*ones(1,25)];
yzero = zeros(1,25);
subplot(2,2,1);
stem(n,f1n);hold on; plot(n,yzero);
ylabel('Amplitude [f1(n)]');xlabel('n');
axis([-1 25 -1.3 1.3])
title ('f1(n) vs. n');
% part (b)
nn = 0:0.1*pi:0.9*pi;
Y = sin(nn);
f2n= [Y zeros(1,10) Y zeros(1,10) Y zeros(1,10)]; m=0:.1:5.9;
subplot(2,2,2);stem(m, f2n);axis([-.1 6 -.1 1.3]);
ylabel('Amplitude [f2(n)]'); xlabel('n');
title('f2(n) vs. n');
% part c
f3 = -4:1:4; f3n = [f3 f3 f3 f3];
nnn = linspace(-4,28,36);
subplot(2,2,3); yzer =zeros(1,length(nnn));
stem(nnn, f3n);hold on; plot(nnn,yzer);
xlabel('n');axis([-5 30 -5 5]);
ylabel('Amplitude [f3(n)]');
title('f3(n) vs. n')
% part d
f4 = 10*rand(1,10) ;f4n=[f4 f4 f4 f4];
n4 =linspace(-1,3,40);
subplot (2,2,4);yze=zeros(1,length(n4));
stem(n4,f4n);hold on; plot(n4,yze);
axis ([-1 3 -1 11]);xlabel('n');
ylabel('Amplitude [f4(n)]');
title('f4(n) vs. n')
```

The script file *period_seq* is executed and the results are shown in Figure 1.59.

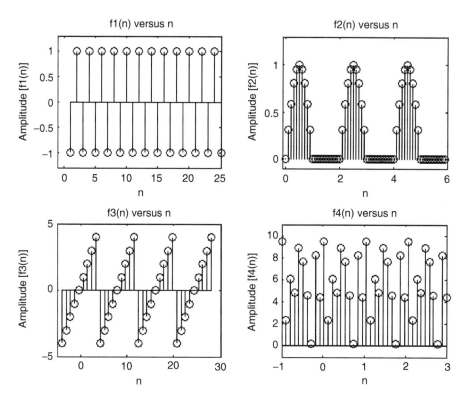

FIGURE 1.59
Plots of $f_1(n)$, $f_2(n)$, $f_3(n)$, and $f_4(n)$ of Example 1.10.

Example 1.11

Given the discrete transfer function

$$H(z) = \frac{0.8 - 0.45z^{-1} + 0.35z^{-2} + 0.01z^{-3}}{1 + 0.85z^{-1} - 0.43z^{-2} - 0.58z^{-3}}$$

Create the script file *disc_tranf_func* that returns the following:

a. The discrete impulse and step response plots by creating an impulse and step sequence as inputs
b. Repeat part a by using the discrete MATLAB functions *dimpulse* and *dstep*, and compare the result obtained with the results of part a

MATLAB Solution
```
% Script file: disc _ tranf _ func
n = 30;
P = [0.8 -0.45 0.35 0.01];Q = [1 0.85 -0.43 -0.58];
I = [1 zeros(1, n-1)];                    % impulse sequence
S = [ones(1, n)] ;                        % step sequence
```

```
figure(1)
subplot(2,2,1);
Y1 = filter(P, Q, I); n=0:29; yzero=zeros(1,30);
stem(n, Y1);hold on; plot (n,yzero);
title('impulse response using a sequence');
ylabel ('Amplitude imp. resp.');
subplot (2,2,2);
Y2 = filter(P,Q,S);
stem (n, Y2);hold on; plot(n,yzero);
title('step response using a sequence');
ylabel('Amplitude step resp.');
subplot(2,2,3);
Y3 = dimpulse(P, Q, n);
stem(n, Y3);hold on; plot(n,yzero);
title('impulse response using dimpulse');
ylabel ('Amplitude imp.resp.');
xlabel ('time index n');
subplot (2,2,4);
Y4 = dstep(P, Q, n);
stem(n, Y4); hold on; plot(n,yzero);
title('step response using dstep');
ylabel('Amplitude step resp.');
xlabel('time index n')
```

The script file *disc_tranf_func* is executed and the results are shown in Figure 1.60.
Note that the results of part a fully agree with the results obtained in part b.

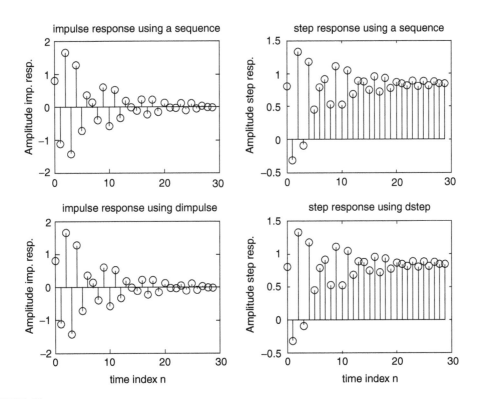

FIGURE 1.60
Plots of system responses of Example 1.11.

Example 1.12

The set of signals given by

$$f(t) = e^{jnw_o t} \quad \text{for } n = 0, \pm 1, \pm 2, ..., \pm \infty$$

constitutes an orthogonal family for any arbitrary w_o (where w_o is referred to as the fundamental frequency, see Chapter 4 for more details), over the time interval given by $T = 2\pi/w_o$, if the following relation is satisfied:

$$\int_{-\pi/w_o}^{\pi/w_o} \exp(jnw_o t) \cdot \exp(-jmw_o t)dt = \begin{cases} 2\pi/w_o & \text{for } n = m \\ 0 & \text{for } n \neq m \end{cases}$$

Create the MATLAB script file *ortog* that verifies the preceding relation for the following arbitrary values of w_o, n, and m:

a. $w_o = 2, n = 5,$ and $m = 5$
b. $w_o = 2, n = 5,$ and $m = 6$
c. $w_o = 7, n = 8,$ and $m = 10$
d. $w_o = 7, n = 10,$ and $m = 10$

MATLAB Solution
```
% Script file: ortog
syms check_a check_b check_c check_d expon t ;
expon = exp(j*5*2*t)*exp(-j*5*2*t);%wo=2,n=5,m=5
check_a = int(expon,-pi/2,pi/2);
% f = vpa(check_a)
expon = exp(j*5*2*t)*exp(-j*6*2*t);%wo=2,n=5,m=6
check_b = int(expon,-pi/2,pi/2);
expon = exp(j*8*7*t)*exp(-j*10*7*t);%wo=7,n=8,m=10
check_c = int(expon,-pi/7,pi/7);
expon = exp(j*10*7*t)*exp(-j*10*7*t);%wo=7,n=10,m=10
check_d = int(expon,-pi/7,pi/7);
disp('***************RESULTS*********************************')
disp('The results for parts (a),(b),(c) and (d) are given by :')
results = [check_a check_b check_c check_d];
disp(results)
disp('**********************************************************')
```

Back in the command window the script file *ortog* is executed and the results are shown as follows:

```
>> ortog

    *********************** RESULTS ***************************
       The results for parts (a), (b), (c) and (d) are given by:
                  [    pi,    0,    0,    2/7*pi ]
    **********************************************************
```

Observe that the results obtained clearly confirm that the exponential family of functions, given by $f(t) = e^{jnw_o t}$, for $n = 0, \pm 1, \pm 2, ..., \pm \infty$ constitute an orthogonal family.

Example 1.13

Repeat Example 1.12, by creating the script file *ortho_sin* that verifies that the sinusoids constitute an orthogonal family by evaluating the following cases:

a. *sin(nwot), sin(mwot)*

b. *sin(nwot), cos(mwot)*

c. *cos(nwot), cos(mwot)*

d. *sin(nwot), sin(nwot)*

e. *sin(nwot), cos(nwot)*

f. *cos(nwot), cos(nwot)*

for an arbitrary $n = 5, m = 7$, and $w_o = 2$ over the period $T = 2\pi/w_o$.

MATLAB Solution

```
% Script file: ortho _ sin
syms check _ a check _ b check _ c check _ d check _ e check _ f sins   t
sins=sin(5*2*t)*sin(7*2*t);%wo=2,n=5,m=7
check _ a=int(sins,-pi/2,pi/2);
sins=cos(5*2*t)*sin(7*2*t);
check _ b=int(sins,-pi/2,pi/2);
sins=cos(5*2*t)*cos(7*2*t);
check _ c=int(sins,-pi/2,pi/2);
sins=sin(2*5*t)*sin(2*5*t);%wo=2,n=5,m=5
check _ d=int(sins,-pi/2,pi/2);
sins=sin(2*5*t)*cos(2*5*t);
check _ e=int(sins,-pi/2,pi/2);
sins=cos(2*5*t)*cos(2*5*t);
check _ f=int(sins,-pi/2,pi/2);
disp('****************RESULTS********************************')
disp('The results for parts (a),(b),(c),(d),(e) and (f) are given below :')
results=[check _ a check _ b check _ c check _ d check _ e check _ f];
disp(results)
disp('*********************************************************')
```

Back in the command window, the script file *ortho_sin* is executed and the results are indicated as follows:

```
>> ortho _ sin

*************************** RESULTS ****************************
  The results for parts (a), (b), (c), (d), (e)   and (f) are given below:
                  [   0,  0,   0, 1/2*pi, 1/2*pi, 1/2*pi]
****************************************************************
```

Note that the preceding results clearly indicate that the sinusoids constitute an orthogonal family. Not a surprising result, since the complex exponentials constitute an orthogonal family (Example 1.12), and are related to the sinusoidals, by the Euler's identities.

Example 1.14

Let *f(t)* be defined as

$$f(t) = \begin{cases} 0 & \text{for } -10 \leq t < -5 \\ 10 & \text{for } -5 \leq t < 0 \\ -t + 10 & \text{for } 0 \leq t \leq 10 \end{cases}$$

Create the script file *even_odd* that returns, using 201 points over the range $-10 \le t \le 10$, the following:

    a. Plots of $f(t)$ versus t, $f_e(t)^*$ versus t, and $f_o(t)$ versus t

    b. Verifies graphically that $f(t) = f_e(t) + f_o(t)$

    c. The energies of $f(t)$, $f_e(t)$, and $f_o(t)$

    d. Verifies that the energy $\{f(t)\} = energy\{f_e(t)\} + energy\{f_o(t)\}$

MATLAB Solution

```
% Script file: even _ odd
t = -10:0.1:10;                        % 201points
yzero = zeros(1,201);
y1 = [zeros(1,50) 10*ones(1,50)];
y2 = [10:-.1:0];
y = [y1 y2];

figure(1)
subplot(3,1,1)
plot(t,y);axis([-10 10 -5 13]);
title(' f(t) vs. t ')
ylabel('Amplitude [f(t)]');xlabel('time');
flipft = fliplr(y);
feven =.5*(y+flipft);
fodd =.5*(y-flipft);
subplot(3,1,2);
plot(t,feven); axis([-10 10 -5 13]);
ylabel('Amplitude [feven(t)]');xlabel('time')
title('feven(t) vs.t')
subplot(3,1,3)
plot(t,fodd,t,yzero);
ylabel('Amplitude [fodd(t)]'); xlabel('time')
title('fodd(t) vs.t')

figure(2)
check = feven+fodd;
ysquare = y.^2;
energyft = trapz(t,ysquare);
poweraveft = (1/200)*energyft;
fevensq = feven.^2;
energyfeven = trapz(t,fevensq);
foddsq = fodd.^2;
energyfodd = trapz(t,foddsq);
disp('**********************************************')
disp('*********** ENERGY ANALYSIS **************')
disp('**********************************************')
disp('The energy of f(t) =')
disp(energyft)
disp('The energy of feven(t) =')
disp(energyfeven)
disp('The energy of fodd(t) =')
```

---

* Recall that $f_o(t)$ stands for the odd portion of $f(t)$, whereas $f_e(t)$ stands for the even portion of $f(t)$.

```
disp(energyfodd)
disp(' (in joules)')
disp('*********************************************')
disp('*********************************************')
plot(t,check);
title('[feven(t) + fodd(t)] vs. t )');
axis([-10 10 -5 15]);xlabel('time');
ylabel('Amplitude [feven(t) + fodd(t)]')
```

Back in the command window the script file *even_odd* is executed and the results are shown as follows and in Figures 1.61 and 1.62.

```
>> even _ odd

**************************************************
************** ENERGY ANALYSIS *****************
**************************************************
          The energy of f(t) =
                          838.3500
          The energy of feven(t) =
                             796.6750
          The energy of fodd(t) =
                          41.6750

          (in joules)
**************************************************
**************************************************
```

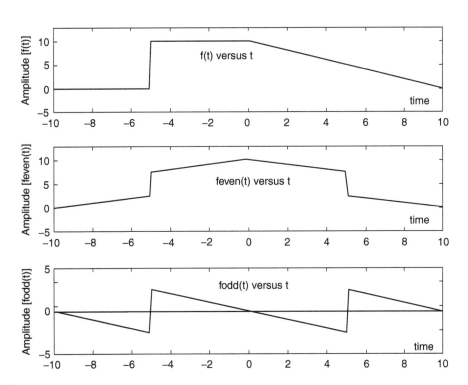

FIGURE 1.61
Plots of even and odd parts of *f(t)* of Example 1.14.

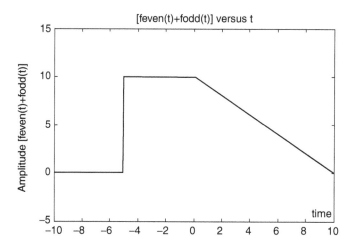

FIGURE 1.62
Plot of adding the even and odd parts of *f(t)* of Example 1.14.

Example 1.15

Let the discrete sequence

$$f(n) = 4n0.8^n u(n)$$

be contaminated by a random noisy signal with a magnitude less than 1.
    Create the script file *averages* that returns the following plots:

a. *f(n)* versus *n*
b. *noise signal[n(t)]* versus *n*
c. *[f(n) + n(t)]* versus *n*
d. *[f(n) + n(t)]* moving average using two terms versus *n*
e. *[f(n) + n(t)]* moving average using three terms versus *n*
f. *[f(n) + n(t)]* moving average using four terms versus *n*
g. Estimate the best moving average approximation for *[f(n) + n(t)]* versus *n*, by executing the script file at least three times and observing and recording the results

Where the *L* point moving average is defined by

$$\{moving \ aver. \ L\} = \frac{1}{L}\sum_{k=0}^{L-1} f(n-k)$$

for *L* = 2, 3, and 4, over the range *0 ≤ n ≤ 20*.

MATLAB Solution
```
% Script file: averages
n = 0:1:20;

figure(1)
subplot(3,1,1)
signal = 4.*n.*(.8.^n);stem(n,signal);hold on;
plot(n,signal,n,signal,'o');
title('Signal vs. n')
ylabel('Amplitude');xlabel('time index n');
subplot(3,1,2)
noise = 2.*rand(1,21)-1.0;y = zeros(1,21);
```

```
plot(n,y);title('Noise vs. n');hold on;
stem(n,noise);hold on;plot(n,y,n,noise);
ylabel('Amplitude');xlabel('time index n');
axis([0 20 -1.3 1.3]);
subplot(3,1,3);
signoi =signal+noise;
stem(n,signoi);title('[Signal + Noise] vs. n');hold on;
plot(n,signoi);ylabel('Amplitude[ signal+noise]');
xlabel(' time index n')

figure(2)
subplot(3,1,1)
N = [.5 .5];
D = 1;
movave2 = filter(N,D,signoi);
plot(n,signal,n,signoi,'s--',n,movave2,'o-.');
legend('signal','signal+noise','moving ave/2term')
title('Various moving averages');ylabel('magnitude')
subplot(3,1,2)
N=[.33 .33 .33];
D=1;
movave3=filter(N,D,signoi);
plot(n,signal,n,signoi,'s--',n,movave3,'o-.');ylabel('magnitude')
legend('signal','signal+noise','moving ave/3term')
subplot(3,1,3)
N = [.25 .25 .25 .25];
D = 1;
movave4 = filter(N,D,signoi);
plot(n,signal,n,signoi,'s--',n,movave4,'o-.');
legend('signal','signal+noise','moving ave/4term')
ylabel('magnitude');xlabel(' time index n')

figure(3)
plot(n,signal,n,signoi,'ks--',n,movave2,'ko-.');
legend('signal','signal+noise','moving ave/2term');
title('Best approximation using moving averages');
ylabel('magnitude');xlabel(' time index n')

figure(4)
err2 = signal-movave2;
stem(n,err2);hold on;plot(n,y,n,err2);
title('error=[mov.ave/2terms ] vs t');
ylabel('error');xlabel(' time index n')
error1= sum(abs(signal-signoi)/21);
error2 = sum(abs(signal-movave2)/21);
error3 = sum(abs(signal-movave3)/21);
error4 = sum(abs(signal-movave4)/21);
disp('**************************************************')
disp('***************ANALYSIS OF ERROR*****************')
disp('**************************************************')
disp('  no ave  2 term ave  3 term ave  4 term ave ');
disp([error1 error2 error3 error4])
disp('**************************************************')
disp('**************************************************')
```

Back in the command window the script file *averages* is executed three times, and the results are shown as follows and in the plots of Figures 1.63 through 1.66.

```
>> averages

***********************************************************
*******************ANALYSIS  OF  ERROR*******************
***********************************************************
    no ave      2 term ave      3 term ave      4 term ave
    0.5389      0.5276          0.7011          0.9793
***********************************************************
***********************************************************

>> averages

***********************************************************
*******************ANALYSIS  OF  ERROR*******************
***********************************************************
    no ave      2 term ave      3 term ave      4 term ave
    0.5272      0.4794          0.6574          0.9033
***********************************************************
***********************************************************

>> averages

***********************************************************
*******************ANALYSIS  OF  ERROR*******************
***********************************************************
    no ave      2 term ave      3 term ave      4 term ave
    0.4144      0.3750          0.5572          0.8223
***********************************************************
***********************************************************
```

Observe that the results obtained by executing the script file *averages* three times are not the same, but in each case the best results are achieved when the moving average uses two terms.

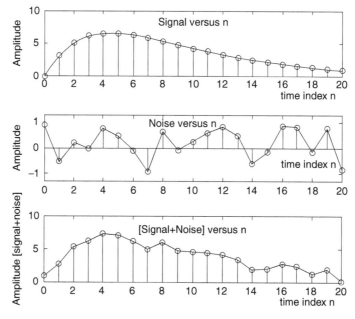

FIGURE 1.63
Plots of $f(n)$, *noise signal[n(t)]*, and *[f(n) + n(t)]* of Example 1.15.

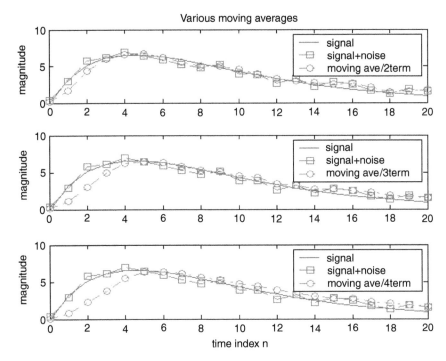

FIGURE 1.64
Plots of various moving averages of Example 1.15.

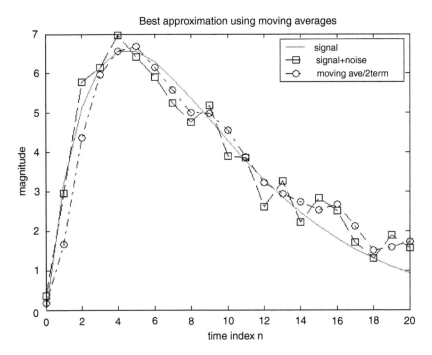

FIGURE 1.65
Plots of $f(n)$, $[f(n) + n(t)]$, and best moving average of Example 1.15.

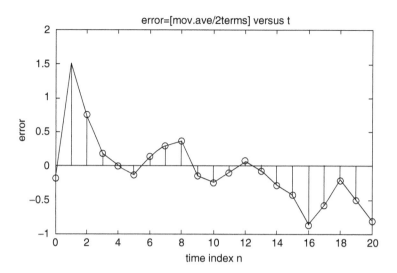

FIGURE 1.66
Error plot of moving average using two terms of Example 1.15.

Example 1.16

Let $f(t) = te^{-1000t}u(t)$

Create the script file *sample_data* that returns the following plots:

a. $f(t)$ versus t, over the range $0 \le t \le 5$ ms
b. $f(nT)$ versus nt, for $T = 0.2$ ms, over the range $1 \le n \le 26$
c. Reconstruct $f(t)$ from $f(nT)$ using
    i. Summations of *sinc* functions (emulating a low-pass filter)
    ii. Stair function
    iii. Spine function
d. *error(t)* versus t, for the reconstruction process when a low-pass filter is used, where

$$error(t) = \left| f(t) - \sum_{n=1}^{N=26} fn(nT_s)\operatorname{sinc}[F_s(t - nT_s)] \right|, \quad \text{for } F_S = \frac{1}{T_S}$$

MATLAB Solution
```
% Script file: sample _ data
t = 0:0.0005:0.005;          %  26 time samples
fa = t.*exp(-1000.*t);

figure(1)
subplot(2,1,1)
plot(t,fa);title('f(t) vs t')
ylabel('Amplitude');xlabel('time t (msec)');
subplot(2,1,2)
Ts = 0.0002; n = 0:1:25; Fs=1/Ts;
nTs = n*Ts;
fd = nTs.*exp(-1000*nTs);
stem(nTs,fd);hold on;
plot(nTs,fd);
title('f(t) sample with Ts=.2msec.');
ylabel('Amplitude'); xlabel('time index n (msec)')
```

```
figure(2)                     % reconstructions
f25 = 0;
for k = 1:1:26;
    fr25 = fd(k)*sinc(Fs*(t-k*Ts));
    f25 = f25+fr25;
end
subplot(3,1,1)
plot(t,f25,'ko-',t,fa,'ks-.'); legend('sinc-reconst','f(t)');
title('Reconstruction of f(t) ')
ylabel('Amplitude')
subplot(3,1,2)
stem(nTs,fd);hold on;
stairs(nTs,fd); ylabel('Amplitude');legend('stairs')
subplot(3,1,3)
y = spline(nTs,fd,t);
plot(nTs(1:2.5:26.5),y); legend('spline');
ylabel('Amplitude')
xlabel('time (sec)')

figure(3)
error=abs(fa-f25);
plot(t,error);ylabel('magnitude')
title('error(t) = abs [f(t) - Reconstruction (Sums(sinc))]')
xlabel('time(sec)')
```

Back in the command window the script file *sample_data* is executed and the results are shown in Figures 1.67 through 1.69.

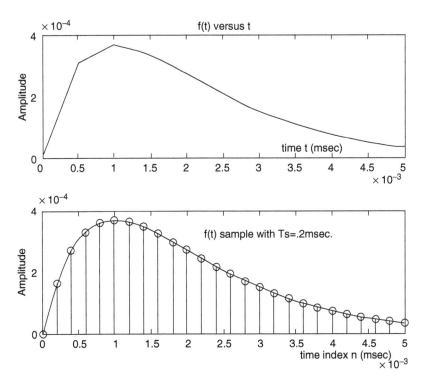

FIGURE 1.67
Plots of *f(t)* and *f(nT)* of Example 1.16.

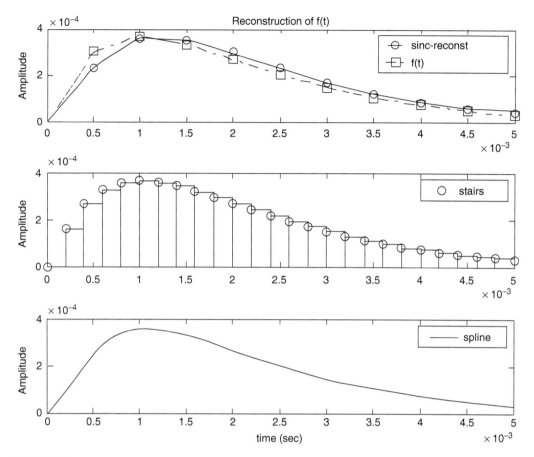

FIGURE 1.68
Reconstruction plots for *f(t)* of Example 1.16.

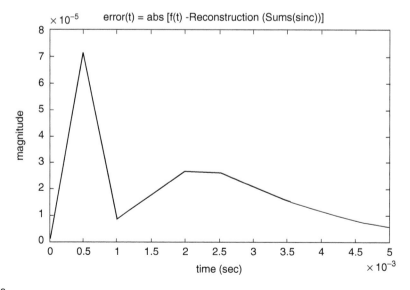

FIGURE 1.69
Reconstruction plot error for *f(t)* of Example 1.16.

Example 1.17

Create the script file *up_down_samples* that returns the approximation plots of the following time discrete function:

$$f(n) = \cos(2\pi\,0.05n) + 2\sin(2\pi 0.03n)$$

for a length of $N = 100$ samples, for the following cases:

1. Down-sample or decimate the sequence $f(n)$ with the integer factors of $M = 2, 4,$ and 10.
2. Up-sample or interpolate the sequence $f(n)$ with the integer factors of $L = 2, 4,$ and 10.
3. Resample $f(n)$ by the ratio of the two integers given by L/M, for the following cases: $L = 3, M = 2,$ and $L = 2, M = 3$.

MATLAB Solution

```
% Script file: up _ down _ samples
n = 0:1:99;
f = cos(2*pi*.05*n)+2*sin(2*pi*.03*n);
yzero = zeros(1,500);


figure(1);                      % down-sample with M=2, 4, 10
subplot(2,2,1)
stem(n,f);hold on; plot(n,f,n,yzero(1:100))
title('f(n) vs n'); ylabel('Amplitude');
xlabel('time index n');
subplot(2,2,2)
gm2 = decimate(f,2,'fir');
nm2 = 0:100/2-1;
stem(nm2,gm2(1:50)); hold on; plot(nm2,yzero(1:50));
title('f(n) is decimated with M=2'); ylabel('Amplitude');
xlabel('time index n');
subplot(2,2,3)
gm4 = decimate(f,4,'fir');
nm4 = 0:100/4-1;
stem(nm4,gm4(1:25));hold on;
plot(nm4,gm4(1:25),nm4,yzero(1:25));
ylabel('Amplitude'); xlabel('time index n');
title('f(n) is decimated with M=4')
subplot(2,2,4)
gm10 = decimate(f,10,'fir');
nm10=0:100/10-1;
stem(nm10,gm10(1:10));hold on;
plot(nm10,gm10(1:10),nm10,yzero(1:10));
```

```
ylabel('Amplitude');xlabel('time index n')
title('f(n) is decimated with M=10')


figure(2);                          % up-sample with L = 2, 4, 10
subplot(2,2,1)
stem(n,f); hold on; plot(n,f,n,yzero(1:100))
title('f(n) vs n');ylabel('Amplitude');
xlabel('time index n');
subplot(2,2,2)
gL2 = interp(f,2);
nL2 = 0:100*2-1;
stem(nL2,gL2(1:200));hold on; plot(nL2,yzero(1:200));
title('f(n) is upsampled with L=2'); ylabel('Amplitude');
xlabel('time index n');
subplot(2,2,3)
gL4 =  interp(f,4);
nL4 = 0:100*4-1;
stem(nL4,gL4(1:400)); hold on;
plot (nL4,yzero(1:400));
ylabel('Amplitude');xlabel('time index n');
title('f(n) is upsampled with L=4')
subplot(2,2,4)
gL10= interp(f,10);
nL10 = 0:100*10-1;
stem(nL10,gL10(1:1000));hold on;
ylabel('Amplitude'); xlabel('time index n')
title('f(n) is upsampled with L=10')


figure(3)
                  % re-sample by ratio of L=3/M=2 & L=2/M=3
subplot(3,1,1)
stem(n,f);hold on; plot(n,f,n,yzero(1:100))
title('f(n) vs n');
ylabel('Amplitude'); xlabel('time index n');
subplot(3,1,2)
gr32= resample(f,3,2);
nr32 = 0:100*3/2-1;yzeros = zeros(1,length(nr32));
stem(nr32,gr32(1:100*3/2));hold on;plot(nr32,yzeros);
ylabel('Amplitude'); xlabel('time index n');
subplot(3,1,3)
gr23 = resample(f,2,3);
nr23 = 0:100*2/3-1;
yzero = zeros(1,length(nr23));
stem(nr23,gr23(1:100*2/3));
hold on;plot(nr23,yzero);
axis([0 65 -5 5]);
ylabel('Amplitude'); xlabel('time index n')
```

Back in the command window, the script file *up_down_samples* is executed and the results are shown in Figures 1.70 through 1.72.

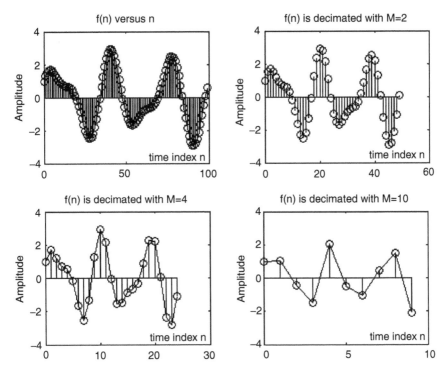

FIGURE 1.70
Decimation plots of *f(t)* of Example 1.17.

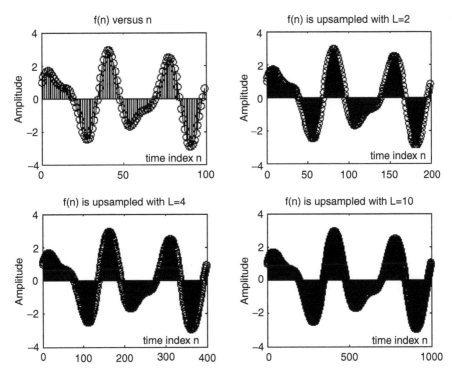

FIGURE 1.71
Interpolation plots of *f(t)* of Example 1.17.

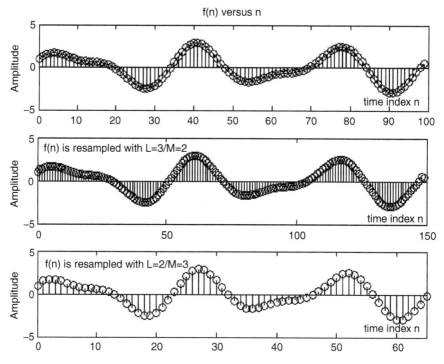

FIGURE 1.72
Plots of *f(t)* and resampled *f(t)* with notes given by *L/M* of Example 1.17.

Example 1.18

An analog communication system consists of an information signal or message given by $m(t) = 3\cos(5t)$, and a high-frequency carrier given by $f_c(t) = \cos(50t)$.

1. Create the script file *modulation* that returns the following plots:
   a. *m(t)* versus *t*
   b. $f_c(t)$ versus *t*
   c. Amplitude modulated *(AM) signal* versus *t*
   d. Angle modulated *(FM/PM) signal* versus *t*
2. Let us assume now that the message signal *m(t)* is a binary-periodic signal with period $T = 2$, defined as follows:

$$m(t) = \begin{cases} 3 & 0 < t \le 1 \\ 1 & 1 < t \le 2 \end{cases}$$

Obtain the plots of the following communication signals:

a. *ASK* versus *t*
b. *FSK* versus *t*
c. *PSK* versus *t*

MATLAB Solution

```
% Script file: modulation

figure(1)
subplot(2,2,1);                        % analog signals
fplot('3*cos(5*t)',[0 3 -6 6]);
title('Information signal: 3*cos(5*t)');
ylabel('Amplitude');xlabel('time');
subplot(2,2,2)
fplot('cos(50*t)',[0 0.5 -1.5 1.5]);
title('Carrier signal: cos(50*t)');
ylabel('Amplitude');xlabel('time');
subplot(2,2,3)
fplot('3*cos(50*t)*cos(5*t)',[0 3 -6 6]);
title('AM (Amplitude Modulated) signal');
ylabel('Amplitude');
xlabel('time');
subplot(2,2,4)
fplot('3*cos((50*t)+3*cos(5*t))',[0 2 -6 6]);
title('Angle Modulated signal');
xlabel('time');
ylabel('Amplitude');


figure(2)
a=3*ones(1,50);
b=ones(1,50);
clock = [a b a b];                     % discrete/binary signals
t = linspace(0,4,200);
carrier = cos(50*t);
subplot(2,2,1)
plot(t,clock);title('Binary information signal');
axis([0 4 0 4]);ylabel('Amplitude');
xlabel('time');
subplot(2,2,2)
ASK= carrier.*clock;plot(t,ASK);
title('ASK signal')
axis([0 4 -4 4]);ylabel('Amplitude');
xlabel('time');
subplot(2,2,3)
FSK=cos(20*t.*clock);
plot(t,FSK);xlabel('time')
axis([0 4 -1.3 1.3]);ylabel('Amplitude')
title('FSK signal')
subplot(2,2,4)
PSK=cos(50*t+clock);
plot(t,PSK);
title('PSK signal')
xlabel('time'); ylabel('Amplitude');
axis([0 1.5 -1.3 1.3]);
```

Back in the command window the script file *modulation* is executed and the results are shown in Figures 1.73 and 1.74.

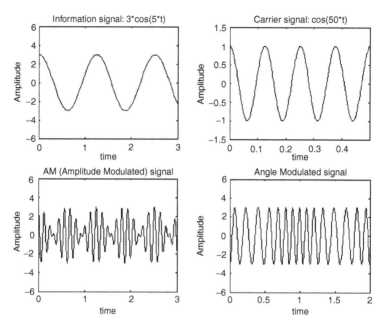

FIGURE 1.73
Plots of standard analog telecommunication signals of Example 1.18.

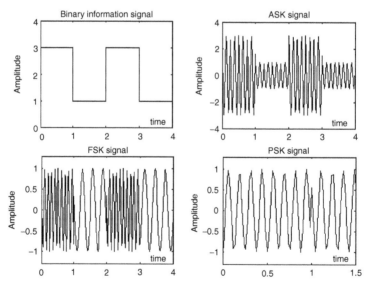

FIGURE 1.74
Plots of ASK, FSK, and PSK telecommunication signals for a binary information signal of Example 1.18.

Example 1.19

Given the following time functions:

$$f_1(t) = cos(2t) + 2\,sin(3t)$$

and

$$f_2(t) = 10\,sin(2t)e^{-t/\pi}$$

over the range $0 \le t \le 2\pi$.

Create the script file *multiplex* that returns the following plots:

a. $f_1(t)$ and $f_2(t)$ are sampled with a sampling rate of $T_s = 2\pi/100$.
b. The multiplex signals of part a.

MATLAB Solution

```
t = linspace(0,2*pi,100);
f1 = cos(2*t)+2*sin(3*t);
f2 = 10*sin(2*t).*exp(-t/pi);
yzero = zeros(1,100);

figure(1)
subplot(2,2,1)
plot(t,f1,'dk',t,f2,'sk',t,yzero);
axis([0 2*pi -6 10])
legend('f1(t)','f2(t)')
title('f1(t) vs. n and f2(t) vs. t')
ylabel('Amplitude'); xlabel('time index n')
subplot(2,2,2)
a = [1 0 1 0 1 0 1 0 1 0];
aa = [a a a a a a a a a a];
b = [0 1 0 1 0 1 0 1 0 1];
bb = [b b b b b b b b b b];
f1samp = f1.*aa;
f2samp = f2.*bb;
stem(t(1:2:100),f1(1:2:100));hold on; plot(t,yzero);
title('f1(n) vs. n')
axis([0 2*pi -4 5]);ylabel('Amplitude');
xlabel ('time index n');
subplot(2,2,3)
stem(t(1:2:100),f2(1:2:100));hold on;plot(t,yzero)
title('f2(n)vs. n')
axis([0 2*pi -6 10]);ylabel('Amplitude')
xlabel('time index n')
subplot(2,2,4);
stem(t(1:2:100),f1(1:2:100),'d');hold on;
stem(t(1:2:100),f2(1:2:100),'s');
hold on; plot(t,yzero);
axis([0 2*pi -6 10]);
title(' [Multiplexed samples of f1(n) and f2(n)] vs. n');
legend('f1(t)','f2(t)')
xlabel('time index n')

figure(2)
plot(t(1:2:50),f1(1:2:50),t(1:2:50),f2(1:2:50),t,yzero);
hold on; stem(t(1:2:50),f1(1:2:50),'d'); hold on;
stem(t(2:2:50),f2(2:2:50),'s');
title('[Enlarge alternating (multiplexed) samples from f1(n)
      and f2(n)] vs. n ')
axis([0 3.1 -6 10]);ylabel('Amplitude');
xlabel('time index n')
```

Back in the command window the script file *multiplex* is executed and the results are shown in Figures 1.75 and 1.76.

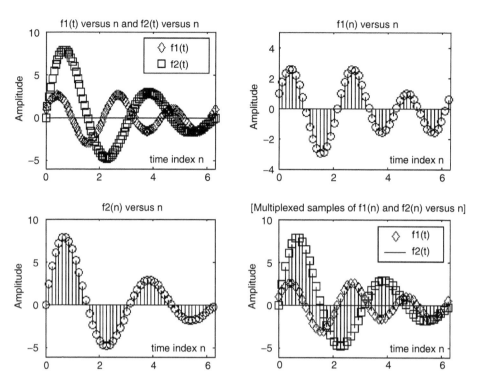

FIGURE 1.75
Plots of discrete telecommunication signals of Example 1.19.

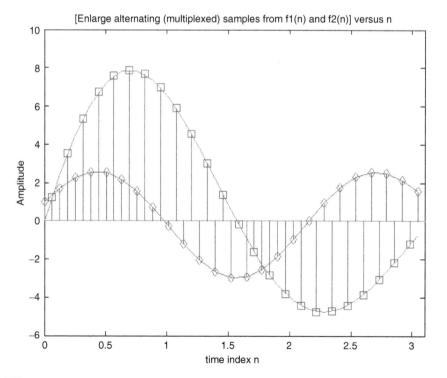

FIGURE 1.76
Enlarge plot of multiplexed signals of Example 1.19.

Example 1.20

Create the script file *explore_window* that returns the plots of the truncated or windowed function $f(t) = cos(2\pi t)$ using the following window types:

1. Hamming
2. Hanning
3. Blackman
4. Kaiser (with $\beta = 3.4$)
5. Triangular
6. Boxcar
7. Bartlett

Let us define $f(t) = cos(2\pi t)$ by using 301 points over the range $-15 \le t \le 15$, and limit $f(t)$ using the above-mentioned windows over the range $-10 \le t \le 10$.

Display the plots of the windowed function $f(t)$ over the range $-15 \le t \le 15$, for the first three cases and $-10 \le t \le 10$, for the remaining four.

MATLAB Solution

```
% Script file: explore _ window
t = -15:0.1:15;                    % returns a vector with 301 elements
f = cos(2*pi*t);                   % returns 301 element for f(t)
winpoints = -100:1:100;            % 201 window points
WHam=hamming(201);
WHan= hanning(201);
WBlac= blackman(201);
WKai= kaiser(201, 3.4);            % Beta = 3.4
WTrian= triang(201);
WRect= boxcar(201);
WBar= bartlett(201);
W1Ham= [zeros(1,50) WHam' zeros(1,50)];    % Hamming window with 301
                                             points
W1Han = [zeros(1,50) WHan' zeros(1,50)];   % Hanning window with 301
                                             points
W1Blac = [zeros(1,50) WBlac' zeros(1,50)]; % Blackman window with 301
                                             points
W1Kai = [ zeros(1,50) WKai' zeros(1,50)] ; % Kaiser window with 301
                                             points
W1Tria = [ zeros(1,50) WTrian' zeros(1,50)]; % Triangular window with
                                               301 points
W1Box = [ zeros(1,50) WRect' zeros(1,50)]; % Boxcar window with 301
                                             points
W1Bar = [ zeros(1,50) WBar' zeros(1,50)] ; % Bartlett window with 301
                                             points
Hamwin= W1Ham.*f;                  % f(t) is windowed
Hanwin= W1Han.*f;
Blackwin =W1Blac.*f;
Kaiwin=W1Kai.*f;
Triwin=W1Tria.*f;
Boxwin=W1Box.*f;
Barwin=W1Bar.*f;

figure(1)
subplot(3,1,1)
plot(t, Hamwin);ylabel('Amplitude');
title('[f(t)*Hamming widow] vs.t')
```

```
axis([-15 15 -1.2 1.2]);
subplot(3,1,2);
plot(t, Hanwin);
axis([-15 15 -1.2 1.2]);
ylabel('Amplitude');title(' [f(t)*Hanning widow] vs. t');
subplot(3,1,3);
plot(t, Blackwin);
axis([-15 15 -1.2 1.2]);xlabel('t');title('[f(t)*Blackman widow] vs.t')
ylabel('Amplitude')

figure(2)
subplot(2,2,1)
plot(t, Kaiwin);ylabel('Amplitude')
title('[f(t)*Kaiser window] vs. t')
axis([-10 10 -1.2 1.2]);
subplot(2,2,2)
plot(t,Triwin );ylabel('Amplitude');
title('[f(t)*Triang. window] vs. t');axis([-11 11 -1.2 1.2]);
subplot(2,2,3)
plot(t,Boxwin );
ylabel('Amplitude');title('[f(t)*Boxcar window] vs. t');
axis([-11 11 -1.2 1.2]);xlabel('t')
subplot(2,2,4)
plot(t,Barwin );
ylabel('Amplitude')
title('[f(t)*Barlett window] vs. t')
axis([-11 11 -1.2 1.2]);xlabel('t')
```

Back in the command window, the script file *explore_window* is executed and the results are shown in Figures 1.77 and 1.78.

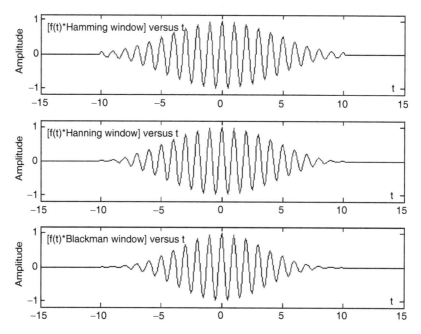

FIGURE 1.77
Plots of the windowed function *f(t)* using the *Hamming, Hanning,* and *Blackman* windows of Example 1.20.

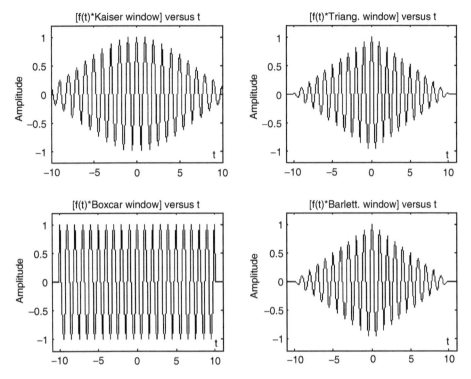

FIGURE 1.78
Plots of the windowed function *f(t)* using the *Kaiser, Triangular, Boxcar,* and *Bartlett* windows of Example 1.20.

1.5 Application Problems

P.1.1 Sketch by hand, over the range $-3 \le t \le 6$, the following analog functions:

a. $f_1(t) = 5u(t - 2)$

b. $f_2(t) = 3u(t - 1) - 3u(t - 3)$

c. $f_3(t) = t[u(t - 1) - u(t - 2)]$

d. $f_4(t) = \delta(t + 1) + 2\delta(t - 1) + u(t - 2)$

e. $f_5(t) = t[u(-t) * u(t + 3)]$

f. $f_6(t) = u(t + 1) - u(t) + 3\delta(t - 3)$

P.1.2 Sketch by hand, over the range $-3 \le n \le 6$, the following discrete sequences:

a. $f_1(n) = 3(n - 2)u(n)$

b. $f_2(n) = n\,u(n)$

c. $f_3(n) = (-n)^2\,u(n)$

d. $f_4(n) = -3\delta(n + 2) + 2\delta(n) + u(n) + u(n - 1) + 2u(n - 1)$

e. $f_5(n) = 2\eta[u(n) - u(n - 2)]$

f. $f_6(n) = n[u(n + 2) \cdot u(-n)]$

P.1.3   Create a script file that returns the plots of each of the functions defined in P.1.1.

P.1.4   Create a script file that returns the plots of each of the sequence defined in P.1.2.

P.1.5   Sketch by hand, over the range $-1 \leq n \leq 8$, the sequence $f(n) = (0.32)^n [u(n) - u(n - 6)]$.

P.1.6   Let $f(n) = (0.32)^n [u(n) - u(n - 6)]$, over $-1 \leq n \leq 8$.
         Write a script file that returns the following plots:

   a. $f(n)$ versus n

   b. $f(2n)$ versus n

   c. $f(n/2)$ versus n

   d. $f(n - 3)$ versus n

P.1.7   Given the sequence $f(n) = \sum\limits_{m=0}^{+\infty} (1/2)^m \delta(n - m)$

   a. Write a program that returns the plot of $f(n)$ versus n, over the range $0 \leq n \leq 30$.

   b. Evaluate the energy and power of $f(n)$.

P.1.8   Given the sequence $f(n) = 15(0.75)^n u(n)$. Create the script file that returns the plot of $f(n)$ versus n, over the range $-10 \leq n \leq 30$.
         Observe and discuss if the sequence $f(n)$ diverges or converges.

P.1.9   Given the sequence $f(n) = 0.2(1.1)^n u(n)$, create a script file that returns the plots of $f(n)$ versus n, over the range $-10 \leq n \leq 30$.
         Discuss if the sequence $f(n)$ converges or diverges.

P.1.10   Given the following analog signals:

   1. $f_1(t) = 4e^{-2t} u(t)$

   2. $f_2(t) = 5e^{-1.5t} \cos(5t - \pi/2)u(t)$

   3. $f_3(t) = 6(1 - e^{-2t})u(t)$

   4. $f_4(t) = e^{-t}\cos(5t - \pi/4)u(t)$

   (a) Sketch by hand each of the given functions versus t. (b) Write a program that returns the plots of each of the functions of part (a), over the range $-1 < t < 6$.

P.1.11   Write a MATLAB program that returns the plots of the continuous function $f(t) = 3\cos(0.15\pi t) + 2\sin(0.20\pi t)$, and $f(t)$ sampled with $T_S = 0.1\pi$ over the range $0 < t < 15\pi$.

P.1.12   In general, a random noisy signal of length N can be generated by using the following MATLAB command:

```
Noise=rand(1,N)
```

   Likewise, the sequence

```
Noisen=randn(1,N)
```

   returns a random sequence of length N, normally distributed with zero mean and unit variance. Write a program that returns the plots, the average value, the maxima, and minima of each of the noisy sequences defined earlier for $N = 100$ and 200.

P.1.13   Plot the functions indicated as follows over the range $-3 \leq t \leq 3$:

$$f_1(t) = t^3$$
$$f_2(t) = 3 + t^3$$
$$f_3(t) = 2t^3 + t^5 + 3$$
$$f_4(t) = tan(2t)$$
$$f_5(t) = sin(2t)$$

and verify that all the preceding functions present odd symmetry, that is, $f(t) = -f(t)$.

P.1.14   Plot the functions indicated as follows for the range $-3 \leq t \leq 3$:

$$f_1(t) = t^2$$
$$f_2(t) = 3 + t^2$$
$$f_3(t) = 2t^4 + t^6 + 3$$
$$f_4(t) = cos(2t)$$
$$f_5(t) = |t|$$

and verify that all the preceding functions present even symmetry, that is, $f(t) = f(-t)$.

P.1.15   Write a MATLAB function file that analyzes a given arbitrary function $f(t)$ over the range $-3 \leq t \leq 3$, and returns a message indicating if $f(t)$ presents even or odd symmetry. Repeat the preceding problem over any given range.

P.1.16   Given the functions shown in Figure 1.79, sketch by hand for each function its even and odd parts.

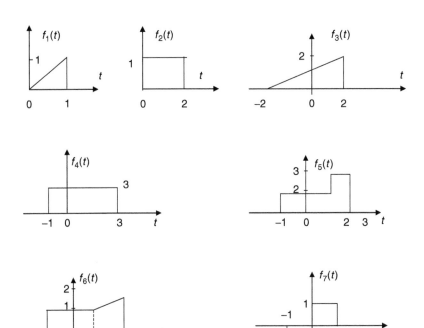

FIGURE 1.79
Plots of P.1.16.

P.1.17 Write a program that returns the plots over three cycles of the following periodic functions, defined as follows over one period ($T = 2$):

$$f_1(t) = t, \qquad \text{for } 0 < t < 2$$

$$f_2(t) = -t + 2, \quad \text{for } 0 < t < 2$$

$$f_3(t) = \begin{cases} 2 & 0 < t < 1 \\ 0 & 1 < t < 2 \end{cases}$$

$$f_4(n) = \begin{cases} 1 & 0 < t < 1 \\ 2 & 1 < t < 2 \end{cases}$$

$$f_5(t) = 2 + t, \qquad 0 < t < 2$$

$$f_6(t) = -t + 4, \quad 0 < t < 2$$

P.1.18 Sketch by hand each of the functions defined in P.1.13.

P.1.19 Sketch by hand the even and odd portions of each of the functions defined in P.1.13.

P.1.20 Let $f(t)$ be a periodic function with period $T = 2$, defined as follows by

$$f(t) = \begin{cases} t & 0 \le t \le 1 \\ 1 & 1 < t \le 2 \end{cases}$$

    a. Manually sketch $f(t)$ over three periods

    b. Write a program that returns the plots of part a

    c. Sketch by hand $f(-t)$

    d. Write a program that returns the plot of part c

    e. Sketch by hand $[f_1(t) = -f(-t) + f(t)]$ versus t, over the range $-3 \le t \le +3$

    f. Write a program that returns the plot of part e

    g. Sketch $f_2(t) = \frac{1}{2}[f(t) + f(-t)]$ versus t and $f_3(t) = \frac{1}{2}[f(t) - f(-t)]$ versus t, over the range $-3 \le t \le +3$

    h. Sketch by hand $[f_4(t) = f_2(t) + f_3(t)]$ versus t

P.1.21 Modify the script file *even_odd.m* (of Example 1.14) into a function file where the input is the function $f(t)$, over a given range r, and its outputs are the plots of the even and odd parts of $f(t)$.

P.1.22 Test the function file *even_odd.m* created in P.1.21, over the range $-10 \le t \le 10$, using the following function:

    a. $f_1(t) = cos(3t + pi/3)u(t)$

    b. $f_2(t) = t\,u(t)$

    c. $f_3(t) = t\,u(t) - t\,u(t - 5)$

P.1.23 Let $f(t) = 3r(t) - 3r(t - 1) + 3r(t - 2)$, sketch and obtain MATLAB plots, over the range $-5 \le t \le 5$, for the following functions:

    a. $f(t)$ versus t

    b. $f_e(t)$ versus t

    c. $f_o(t)$ versus t

P.1.24   Let $f_1(t) = 3e^{2t/3} + 3$ and $f_2(t) = cos(\pi t)$. Create a script file that returns the energy and power of $f_1(t)$ and $f_2(t)$.

P.1.25   Repeat P.1.24 for the following discrete sequences: $f_1(n) = 3e^{2n/3} + 3$ and $f_2(n) = cos(0.1\pi n)$, over the range $-10 \le n \le 10$.

P.1.26   Evaluate by hand which of the following discrete sequences are periodic, and if periodic, evaluate its period.

    a. $cos(0.2n)$

    b. $cos(0.2\pi n)$

    c. $sin(2\pi n/3)$

    d. $sin(9.1n)$

P.1.27   Verify the following equalities:

    a. $Imp[(n)] = \sum\limits_{k=-\infty}^{+\infty} \delta(n - k)$

    b. $u[(n)] = \sum\limits_{k=0}^{+\infty} \delta(n - k)$

    c. $r(n) = \sum\limits_{k=0}^{+\infty} k\delta(n - k)$

P.1.28   Evaluate the following integrals:

    a. $\int_{-\infty}^{\infty} (2t + 3)\delta(t - 1)dt =$

    b. $\int_{-\infty}^{\infty} (2t + 3)\delta((3t/2) - 1)dt =$

    c. $\int_{-\infty}^{\infty} sin(wt)\delta((3t/2) - (2/3))dt =$

    d. $\int_{-\infty}^{\infty} e^{(-3t1+1)}\delta(t - (2/3))dt =$

P.1.29   Analyze and draw a flow chart of Example 1.15 and indicate in each average approximation the filter used as well as the effect of the filter.

P.1.30   Evaluate the first and second derivative of the following expressions:

    a. $f_1(t) = u(t) + 7u(t - 5) - 2u(t - 7)$

    b. $f_2(t) = tu(t) + e^{3t}u(t - 1) + 3\delta(t - 3)$

    c. $f_3(t) = r(t)u(t) - r(t - 1)u(t - 1)$

P.1.31   Let $H(s) = \dfrac{s + 3}{s^3 + s^2 + 5s + 10}$ be the transfer function of a given analog system. Write a MATLAB program that returns

    a. The Bode plot of $H(s)$, magnitude and phase

    b. The impulse as well as the step responses

    c. The zero/pole diagram

    d. The system differential equation

P.1.32   Let

$$H(z) = \frac{0.32 + 0.43z^{-1} - 0.85z^{-2} + 0.2z^{-3}}{1 + 0.802z^{-1} - 0.42z^{-2} + 0.62z^{-3}}$$

be the transfer function of a discrete system. Write a MATLAB program that returns

a. The impulse as well as the step response
b. The zero/pole diagram
c. The system output if its input is given by

$$x(n) = 0.6^n cos((2 * pi * n)/(256)), \quad for\ n = 0, 1, 2, 3, \ldots, 256$$

d. The plot of the output if the input $x(n)$ given in part c is truncated by a triangular window with size of $N = 128$
e. Repeat part d for the *Hamming, Hanning,* and *Blackman* windows

P.1.33 Write a program (or programs) that returns the sequences and plots shown in Figure 1.80.

P.1.34 The Parabolic or Parzen window is defined by the following equation:

$$w(n) = 1 - \left\{ \frac{n - M}{M} \right\}^2$$

Write a program that returns the plots of $w(n)$ versus n, for $n = 31$ and $M = 0.5, 1,$ and 5, and discuss the effect of M.

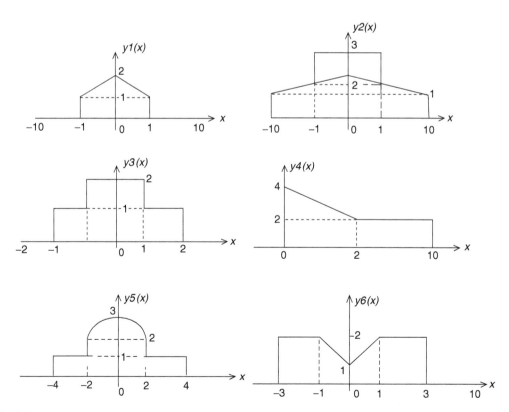

FIGURE 1.80
Plots of P.1.33.

P.1.35 The Cauchy window is defined by the following equation:

$$w(n) = \frac{M^2}{M^2 + \alpha^2(n - M)^2}$$

Write a program that returns the plots of $w(n)$ versus n, for $n = 31$ and $\alpha = 0.5, 1,$ 3, and 5, and discuss the effect of α.

P.1.36 The Gaussian window is defined by the following equation:

$$w(n) = \exp\left\{-0.5\alpha^2\left(\frac{n - M}{M}\right)^2\right\}$$

Write a program that returns the plots of $w(n)$ versus n, for $n = 31$, $M = 0.5$, and $\alpha = 1, 3,$ and 5, and discuss the effect of α.

P.1.37 Consider the following discrete time sequences:

$$f_1(n) = \left\{\frac{\sin[0.3\pi(n - 64)]}{0.3\pi(n - 64)}\right\}^2$$

$$f_2(n) = \left\{\frac{\sin[0.2\pi(n - 64)]}{0.2\pi(n - 64)}\right\}^2$$

Write a program that returns

a. The plots of $f_1(n)$ versus n and $f_2(n)$ versus n, over $0 \le n \le 128$ using the *sinc* function, and appropriately label the axis.

b. The plots of the expanded signals $f_1(n)$ and $f_2(n)$ by a factor L, by inserting $L - 1$ zeros between each one of the samples of $f_1(n)$ and $f_2(n)$, for $L = 2, 3,$ and 4. Therefore, each signal $f_1(n)$ and $f_2(n)$ is expanded by a factor of 2, 3, and 4.

c. The plots of the down-sampled signals $f_1(n)$ and $f_2(n)$ by a factor of $M = 2, 3,$ and 4.

P.1.38 Truncate or limit the length of each of the functions $f_1(n)$ and $f_2(n)$ defined in P.1.37 by using the following windows:

a. Parabolic

b. Cauchy

c. Gaussian

defined in P.1.34/36 over a length of $N = 31$.

P.1.39 Obtain the multiplex signal $mult[f_1(n), f_2(n)]$; for the sequences $f_1(n)$ and $f_2(n)$ given by

$$f_1(n) = \left\{\frac{\sin[0.3\pi(n - 64)]}{0.3\pi(n - 64)}\right\}^2$$

$$f_2(n) = \left\{\frac{\sin[0.2\pi(n - 64)]}{0.2\pi(n - 64)}\right\}^2$$

P.1.40 Consider the analog time functions

$$f_1(t) = [5\sin(3\pi t)/(3\pi t)]^2$$

$$f_2(t) = [3\sin(2\pi t)/(2\pi t)]^2$$

Write a program that returns

a. The plots of $f_1(t)$ versus t and $f_2(t)$ versus t, over the range $-2\pi \le t \le 2\pi$, using the *sinc* function

b. The plots of the amplitude modulated (AM) signals for $f_1(t)$ and $f_2(t)$ versus t, over the range $0 \le t \le 3$, if the carrier signal is $f_c(t) = 10\cos(20\pi t)$

c. Repeat part b for the case of angular modulation (FM or PM)

P.1.41   Let $f(t) = A\cos(2\pi w_o t + \varphi)$ be a continuous time function.
How many samples are then required to uniquely determine A, w_o, and φ.
Is it always possible to solve for A, w_o, and φ?

P.1.42   Let $A = 5$, $w_o = 2$ rad/s, and $\varphi = 30°$, for the signal defined in P.1.41.
Write a program that returns the following plots:

a. $f(t)$ versus t

b. $f(nT)$ versus nT, for $T = 0.1$

c. Reconstruct $f(t)$ from $f(nT)$ using the summation of *sinc* functions (by applying the sampling theorem)

P.1.43   Create the symbolic expression for $f(t)$ defined in P.1.42 and obtain the following plots:

a. $f(t)$ versus t, using *ezplot*.

b. Let $A = 7$, $w_o = 3$ rad/s, and $\varphi = 45°$ by redefining the symbolic expression of part a, by using the *subs* symbolic function.

c. AM and FM/PM signals using the symbolic approach, if the carrier is $f_c(t) = 10\cos(20\pi t)$ and the information is $f(t)$.

2

Direct Current and Transient Analysis

> I often say that when you can measure what you are speaking about, and express it in numbers, you know something about it; but when you cannot express it in numbers, your knowledge is of meager and unsatisfactory kind.
>
> **Lord Kelvin**

2.1 Introduction

This chapter deals with the basic concepts and principles of electricity as well as the analysis of simple electric circuits. A brief and compressed history of electricity is also presented to introduce the reader to how this important field of science and engineering evolved over time.

The simplest manifestation of electricity in nature is the phenomenon of magnetism and static electricity. Static electricity was first observed by the ancient Greeks, and back then they called this phenomenon *elektron*. The curious Greeks observed and studied the effects of *elektron*, but no written records about the subject exist.

The first recorded observations about electricity and magnetism date back to Thales (640–546 BC), a famous philosopher and mathematician who lived on the west coast of Asia Minor. Years later, the Chinese military commanders, during the Hun dynasty (AD 206–220) are believed to have used the magnet's properties to implement the first compass to indicate direction. It took about 900 years for the Europeans to incorporate the compass in navigation. No significant recorded scientific progress was made in electricity and magnetism until the 1600s, when William Gilbert (1540–1603), Queen Elizabeth's physician, recorded experiments he performed with magnetic materials. He thought that magnetism could have healing effects on the human body.

Probably the first major modern scientific contribution was the Leyden jar developed by Pieter van Musschenbrock and Benjamin Franklin. The jar was a device capable of storing an electric charge (static electricity), which could be discharged producing an electric arc or spark, thus emulating lighting.

Years later, it was discovered that lightning was in essence an electric discharge. The most important scientific contributions occurred during the late 1700s through the late 1800s, when the effects of electricity and magnetism were observed, recorded, and seriously studied. Concepts and principles once understood, tested, and proven became rules, laws, and theories.

A number of electric units bear the name of those early pioneers, such as

- *Coulomb.* After Charles Coulomb (1736–1806), Frenchman; observed, measured, and quantized the amount of electricity (electric charge)

- *Ampere.* After Andres Maria Ampere (1775–1836), Frenchman; studied the current induced by a magnetic field

- *Volt.* After Count Alexander Volta (1747–1827), Italian; developed the voltaic cell
- *Ohm.* After Georg Ohm (1787–1854), German; developed a relationship among the current, voltage, and resistance in an electric circuit
- *Faraday.* After Michael Faraday (1791–1867), Englishman; developed the basis of the electromagnetic theory
- *Hertz.* After Heinrich Rudolph Hertz (1857–1894), German; experimented with electromagnetic waves

Two additional names are associated with the early developments of electric theories and discoveries, and their work constitutes the basis of electric circuits. They are:

James Clark Maxwell (1831–1879), Scottish. Maxwell developed the relationship between electricity and magnetism, as well as the electromagnetic theory of light (1862).

Gustav Robert Kirchhoff (1824–1887), Russian. Kirchhoff formulated the basic circuit laws (the network current and voltage laws).

The theoretical fundamentals of electricity and magnetism were observed, studied, and firmly established, predominantly by European physicists, engineers, and mathematicians.

American scientists and engineers are distinguished as the great inventors, patent holders, and industrialists who brought the electric discoveries into the market place. The most important Americans, just to mention a few, are

- Thomas Edison (1847–1931)
- Alexander Graham Bell (1847–1922)
- Lee De Forest (1873–1961)

Today we know that electricity is produced by electrons. An electron is the smallest unit of electric charge, and its quantity is known as "1 esu." An electron is not easily found in nature, but it can be generated, and it constitutes the most important source of energy in modern societies. In its simplest version, the concepts of current and voltage constitute the bases of electricity.

Let us define current and voltage using the simplest possible terms. Current is the flow (movement) of electrons, whereas voltage is the force that makes this flow possible.

This chapter deals with DC electrical systems and transients. The term "DC" is an abbreviation of direct current, which refers to electrical sources that employ unidirectional current. These sources of voltage or current provide a fixed or constant output.

A current that varies with respect to time reversing directions periodically is called alternating current, abbreviated as "AC," and is the subject of Chapter 3.

DC voltage sources consist of batteries, generators, and power supplies. These devices employ different technologies and modes of operations, but the end result is to supply a constant terminal voltage, regardless of the current demand and load connected to its terminals.

A DC current source is analogous to a voltage source in the sense that it supplies a constant current flow, whereas the voltage demand may vary across its terminals.

DC is produced either by a voltage or current source that delivers a constant output, denoted mathematically as $e(t) = E$ V (volts) or $i(t) = I$ A (amp) respectively, and the electrical responses (steady state) in any of its branches of an electrical network are also constant.

When a circuit switches from one condition to a new set of conditions, there is a transition period where the currents and voltages adjust to the new circuit conditions. The transition time is referred to as the transient response or simply as transient, and is usually defined by a set of differential equations.

The basic passive circuit elements such as resistors, capacitors, and inductors are introduced, defined, and used in this chapter. Circuit variables such as voltage, current, power, and energy are also defined in terms of the circuit elements and its excitations; and transient conditions for the standard electrical configurations are stated, explored, and analyzed.

2.2 Objectives

On reading this chapter, the reader should be able to

- Distinguish between DC and AC
- State the laws of attraction and repulsion (Coulomb's law) at the subatomic level
- Define ampere (the unit of electric current) and volt (the unit of electric voltage) quantitatively and qualitatively
- Relate current, charge, and time
- Define the concept of conductor, superconductor, semiconductor, and insulator
- Define and know Ohm's law
- Define an electrical short and open and its electrical equivalents
- Define and know how to use the basic electric concepts and terminology such as voltage drop, current flow, and resistance
- Recognize a series and parallel connection
- Draw schematic diagrams of simple electric circuits
- State Kirchhoff's current law (KCL) and Kirchhoff's voltage law (KVL)
- State the voltage divider rule
- State the current divider rule
- Know how and when to apply the voltage or current divider rules
- Define electric power and energy
- State the equations that define the energy storing devices such as the inductor (L) and capacitor (C)
- Combine resistors, capacitors, and inductors in a series and parallel configuration
- Calculate the equivalent capacitance and inductance
- State the conditions for the steady-state and transient responses
- Define the time constant of an electric circuit and its meaning and applications
- Present the equation and plots of the charging and discharging for the simple circuits known as resistor capacitor (RC) and resistor inductor network (RL)
- Set up the circuit equations for simple circuits
- State the differential equations and the exponential form of the solution for the RC and RL cases

- State and analyze the transient solution for the parallel and series resistor capacitor network (RLC)
- Define the system loop and node equations
- Use node and loop equations to solve circuit problems
- Define and use the superposition theorem when dealing with multiple sources
- Recognize that an independent voltage source set to zero is equivalent to a short circuit
- Recognize that an independent current source set to zero is equivalent to an open circuit
- Recognize that a voltage source can be transformed into an equivalent current source and vice versa
- Define and determine the Thevenin's equivalent circuit of an electrical network
- Recognize that the Thevenin's equivalent circuit can be evaluated for any load of interest
- Define and determine the Norton's equivalent circuit given an electrical network
- Recognize that the relation between Thevenin's and Norton's equivalent circuits consist of a simple source transformation
- Use MATLAB® as a tool in the analysis of electric circuits

2.3　Background

R.2.1　The idea that matter is composed of atoms is an old concept that was first proposed by the Greek philosophers Empedocles and Democritus around 500 and 400 BC.

R.2.2　The scientific community accepted the existence of the atom in modern times, initially proposed by the chemist Dalton in the nineteenth century, and supported and consolidated by Cannizzaro 50 years later.

R.2.3　An atom is too small to be seen directly or even with the help of modern and sophisticated devices such as a microscope, because its diameter is estimated to be in the order of 10^{-10} m.

R.2.4　Atoms are composed of protons, neutrons, and electrons.

R.2.5　In modern times, J. J. Thomson, around 1897, is credited with the discovery and study of the electron.

R.2.6　The mass of a neutron is slightly larger than the mass of a proton, and the mass of an electron is much smaller than either of them.

R.2.7　An electron constitutes the unit of a negative electric charge.

R.2.8　A proton constitutes the unit of a positive electric charge.

R.2.9　Atoms move in nature in a perfectly random way.

R.2.10　The effect of an electron exactly cancels the effect of a proton.

R.2.11　An atom contains a certain number of protons in a nucleus and an equal number of electrons orbiting the nucleus.

R.2.12　Most bodies in nature are electrically neutral, because they contain equal amounts of positive and negative charges that exactly cancel one another.

R.2.13 By a process that is known to scientists, but is beyond the scope of this discussion, energy may be absorbed by some electrons in the outer orbits and migrate outside its natural orbits, and in this way, become free electrons.

R.2.14 In metals such as silver and copper, some electrons are very loosely held, since they are not strongly attached to any atom, and they can be shaken off to become free electrons.

R.2.15 In nature as well as in the physical sciences, only forces are capable of making or creating changes.

R.2.16 Gravity is nature's force.

R.2.17 Recall from physics (basic mechanics) that force (Newtons) = mass (kilograms) * acceleration (meters/second^2).

R.2.18 An electric charge may induce or create an electrical force.

R.2.19 Coulomb's law states that the force F between two electrically charged points Q_1 and Q_2 is given by

$$F = \frac{k * Q_1 * Q_2}{r^2}$$

where F is given in Newton (N) (1 N = 1 kg * m/s^2), Q in coulombs (C; where a coulomb is a measure of the amount of electric charges given in terms of electrons; the relationship is 1 C = 2.64 * 10^{18} electrons), and k a proportionality constant given by

$$k = \frac{1}{4\pi\epsilon_0} = 9 * 19^9 \frac{N * m^2}{C^2}$$

where $\epsilon_0 = 8.85 * 10^{-12}$ C^2/N · m^2 is the permittivity of free space, and r is the distance between Q_1 and Q_2 in meters. In this discussion, it is assumed for simplicity Q_1 and Q_2 are stationary, otherwise additional forces must be considered.

R.2.20 The charged electric points Q_1 and Q_2 induce a radial electric field around itself. The electric field is known as a force field.

R.2.21 The following can be said about the nature of the induced electric force F:

a. Like charges repel each other (creating a repulsion force).

b. Unlike charges attract each other (creating an attraction force).

c. The closer the charges the stronger the force.

d. The larger the charges the stronger the force.

R.2.22 As mentioned earlier, the smallest charged (negative) particle in nature is the electron. The charge of a single electron is given by

$$1 \text{ electron} = 1.602 * 10^{-19} \text{ C}$$

Therefore, it takes 6.242 * 10^{18} electrons to generate a charge of 1 C.

R.2.23 Some materials (in particular metals) contain free electrons in their natural state. The movement or flow of free electrons creates an electric current.

R.2.24 Materials can be classified according to their capacity to conduct current or possess free electrons as

• Conductors (including superconductors)

• Semiconductors

• Insulators

R.2.25 Conductors are materials in which an electric charge can be moved from one point to another over a finite interval of time with little resistance.

A superconductor is a conductor that has no resistance to the flow of electric current (no energy is converted to heat).

R.2.26 Insulators are materials that resist the flow of electrons by presenting a large resistance to its current flow, and they require a large external amount of energy to produce a measurable current.

R.2.27 Semiconductors are materials that exhibit characteristics that in some cases behave as a conductor, whereas in other cases, act as insulators depending on the intensity and polarity of an external excitement such as temperature, light, or voltage. Examples of semiconductor materials are silicon (Si), germanium (Ge), and gallium arsenate (GaAs).

R.2.28 An electric current, denoted by I, is defined as the rate of flow of an electric charge (made up of electrons). Analytically, current is defined as

$$I\,(\text{A}) = \frac{Q\,(\text{Charge})}{t\,(\text{time})} = \frac{dQ}{dt}\,(\text{C/s})$$

The unit of an electric current is the ampere (denoted by A or amp), defined as the rate of flow of an electric charge of 1 C/s (coulombs per second). Since current involves motion, magnitude and direction must be indicated. The positive direction for current is defined as the flow of a positive electric charge. Electrons consist of negative charges and constitute the bulk of the electric charge. The direction of a positive current is then defined as the opposite of the electron flow.

R.2.29 The work required to move one unit of charge from one point to another is measured in volts. Therefore,

1 V (volt) = 1 J/C (Joule/Coulomb)

where J is the unit of energy. Electric voltage is measured between two points, which is also referred as electrical potential or potential difference and is the electrical force that causes electrons to flow creating an electric current. Observe that voltage is the potential to move a charge even when no charge is moved. When charges (current) move through the circuit's elements, energy is transferred. The volt is the energy transferred per unit of charge through the affected elements. A voltage is always defined by assigning a polarity (positive–negative) indicating the direction of energy flow.

If a positive charge moves from the positive polarity through the element toward the negative polarity, then the device generates or supplies energy. However, if a positive charge moves from the negative toward the positive polarity, then the device absorbs energy.

R.2.30 An ammeter is an instrument used to measure an electric current. To measure current, an ammeter must be inserted in the path of the current flow (electrons) by physically opening the circuit in order to insert the ammeter.

R.2.31 A voltmeter is an instrument used to measure an electric potential between two arbitrary points in an electric circuit. To measure a voltage difference, a voltmeter must be inserted (attached) to the two referred points.

R.2.32 In its simplest version, a current flow can be generated in a circuit if a close path or loop is established and a source is present in the loop.

R.2.33 Electric power is defined as the rate of generating or consuming (dissipating) energy. Electric power (denoted by P) is defined as the product of a voltage across, and the resulting current through an element. The unit of electric power is watt (W). Then

$$P \text{ (W)} = \text{current (A)} * \text{voltage (V)}$$

or

$$P \text{ (W)} = \text{energy/time (J/s)}$$

The power of 1 W implies a rate of generating or consuming 1 J of energy per second (s).

R.2.34 Decibel (denoted by dB) is the unit used to express powers, voltages, and currents as ratios. Decibel is defined as

$$dB = 10 \log_{10}[P_O/P_I]$$

$$dB = 20 \log_{10}[V_O/V_I]$$

and

$$dB = 20 \log_{10}[I_O/I_I]$$

Strictly speaking, a decibel can be used to express only ratios, which is a measure of losses, gains, or no changes. The subscripts "O" and "I" denote output and input, respectively. For example, P_O/P_I denotes output power divided by input power referred to as power gain.

R.2.35 Electrical energy (denoted by W) is the amount of power generated or consumed during a given interval of time $[t_0, t_1]$ of interest.
Therefore,

$$W \text{ (J)} = \int_{t_0}^{t_1} p(t)dt$$

is often expressed in terms of watts $*$ hours (W h), especially by the utilities companies.*

R.2.36 The principle of conservation of energy states that

energy can be transformed, but never destroyed.

Electrical energy can be transformed into mechanical energy (motor) and vice versa (generator) or any other form of energy such as nuclear, thermal, and hydraulic.
The following relations indicate the equivalency between mechanical and electrical units of energy often used in practice:

$$1 \text{ J} = 0.737 \text{ ft} \cdot \text{lb}_f$$

$$1 \text{ft} \cdot \text{lb}_f = 1.357 \text{ J}$$

$$1 \text{ Btu} = 1055 \text{ J}$$

$$1 \text{ Hp} = 746 \text{ W} = 550 \text{ ft} \cdot \text{lb}_f/s$$

---

* The reader should not confuse P (W), that is the power given in watts with W (J), that is the energy given in Joules, where W = J/s (Watt = Joule/sec).

where lb_f denotes pound-force. The law or principle of conservation of energy for an electric circuit can simply be restated as one of the following:

The total energy in a closed system is zero.

Energy in, equals energy out.

In a close loop system the energy consumed is equal to the energy generated.

R.2.37 When electrical energy is supplied to a device, and the device dissipates that energy in the form of heat, then the device is called a "resistor" (denoted by R), with the unit given in ohms, denoted by the Greek character Ω.

R.2.38 When electrical energy is supplied to a device, and the device stores that energy in its magnetic field, then the device is an inductor (denoted by L), with the unit given in henries (H).

R.2.39 When electrical energy is supplied to a device, and the device stores that energy in its electric field, then the device is a capacitor (denoted by C), with units given in faradays (F).

R.2.40 A practical device usually has more than one way to deal with the energy it receives. There are no pure inductors in nature, rather a combination of a resistor and an inductor (RL circuit). The same statement is valid for capacitors. There are no pure capacitors in real life, rather a combination of a resistor with a capacitor (RC circuit).

R.2.41 The standard and widely accepted symbols used to represent resistors, capacitors, inductors, and sources are shown in Table 2.1.

R.2.42 Ohm's law relates the voltage (V) across and the current (I) through a resistor (R) by means of the following equation:

$$V = R * I$$

TABLE 2.1

Electrical Symbols

Variable	Symbol	Units
R	⎓Ⱳ⎓	Ohms (Ω)
C	⊣⊢	Farads (F)
L	⟋⟋⟋⟋	Henries (H)
V		DC volts (V)
$v(t)$		AC volts (V)
I		DC amp (A)
$i(t)$		AC amp (A)

The voltage across a resistor presents the following polarities: it is always positive at the terminal at which the current enters the resistor and negative at the point where it leaves.

The inverse of a resistance is called a conductance (denoted by G), where $G = 1/R$ with the units given in *siemens* (*sie*), where 1 sie $= 1/\Omega$.

R.2.43 The resistance of electrical material (including a wire) is a function of its dimensions as well as its physical properties, and is given by the following equation:

$$R = \alpha * l/a$$

where α is called the resistivity factor of the material in $\Omega *$ meters, l is its length in meters, and a is its cross-sectional area in meters squared.

The resistivity factor α of a good conductor is bounded by the following (resistivity) limits $1.6 * 10^{-6}\,\Omega$ cm $< \alpha < 2.5 * 10^{-6}\,\Omega$ cm, whereas the resistivity factor of an insulator is given by

$$\alpha > 200 * 10^{-6}\,\Omega\,\text{cm}$$

For example, the resistivity factor of copper is $\alpha = 1.7 * 10^{-6}\,\Omega$ cm (conductor), whereas the resistivity factor of carbon is $\alpha = 3500 * 10^{-6}\,\Omega$ cm (insulator). The equation $R = \alpha * l/a$ is especially useful when evaluating the resistance of a segment of a wire.

R.2.44 The current through and voltage across an inductor* L are related by

$$v_L(t) = L\frac{di_L(t)}{dt}$$

and

$$i_L(t) = \frac{1}{L}\int v_L(t)dt$$

R.2.45 The current through and the voltage across a capacitor C are related by the following equations:

$$i_C(t) = C\frac{dv_C(t)}{dt}$$

and

$$v_C(t) = \frac{1}{C}\int i_C(t)dt$$

---

* An inductor L is a coil of wire, which when connected across a voltage source, produces a magnetic field in the coil that opposes current changes. According to Faraday's law, the induced voltage is proportional to the change in flux linkage which in turn is proportional to its current.

R.2.46 A capacitor (C) basically consists of two conductor plates called electrodes, separated by insulating material referred to as dielectric. When a voltage V is applied across the plates, an electric field is established and the capacitor C *charges*. The capacitance C is defined in terms of its electrical variables by

$$C = \frac{Q}{V}$$

The capacitance C, in terms of its physical dimensions is defined by

$$C = \epsilon_0 \frac{A}{d} \text{(farads)}$$

where $\epsilon = \epsilon_r * \epsilon_0$, $\epsilon_0 = 8.85 * 10^{-12}$ F/m (permittivity of vacuum), and ϵ_r is the relative permittivity or dielectric constant of a particular material.

For example, ϵ_r (vacuum) = 1, ϵ_r (mica) = 5, and ϵ_r (glass) = 7.5.

$$A = \text{area of the plates} = \text{width} * \text{height (in meters}^2)$$

and

$$d = \text{distance between the plates (in meters)}$$

R.2.47 The behavior of the voltages and currents in an electrical network is governed by KCL and KVL stated in R.2.48 and R.2.49.

R.2.48 KCL states that the sum of the currents entering a junction (also referred to as a node) must equal the sum of the currents leaving that junction (node), or the sum of the currents at any node is always equal to zero, illustrated in Figure 2.1.

R.2.49 KVL states that the sum of the voltage raises (gains) are equal to the sum of the voltage drops (losses) in any close electrical loop (which is a consequence of the law of conservation of energy) illustrated in Figure 2.2.

KVL must be applied around a closed electric loop in which the referred loop may or may not have current (open circuit), but the sum of all the voltage drops around it (loop) is always zero.

R.2.50 When two (or more) electrical elements, denoted by A and B, have the same current flow and are connected in cascade, as shown in Figure 2.3, they are referred to as a series connection.

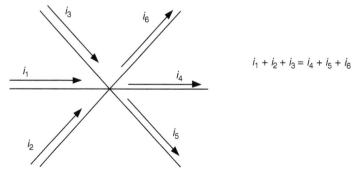

$$i_1 + i_2 + i_3 = i_4 + i_5 + i_6$$

FIGURE 2.1
An electric node (KCL).

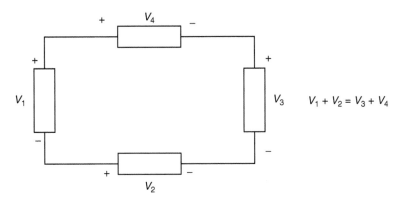

FIGURE 2.2
An electric loop (KVL).

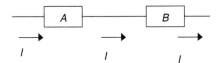

FIGURE 2.3
Series connection.

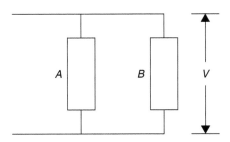

FIGURE 2.4
Parallel connection.

R.2.51 When two (or more) electrical elements, denoted by A and B, share the same voltage, that is, they are connected between the same two points, they are referred to as connected in parallel, as illustrated in the Figure 2.4.

R.2.52 A current I_T divides between two resistances R_1 and R_2 connected in parallel into I_1 and I_2, the respective currents as indicated in Figure 2.5. Then the branch currents, I_1 and I_2, can be expressed in terms of the resistances R_1, R_2, and the current I_T by the following relation known as the current divider rule:

$$I_T = I_1 + I_2$$

$$I_1 = \frac{R_2}{R_1 + R_2} I_T$$

$$I_2 = \frac{R_1}{R_1 + R_2} I_T$$

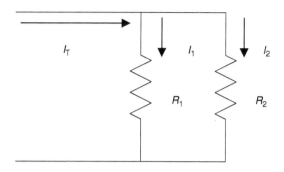

FIGURE 2.5
Current divider network.

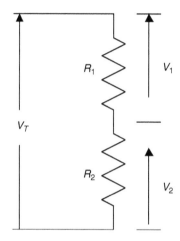

FIGURE 2.6
Voltage divider network.

Observe that indeed

$$I_T = I_1 + I_2 = \frac{R_2}{R_1 + R_2} I_T + \frac{R_1}{R_1 + R_2} I_T \text{ (KCL)}$$

R.2.53 An applied voltage V_T across a series connection consisting of two resistors R_1 and R_2 divides the voltage V_T into two voltage drops V_1 and V_2 as indicated in Figure 2.6. Each voltage is a function of R_1, R_2, and V_T referred to as the voltage divider rule given by the following relation:

$$V_1 = \frac{R_1}{R_1 + R_2} V_T$$

$$V_2 = \frac{R_2}{R_1 + R_2} V_T$$

where indeed $V_T = V_1 + V_2 = \frac{R_1}{R_1 + R_2} V_T + \frac{R_2}{R_1 + R_2} V_T \text{ (KVL)}$

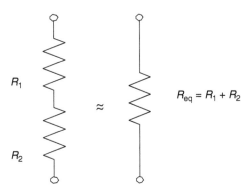

FIGURE 2.7
Series equivalent resistor.

R.2.54 The equivalent resistance R_{eq} of a set of two resistors R_1 and R_2 connected in series, as illustrated in Figure 2.7, is given by its sum.

If n resistors are connected in series ($n > 2$) denoted by R_1, R_2, R_3, ..., R_n, then the equivalent resistance R_{eq} is given by

$$R_{eq} = \sum_{i=1}^{n} R_i$$

R.2.55 The equivalent resistance R_{eq} of two resistors R_1 and R_2 connected in parallel is given by

$$R_{eq} = \frac{R_1 * R_2}{R_1 + R_2}$$

and for the case of the three resistors, the equivalent resistance is given by

$$R_{eq} = \frac{R_1 * R_2 * R_3}{R_1 * R_2 + R_2 * R_3 + R_3 * R_1}$$

For the case of n ($n > 3$) resistors (R_1, R_2, R_3, ..., R_n) connected in parallel, the equivalent resistance is given by

$$\frac{1}{R_{eq}} = \frac{1}{R_1} + \frac{1}{R_2} + \cdots + \frac{1}{R_n}$$

$$\frac{1}{R_{eq}} = \sum_{i=1}^{n} \frac{1}{R_i}$$

or

$$R_{eq} = \frac{1}{\sum_{i=1}^{n} \frac{1}{R_i}}$$

R.2.56 For the case of two capacitors C_1 and C_2 connected in parallel, the equivalent capacitance C_{eq} is given by

$$C_{eq} = C_1 + C_2$$

For the case of n capacitors denoted by $C_1, C_2, C_3, \ldots, C_n$ connected in parallel, the equivalent capacitance C_{eq} is given by

$$C_{eq} = \sum_{i=1}^{n} C_i$$

R.2.57 For the two capacitors C_1 and C_2 connected in series, the equivalent capacitance C_{eq} is given by

$$C_{eq} = \frac{C_1 * C_2}{C_1 + C_2}$$

The equivalent capacitance C_{eq} of n capacitors denoted by $C_1, C_2, C_3, \ldots, C_n$ connected in series is given by

$$\frac{1}{C_{eq}} = \sum_{i=1}^{n} \frac{1}{C_i}$$

or

$$C_{eq} = \frac{1}{\sum_{i=1}^{n} (1/C_i)}$$

R.2.58 The equivalent inductor L_{eq} for the case of two inductors L_1 and L_2 connected in series is given by

$$L_{eq} = L_1 + L_2$$

For the case of n inductors $L_1, L_2, L_3, \ldots, L_n$ connected in series, the equivalent inductor L_{eq} is given by

$$L_{eq} = \sum_{i=1}^{n} L_i$$

R.2.59 For the case of the two inductors L_1 and L_2 connected in parallel, the equivalent inductor L_{eq} is given by

$$L_{eq} = \frac{L_1 * L_2}{L_1 + L_2}$$

For the case of n inductors L_1, L_2, L_3, ..., L_n connected in parallel, the equivalent inductor L_{eq} is given by

$$\frac{1}{L_{eq}} = \sum_{i=1}^{n} \frac{1}{L_i}$$

or

$$L_{eq} = \frac{1}{\sum_{i=1}^{n} (1/L_i)}$$

R.2.60 Recall that the electric power dissipated by a resistor R, with voltage V across it, and a current I flowing through it is given by

$$P = I * V = I^2 * R = V^2/R \text{ (W)}$$

During an interval of time T, the energy (W_R) dissipated by a resistor R is given by

$$W_R = P * T = I * V * T = (V^2/R) * T = I^2 * R * T \text{ (J)}$$

R.2.61 An ideal inductor with no (zero) resistance cannot dissipate energy; it can only store energy. The energy stored in an inductor L is given by

$$W_L(t) = \frac{1}{2} L * i^2(t) \text{ (J)}$$

For the DC case (constant current I), the energy is given by

$$W_L = \frac{1}{2} * L * I^2 \text{ (J)}$$

R.2.62 An ideal capacitor with no (zero) resistance cannot dissipate energy; it can only store energy in its electric field. Its energy is given by

$$W_C = \frac{1}{2} * C * v^2(t) \text{ (J)}$$

For the DC case (constant voltage V), the energy is given by

$$W_C = \frac{1}{2} C * V^2 \text{ (J)}$$

R.2.63 In a DC circuit, the steady-state voltage drop across an inductor L is 0 V since

$$v_L(t) = L \frac{di(t)}{dt}$$

and

$$\frac{di(t)}{(dt)} = 0$$

since $i(t)$ is a constant, then $v_L(t) = 0$ V.

R.2.64 In a DC circuit, the steady-state current* through a capacitor (C) is always zero (A) since

$$i_C(t) = C \frac{dv_C(t)}{dt}$$

and

$$\frac{dv_C(t)}{dt} = 0$$

since $v_C(t)$ is a constant, then $i_C(t) = 0$ A.

R.2.65 Note that R.2.63 and R.2.64 imply that in a DC circuit during the steady-state response, that is, for $t \geq 5\ \tau$,[†] a pure inductor and a pure capacitor can be replaced by a short and an open circuit, respectively.

R.2.66 The efficiency of a system denoted by η is defined as the ratio of the output power divided by its input power. Then

$$\eta = (P_O/P_I) * 100\%$$

where P_O denotes the output power and P_I its input power, both in watts.

R.2.67 Mesh or loop analysis (first proposed by Maxwell) refers to a procedure in which, given an electrical network that consists for simplicity of resistors and voltage sources can be expressed as a set of equations in terms of its loop currents by applying KVL around each of the independent loops of the network. The resulting set of equations is sufficient to determine all the network currents.

R.2.68 The steps involved in obtaining the set of loop equations are

a. Assign a clockwise direction to each of the loop currents of the n loops of an electrical network, and label the unknown loop currents $I_1, I_2, I_3, I_4, ..., I_n$.

b. For each one of the independent n loops, write the corresponding loop equation by applying KVL around the loop.

c. A set of n equations in terms of the n unknown currents are then obtained.

d. Solve the set of n equations for the n unknown currents ($I_1, I_2, I_3, I_4, ..., I_n$).

e. A branch of the electrical network, which is a part of two adjacent loops, labeled x and y, with loop currents I_x and I_y respectively, results in a net branch current that is the algebraic sum of the two currents I_x and $I_y (I_x - I_y)$ where $1 < x \leq n$ and $1 < y \leq n$.

---

* The concept of steady state is introduced later on in this chapter. At this point, consider steady state as the stable or final current.

† τ is referred to as the time constant of the circuit introduced and discussed later on in this chapter. For simple first-order circuits, τ is either RC or L/R.

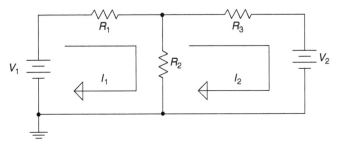

FIGURE 2.8
Electrical network with two independent loop currents.

R.2.69 The following example illustrates the procedure used to obtain the set of two independent loop equations for the circuit shown in Figure 2.8.

$$V_1 = (R_1 + R_2) * I_1 - R_2 * I_2 \qquad \text{(loop\# 1)}$$

$$-V_2 = -R_2 * I_1 + (R_2 + R_3) * I_2 \qquad \text{(loop\# 2)}$$

R.2.70 Observe that to obtain the loop equations set, the electric circuit must first be drawn as a planar network. Also observe that the circuit shown in Figure 2.8 presents two independent loops (from a total of three loops) in which two equations in terms of two unknown currents (I_1 and I_2) are obtained.

R.2.71 Note that the resulting branch current through R_2 in the circuit of Figure 2.8 is given by $I_1 - I_2$.

R.2.72 Note that the structure (and format) of each of the loop equations is given by the following resulting expression:

for any arbitrary loop x

$$\sum[\text{voltage sources in loop } x] = \sum[\text{resistances in } x] * I_x$$

$$- \sum[\text{resistances in branch } xy] * I_y \quad \text{for all possible } ys$$

where the voltage sources (left-hand side of the preceding equation) that generate power are considered positive (otherwise negative if the source consumes power), and the branch xy consists of the elements common to loops x and y that carry the opposing currents I_x and I_y. The branch current in xy is then $I_x - I_y$.

R.2.73 Node analysis refers to a procedure where a given electrical network that consists for simplicity of resistors and current sources are expressed in terms of all the nodal voltages with respect to an arbitrary reference node label ground (with zero potential). Thus, in a circuit with n nodes, one node is designated as the reference node (ground), and for the remaining $n - 1$ nodes (applying KCL), $n - 1$ equations can then be expressed in terms of the $n - 1$ unknown nodal voltages labeled $V_1, V_2, V_3, V_4, ..., V_{n-1}$.

For each one of the nodes, assume that the unknown current directions are toward the reference node. Then solve the $n - 1$ nodal equations simultaneously for all the unknown nodal voltages. Once all the nodal voltages are known, then all the network voltages as well as all the network currents can be easily evaluated.

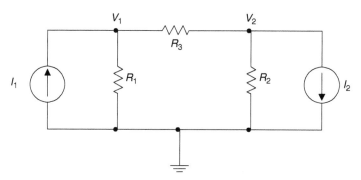

FIGURE 2.9
Electrical network requiring two node equations.

R.2.74 Nodal analysis is more powerful than mesh analysis in the sense that it applies equally well to both planar and nonplanar networks.
        For any arbitrary node x, the structure (and format) of the node equation is given by

$$\sum [\text{current sources at } x] = \sum [\text{admittances connected between } x \text{ and } y] * V_x$$

$$- \sum [\text{admittances between } x \text{ and } y] * V_y \quad \text{for all possible } ys$$

where the current sources (left side of equation) are considered positive if its polarity points towards x and negative otherwise, and y is a node connected to x through an element (or group of elements) for any y, $1 < y \leq n-1$.

R.2.75 The following example, a network consisting of three nodes as shown in Figure 2.9, is analyzed using node equations.

ANALYTICAL Solution
The node equations are
For node V_1

$$I_1 = \left(\frac{1}{R_1} + \frac{1}{R_3} \right) * V_1 - \frac{1}{R_3} * V_2$$

For node V_2

$$-I_2 = -\frac{1}{R_3} * V_1 + \left(\frac{1}{R_2} + \frac{1}{R_3} \right) * V_2$$

R.2.76 Source transformation refers to the fact that any voltage source V in series with a resistor R_s can be replaced by a current source $I = V/R_s$, in parallel with the same resistor R_s and vice versa as illustrated in Figure 2.10.

R.2.77 An example of source transformation is shown in Figure 2.11, where a voltage source $V = 10$ V, in series with a resistance $R_s = 5\ \Omega$, can be transformed into an equivalent current source $I = (10/5) = 2$ A in parallel with $R_s = 5\ \Omega$.

R.2.78 The superposition theorem states that in a linear DC network, the current through, or voltage across any element is given by the algebraic sum (contributions) of the

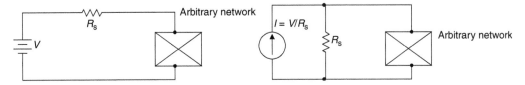

FIGURE 2.10
Equivalent circuits obtained by source transformation.

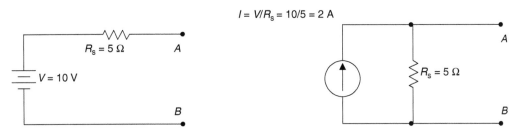

FIGURE 2.11
Equivalent source transformation circuits as seen through terminals A and B.

currents or voltages produced independently by each source (current or voltage). The concept of independence means that only one source is considered active, whereas all the other sources (voltage or current) are set to zero.

Recall that when a voltage source is set to zero, the source can be replaced by a short circuit, and similarly, when a current source is set to zero, the source can be replaced by an open circuit.

The superposition theorem states then that in an n source electrical network, the current through or the voltage drop across any arbitrary network element can be obtained by solving n single source networks (where the single source network refers to the original network with only one source at the time whereas all the remaining sources are set to zero) for the variable of interest, which can be a current or a voltage. The solution is the algebraic sum of the partial solutions (of the single source networks).

Note that the voltage and current polarities (directions) are important when solving each single source network, since the solution is the algebraic sum of the (solutions) contribution of each source.

R.2.79 Thevenin's theorem states that any two-terminal linear DC network across a load R_L can be replaced by two elements: a voltage source called the Thevenin voltage V_{TH} and a resistor placed in series called the Thevenin resistance R_{TH}. The Thevenin resistance R_{TH} is calculated by setting all the sources to zero, removing the arbitrary load R_L (replace it by an open), labeling its terminals aa', and either calculating or measuring the resistance looking into the terminals (aa').

The Thevenin voltage V_{TH} is the open-circuit voltage across terminals aa', after having removed the arbitrary load R_L from the original circuit, while preserving all the sources.

R.2.80 Norton's theorem states that any two-terminal linear DC network can be replaced by two elements, a current source in parallel with a resistor, where the current source is the Norton short-circuit current denoted by I_N, and the resistance is the Thevenin resistance R_{TH}.

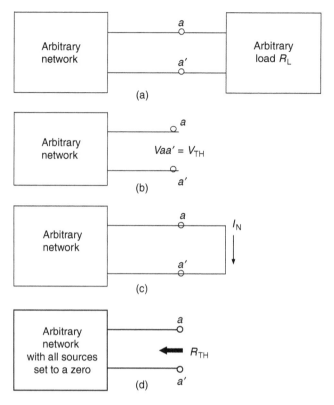

FIGURE 2.12
(a) A network with a connected load, (b) Thevenin model for V_{TH}, (c) Norton model for I_N, (d) Thevenin–Norton model for R_{TH}.

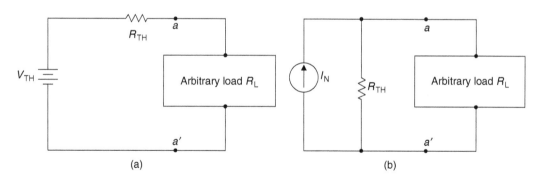

FIGURE 2.13
(a) Thevenin's equivalent circuit, (b) Norton's equivalent circuit.

The Norton's short-circuit current I_N is the current through terminals aa', obtained by replacing R_L by a short-circuit while preserving all sources.

R.2.81 The equivalent circuits used to evaluate the values of V_{TH}, I_N, and R_{TH} are shown in Figure 2.12.

   The Thevenin's and Norton's equivalent circuits are shown in Figure 2.13.

R.2.82 Note that the Thevenin's and Norton's equivalent circuits are related by a simple source transformation as illustrated in Figure 2.14.

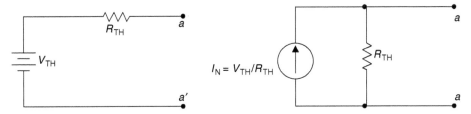

FIGURE 2.14
Thevenin's source transformation into a Norton's equivalent circuit.

R.2.83 Let us gain some experience by using the different network theorems and techniques just presented by solving the following example.

For the circuit shown in Figure 2.15, find the voltage across $R_L = 5\ \Omega$ (*Vaa'*) by hand calculation by using

a. Loop equations

b. Node equations

c. Source transformation—voltage to current source

d. Source transformation—current to voltage source

e. Superposition

f. Thevenin's theorem

g. Norton's theorem

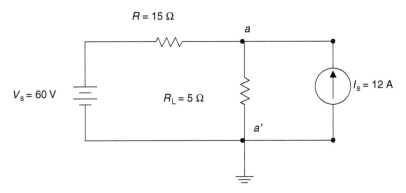

FIGURE 2.15
Network of R.2.83.

ANALYTICAL Solutions

a. The loop equation solution is illustrated in the circuit diagram of Figure 2.16. The loop currents I_1 and I_2 are indicated in Figure 2.16. Then the loop equation for loop # 1 is given by

$$60\ \text{V} = 20I_1 - 5I_2$$

and since

$$I_2 = -12\ \text{A}$$

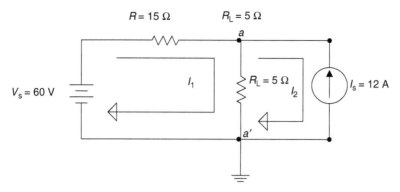

FIGURE 2.16
Loop model of network of R.2.83.

Then

$$60 \text{ V} = 20I_1 - 5(-12)$$

$$60 = 20I_1 + 60$$

$$\frac{60 - 60}{20} = I_1 = 0$$

Then

$$Vaa' = 5 \ \Omega * 12 \text{ A} = 60 \text{ V}$$

b. The node equation solution is referred to as the circuit diagram of Figure 2.16. The node equation for node a shown in Figure 2.16 is given below while the reference node is a' (grounded).

$$12 \text{ A} = \left(\frac{1}{5} + \frac{1}{15}\right)Va - \frac{1}{15}60$$

$$12 = \frac{4}{15}Va - 4$$

$$12 + 4 = \frac{4}{15}Va$$

$$16\frac{15}{4} = Va$$

$$Va = 60 \text{ V or } Vaa' = 60 \text{ V}$$

c. The voltage source transformation into a current source is shown by the circuit diagram of Figures 2.17 and 2.18, where first the transformation is illustrated (Figure 2.17), followed by the currents' source combinations (Figure 2.18).

Then

$$Vaa' = (15 \mid\mid 5) 16$$

$$Vaa' = \frac{(15 * 5 * 16)}{15 + 5} = 60 \text{ V}$$

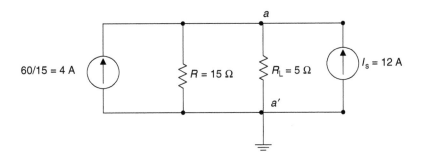

FIGURE 2.17
The voltage source is transformed in the network of R.2.83.

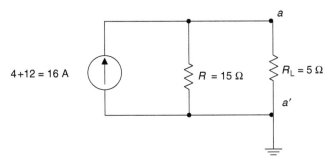

FIGURE 2.18
Current sources are added (combine into an equivalent source) in the network of Figure 2.17.

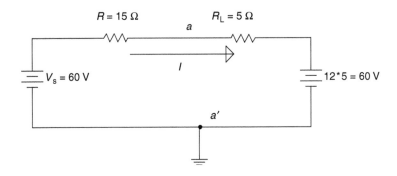

FIGURE 2.19
Current source transforms into a voltage source in the network of R.2.83.

d. The current source transformation into a voltage source is illustrated by the circuit diagram of Figure 2.19. Then

$$I = \frac{60 - 60}{20} = 0 \text{ A}$$

therefore

$$Vaa' = 60 \text{ V}$$

e. The superposition solution is illustrated by the two circuit diagrams shown in Figure 2.20.

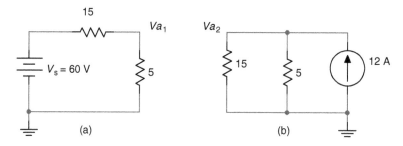

FIGURE 2.20
Superposition models of the network of R.2.83. (a) Circuit of Figure 2.15 with $I_s = 0$, (b) Circuit of Figure 2.15 with $V_s = 0$.

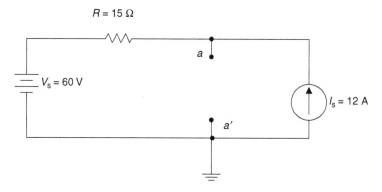

FIGURE 2.21
Thevenin's models of the network of R.2.83.

The two single source networks of Figure 2.20 are solved for the voltage drop across $R = 5\,\Omega$ labeled Va_1 and Va_2 and the voltage Vaa' is then the algebraic sum $(Va_1 + Va_2)$. The steps involved are indicated as follows:

a. Circuit of Figure 2.15 with $I_s = 0$
b. Circuit of Figure 2.15 with $V_s = 0$

The voltage Va_1 for the circuit shown in Figure 2.20a is given by

$$Va_1 = \frac{(5 * 60)}{20} = 15\ \text{V} \quad \text{(using the voltage divider rule)}$$

The voltage Va_2 for the circuit shown in Figure 2.20b is given by

$$Va_2 = \frac{(15 * 5 * 12)}{15 + 5} = 45\ \text{V}$$

Since Va_1 and Va_2 have the same polarity

$$Va = Va_1 + Va_2 = 15\ \text{V} + 45\ \text{V} = 60\ \text{V}$$

6. The solution using Thevenin's theorem is indicated by the circuit diagram of Figure 2.21. (Note that the load is $R_L = 5\,\Omega$.)

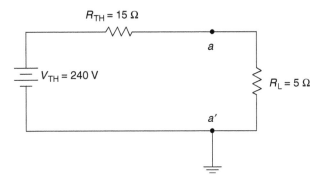

FIGURE 2.22
The Thevenin's equivalent circuit of the network of R.2.83.

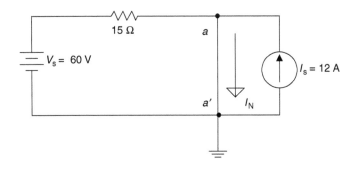

FIGURE 2.23
The Norton's equivalent model of the network of R.2.83.

Note that the load R_L is removed from the circuit of Figure 2.21 and the open voltage Vaa' is V_{TH}. The evaluation of V_{TH} is given by

$$Vaa' = V_{TH} = 60 \text{ V} + 15 \text{ } \Omega * 12 \text{ A} = 60 \text{ V} + 180 \text{ V} = 240 \text{ V}$$

Then $R_{TH} = 15 \text{ } \Omega$ is obtained by setting all the sources to zero.
The resulting Thevenin's equivalent circuit is shown in Figure 2.22.

Finally, Vaa' is given by

$$Vaa' = \frac{(5 * 240)}{15 + 5} = 60 \text{ V}$$

g. Norton's solution is indicated by the equivalent circuit diagram model of Figure 2.23. The short current I_N is then evaluated below

$$I_N = 12 \text{ A} + \frac{60 \text{ V}}{15 \text{ } \Omega} = 12 \text{ A} + 4 \text{ A} = 16 \text{ A}$$

The resulting Norton's equivalent circuit is shown in Figure 2.24.

$$V_{R_L} = Vaa' = (15 \,\|\, 5) I_N$$

$$V_{R_L} = Vaa' = \frac{(15 * 5 * 16)}{15 + 5} = 60 \text{ V}$$

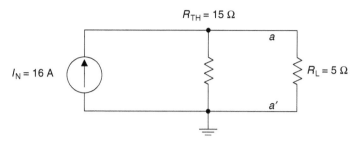

FIGURE 2.24
The Norton's equivalent circuit of the network of R.2.83.

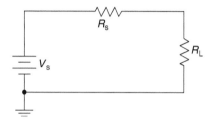

FIGURE 2.25
Maximum power delivered to R_L when $R_L = R_s$.

R.2.84 The maximum power transfer theorem states that maximum power is delivered to a load R_L connected to an ideal voltage source V_s connected to a series resistor R_s when the load resistance R_L is equal to the resistance R_s as illustrated in the circuit diagram of Figure 2.25.

(Maximum power is delivered to R_L when $R_L = R_s$.)

R.2.85 For a load R_L connected to an arbitrary network, maximum power is delivered to the load R_L, where R_L is calculated in the following way:

a. The load is removed and replaced by an open circuit (with terminals aa').

b. Calculate the Thevenin's equivalent resistance R_{TH} by looking into the open terminals (aa').

c. Maximum power is delivered to the load by adjusting R_L to be equal to R_{TH}.

R.2.86 Note that the maximum power transfer theorem holds for the case where R_{TH} is fixed and R_L is allowed to vary. If R_L is fixed but R_{TH} is allowed to vary, then maximum power delivered to R_L will not occur when R_L is equal to R_{TH}, but rather when $R_{TH} = 0$.

R.2.87 For the example presented in R.2.83 referred to as Figure 2.15, maximum power is delivered to the load R_L (5 Ω) by changing (increasing) R_L to 15 Ω. The resulting Thevenin's equivalent circuit of the given network as well as the new load R_L is shown in Figure 2.26.

The maximum power delivered to R_L is then given by

$$P_{RL\text{-max}} = \frac{(30 \text{ V})^2}{15 \text{ Ω}} = 60 \text{ W}$$

R.2.88 If a circuit is purely resistive with no energy-storing elements (L or G) and switching occurring, then there will be no transient behavior, and current and voltages

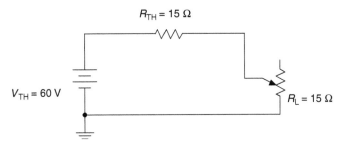

FIGURE 2.26
Load R_L is increased to 15 Ω from 5 Ω (potentiometer) for maximum power transfer of the network of R.2.83.

reach a steady state instantaneously. In contrast, if storing energy elements are present in a circuit, such as inductors or capacitors, and switching occurs, then transient takes place due to the trapped network voltages and currents, and it takes time to settle into the new stable values.

In brief, transients exist in a circuit if and only if, at any time, a sudden change is made (switched) and energy-storing devices are present.

R.2.89 Transient solutions involve differential equations in which initial conditions must be considered. These solutions require the inclusion of the energy-storing devices (capacitors and inductors) as well as the trapped conditions.

R.2.90 Transient solutions involve either growth or decay time exponentials, and are dependent on the exponential coefficient referred as the network time constant (τ).

R.2.91 The time constant can be either RC or L/R or a combination or superposition of time constants for the case of more complex circuits.

R.2.92 The time constant (τ) is the time the circuit requires to complete 63.2% of the change (or discharge) that ultimately takes place. Practical considerations are associated with the time constant, for example, four or five time constants is the time that a circuit needs to reach the steady-state conditions, and after that time, all transients die out or are no longer present.

R.2.93 The transient solution of a given circuit depends on the number of independent energy-storing elements (capacitors and inductors), and not on the number of the network loops or nodes. The reader should observe that capacitors or inductors in series or parallel are not independent, since they can easily be combined. Similarly, a delta (Δ) or Y connection consisting of three inductors or three capacitors, one in each branch, leads to just two independent energy-storing elements, since the third element depends on the value of the other two.

R.2.94 Electric circuits with one capacitor or inductor lead to first-order ordinary differential equations, regardless of the structure or complexity of the circuit. Complex circuits with only one type of energy-storing device can be analyzed by using Thevenin's theorem, assuming that the load is the energy-storing element.

R.2.95 Let us start the transient analysis by considering the RL series circuit shown in Figure 2.27. The switch shown in the circuit closes at $t = 0$ (it is open for $t < 0$), and let us assume that the coil is discharged, that is, $i(0) = 0$. Then by applying KVL, a first-order linear differential equation is obtained, and then solved in terms of the unknown current $i(t)$, illustrated as follows:

$$Ri(t) + L\frac{di(t)}{dt} = V$$

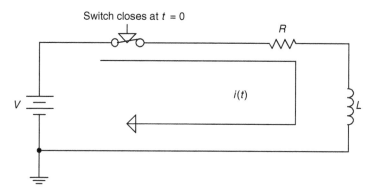

FIGURE 2.27
RL series circuit.

solving for $i(t)$ results in

$$i(t) = \frac{V}{R}(1 - e^{-t/\tau}) \quad \text{for } t \geq 0 \quad \text{with } \tau = \frac{L}{R}$$

R.2.96 When a sudden voltage V is applied to a simple-series RL circuit, a transient condition results (recall that the inductor controls the current flow), and the general current and voltage relations are given by

$$i(t) = \frac{V}{R}(1 - e^{-t/\tau})$$

and the voltage across the resistor R is given by

$$v_R(t) = V(1 - e^{-t/\tau})$$

and the voltage across the inductor L is given by

$$v_L(t) = V - v_R(t) = V - (V - Ve^{-t/\tau}) = Ve^{-t/\tau}$$

where $\tau = L/R$ (s) is the time constant of the circuit. Note that $V = v_R(t) + v_L(t)$.

R.2.97 In an RL circuit in which the inductor is charged and its initial current is I_o and initial voltage across V_o (with no source), the inductor discharges and the current and voltage relations are given by

$$i(t) = I_o * e^{-t/\tau} = \frac{V_o}{R} * e^{-t/\tau}$$

and

$$v_L(t) = V_o * e^{-t/\tau}$$

R.2.98 Let us now turn our attention to the RC series circuit shown in Figure 2.28. The switch is open for $t < 0$, and closes at $t = 0$, assuming that the capacitor is discharged, that is, $v_C(0) = 0$. Then applying KVL (around the loop), the

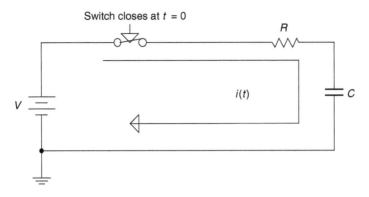

FIGURE 2.28
RC series circuit.

following first-order linear differential equation is obtained in terms of the unknown loop current $i(t)$ illustrated as follows:

$$\frac{1}{C}\int i(t)dt + Ri(t) = V$$

then

$$i(t) = \frac{V}{R}e^{-t/\tau}$$

where $\tau = RC$.

$$v_R(t) = Ve^{-t/\tau} \quad \text{for } t > 0$$

and

$$v_C(t) = V - v_R(t) = V - Ve^{-t/\tau} \quad \text{for } t > 0$$

R.2.99   When a sudden DC voltage V is applied to a simple series RC circuit (with $v_c(0) = 0$ V), the capacitor voltage charges up to

$$v_C(t) = V(1 - e^{-t/\tau})$$

and the current through C is then given by

$$i_C(t) = \frac{V}{R}e^{-t/\tau}$$

where $\tau = RC$ (s) is the time constant of the circuit.

R.2.100   A charged capacitor in an RC circuit with an initial voltage $V(0)$ and no sources, as indicated in Figure 2.29, discharges with the following current and voltage relations:

$$v_C(t) = V_0 e^{-t/\tau}$$

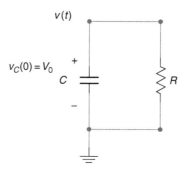

FIGURE 2.29
Charged capacitor in an RC series circuit.

where $\tau = RC$, and

$$i(t) = \frac{v_C(t)}{R} = \frac{V_0}{R} e^{-t/\tau} \quad \text{for } t \geq 0$$

R.2.101 When a circuit switches from one set of conditions to a new set of conditions, there is a transition period where voltages and currents are adjusting to the new conditions. The term *transient response* is the adjusting transition time from its initial to its final voltage and current values. The transition takes place in the interval $0 \leq t \leq 5\tau$; and the steady-state response refers to the voltages and currents for $t > 5\tau$.

R.2.102 Let us now review and summarize the inductor's characteristics.

- There is no voltage drop across an inductor if the current through it is not changing with time. An inductor is, therefore, a short circuit for DC, and $v_L(t) = 0$, for $t > 5\tau$.

- A finite amount of energy can be stored in an inductor even if the voltage across the inductor is zero.

- An inductor resists abrupt changes in current.

- An inductor never dissipates energy; it is only capable of storing energy. This statement is true for the ideal inductor; however, it is false for the real inductor, which always possesses some internal resistance.

R.2.103 The capacitor's characteristics are summarized as follows:

- The current through the capacitor is zero, if the voltage across it is not changing with time (DC).

- A finite amount of energy can be stored in a capacitor even if the current through the capacitor is zero.

- A capacitor resists abrupt change in voltage across it.

- A pure capacitor cannot dissipate energy; it can only store energy.

R.2.104 For example, analyze the circuit shown in Figure 2.30, and write the differential equation for $t \geq 0$, if the switch opens at $t = 0$ after being closed for a long time. Obtain expressions for $v_C(t)$ and $i_R(t)$ for $t > 0$.

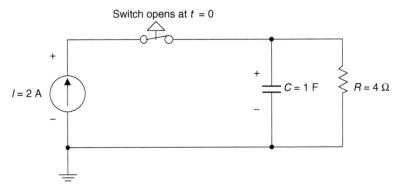

FIGURE 2.30
RC network of R.2.104.

ANALYTICAL Solution

For $t < 0$, $v_C(0) = 2$ A $* 4$ $\Omega = 8$ V.
For $t > 0$, the nodal differential equation is given by

$$C\frac{dv_C(t)}{dt} - \frac{v_C(t)}{R} = 0$$

where $\tau = RC$. The solution to the preceding differential equation is of the form $v_C(t) = Ae^{-t/\tau}$, and by satisfying the initial conditions at $t = 0$, the following is obtained: $v_C(0) = 8 = A$, then $v_C(t) = 8e^{-t/\tau}$ V and $i_R(t) = v_C(t)/R = 2e^{-t/\tau}$ (amp).

R.2.105 The RC and RL transient responses are modeled by first-order differential equations. The RLC transient response leads to a second-order differential equation where the roots of its characteristic equation may be complex numbers. Those roots are referred to as the complex natural frequencies of the circuit. The location of the roots on the complex plane determines the form of the transient response of the electric network. Roots located on the left half of the plane represent decaying exponentials, whereas roots on the right half of the plane represent growing exponentials.

Complex conjugate roots can be associated with oscillations (sinusoids); and if they are located in the left half of the plane, they are decaying; and in the right half of the plane, they represent growing sinusoids. When the roots are purely imaginary, they are located on the imaginary (jw) axis of the complex plane and represent sustained oscillations.

R.2.106 Oscillations occur in an RLC circuit when R is small, or the ideal case is when $R = 0$, then oscillations are sustained. Observe that the resistance R controls the energy dissipated (losses) in the form of heat (friction) and is commonly referred to as the damping coefficient.

R.2.107 Three distinct cases are encountered in the solution of second-order systems. They are

- Overdamped
- Critical damped
- Underdamped

These cases are analyzed next for both the series and parallel RLC configurations.

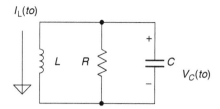

FIGURE 2.31
Source free parallel RLC of R.2.109.

R.2.108 Systems higher than the second order are not considered in this section, but in general, the solution follows the same steps outlined for the second-order system. Higher-order systems arise when there are more than two independent energy-storing elements. The transient response is obtained by solving for the roots of the network characteristic equation that consist of either real numbers, in which case the response consists of exponentials, or complex conjugate, in which case the response consists of sinusoids (decaying or growing).

R.2.109 Let us analyze the source-free parallel RLC circuit shown in Figure 2.31. The nodal differential equation (KCL) is given by

$$\frac{v(t)}{R} + \frac{1}{L}\int_{to}^{t} v(\lambda)d\lambda - i_L(to) + C\frac{dv(t)}{dt} = 0$$

Then differentiating each term with respect to t yields

$$C\frac{d^2v(t)}{dt^2} + \frac{1}{R}\frac{dv(t)}{dt} + \frac{1}{L}v(t) = 0$$

The auxiliary or characteristic equation becomes (see Chapter 7 of *Practical MATLAB® Basics for Engineers*)

$$Cs^2 + \frac{s}{R} + \frac{1}{L} = 0$$

where $s = d/dt$, and solving for the two roots yields

$$s_{1,2} = \frac{-1}{2RC} \pm \sqrt{\left(\frac{1}{2RC}\right)^2 - \frac{1}{LC}}$$

Let

$$w_0 = \frac{1}{\sqrt{LC}}$$

where w_0 is called the resonant frequency, and let

$$\alpha = \frac{1}{2RC}$$

where α is referred to as the neper frequency,

and s_1 and s_2 are referred as the complex frequencies given by

$$s_{1,2} = -\alpha \pm \sqrt{\alpha^2 - w_0^2}$$

R.2.110 As mentioned, the natural response of a parallel RLC circuit results in one of the following three cases:

a. Overdamped

b. Critical damped

c. Underdamped

Observe that the elements define the specific case.

R.2.111 Let us analyze each case, starting with the overdamped parallel configuration that occurs for the following condition:

$$\sqrt{\alpha^2 - w_0^2} < \alpha$$

and since

$$s_{1,2} = -\alpha \pm \sqrt{\alpha^2 - w_0^2}$$

then

$$\left[-\alpha - \sqrt{\alpha^2 - w_0^2}\right] < \left[-\alpha + \sqrt{\alpha^2 - w_0^2}\right] < 0$$

Note that both s_1 and s_2 are real, distinct, and negative. Then the solution of the differential equation of R.2.109 is of the form

$$v(t) = A_1 e^{-s_1 t} + A_2 e^{-s_2 t}$$

where A_1 and A_2 are constants that can be evaluated from the network initial conditions.

R.2.112 The critical-damped parallel case occurs for the following condition:

$$\sqrt{\alpha^2 - w_0^2} = 0$$

Therefore both s_1 and s_2 are equal to $-\alpha$, and α is real and negative. The solution of the differential equation of R.2.109 is then of the form

$$v(t) = A_1 e^{-\alpha t} + A_2 t e^{-\alpha t}$$

where A_1 and A_2 are constants that can be evaluated from the network initial conditions.

R.2.113 The underdamped parallel case occurs for the following condition:

$$\alpha^2 - w^2 < 0$$

Then s_1 and s_2 become complex conjugate frequencies, and the response of the differential equation of R.2.109 is of the following form:

$$v(t) = e^{-\alpha t}[A_1 \cos(w_d t) + A_2 \sin(w_d t)]$$

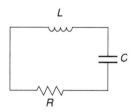

FIGURE 2.32
Source free series RLC of R.2.114.

where

$$w_d = \sqrt{w_o^2 - \alpha^2}$$

where A_1 and A_2 are constants that can be evaluated from the initial network conditions.

R.2.114 Let us now turn our attention to the source-free RLC series circuit shown in Figure 2.32, assuming for simplicity that all the initial conditions are zero.

The loop differential equation of the circuit of Figure 2.32 is given by

$$\frac{1}{C}\int i(t)\,dt + L\frac{di(t)}{dt} + R\,i(t) = 0.$$

Differentiating every term of the preceding equation with respect to t yields

$$L\frac{d^2i(t)}{dt^2} + R\frac{di(t)}{dt} + \frac{1}{C}i(t) = 0$$

or

$$\frac{d^2i(t)}{dt^2} + \frac{R}{L}\frac{di(t)}{dt} + \frac{1}{CL}i(t) = 0.$$

The preceding equation is a second-order, linear, homogeneous differential equation, and the auxiliary equation is given by

$$s^2 + \frac{R}{L}s + \frac{1}{LC} = 0$$

and the roots of this equation are

$$s_{1,2} = \frac{-R}{2L} \pm \sqrt{\left(\frac{R}{2L}\right)^2 - \frac{1}{LC}}$$

where $\alpha = \frac{R}{2L}$ is referred as the neper frequency, $w_o = \frac{1}{\sqrt{LC}}$ is the resonant frequency, then $s_{1,2} = -\alpha \pm \sqrt{\alpha^2 - w_o^2}$ are referred to as the complex network frequencies.

Observe that the term (similar to the parallel case) $\alpha^2 - w_o^2$ can be positive, zero, or negative. Then the respective solutions are

- Overdamped
- Critical damped
- Underdamped

R.2.115 The natural overdamped RLC series circuit response occurs for the condition $\alpha^2 > w_o^2$. Then the roots s_1 and s_2 are real, unequal, and negative numbers, resulting in a solution of the form $i(t) = A_1 e^{-s1t} + A_2 e^{-s2t}$, where the constants A_1 and A_2 depend on the initial network conditions $\left(\text{usually given by } i(t = 0) \text{ and} \dfrac{di(t)}{dt}\bigg|_{t=0}\right)$.

R.2.116 The natural critical-damped RLC series circuit response occurs for the condition $\alpha^2 = w_o^2$. Then the roots s_1 and s_2 are real $(-\alpha)$ and repeated, resulting in a solution of the form $i(t) = A_1 e^{-\alpha t} + A_2 t e^{-\alpha t}$, where the constants A_1 and A_2 can be evaluated from the system's initial conditions.

R.2.117 The natural underdamped RLC series circuit response occurs for the condition $\alpha^2 < w_o^2$. Then the roots s_1 and s_2 are a complex conjugate pair, resulting in a solution of the form

$$i(t) = e^{-\alpha t}[A_1 \cos(w_d t) + A_2 \sin(w_d t)]$$

where $w_d = \sqrt{w_o^2 - \alpha^2}$ and A_1 and A_2 are constants that depend on the initial network conditions.

R.2.118 The transient analysis of a circuit that contains more than one loop can be described by a set of differential equations. The set of differential equations may consist of either loop or node equations where the unknowns may be either the loop currents or the node voltages. For example, the set of two loop differential equations (using KVL) for the circuit shown in Figure 2.33 is illustrated, where the two loop currents define the system's transient response.

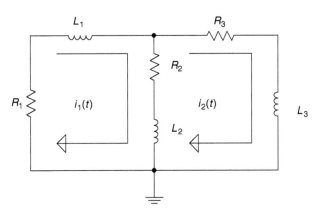

FIGURE 2.33
Transient electrical network of R.2.118.

ANALYTICAL Solution

The set of loop differential equations are

$$\text{Loop\#1:}\quad (L_1 + L_2)\frac{di_1(t)}{dt} + (R_1 + R_2)i_1(t) - \left(R_2 i_2(t) + L_2 \frac{di_2(t)}{dt} \right) = 0$$

$$\text{Loop\#2:}\quad -L_2 \frac{di_1(t)}{dt} - R_2 i_1(t) + (L_2 + L_3)\frac{di_2(t)}{dt} + (R_2 + R_3)i_2(t) = 0$$

The set of initial (or boundary) current conditions such as $i_1(t_0) = I_1$ and $i_2(t_0) = I_2$ must be known to completely specify its response.

R.2.119 Recall that capacitors and inductors can be combined and represented by a single equivalent element depending on the type of connection (series or parallel).
The voltage and current divider rules for capacitors and inductors are stated below.

R.2.120 The voltage divider rule is used to evaluate the voltage drops $VL1$ and $VL2$ across the inductors L_1 and L_2 when connected in series with and applied voltage V_o across them as shown in Figure 2.34.

$$VL_1 = \frac{L_1}{L_1 + L_2} V_o$$

$$VL_2 = \frac{L_2}{L_1 + L_2} V_o$$

R.2.121 Let us now consider two inductors L_1 and L_2 connected in parallel, with a current (source) I_o feeding the parallel combination as indicated in Figure 2.35.

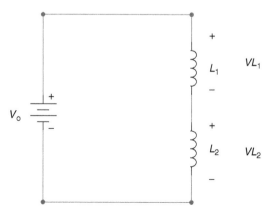

FIGURE 2.34
Voltage divider rule across inductors in series of R.2.120.

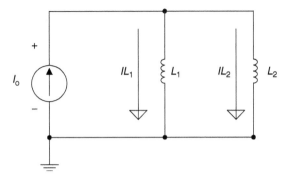

FIGURE 2.35
Current divider rule for inductors in parallel.

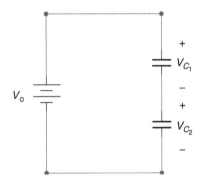

FIGURE 2.36
Voltage divider rule across capacitors in series.

Then the current divider rule states that the currents IL_1 and IL_2 are given by

$$IL_1 = \frac{L_2}{L_1 + L_2} I_o$$

$$IL_2 = \frac{L_1}{L_1 + L_2} I_o$$

R.2.122 Now consider two capacitors C_1 and C_2 connected in series as shown in Figure 2.36 with and applied voltage V_o across them.

Then the voltages V_{C1} and V_{C2} are evaluated by using the voltage divider rule as follows:

$$V_{C_1} = \frac{C_2}{C_1 + C_2} V_o$$

$$V_{C_2} = \frac{C_1}{C_1 + C_2} V_o$$

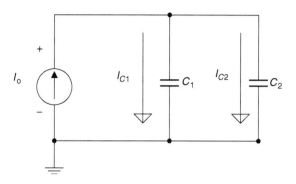

FIGURE 2.37
Current divider rule for capacitors in parallel.

R.2.123 Finally, consider the two capacitors C_1 and C_2 connected in parallel with a current (source) I_o feeding the parallel combination as shown in Figure 2.37.

The currents I_{C1} and I_{C2} are evaluated by using the current divider rule as follows:

$$I_{C1} = \frac{C_1}{C_1 + C_2} I_o$$

$$I_{C2} = \frac{C_2}{C_1 + C_2} I_o$$

2.4 Examples

Example 2.1

1. Determine the charge induced when a current of 1 A passes through a point during 2 min
2. Determine the number of electrons required to build up the charge of part 1

Solve this problem analytically and by using MATLAB.

ANALYTICAL Solution

$$\text{Charge } Q \text{ (coulombs)} = I*t = 1 \text{ A} * 2 \text{ min} * \frac{60 \text{ s}}{\text{min}}$$

$$Q = 120 \text{ A s} = 120 \text{ C}$$

The number of electrons required to build up a charge of $Q = 120$ C is given by

$$\text{Number of electrons} = 120 \text{ C} * \frac{1 \text{ electron}}{1.602 \times 10^{-19} \text{ C}} = 7.49 \times 10^{-19} \text{ electrons}$$

MATLAB Solution
```
% Script file: charge
I = 1;
T = 2*60;
Q = I*T;
disp('****************************************************')
disp('The charge of 1 amp passing through a point during 2 minutes is :');
disp(Q);
disp('Coulombs')
format long
Ne=Q*(1/(1.602*10e-19));
disp('The number of electrons is:'); disp(Ne)
disp('****************************************************')

>> charge
```

```
*********************************************************************
The charge of 1 amp passing through a point during 2 minutes is :
                120   Coulombs
The number of electrons is:
                            7.490636704119849e+019
*********************************************************************
```

Example 2.2

Create the MATLAB script file *series* that returns the equivalent resistance (across terminals *AB*) shown in Figure 2.38, where $R_1 = 10\,\Omega$, $R_2 = 10\,\Omega$, $R_3 = 20\,\Omega$, and $R_4 = 30\,\Omega$.

MATLAB Solution
```
% Script file: series
Resist = [10 10 20 30];
Requi = sum(Resist);
disp('*******************************')
disp('The equivalent resistance is='); disp(Requi);
disp('                      Ohms')
disp('*******************************')

>> series
```

```
            *******************************************
            The equivalent resistance is =
                                        70 Ohms
            *******************************************
```

FIGURE 2.38
Series connections of R_1, R_2, R_3, and R_4 of Example 2.2.

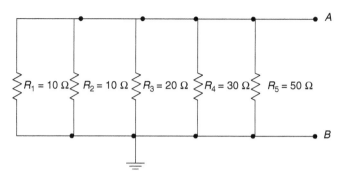

FIGURE 2.39
Parallel connection of R_1, R_2, R_3, R_4, and R_5 of Example 2.3.

Example 2.3

Create the script file *parallel* that returns the equivalent resistance R_{eq} across terminals AB of the circuit shown in Figure 2.39, where $R_1 = 10\ \Omega$, $R_2 = 10\ \Omega$, $R_3 = 20\ \Omega$, $R_4 = 30\ \Omega$, and $R_5 = 50\ \Omega$.

MATLAB Solution
```
% Script file: parallel
Resist = [10 10 20 30 50];
ones =[1 1 1 1 1];
Re = ones./Resist
disp('***************************************************')
disp('The admittances (in sie.) are=');disp(Re)
Res = sum(Re);
Req = 1/Res;
disp ('The equivalent resistance is given by:'); disp(Req)
disp('      Ohms      ')
disp('***************************************************')

>> parallel

***************************************************
The admittances (in sie.) are =
              Re =
                    0.1000    0.1000    0.0500    0.0333    0.0200
The equivalent resistance is given by:
                    3.29670329670330  Ohms
***************************************************
```

Example 2.4

Color coding is used to specify the resistance values (in ohms) as well as its statistical performance (tolerance and reliability).

The color code of a resistor consists of five color bands, as shown in Figure 2.40, where the first three color bands are used to specify the resistance value and the remaining two bands are used to specify their statistical performance.

Each color band represents a numerical value defined in Table 2.2.

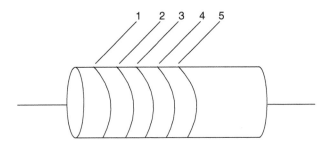

FIGURE 2.40
Resistor color bands.

TABLE 2.2

Band 1 (First Digit)	Band 2 (Second Digit)	Band 3 (Multiplying Factor)	Band 4 (Tolerance)	Band 5 (Reliability)
Brown = 1	Black = 0	Silver = −2	Gold = 0.05	Brown = 0.01
Red = 2	Brown = 1	Gold = −1	Silver = 0.10	Red = 0.001
Orange = 3	Orange = 3	Black = 0	None = 0.20	Orange = 0.0001
Yellow = 4	Yellow = 4	Brown = 1		Yellow = 0.00001
Green = 5	Green = 5	Red = 2		
Blue = 6	Blue = 6	Orange = 3		
Violet = 7	Violet = 7	Yellow = 4		
Gray = 8	Gray = 8	Green = 5		
White = 9	White = 9	Blue = 6		
		Violet = 7		
		Gray = 8		
		White = 9		

The value of an arbitrary color-coded resistor R is specified by the following rules:

1. The first (A) and the second (B) band represent the first (A) and the second digit (B), of the value R.
2. The third color band (C) represents a multiplication factor of 10 raised to the power C (10^C), with unit ohms (Ω).
3. The fourth band (D) gives an index of the manufacturer's tolerance given by [$\pm(AB) * 10^C * D$].
4. The fifth color band (E) gives an index of the manufacturer's component reliability, which represents the probability of failure of R per the first 1000 hours of use.

The nominal value of an arbitrary color-coded resistor R is given by

$$R = (AB) * 10^C \ \Omega$$

The tolerance of the resistor R would be bounded by

$$\{(AB) * 10^C - (AB) * 10^C * D\} \leq R \leq \{(AB) * 10^C + (AB) * 10^C * D\}$$

Finally, the reliability of the resistor R is given either by E or $E\%$, and that quantity indicates the probability of failure of R during the first 1000 hours of usage.

For example, let a resistor R be defined by the following color bands:

Band 1 = Brown (A)
Band 2 = Black (B)
Band 3 = Orange (C)
Band 4 = Silver (D)
Band 5 = Brown (E)

ANALYTICAL Solution

From Table 2.2, the nominal value of R is given by

$$R = 10 * 10^3 \, \Omega = 10 \, k\Omega$$

and the range of the resistor R (tolerance) is given by

$$(10 * 10^3) - (10 * 10^3 * 0.1) \leq R \leq (10 * 10^3) + (10 * 10^3 * 0.1)$$

then

$$9{,}000 \, \Omega \leq R \leq 11{,}000 \, \Omega$$

The fifth color band indicates the reliability of R, that is, 0.01. In the present case, it indicates that one in every 100 resistors would fail during the first 1000 hours of usage.

Having defined the color code of resistors, we next proceed to write the MATLAB script file *color_code*, which returns the nominal value of a given resistor R, its tolerance and reliability factor, given its color-coded specs. The program is then tested for the resistor with the following color specs: brown, green, orange, silver, and red.

MATLAB Solution
```
% Script file: color _ code
% Color Code Program
Color1 = char('Brown=1', 'Red=2', 'Orange=3', 'Yellow=4', 'Green=5','Blue=6',
                                        'Violet=7', 'Gray=8','White=9');
disp ('The first color band, with its numerical values are given below:');
disp (Color1)
A=input('Enter the numerical value of the first band, A=');
disp('  ')
Color2=char('Black=0', Color1);
disp('The second color band, with its numerical values are given below:');
disp(Color2)
B=input('Enter the numerical value of the second ban, B=');
disp('  ')
Color3=char('Silver=-2', 'Gold=-1', Color2);
disp('The third color band, with its numerical values are given below:');
disp(Color3)
C=input('Enter the numerical value of the third band, C=');
disp('  ')
Color4=char('Gold=0.05', 'Silver=0.1','Noband=0.2');
disp('The forth color band, with its numerical values are given below:');
disp(Color4)
D=input('Enter the numerical value of the fourth band, D=');
disp('  ')
Color5=char('Brown=0.01', 'Red=0.001','Orange=0.0001','Yellow=0.00001');
disp('The fifth color band, with its numerical values are given below: ');
disp(Color5)
E=input('Enter the numerical value of the fifth band, E=');
disp('  ')
R=(A*10+B)*10^C;
Rmin=R-R*D;Rmax=R+R*D;
```

```
Relia=100000*E;
disp('*************************************************************')
disp('***************R  E  S  U  L  T  S  ****************************')
disp('*************************************************************')
disp(['The nominal value of R is ', num2str(R),'ohms'])
disp(['The range of R (tolerance)is between ',num2str(Rmin),' and ',
  num2str(Rmax),' Ohms'])
disp(['The statistical number of failure (reliability) is
',num2str(Relia),' for every 100,000 resistors per 1000 hours of use'])
disp('*************************************************************')
```

The script file *color_code* is now tested for the resistor R defined by the following color bands: brown, green, orange, silver, and red.

```
>> color _ code

    The first color band, with its numerical values are given below:
    Brown=1
    Red=2
    Orange=3
    Yellow=4
    Green=5
    Blue=6
    Violet=7
    Gray=8
    White=9
    Enter the numerical value of the first band, A=1

    The second color band, with its numerical values are given below:
    Black=0
    Brown=1
    Red=2
    Orange=3
    Yellow=4
    Green=5
    Blue=6
    Violet=7
    Gray=8
    White=9
    Enter the numerical value of the second ban, B=5

    The third color band, with its numerical values are given below:
    Silver=-2
    Gold=-1
    Black=0
    Brown=1
    Red=2
    Orange=3
    Yellow=4
    Green=5
    Blue=6
    Violet=7
    Gray=8
    White=9
    Enter the numerical value of the third band, C=3
    The fourth color band, with its numerical values are given below:
    Gold=0.05
    Silver=0.1
```

```
Noband=0.2
Enter the numerical value of the fourth band, D=0.1

The fifth color band, with its numerical values are given below:
Brown=0.01
Red=0.001
Orange=0.0001
Yellow=0.00001
Enter the numerical value of the fifth band, E=0.001
```

```
*****************************************************************
***************************R E S U L T S*************************
*****************************************************************
The nominal value of R is 15000 ohms
The range of R (tolerance) is between 13500 and 16500 Ohms
The statistical number of failure (reliability) is 100 for every
100,000 resistors per 1000 hours of use
*****************************************************************
```

The reader is encouraged to run and test the preceding program for the three color-coded resistors, given in Table 2.3, and for any arbitrary color-coded resistance.

TABLE 2.3

Examples of Color Coded Resistors

Band 1	Band 2	Band 3	Band 4	Band 5
Yellow	Red	Orange	Silver	Yellow
Gray	White	Violet	Gold	Orange
Red	Green	Black	No band	Red

Example 2.5

Solve by hand and by using MATLAB for the energies stored in the capacitors $C_1 = 1$ F and $C_2 = 2$ F shown in the circuit diagram of Figure 2.41 expressed in the following units:

1. Joules
2. British thermal units
3. Foot pound
4. Watt hour

See R.2.36 for unit conversions.

ANALYTICAL Solution

The equivalent DC circuit of Figure 2.41 is redrawn in Figure 2.42

1. $V_{C1} = 10$ V

    $V_{C2} = V_{R2} = (R_2/(R_1 + R_2)) * 10 = (2 * 10)/5 = 4$ V

    $W_1 = (1/2) * C_1 * V_{C1}^2 = (1/2) * 1 * 10^2 = 50$ J

    $W_2 = (1/2) * C_2 * V_{C2}^2 = (1/2) * 2 * 4^2 = 16$ J

2. $W_1 = 50/1055 = 0.0474$ Btu

    $W_2 = 16/1055 = 0.0152$ Btu

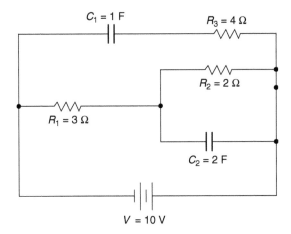

FIGURE 2.41
Electrical network of Example 2.5.

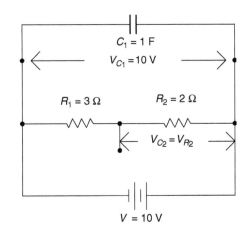

FIGURE 2.42
Equivalent DC circuit of Figure 2.41.

3. $W_1 = 50 * 0.737 = 36.85$ ft · lb$_f$

   $W_2 = 16 * 0.737 = 11.79$ ft · lb$_f$

4. $W_1 = 50/3.6 * 10^{-6} = 0.0000138$ kW h

   $W_2 = 16/3.6 * 10^{-6} = 0.00000444$ kW h

MATLAB Solution
```
>> R1=3; R2=2; C1=1; C2=2; VC=10;
>> VC1=VC; VC2=(R2/(R1+R2))*VC;
>> W1 _ J=0.5*C1*VC1^2,W2 _ J=0.5*C2*VC2^2

   W1 _ J =
           50
   W2 _ J =
           16
```

```
>> W1BTU=W1J/1055, W2BTU=W2J/1055

    W1 _ BTU =
                0.0474
    W2 _ BTU =
                0.0152
>> W1 _ ftlbf =W1J*0.737,W2 _ ftlbf=W2J*0.737
    W1 _ ftlbf =
                36.8500
    W2 _ ftlbf =
                11.7920
>> W1 _ Kwh=W1J/3.6*10^-6, W2 _ Kwh=W2J/3.6*10^-6
    W1 _ Kwh =
                1.3889e-005
    W2 _ Kwh =
                4.4444e-006
```

Example 2.6

Create the script file linearity that verifies the linear relation between the current and voltage in a resistive circuit (Ohm's law), and the nonlinearity of its power for the circuit shown in Figure 2.43, where the applied voltage V_0 varies over the range $10\text{ V} \leq V_0 \leq 1000\text{ V}$ in linear increments of 10 V by obtaining the following plots:

1. V_0 versus I
2. V_0 versus P_R (power dissipated by $R = 1\text{ k}\Omega$)

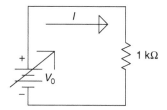

FIGURE 2.43
Network of Example 2.6.

MATLAB Solution
```
%Script file: linearity
R=1.0e3;
V=10:10:1000;
I=V./R;
P=I.*V;
subplot(2,1,1)
plot (V,I*1000)
title('Current vs. Voltage')
xlabel('Voltage (Volts)')
ylabel('Current (mA)')
subplot(2,1,2)
plot (V,P)
title('Power  vs. Voltage')
xlabel('Voltage (Volts)')
ylabel('Power (watts)')
```

The script file linearity is executed and the results are shown in Figure 2.44.

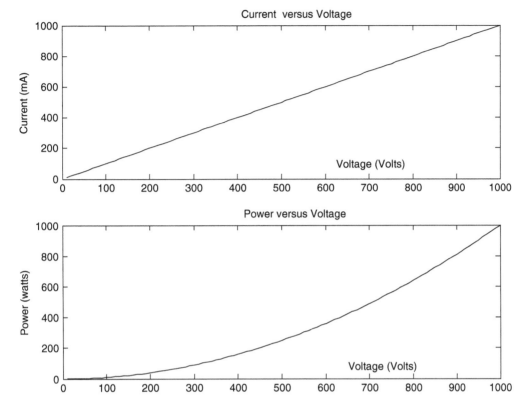

FIGURE 2.44
Plots of Example 2.6.

Example 2.7

Analyze the circuit shown in Figure 2.43 for the case where the applied voltage source is constant at $V_0 = 250$ V, but the resistance (R) varies over the range $200 \ \Omega \leq R \leq 2 \ k\Omega$ in linear increments of 10 Ω. Obtain plots of

1. I versus R
2. P_R versus R

MATLAB Solution

```
>> V = 250;
>> R = 200:10:2000;
>> I = V./R;
>> P = I.*V;
>> plot(R,I*1000,R,P)          % current in mA
>> title('Current and Power vs Resistance')
>> xlabel('Resistance (Ohms)'), ylabel('Current (mA), Power(Watts)')
>> grid on
>> gtext ('Current')           % place the text 'Current' and 'Power'
>> gtext ('Power')             % control text by mouse
>> % plots are shown in Figure 2.45
```

Note that by increasing R, the current I decreases. Since $P_R = I^2R$, increasing R decreases I, where I is the dominant variable. Then by increasing R, the power P_R decreases.

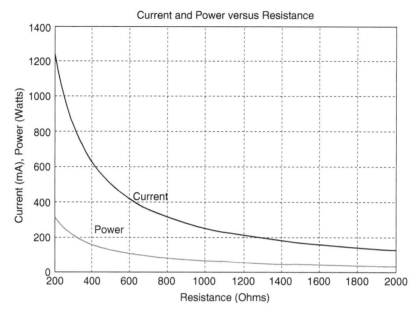

FIGURE 2.45
Plots of Example 2.7.

Example 2.8

Analyze the circuit diagram shown in Figure 2.46, over the range $0 \le R_2 \le 10$ kΩ, in linear increments of 10 Ω by returning the following plots:

1. R_2 versus I
2. R_2 versus V_2
3. R_2 versus V_1

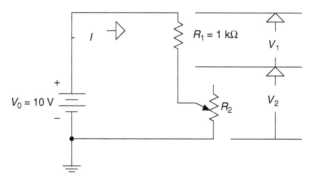

FIGURE 2.46
Network of Example 2.8.

MATLAB Solution
```
>> R1 = 1.0e3;
>> V = 10;
>> R2 = 0:10:10000;
>> I = V./(R1+R2);
>> subplot (3,1,1)
```

```
>> plot (R2,I*1.0e3)                     % current in ma
>> title ('Current I, Voltage V2, and Voltage V1 vs R2');
>> ylabel ('Current I (mA)')
>> grid on; V2 = R2.*I;
>> subplot(3,1,2)
>> plot (R2,V2)
>> ylabel ('Voltage V2 (Volts)')
>> grid on
>> subplot (3,1,3)
>> V1 = R1*I; plot(R2,V1)
>> xlabel ('Resistance R2(Ohms)'),
>> ylabel ('voltage V1(Volts)')
>> grid on;
>> % plots are shown in Figure 2.47
```

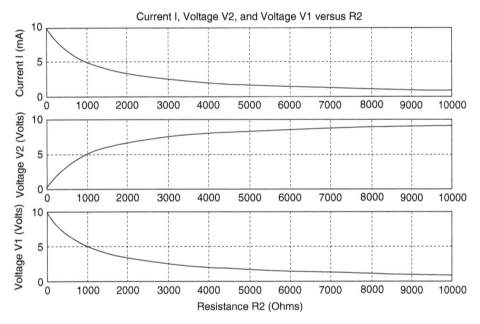

FIGURE 2.47
Plots of Example 2.8.

It is left as an exercise for the reader to verify that $V_0 = 10\ \text{V} = V_1 + V_2$, for any value of R_2 satisfying KVL.

Example 2.9

Write a program that analyzes the circuit shown in Figure 2.48, over the range $0\ \Omega \le R_3 \le 500\ \Omega$, in linear increments of $10\ \Omega$ by returning the following plots:

1. R_3 versus I
2. R_3 versus I_2
3. R_3 versus I_3
4. From the preceding plots, indicate which current is the least and the most affected by the variations of (potentiometer) R_3

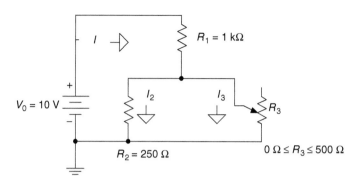

FIGURE 2.48
Network of Example 2.9.

```
MATLAB Solution
>> V = 10;
>> R1 = 1.0e3;
>> R2 = 250;
>> R3 = 0:10:500;
>> RP = (R2*R3)./(R2+R3);        % equivalent resist. of R2 and R3
>> RT = R1+RP;                   % total resistance
>> I = V./RT;
>> I2 =( I.*R3)./(R2+R3);
>> I3 = (I.*R2)./(R2+R3);
>> plot (R3,I*1.0e3,'*', R3, I2*1.0e3, 'o',R3,I3*1.0e3,'+')
>> title ('Currents vs Resistance R3')
>> xlabel ('Resistance R3 in Ohms')
>> ylabel ('Currents in Milli-Amperes')
>> legend ('I','I2','I3')
>> grid on
```

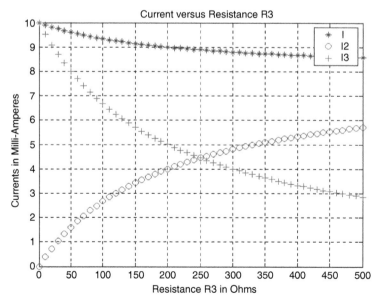

FIGURE 2.49
Plots of Example 2.9.

Observe from Figure 2.49 that when $R_3 = 250\ \Omega$, $I_2 = I_3$. Also observe that the current I is the least affected, whereas the currents I_3 and I_2 are equally affected by the changes in R_3, over the range 0–500 Ω.

Example 2.10

Write a MATLAB program that returns in a tablelike format the resistance values (R) versus length L of a copper wire with a diameter of $\sqrt{300}$ mills (the cross-sectional area: 300 CM [circular mills]), over the range 50 ft $\le L \le$ 500 ft in linear increments of 50 ft, assuming that the temperature remains constant at 20°C.

ANALYTICAL Solution

Recall that the resistance R is given by

$$R = \frac{rho * L}{A} \quad \text{(at a room temperature of 20°C)}$$

where $rho = \rho$ denotes resistivity in CM Ω/ft, and the *rho* of copper is 10.37 CM Ω/ft, at 20°C; L defines its length in feet; A defines the cross-sectional area in CM (circular mills); A (area of a circle) = pi $* r^2 = \pi * d^2/4$; 1000 mills = 1 in. and $\pi/4$ sq. mills = 1 CM, then if d is given in mills, $A = d^2$ (in CM).

MATLAB Solution
```
>> L = 50:50:500;
>> Rho =10.37;
>> A = 300;
>> R=Rho.*L./A;
>> disp('************************')
>> disp('Length (ft)  Resit (Ohms)'); results = [L' R'];
>> disp('************************')
>> disp(results)
>> disp('************************')

        ***************************
        Length (ft)    Resit (Ohms)
        ***************************
           50.0000        1.7283
          100.0000        3.4567
          150.0000        5.1850
          200.0000        6.9133
          250.0000        8.6417
          300.0000       10.3700
          350.0000       12.0983
          400.0000       13.8267
          450.0000       15.5550
          500.0000       17.2833
        ***************************
```

Example 2.11

Write a program that returns in a tablelike format the resistance of a copper wire of 100 ft long with a fixed cross-sectional area $A = 1$ CM as a function of temperature variations (ΔT), over the range 0–100°C, in linear increments of $\Delta T = 10$°C.

ANALYTICAL Solution

Recall that

$$R = \rho \frac{L}{A}(1 + TC * \Delta T)$$

where ρ is the Greek letter *rho*, TC = 0.00393 (the temperature coefficient of copper), and $R = \rho(100/1)(1 + 0.00393 * \Delta T)$.

MATLAB Solution

```
>> T = 0:10:100;
>> TC = 0.00393;
>> L =100;
>> Rho =10.37;
>> A=1;
>> R= (Rho*100./A)*(1+TC.*T);
>> result = [T' R'];
>>disp('********************************')
>> disp('Temperature (°C)****Resistance (in Ohms)');
>>disp('********************************')
>>disp(result)
>>disp('********************************')

*********************************************************
      Temperature (°C)****Resistance (in Ohms)
*********************************************************
           1.0e+003

                0    1.0370
           0.0100    1.0778
           0.0200    1.1185
           0.0300    1.1593
           0.0400    1.2000
           0.0500    1.2408
           0.0600    1.2815
           0.0700    1.3223
           0.0800    1.3630
           0.0900    1.4038
           0.1000    1.4445
*****************************************************
```

Example 2.12

Let the resistor R_L shown in the circuit of Figure 2.50 vary over the range $0\,\Omega \le R_L \le 10\,\Omega$ in linear incremental steps of 0.25 Ω.

Write a program that returns the following plots:

1. I_L versus R_L
2. V_L versus R_L
3. $[P_{RL} = V_L * I_L]$ versus R_L

For each of the plots, determine the maximum of I_L, V_2, and P_{RL} over the range $0\,\Omega \le R_L \le 10\,\Omega$, and verify that $P_{RL\text{-max}} \neq I_{L\text{-max}} * V_{L\text{-max}}$ and $P_{RL\text{-max}} < I_{L\text{-max}} * V_{L\text{-max}}$.

MATLAB Solution

```
>> VS =10;
>> RL = 0:0.25:10;
>> IL = VS./(5+RL);
>> VL = IL.*RL;
>> subplot (2,2,1)
```

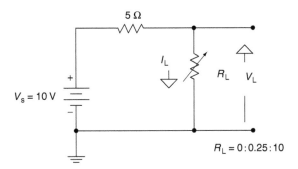

FIGURE 2.50
Network of Example 2.12.

```
>> plot (RL, IL)
>> grid on
>> xlabel('Load Resistance RL'), ylabel('Current IL')
>> title('IL vs. RL');
>> subplot(2,2,2)
>> plot(RL,VL)
>> grid on
>> xlabel('Load Resistance RL'), ylabel('Voltage VL')
>> title('VL vs. RL');
>> subplot(2,2,3)
>> P = IL.*VL;
>> plot(RL,P)
>> grid on
>> xlabel('Load Resistance RL'), ylabel('Power of RL')
>> title('Power vs. RL');
>> subplot(2,2,4)
>> plot(RL,IL,'o',RL,VL,'+',RL,P,'*')
>> grid on
>> xlabel('Load Resistance RL'), ylabel('IL,VL,P=IL*VL')
>> title('IL, VL, P vs. RL');
>> legend('IL','VL','P');
>> Imax = max(IL);
>> Vmax = max(VL);
>> Pmax = max(P);
>> Result = [Imax Vmax Pmax];
>> disp('        ****** R E S U L T S ******")
>> disp('*************************************************')
>> disp('The maximum values for IL, VL, and PL for 0<RL<10 Ohms are:');
>> disp (Result) % the plots are shown in Figure 2.51.
>> disp('     amps    volts    watts')
>> disp('*************************************************')
*********************** R E S U L T S *******************
*************************************************************
   The maximum values for IL, VL, and PL for 0<RL<10 Ohms are:
          2.0000     6.6667     5.0000
            amps      volts      watts
*************************************************************
```

Clearly the results indicate that $P_{RL\text{-max}} \neq I_{L\text{-max}} * V_{L\text{-max}}$ {5 W $\neq$ (2 A) * (6.6667 V)} and $P_{RL\text{-max}} = 5$ W$< I_{L\text{-max}} * V_{L\text{-max}} = Z * 6.6667$ W; the resulting plots are shown in Figure 2.51.

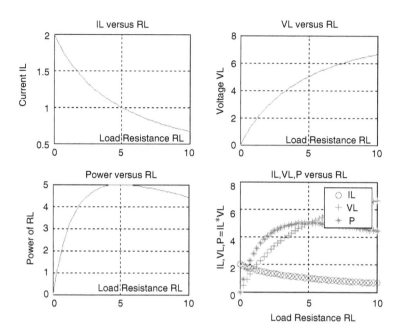

FIGURE 2.51
Plots of Example 2.12.

Example 2.13

Using two loop equations and one node equation, and the matrix operations $I = inv(R) * V$ (where R is the [equivalent] impedance matrix of the network and V the voltage vector), solve for the three branch currents—I_1, I_2, and I_3—shown in the circuit diagram of Figure 2.52.

ANALYTICAL Solution

The node and the two loop equations are shown as follows:

Node A; $-I_1 + I_2 + I_3 = 0$
Loop# 1; $10 * I_1 + 5 * I_2 = 10$
Loop# 2; $-5 * I_2 + (10 + 20) * I_3 = 0$

The preceding equations in matrix form is given by

$$\begin{bmatrix} -1 & 1 & 1 \\ 10 & 5 & 0 \\ 0 & -5 & 30 \end{bmatrix} * \begin{bmatrix} I_1 \\ I_2 \\ I_3 \end{bmatrix} = \begin{bmatrix} 0 \\ 10 \\ 0 \end{bmatrix}$$

Then the 3×3 matrix becomes the resistance matrix R, and the voltage V is given by the column vector $[0 \ 10 \ 0]^T$ as illustrated by the following matrix equation:

$$[R] * [I] = [V]$$

then

$$I = inv(R) * V$$

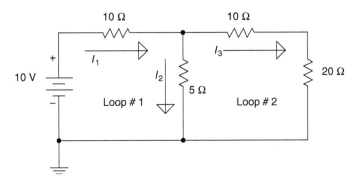

FIGURE 2.52
Network of Example 2.13.

```
MATLAB Solution
>> R = [-1 1 1;10 5 0;0 -5 30];
>> V = [0;10;0];
>> I = inv(R)*V;   % Solves for the loop currents
>> Result = [I(1) I(2) I(3)];
>> % Results are printed
>> disp('*******************************');
>> disp('*********R E S U L T S********');
>> disp('*******************************');
>> disp('   The currents I1, I2, I3 are:');
>> disp(Result)
>> disp('   amp.   amp.   Amp.');
>> disp('*******************************');

        ***********************************
        *********R E S U L T S***********
        ***********************************
           The currents I1, I2, I3 are:
              0.7000    0.6000    0.1000
                 amp.      amp.      amp.
        ***********************************
```

Example 2.14

Solve for the loop currents I_1 and I_2 shown in the circuit diagram of Figure 2.53 (Example 2.13) by using

  i. The matrix operation $I = R\backslash V$
 ii. The symbolic method
iii. Compare the results of i and ii with the solution obtained in Example 2.13

ANALYTICAL Solution

The loop equations are

For loop# I, $15 * I_1 - 5 * I_2 = 10$
For loop# II, $-5 * I_1 + (5 + 10 + 20) * I_2 = 0$

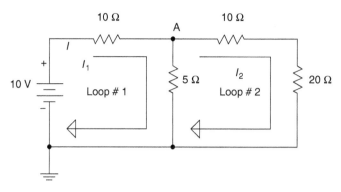

FIGURE 2.53
Network of Example 2.14.

The matrix equation is then given by

$$\begin{bmatrix} 15 & -5 \\ -5 & 35 \end{bmatrix}\begin{bmatrix} I_1 \\ I_2 \end{bmatrix} = \begin{bmatrix} 10 \\ 0 \end{bmatrix}$$

where the 2 × 2 matrix is the resistance matrix R and $[10\ 0]^T$ represents the column vector V, as indicated by

$$[R] * [I] = [V]$$

then

$$[I] = [R] \setminus [V]$$

MATLAB Solution
```
>> % part (i) matrix solution
>> R = [15 -5;-5 35];
>> V = [10;0];
>> I = R\V;
>> Result = [I(1) I(2) I(1)-I(2)];              % results are printed
>> disp('*************************************');
>> disp('**********R E S U L T S*************');
>> disp('   Matrix current solutions');
>> disp('*************************************');
>> disp ('The currents I1, I2, I3 are given by:'); disp(Result)
>> disp ('   amp.    amp.     amp.');
>> disp('*************************************')
*************************************************

***************R E S U L T S******************
      Matrix current solutions
*************************************************
   The currents I1, I2, I3 are given by:
         0.7000     0.1000     0.6000
           amp.       amp.       amp.
*************************************************
>> % Part (ii); symbolic solution
>> sym I1 I2 I3;
>> disp('*************************************')
```

```
>> disp('The Symbolic current solutions are given by:')
>> [I1 I2 I3] = solve('15*I1-5*I2=10','-5*I1+35*I2=0','I1-I2=I3')
>> disp('*******************************')

        **************************************************
        The Symbolic current solutions are given by:
        I1 = 7/10
        I2 = 1/10
        I3 = 3/5
        **************************************************
```

Note that the results obtained for parts i and ii are equivalent to the solutions obtained in Example 2.13. Also note that the current I_3 in Example 2.13 is labeled I_2 in Example 2.14.

Example 2.15

Show that the circuit diagrams in Figure 2.54 are equivalent, as seen through the terminals labeled *a* and *b*.

ANALYTICAL Solution

The circuit diagrams in Figures 2.54a and 2.54b are equivalent due to the equivalent substitutions shown in Figure 2.55.

The circuit diagram shown in Figure 2.54b is equivalent to the circuit diagram shown in Figure 2.54c due to the equivalent substitutions indicated in Figure 2.56.

The circuit diagram in Figure 2.54c is equivalent to the diagram shown in Figure 2.54d due to the equivalencies shown in Figure 2.57.

Finally, the diagram in Figure 2.54d is equivalent to the diagram in Figure 2.54e due to the substitution indicated in Figure 2.58.

The equivalencies shown can be evaluated by using the MATLAB programs of Examples 2.2 and 2.3 and are left as an exercise for the reader to verify.

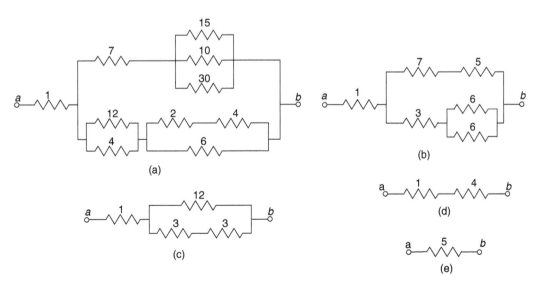

FIGURE 2.54
Diagrams of Example 2.15.

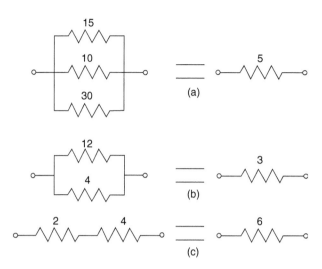

FIGURE 2.55
Equivalent substitutions of Example 2.15 (from Figures 2.54a to 2.54b).

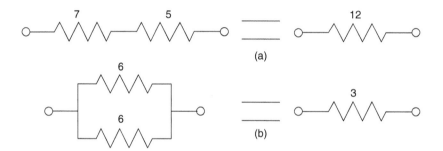

FIGURE 2.56
Equivalent substitutions of Example 2.15 (from Figures 2.54b to 2.54c).

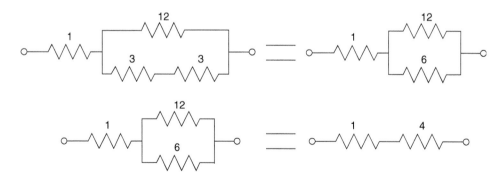

FIGURE 2.57
Equivalent substitutions of Example 2.15 (from Figures 2.54c to 2.54d).

FIGURE 2.58
Equivalent substitutions of Example 2.15 (from Figures 2.54d to 2.54e).

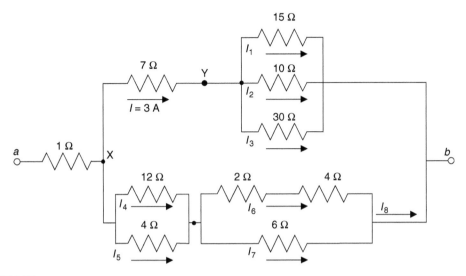

FIGURE 2.59
Network of Example 2.16.

Example 2.16

Create the script file *current_voltage* that returns the currents I_1, I_2, I_3, I_4, I_5, I_6, I_7, I_8, and the voltage V_{ab} shown in the circuit diagram of Figure 2.59 (used in Figure 2.54), given the fact that the current $I = 3$ A through the 7 Ω resistor.

MATLAB Solution

```
% Script file: current _ voltage
clear;clc;
I = 3;
Vyb = 3*5;                      % from Figure 2.54a
Vxb = 12*3;                     % from Figure 2.54b
I1 = Vyb/15;
I2 = Vyb/10;
I3 = Vyb/30;
I8 = Vxb/6;                     % from Figure 2.54c
I4 = (4/16)*I8;                 % current divider
I5 = (12/16)*I8;
I6 = I8/2;
I7 = I6;
Vab = (I+I8)*5;                 % from Figure 2.54e
result = [I1 I2 I3 I4 I5 I6 I7 I8];
disp('*******************************');
disp('**********R E S U L T S*********');
disp('*******************************');
disp('  ');
disp('The currents I1 I2 I3 I4 I5 I6 I7 and I8 (in amps) are given by:');
disp(result');
disp('*******************************');
disp('The voltage Vab is:'); disp(Vab); disp('volts')
disp('*******************************');
```

The script file *current_voltage* is executed and the results are shown as follows:

```
>> current _ voltage

**************************************************************
*************************R E S U L T S************************
**************************************************************
 The currents I1 I2 I3 I4 I5 I6 I7 and I8 (in amps) are given by:
   1.0000
   1.5000
   0.5000
   1.5000
   4.5000
   3.0000
   3.0000
   6.0000
**************************************************************
 The voltage Vab is:
                   45
                   volts
**************************************************************
```

Example 2.17

For the circuit diagram shown in Figure 2.60

1. Write the three system mesh (loop) equations
2. Arrange the result of part 1 into a matrix equation
3. Using MATLAB, solve for the three loop currents (I_1, I_2, and I_3)

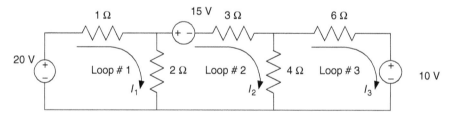

FIGURE 2.60
Network of Example 2.17.

ANALYTICAL Solution

The loop equations are

Loop 1: $20 = 3I_1 - 2I_2 + 0I_3$
Loop 2: $-15 = -2I_1 + 9I_2 - 4I_3$
Loop 3: $-10 = 0I_1 - 4I_2 + 10I_3$

The resulting matrix equation is

$$\underbrace{\begin{bmatrix} 3 & -2 & 0 \\ -2 & 9 & -4 \\ 0 & -4 & 10 \end{bmatrix}}_{R} * \underbrace{\begin{bmatrix} I_1 \\ I_2 \\ I_3 \end{bmatrix}}_{I} = \underbrace{\begin{bmatrix} 20 \\ -15 \\ -10 \end{bmatrix}}_{V}$$

MATLAB Solution
```
>> R = [3 -2 0; -2 9 -4; 0 -4 10];
>> V = [20; -15; -10];
>> I = inv(R)*V;
>> Result = [I(1) I(2) I(3)];
>> disp('*********************************************')
>> disp('The loop currents I1, I2 and I3(in amp) are:');
>> disp(Result');
>> disp('*********************************************')
```

```
*************************************************
The loop currents I1, I2 and I3 (in amp) are:
      6.0440
     -0.9341
     -1.3736
*************************************************
```

Example 2.18

For the circuit diagram shown in Figure 2.61

1. Write the two node equations (for nodes X and Y)
2. Arrange the result of part 1 into a matrix equation
3. Use MATLAB to solve for the two voltages (V_X and V_Y)

ANALYTICAL Solution

1. The node equations are

For node X: $\quad I_1 = \left(\dfrac{1}{R_1} + \dfrac{1}{R_2}\right) V_X - V_Y \quad$ or $\quad 2 = (1 + 2) V_X - V_Y$

For node Y: $\quad I_2 = -V_X + \left(\dfrac{1}{R_3} + \dfrac{1}{R_1}\right) V_Y \quad$ or $\quad 3 = -V_X + (3 + 1)V_Y$

The two simplified and rearranged node equations are

$$2 = 3V_X - V_Y$$

$$3 = -V_X = +4V_Y$$

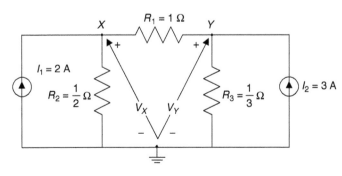

FIGURE 2.61
Network of Example 2.18.

2. The resulting system nodal matrix equation is given by

$$\underbrace{\begin{bmatrix} 3 & -1 \\ -1 & 4 \end{bmatrix}}_{Y} * \underbrace{\begin{bmatrix} V_X \\ V_Y \end{bmatrix}}_{V} = \underbrace{\begin{bmatrix} 2 \\ 3 \end{bmatrix}}_{I}$$

where Y is the admittance matrix of the network, and the current vector I is given by

$$I = \begin{bmatrix} 2 \\ 3 \end{bmatrix}$$

MATLAB Solution
```
>> Y = [3 -1;-1 4];
>> I = [2;3];
>> V= inv(Y)*I;
>> VX = V(1);
>> VY =V(2);
>> Result = [VX VY];
>> disp ('*********************************')
>> disp ('The voltage drops Vx and Vy (in volts) are given by:');
>> disp (Result)
>> disp ('*********************************')
```

```
        ************************************************************
          The nodal voltages Vx and Vy (in volts) are given by:
                                1.0000    1.0000
        ************************************************************
```

Example 2.19

The switch shown in the circuit diagram in Figure 2.62 has been in position *a* for a long time. At $t = 0$, the switch is moved to position *b* where it remains for 2 s and then moves back to position *a*, where it remains indefinitely.

1. Obtain analytical expressions for $v_C(t)$ and $i_C(t)$ for all t
2. Use MATLAB to obtain plots over the range $0\ s \le t \le 10\ s$ of
   a. The voltage $v_C(t)$ versus t
   b. The current $i_C(t)$ versus t

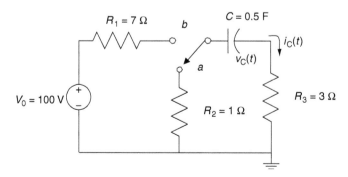

FIGURE 2.62
Network of Example 2.19.

ANALYTICAL Solution

For $t < 0$, the switch is in position a. Therefore,

$$v_c(t) = 0$$

and

$$i_c(t) = 0$$

For $0 < t < (t_1 = 2)$ seconds, the switch is in position b. Then,

$$v_C(t) = V_0(1 - e^{-t/\tau_1})$$

and

$$i_C(t) = \frac{V_0 - v_C(t)}{R_1 + R_3}$$

where $\tau_1 = (R_1 + R_3) * C = 10\ \Omega * 0.5\ \text{F} = 5\ \text{s}$.
At time $t = t_1 = 2$ s, the voltage across the capacitor is

$$v_C(t_1) = V_{C\max} = V_0(1 - e^{-2/\tau_1})$$

For $t > (t_1 = 2\ \text{s})$, the switch is back in position a. Then

$$v_C(t) = V_{C\max} * e^{-(t-t_1)/\tau_2}$$

and

$$i_C(t) = \frac{-v_C(t)}{R_3 + R_2}$$

where $\tau_2 = (R_2 + R_3) * C = 4\ \Omega * 0.5\ \text{F} = 2\ \text{s}$.

MATLAB Solution

```
>> V = 100; R1 =7; R2 =1; R3=3; C =0.5;      % circuit elements
>> tau1 = (R1+R3)*C                          % time constant #1

    tau1 =
          5
>> tau2 = (R2+R3)*C                           % time constant #2

    tau2 =
          2
>> for k =1:40
        t(k)  = k/20;   % 0 < t < 2
        v(k)  = V*(1-exp(-t(k)/tau1));
        i(k)  = (V-v(k))/(R1+R2);
    end
```

```
>> Vmax = v(40)

   Vmax =
         32.9680
>> for k=41:200
     t(k)=k/20;   % 2 < t < 10
     v(k)=Vmax*exp(-(t(k)-t(40))/tau2);
     t(40)=40/20=2sec=t1
     i(k)= -v(k)/(R2+R3);
   end
>>                              % plot the Voltage VC(t)
>> subplot(2,1,1)
>> plot(t,v)
>> axis([0 10 0 40]);
>> title ('Transients in the circuit of Fig 2.62')
>> xlabel('time in sec'), ylabel('Voltage vc(t)')
>> grid on
>>                              % plot the current IC (t)
>> subplot(2,1,2)
>> plot(t,i)
>> axis([0 10 -13 15]);
>> xlabel('time in sec'),ylabel('Current ic(t)')
>> grid on
```

The plots of the Voltage vc(t) and the Current ic(t) versus t are shown in Figure 2.63.

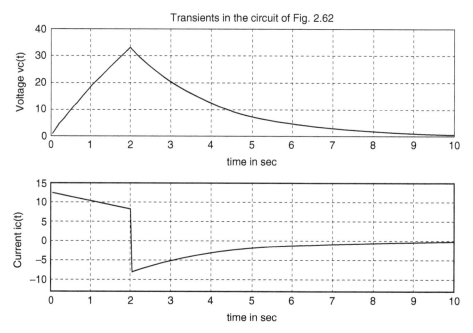

FIGURE 2.63
Transient plots of Example 2.19.

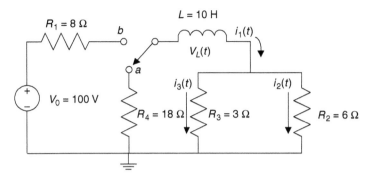

FIGURE 2.64
Network of Example 2.20.

Example 2.20

The switch shown in Figure 2.64 has been at position *a* for a very long time. At $t = 0$, the switch is moved to position *b* where it remains for 2 s and then moves back to position *a* where it remains indefinitely.

1. Obtain analytical expressions of $i_1(t)$, $i_2(t)$, $i_3(t)$, and $v_L(t)$ for all t.
2. Create the script file *analysis_exam* that returns the following:
   a. The network time constants
   b. The maximum currents I_1, I_2, and I_3
   c. The maximum voltage drop across L
   d. The current plots of $i_1(t)$ versus t, $i_2(t)$ versus t, and $i_3(t)$ versus t over the range $0 \, s \leq t \leq 5 \, s$
   e. The voltage plot $v_L(t)$ versus t over the range $0 \, s \leq t \leq 5 \, s$
3. If the applied source voltage is $V_0 = 100$ V, can the voltage drop across L exceed 100 V?

ANALYTICAL Solution

For $t < 0$, the switch has been in position *a* for a long time, then

$$i_1(t) = i_2(t) = i_3(t) = 0$$

and

$$v_L(t) = 0$$

During the interval $0 \, s < t < 2 \, s$, the switch is in position *b*. Therefore,

$$i_1(t) = \frac{V_0}{R_{eq1}} * (1 - e^{-t/\tau_1})$$

$$i_2(t) = \frac{R_3}{R_2 + R_3} * i_1(t)$$

$$i_3(t) = \frac{R_2}{R_2 + R_3} * i_1(t)$$

and

$$v_L(t) = V - i_1(t) * R_{eq1}$$

where

$$R_{eq1} = R_1 + (R_2 \parallel R_3) = R_1 + \frac{R_2 * R_3}{R_2 + R_3} = 8 + \frac{6 * 3}{6 + 3} = 8 + 2 = 10\,\Omega$$

and

$$\tau_1 = L/R_{eq1} = 10\,\text{H}/10\,\Omega = 1\,\text{s}$$

At $t = t_1 = 2$ s, the current $i_1(t_1)$ is at its maximum as given by

$$I_{1\text{max}} = i_1(t = 2) = \frac{V_0}{R_{eq1}} * (1 - e^{-2/\tau_1})$$

At $t = 2$ s, the switch moves back to position a, where it remains indefinitely. The branch currents would then be given by

$$i_1(t) = I_{1\text{max}} e^{-(t-t_1)/\tau_2}$$

$$i_2(t) = \frac{R_3}{R_2 + R_3} * i_1(t)$$

$$i_3(t) = \frac{R_2}{R_2 + R_3} * i_1(t)$$

and

$$v_L(t) = i_1(t) * R_{eq2}$$

where

$$R_{eq2} = R_4 + (R_2 \parallel R_3) = R_4 + \frac{R_2 * R_3}{R_2 + R_3} = 18 + \frac{6 * 3}{6 + 3} = 18 + 2 = 20\,\Omega$$

and

$$\tau_2 = L/R_{eq2} = 10\,\text{H}/20\,\Omega = 0.5\,\text{s}$$

MATLAB Solution
```
% Script file: analysis _ exam
V=100; R1 = 8; R2 =3; R3 =6; R4 =18; L=10;        % circuit elements.
Req1 = R1+(R2*R3)/(R2+R3);
```

```
Req2 = R4+(R2*R3)/(R2+R3);
disp('* * * R E S U L T S * * * ')
disp('****************************')
disp('The network time constants are(in sec):')
tau1 = L/Req1                       % time constant #1
tau2 = L/Req2                       % time constant #2
for K=1:40
     t(K) = K/20;                   % 0 < t < 2.
     I1(K) = (V/Req1)*(1-exp(-t(K)/tau1));
     I2(K) = R3*I1(K)/(R2+R3);
     I3(K) = R2*I1(K)/(R2+R3);
     VL(K) = V-I1(K)*Req1;
  end
disp('the maximum currents are:')
I1Max = I1(40);
for K= 41:100
t(K) = K/20;                        % 2 < t < 5.
I1(K) = I1Max*exp(-(t(K)-t(40))/tau2); % t(40) = 40/20 = 2 sec.= t1
I2(K) = R3*I1(K)/(R2+R3);
I3(K) = R2*I1(K)/(R2+R3);
VL(K) = -I1(K)*Req2;
end
                                    % plots the current i1(t)
I1Max= max(I1)
Tmax = max(t);
subplot(2,2,1)
plot(t,I1)
axis([0 5 0 10])
grid on
title('Current i1(t) vs.t'), ylabel('Current in amps');
xlabel('time in sec')               % plots the current i2(t)
I2Max = max(I2)
I3Max=max(I3)
subplot(2,2,2)
plot(t,I2)
axis([0 5 0 6])
grid on
title('Current i2(t) vs.t'), ylabel('Current in amps');
xlabel('time in sec')
subplot(2,2,3)                      % plots the current i3(t)
plot(t,I3)
axis([0 5 0 3])
grid on
title('Current i3(t) vs. t'), ylabel('Current in amps')
xlabel('time in seconds');          % plots the voltage vL(t)
disp(' The magnitude of the maximum voltage across L is:')
VLMax = -max(abs(VL))
subplot(2,2,4)
plot(t,VL)
axis([0 5 -160 100])
grid on
title('Voltage vL(t) vs.t'), xlabel('time in seconds')
ylabel('Voltage in volts ')
disp('**************************************')
```

The script file *analysis_exam* is executed and the results are shown as follows and in Figure 2.65.

```
>> analysis _ exam _ 1020

* * * R E S U L T S * * *
***************************
The network time constants are (in sec):
 tau1 =
        1
 tau2 =
        0.5000
The maximum currents are:
I1Max =
        8.6466
I2Max =
        5.7644
I3Max =
        2.8822
The magnitude of the maximum voltage across L is:
VLMax =
        -156.4762
*****************************************
```

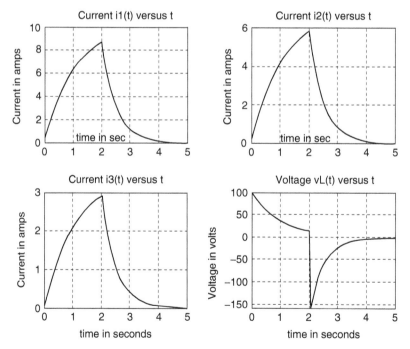

FIGURE 2.65
Transient plots of Example 2.20.

Observe that the voltage across the inductor $v_L(t)$ at $t = 2$ s is $v_L(t = 2) = 156.4762$ V when the applied (source) voltage is $V = 100$ V. Observe also that at $t = 2$ s, $I_{1max} = 8.6466$ A $= I_{2max} + I_{3max} = 5.7644$ A $+ 2.8822$ A verifying KCL.

Example 2.21

Assuming that the load is R_L, determine the Thevenin's equivalent circuit of the electrical network shown in Figure 2.66 by

a. Hand calculations.
b. Creating the script file *Thevenin* that simulates Thevenin's theorem consisting in varying R_L over the range $1 \leq R_L \leq 50\ \Omega$, in steps of $1\ \Omega$, and computing the respective powers for each value of R_L. Recall that the resistance dissipates maximum power when $R_L = R_{TH}$ and its $V_{TH} = 2\ V_{RTH}$.

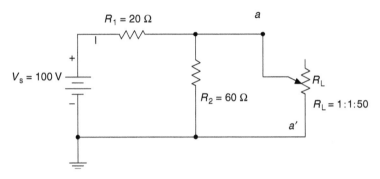

FIGURE 2.66
Network of Example 2.21.

ANALYTICAL Solution

Part a

The V_{TH} and R_{TH} are evaluated by hand.

$$V_{TH} = \frac{60 * 100}{20 + 60} = \frac{6000}{80} = 75\ V$$

and

$$R_{TH} = R_1 \parallel R_2 = \frac{20 * 60}{20 + 60} = \frac{1200}{80} = 15\ \Omega$$

The calculated Thevenin's equivalent circuit, as seen through the terminals aa', is shown in Figure 2.67.

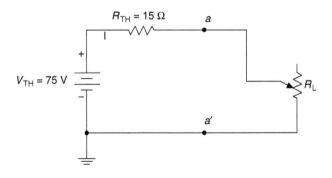

FIGURE 2.67
Calculated Thevenin's equivalent circuit of Example 2.21.

MATLAB Solution

```
% script file: Thevenin
% The Thevenin's equivalent circuit
% can be determined by experimental method
% based on the maximum power theorem
Vs =100;R1 = 20;R2 =60;
RL = 1:1:50; % range of RL
Rpar = (RL.*R2)./(RL+R2);
RT = Rpar + R1;
VL = (Rpar.*Vs)./RT;
PRL = (VL.^2)./RL;
[Pmax,index] = max(PRL);
Rth = RL(index);
Vth = sqrt(4*Rth*Pmax);
%...Display Vth and Rth
disp('********************************');
disp('*********R E S U L T S*********');
disp('********************************');
disp('     ');
fprintf('The Thevenin voltage (in volts) is Vth=%f\n',Vth);
fprintf('The Thevenin resistance (in ohms) is Rth=%f\n',Rth);
disp('     ')
disp('********************************');
```

The script file *Thevenin* is executed, and the results are shown as follows:

```
>>Thevenin
    ********************************************************
    *************** R E S U L T S ***********************
    ********************************************************
    The Thevenin voltage (in volts) is Vth =75.000000
    The Thevenin resistance (in ohms) is Rth =15.000000
    ********************************************************
```

Note that the calculated results completely agree with the simulated experimental results.

Example 2.22

Analyze the series RLC circuit shown in Figure 2.68, where the current source $i(t)$ is given by $i(t) = \left(\frac{2}{\pi}\right)\sum_{n=odd}^{15}\left(\frac{1}{n}\right)\sin(n\pi t)$. The equation for the current $i(t)$ represents the Fourier series approximation of a periodic square wave (see Chapter 4 for additional details).

Create the script file *analysis* that returns the following plots:

1. $i(t)$ versus t (verifies the approximation of the periodic square wave)
2. $i^2(t)$ versus t
3. $\int i(t)\,dt$ versus t (verifies the approximation of the periodic triangular wave)
4. $\frac{di(t)}{dt}$ versus t

5. $V_R(t)$ versus t, where $V_R(t) = Ri(t)$

6. $V_L(t)$ versus t, where $V_L(t) = L\frac{di(t)}{dt}$

7. $V_C(t)$ versus t, where $V_C(t) = \left(\frac{1}{C}\right)\int i(t)dt$

8. $[V_L(t) + V_R(t)]$ versus t

9. $[V_C(t) + V_R(t)]$ versus t

10. $P_R(t)$ versus t, where $P_R(t) = i^2(t)R$

11. $P_L(t)$ versus t, where $P_L(t) = i_L(t)V_L(t)$

12. $P_C(t)$ versus t, where $P_C(t) = i_C(t)V_C(t)$

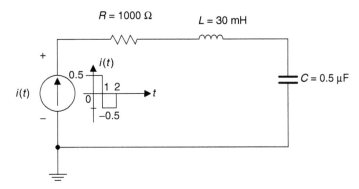

FIGURE 2.68
Network of Example 2.22.

MATLAB Solution
```
% Script file: analysis
% Analysis of RLC Series circuit
% with non DC source
echo off;
syms t;
H1  = 2/pi*sin(pi*t);
H3  = 2/(3*pi)*sin(3*pi*t);
H5  = 2/(5*pi)*sin(5*pi*t);
H7  = 2/(7*pi)*sin(7*pi*t);
H9  = 2/(9*pi)*sin(9*pi*t);
H11 = 2/(11*pi)*sin(11*pi*t);
H13 = 2/(13*pi)*sin(13*pi*t);
H15 = 2/(15*pi)*sin(15*pi*t);
iamps = H1+H3+H5+H7+H9+H11+H13+H15;

figure(1)
ezplot(iamps,[0,3])
xlabel(' time ');ylabel('i(t)in amps');
title('i(t) vs t');grid on;
isquare=iamps^2;

figure(2);%part(c.2)
ezplot(isquare,[0,3])
xlabel(' time ');ylabel('i(t)^2');
title('i(t)^2 vs t');grid on;
```

```
figure(3)
inti = int(iamps);
ezplot (inti,[0,3])
xlabel(' time ');ylabel('integral[i(t)]');
title('integral[i(t)] vs t');grid on;

figure(4)
diffi = diff(iamps);
ezplot(diffi,[0,3])
xlabel(' time ');ylabel('d[i(t)/dt]');
title('d[i(t)/dt] vs t');grid on;

figure(5)
Vr = 1000*iamps;
ezplot(Vr,[0,3])
xlabel(' time ');ylabel('Vr(t)in volts');
title('Vr(t) vs t');grid on;

figure(6)
Vl = 30e-3*diffi;
ezplot(Vl,[0,3])
xlabel(' time ');ylabel('VL(t)in volts');
title('VL(t) vs t');grid on;

figure(7)
Vc=.5e-6*inti;
ezplot(Vc,[0,3])
title('Vc(t) vs t')
xlabel(' time ');ylabel('Vc(t)in volts')
title(' Vc(t) vs t');grid on;

figure(8)
Vlr=Vl+Vr;
ezplot(Vlr,[0,3])
title('[Vr(t)+VL(t)] vs t');xlabel(' time ');
ylabel('[Vr(t)+VL(t)] in volts');grid on;

figure(9)
Vcr=Vc+Vr;
ezplot(Vcr,[0,3])
title('[Vr(t)+Vc(t)] vs t')
xlabel(' time ');ylabel('[Vr(t)+Vc(t)]');
title('[Vr(t)+Vc(t)] vs t');grid on

figure(10)
Pr = iamps*Vr;
ezplot(Pr,[0,3])
title('Pr(t) vs t')
xlabel(' time ')
ylabel('Pr(t) in watts');grid on;

figure(11)
Pl = iamps*Vl;
ezplot(Pl,[0,3])
title('PL(t) vs t'); xlabel(' time ');
ylabel('PL(t)');grid on

figure(12)
Pc = iamps*Vc; ezplot(Pc,[0,3]);
```

```
title('Pc(t) vs t');xlabel(' time ');
ylabel('Pc(t)');grid on;
```

The script file *analysis* is executed and the resulting plots are shown in Figures 2.69 through 2.80.

```
>> analysis
```

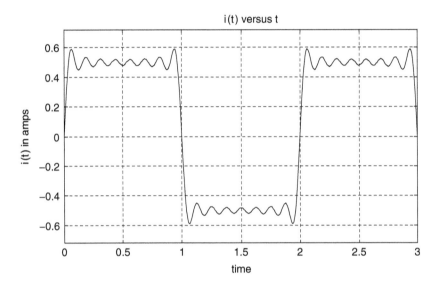

FIGURE 2.69
Plot of i(t) of Example 2.22 MATLAB Figure (1).

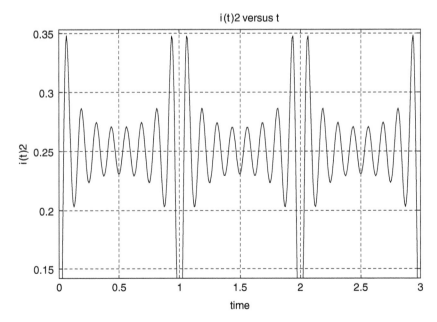

FIGURE 2.70
Plot of i(t)^2 of Example 2.22 MATLAB Figure (2).

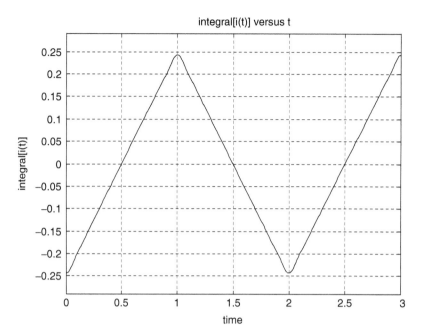

FIGURE 2.71
Plot of $\int i(t)dt$ of Example 2.22 MATLAB Figure (3).

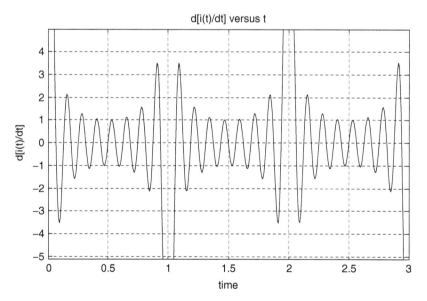

FIGURE 2.72
Plot of $\frac{di(t)}{dt}$ of Example 2.22 MATLAB Figure (4).

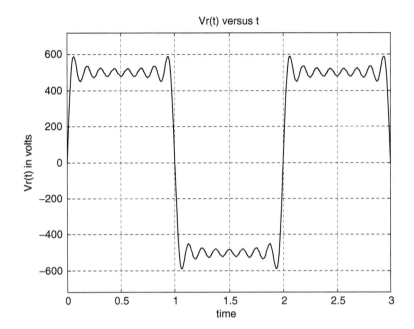

FIGURE 2.73
Plot of the voltage across R of Example 2.22 MATLAB Figure (5).

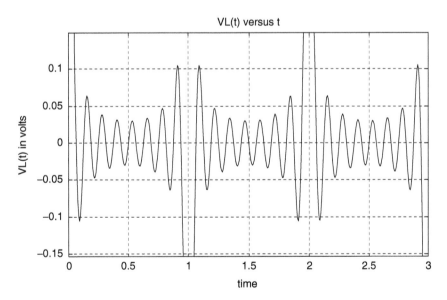

FIGURE 2.74
Plot of the voltage across L of Example 2.22 MATLAB Figure (6).

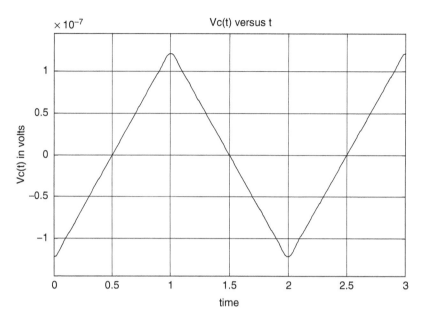

FIGURE 2.75
Plot of the sum of the voltage across *R* and *L* of Example 2.22 Figure (7).

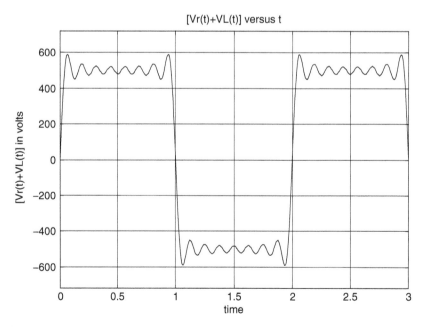

FIGURE 2.76
Plot of the voltage across C of Example 2.22 MATLAB Figure (8).

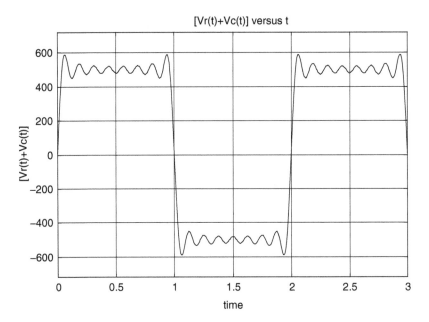

FIGURE 2.77
Plot of the sum of the voltage across R and C of Example 2.22 Figure (9).

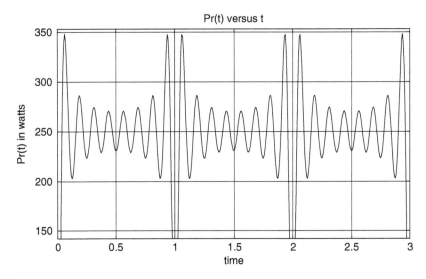

FIGURE 2.78
Plot of the power of R of Example 2.22 MATLAB Figure (10).

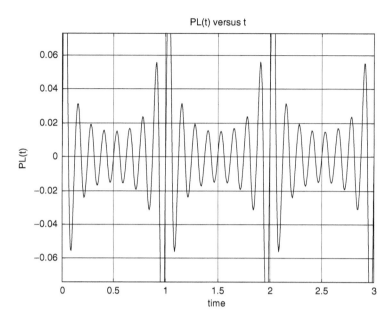

FIGURE 2.79
Plot of the power of L of Example 2.22 MATLAB Figure (11).

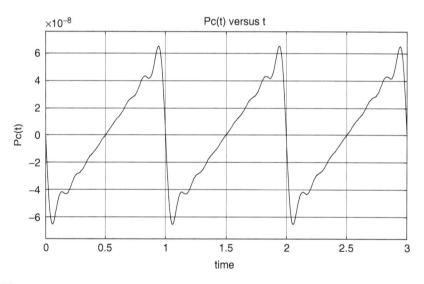

FIGURE 2.80
Plot of the power of C of Example 2.22 MATLAB Figure (12).

Example 2.23

Determine the current $i(t)$ for $t \geq 0$ for each value of R, $R = 1, 2$, and $5\,\Omega$ for the RL circuit diagram shown in Figure 2.81, assuming that the initial current is given by $i_L(0) = 0$ A.

a. By hand calculations
b. By creating the script file *RL* that returns the symbolic MATLAB solution of the loop differential equation, and
c. Each of the corresponding plots *for* $i(t)$ versus t

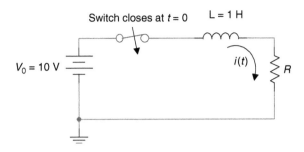

FIGURE 2.81
Network of Example 2.23.

ANALYTICAL Solution

Part a

The loop differential equation, as well as its current solution $i(t)$, is given as follows:

$$L\frac{di(t)}{dt} + Ri(t) = V_0$$

then

$$i(t) = \frac{V_0}{R} - \frac{V_0}{R}e^{-t/\tau} \quad \text{for } t \geq 0 \quad \text{for } R = 1, 2, \text{ and } 5 \ \Omega$$

where $\tau = L/R$.

MATLAB Solution
```
% Script file: RL
% parts ( b and c )
itR _ 1 = dsolve('Dy+y =10','y(0) = 0','t');
itR _ 2 = dsolve('Dy+2*y=10','y(0)=0','t');
itR _ 5 = dsolve('Dy+5*y=10','y(0)=0','t');
disp('**************************');
disp('The solutions i(t) for R=1,2 and 5 (in Ohms) are:');
disp('**************************');
disp('**************************');
itR _ 1
disp('**************************');
itR _ 2
disp('**************************');
itR _ 5
disp('**************************');
disp('**************************');
ezplot(itR _ 5,[0 3]);
hold on;
grid on; ezplot(itR _ 2,[0 3]);
hold on;
ezplot(itR _ 1,[0 3]);
hold on;
title('i(t) vs t for L=1 & R=1,2 and 5')
axis([0 3 0 11]);
xlabel('time t');ylabel('i(t)')
```

The script file *RL* is executed and the results and plots (Figure 2.82) are shown as follows:

```
>> RL
**********************************************************
The solutions i(t) for R=1, 2 and 5 (in Ohms) are :
**********************************************************
**********************************************************
itR_1 =
        10-10*exp(-t)
**********************************************************
itR_2 =
        5-5*exp(-2*t)
**********************************************************
itR_5 =
        2-2*exp(-5*t)
**********************************************************
**********************************************************
```

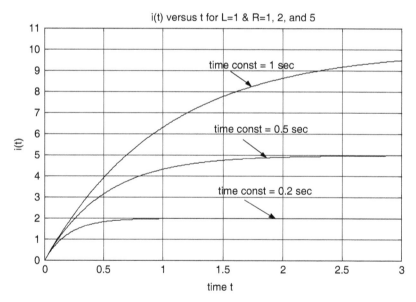

FIGURE 2.82
Transient current plots of Example 2.23.

Example 2.24

Create the script file *RL_vol_plots* that returns

1. The expression of $v_L(t)$ and plot $v_L(t)$ versus t, for $t \geq 0$
2. The expression of $v_R(t)$ and plot $v_R(t)$ versus t, for $t \geq 0$

for the circuit shown in Figure 2.81, for the same three resistors: $R = 1$, 2, and 5 Ω, assuming that the initial current is $i_L(0) = 0$ A.

MATLAB Solution

```
% Script file: RL _ vol _ plots
itR _ 1=dsolve('Dy+y=10','y(0)=0','t');
itR _ 2=dsolve('Dy+2*y=10','y(0)=0','t');
itR _ 5=dsolve('Dy+5*y=10','y(0)=0','t');
disp('********** R E S U L T S **************')
disp('*******************************************');
disp('The  voltages across  R = 1, 2 and 5 (in Ohms) are:');
disp('**************************');
disp('**************************');
vr _ 1=itR _ 1*1
disp('**************************');
vr _ 2=itR _ 2*2
disp('**************************');
vr _ 5=itR _ 5*5
disp('**************************');
figure(1)
ezplot(itR _ 1*1,[0 4]);
hold on
ezplot(itR _ 2*2,[0 4]);
hold on
ezplot(itR _ 5*5,[0 4]);
hold on
grid on;axis([0 3 0 11])
xlabel('time t/sec')
ylabel('voltage ')
title('Voltages across R for, R=1,2 and 5  ')
figure(2)
disp('**************************');
disp('The  voltages across L=1')
disp(' for  R=1,2 and 5 (in Ohms) are:');
disp('**************************');
disp('**************************');
vlr _ 1=10-itR _ 1*1
disp('**************************');
vlr _ 2=10-itR _ 2*2
disp('**************************');
vlr _ 5=10-itR _ 5*5
disp('**************************');
ezplot(10-itR _ 1*1,[0 4]);hold on
ezplot(10-itR _ 2*2,[0 4]);hold on
ezplot(10-itR _ 5*5,[0 4]);
hold on
grid on;axis([0 3 0 11])
xlabel('time t/sec')
ylabel('voltage ')
title('Voltages across L=1, for R=1,2 and 5')
```

The script file *RL_vol_plots* is executed, and the results and plots are shown as follows:

```
>> RL _ vol _ plots
*******************R E S U L T S *******************
****************************************************
The voltages across R = 1, 2 and 5 (in Ohms) are:
****************************************************
```

```
*********************************************
vr _ 1 =
        10 - 10*exp(-t)
*********************************************
vr _ 2 =
        10 - 10*exp(-2*t)
*********************************************
vr _ 5 =
        10 - 10*exp(-5*t)
*********************************************
*********************************************

The voltages across L=1
  for R=1,2 and 5 (in Ohms) are:
*********************************************
*********************************************
vlr _ 1 =
        10*exp(-t)
*********************************************
vlr _ 2 =
        10*exp(-2*t)
*********************************************
vlr _ 5 =
        10*exp(-5*t)
*********************************************
```

See Figures 2.83 and 2.84.

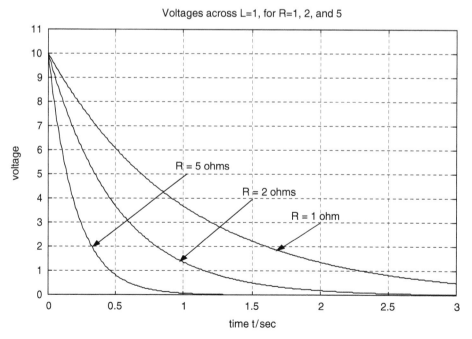

FIGURE 2.83
Transient voltage plots across L of Example 2.24.

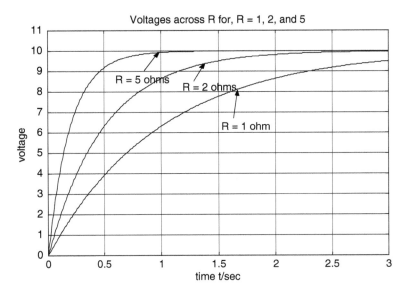

FIGURE 2.84
Transient voltage plots across R of Example 2.24.

Example 2.25

The switch of the circuit diagram shown in Figure 2.85 is open for $t < 0$, and it closes at $t = 0$ where it remains indefinitely. Obtain

1. Analytical expressions for $v_c(t)$, $v_R(t)$, and $i(t)$ assuming that $ic(0) = 0$ A, for $t \geq 0$.
2. Create the script file *RC* that returns the MATLAB symbolic solutions for $v_c(t)$, $v_R(t)$, and $i(t)$ by solving the corresponding loop differential equation for each of the following cases: $R = 1, 2$, and $5 \ \Omega$.
3. The plots of
   a. $v_c(t)$ versus t
   b. $v_R(t)$ versus t
   c. $i(t)$ versus t

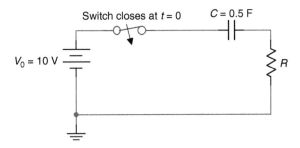

FIGURE 2.85
Network of Example 2.25.

ANALYTICAL Solution

The loop differential equation and its solution are given as follows:

$$RC\frac{dv_C(t)}{dt} + v_C(t) = V_0 \quad \text{and} \quad v_C(t) = V_0 - V_0 e^{-t/\tau} \quad \text{for } t \geq 0$$

then

$$v_R(t) = V_0 - v_C(t) = Ve^{-t/\tau} \quad \text{for } R = 1, 2, \text{ and } 5\ \Omega$$

and

$$i(t) = \frac{v_R(t)}{R} = \frac{V_0}{R}e^{-t/\tau}$$

where $\tau = RC$.

MATLAB Solution
```
% Script file:RC
vctR _ 1=dsolve('0.5*Dy+y=10','y(0)=0','t');
vctR _ 2=dsolve('Dy+y=10','y(0)=0','t');
vctR _ 5=dsolve('2.5*Dy+y=10','y(0)=0','t');
disp('***************************************');
disp('********** R E S U L T S **************')
disp('***************************************');
disp('The voltages across C for R=1,2 and 5(in volta)are:');
disp('***************************************');
vctR _ 1
disp('***************************************');
vctR _ 2
disp('***************************************');
vctR _ 5
disp('***************************************');

figure(1)
ezplot(vctR _ 1,[0 4]);
hold on
ezplot(vctR _ 2,[0 4]);
hold on
ezplot(vctR _ 5,[0 4]);
hold on
grid on;axis([0 3 0 11])
xlabel('time t/sec')
ylabel('voltage ')
title('Voltages across C for R=1,2 and 5 are:')

figure(2)
disp('The voltages across R')
disp(' for R=1,2 and 5 (in volts) are:');
disp('***************************************');
vr1=10-vctR _ 1
disp('***************************************');
vR2=10-vctR _ 2
disp('***************************************');
vR5=10-vctR _ 5
disp('***************************************');
ezplot(10-vctR _ 1,[0 4]);
hold on
ezplot(10-vctR _ 2,[0 4]);
hold on
ezplot(10-vctR _ 5,[0 4]);
hold on
```

```
grid on;axis([0 3 0 11])
xlabel('time t/sec')
ylabel('voltage ')
title('Voltages across R, for R=1,2 and 5')
disp('The current i(t) through R=1,2 and 5 (in amps) are:');
disp('*************************');
iR1=10-vctR _ 1
disp('*************************');
iR2=10-vctR _ 2)/2
disp('*************************');
iR5=(10-vctR _ 5)/5
disp('*************************');

figure(3)
ezplot(10-vctR _ 1,[0 4]);hold on;
ezplot(10/2-vctR _ 2/2,[0 4]);hold on;
ezplot(10/5-vctR _ 5/5,[0 4]);hold on;
grid on;axis([0 3 0 11])
xlabel('time t/sec')
ylabel('currents')
title('Currents i(t) for:R=1,2 and 5')
```

The script file *RC* is executed and the results and plots (Figures 2.86 through 2.88) are shown as follows:

```
>> RC
**********************************************************
************ R E S U L T S ***************************
**********************************************************
The voltages across C for R =1,2 and 5 (in volts) are:
**********************************************************
vctR _ 1 =
          10-10*exp(-2*t)
**********************************************************
vctR _ 2 =
          10-10*exp(-t)
**********************************************************
vctR _ 5 =
          10-10*exp(-2/5*t)
**********************************************************
The voltages across R for  R=1,2 and 5 (in volts) are:
**********************************************************
 vR1 =
       10*exp(-2*t)
**********************************************************
 vR2 =
       10*exp(-t)
**********************************************************
vR5 =
      10*exp(-2/5*t)
**********************************************************
The  current i(t) through R=1,2 and 5 (in amps) are:
**********************************************************
 iR1 =
       10*exp(-2*t)
```

```
************************************************************
iR2 =
     5*exp(-t)
************************************************************
iR5 =
     2*exp(-2/5*t)
************************************************************
```

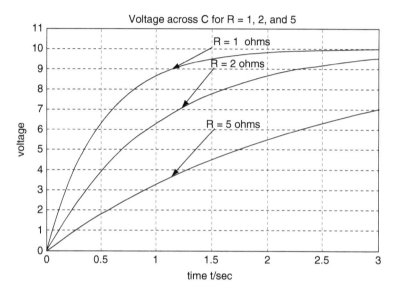

FIGURE 2.86
Transient voltage plots across C of Example 2.25.

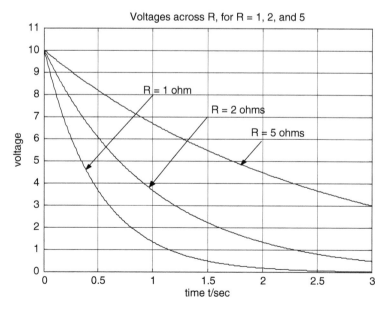

FIGURE 2.87
Transient voltage plots across R of Example 2.25.

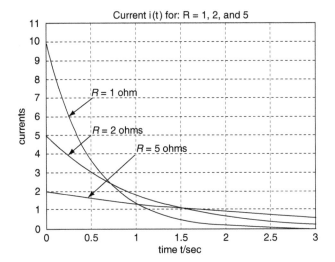

FIGURE 2.88
Transient current plots across R of Example 2.25.

Example 2.26

Steady-state conditions exist in the network shown in Figure 2.89 at $t = 0^-$, when the source $V_2 = 10$ V is connected to the RL circuit. At $t = 0^+$, the switch is moved downward and the source $V_1 = 20$ V is then connected to the RL circuit (while the source $V_2 = 10$ V is disconnected), where it remains for $t \geq 0$.

1. Write and solve the differential loop equation by hand for the current $i(t)$ (through RL), and the voltages $v_R(t)$ and $v_L(t)$, for $t \geq 0$.
2. Create the script file *RL_IC* that returns the solutions for part 1 by using the MAT-LAB symbolic solver *dsolve*.
3. Repeat part 2 by using the MATLAB numerical solver *ode45*.
4. Also obtain the plots of
   a. $i(t)$ versus t
   b. $v_R(t)$ versus t
   c. $v_L(t)$ versus t
   over the range $0 \leq t \leq 2$ s, for parts 2 and 3 and compare their results.
5. By using MATLAB, verify the solution $i(t)$ obtained in part 2 by applying KVL around the loop consisting of V_1, R, and L.

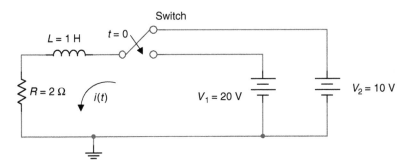

FIGURE 2.89
Network of Example 2.26.

ANALYTICAL Solution

The loop differential equation is given by

$$V_1 = 20 \text{ V} = \frac{di(t)}{dt} + 2i(t) \quad \text{for } t \geq 0$$

with

$$i(0) = \frac{10}{2} = 5 \text{ A} \quad \text{and} \quad i(\infty) = \frac{20}{2} = 10 \text{ A}$$

Assuming a solution for $i(t)$ of the form $i(t) = A + Be^{-t/\tau}$, where $\tau = L/R = 0.5$ s, and by using the initial and final current conditions, the constants A and B can be evaluated. By following this process the current is then $i(t) = 10 - 5e^{-2t}$ A, and the corresponding voltages are $v_R(t) = 20 - 10e^{-2t}$ V and $v_L(t) = 10e^{-2t}$ V, for $t \geq 0$.

MATLAB Solution

```
% Script file: RC _ IC
syms t;
% symbolic solution
it _ 1 = dsolve('Dy+2*y=20','y(0)=5','t');
disp('****************************************');
figure(1)
subplot(3,1,1);
ezplot (it _ 1,[0 2]);ylabel('i(t) in amps');
title('plots of: [i(t), vr(t) and vL(t)] vs t');
subplot(3,1,2);
vtr _ 1 = it _ 1*2;
ezplot(vtr _ 1,[0 2]);ylabel('vr(t)');title('     ');
subplot (3,1,3);
vl _ 1 = 20-vtr _ 1;
ezplot (vl _ 1,[0 2]);ylabel('vL(t)');title('     ');
xlabel ('time in sec.');
disp('**********R E S U L T S ***************');
disp('****************************************');
disp('The i(t) for t >0 is:');
it _ 1
disp('The voltage across R=2 ohms is :')
vtr _ 1
disp('The voltage across L=1 henry is :')
vl _ 1
disp('********Verify solution******************');
echo on
disp('****************************************')
verify = diff(it _ 1,t)+2*it _ 1
echo off
% numerical solution
clear;
% solution using ode45
t0 = 0;R=2;
tf = 2;
ts = [t0 tf];
i0 = 5;   % initil cond.
t =linspace(0,2,100);
[t,it] = ode45('f',ts,i0);
```

```
figure(2)
subplot(3,1,1)
plot(t,it)
title('plots of [i(t), vr(t), and vL(t)] vs t')
ylabel('i(t) in amps')
subplot(3,1,2)
vr=R*it';
plot(t,vr);ylabel('vr(t) in volts');
subplot(3,1,3)
vl=20*ones(1,45)-vr;plot(t,vl);
ylabel('vL(t) in volts ');xlabel('time in sec')
function didt=f(t,i)
didt=20-2*i;
```

The script file *RC_IC* is executed, and the results and plots are shown in Figures 2.90 and 2.91.

```
>> RC _ IC
*************************************
*********R E S U L T S *************
*************************************
The i(t) for t >0 is:
it _ 1 =
          10-5*exp(-2*t)
The voltage across R=2 ohms is :
 vtr _ 1 =
            20-10*exp(-2*t)
The voltage across L=1 henry is :
 vl _ 1 =
          10*exp(-2*t)
********Verify solution***********
verify = diff(it _ 1,t)+2*it _ 1
verify =
          20
*************************************
```

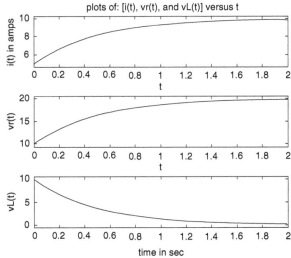

FIGURE 2.90
Plots of symbolic transient solutions of Example 2.26.

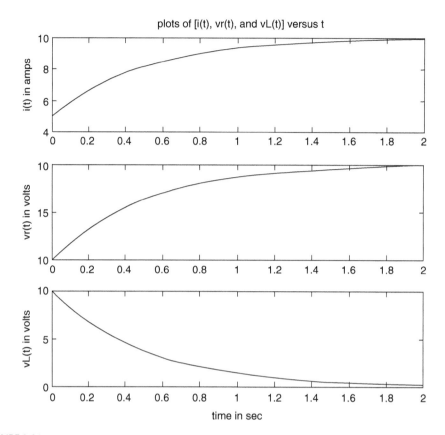

FIGURE 2.91
Plots of numerical transient solutions of Example 2.26.

Note that the symbolic results shown in Figure 2.90 fully agree with the numerical results shown in Figure 2.91.

Example 2.27

Steady-state conditions exist in the network shown in Figure 2.92 at $t = 0^-$, when the $V_1 = 10$ V source is connected to the RC circuit. At $t = 0^+$, the switch moves downward, and the $V_2 = 20$ V source is connected to the RC circuit (while the source $V_1 = 10$ V is disconnected), for $t \geq 0$.

1. Write and solve the loop differential equation by hand, for the current $i(t)$ (through RC) and the voltages $v_R(t)$ and $v_C(t)$, for $t \geq 0$.
2. Create the script file RC_IC that returns the solutions for part 1 using the MATLAB symbolic solver *dsolve*.
3. Repeat part 1 by using the MATLAB numerical solver *ode45*.
4. Also, obtain the plots of
    a. $i(t)$ versus t
    b. $v_R(t)$ versus t
    c. $v_c(t)$ versus t
    over the range $0 \leq t \leq 2$ s, for parts 2 and 3.

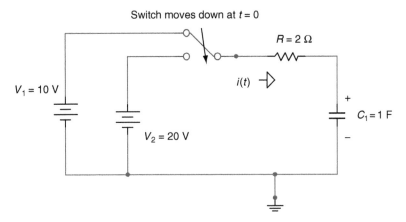

FIGURE 2.92
Network of Example 2.27.

5. By using MATLAB, verify the solution $v_C(t)$ obtained in part 2 by applying KVL around the active loop for $t > 0$.
6. Compare the results of part 1 with the results of part 2 and 3.

ANALYTICAL Solution

The loop differential equation for $t \geq 0$ is given by

$$20 = v_C(t) + 2i(t) \quad \text{for } t \geq 0$$

with $v_C(0) = 10$ V and $v_C(\infty) = 20$ V.

Replacing $i(t) = C \dfrac{dv_c(t)}{dt}$ in the preceding differential equation yields

$$2\frac{dv_C(t)}{dt} + v_C(t) = 20$$

Then assuming a solution for $v_C(t)$ of the form $v_C(t) = A + B\,e^{-t/\tau}$, where $\tau = C*R = 2$ s, and knowing the initial and final voltage values across C, the constants A and B can be evaluated. By following this process the voltage across C is given by $v_C(t) = 20 - 10\,e^{-0.5t}$ V, the voltage across R is $v_R(t) = 10\,e^{-0.5t}$ V, and the current is $i(t) = 5e^{-0.5t}$ A.

MATLAB Solution
```
% Script file: RC_IC
syms t;
% symbolic solution
vct=dsolve('2*Dy+y=20','y(0)=10','t');
disp('*******************************************');
disp('**********R E S U L T S ****************');
disp('*******************************************');
```

```
figure(1)
subplot(3,1,1);
ezplot(vct,[0 2]);ylabel('vc(t) in volts');
title('plots of: [vc(t), vr(t) and i(t)] vs t');
subplot(3,1,2);
vrt=20-vct
ezplot(vrt,[0 2]);ylabel('vr(t)in volts');title('    ');
subplot(3,1,3);
it = vrt/2;
ezplot(it,[0 2]);ylabel('it(t)in amps');title('    ');
xlabel('time in sec.');
disp('The voltage across C, vc(t) for t >0 is:');
vct
disp('The voltage across R=2 ohms is :')
vrt
disp('The current i(t) is :')
it
disp('********Verify solution*******************');
echo on
verify = 2*diff(vct,t)+vct
disp('*****************************************')
% numerical solution
clear;
% solution using ode45
t0 = 0;R = 2;
tf = 2;
ts = [t0 tf];
v0 = 10;                                 % initial condition
t = linspace(0,2,100);
[t,vt] = ode45('f',ts,v0);

figure(2)
subplot(3,1,1)
plot(t,vt)
title('plots of [vc(t), vr(t), and i(t)] vs t')
ylabel('vc(t) in volts')
subplot(3,1,2)
vr = 20*ones(1,41)-vt';
plot(t,vr)
ylabel('vr(t) in volts')
subplot(3,1,3)
it=vr./R;
plot(t,it)
ylabel('i(t) in volts ')
xlabel('time in sec')

function dvtdt=f(t,vt)
dvtdt=10-0.5*vt;
```

The script file *RC_IC* is executed, and the results and plots are shown as follows (Figures 2.93 and 2.94):

```
>> RC _ IC
******************************************
********** R E S U L T S ****************
******************************************
The voltage across C, vc(t) for t >0 is:
 vct =
        20-10*exp(-1/2*t)
The voltage across R=2 ohms is :
 vrt =
        10*exp(-1/2*t)
The current i(t) is :
 it =
        5*exp(-1/2*t)
********Verify solution******************
verify = 2*diff(vct,t)+vct
verify =
        20
******************************************
```

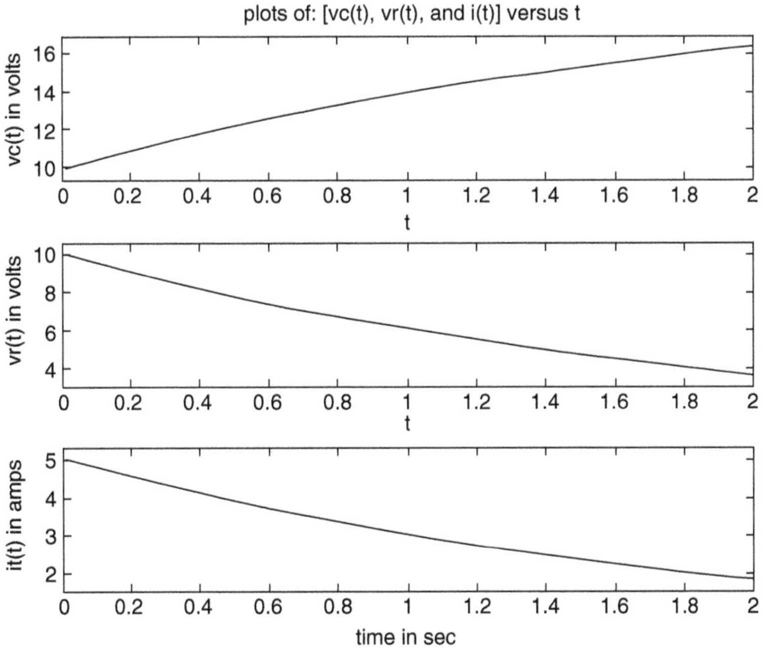

FIGURE 2.93
Plots of symbolic transient solutions of Example 2.27.

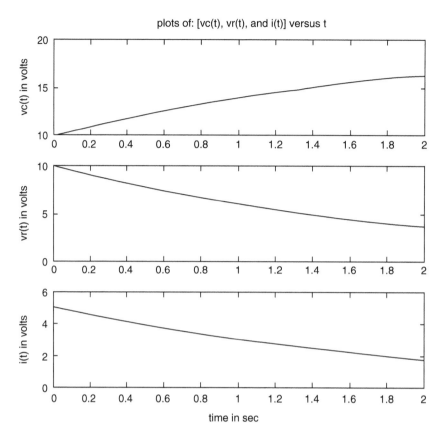

FIGURE 2.94
Plots of numerical transient solutions of Example 2.27.

Note that the analytical solution fully agrees with the symbolic and numerical solutions shown in Figures 2.93 and 2.94.

Example 2.28

Steady-state conditions exist in the network shown in Figure 2.95 at $t = 0^-$, when the $V_0 = 10$ V source is connected to the RCL circuit. At $t = 0^+$, the switch moves downward, and the source $V_0 = 10$ V and resistor R are disconnected from the LC structure, where it remains for $t \geq 0$.

Write the second-order loop differential equation in terms of the current $i(t)$, solve and plot $i(t)$ using MATLAB over the range $0 \leq t \leq 1$ s, and verify that the LC network is an ideal oscillator.

ANALYTICAL Solution

$$CL\frac{d^2i(t)}{dt^2} + i(t) = 0 \quad \text{or} \quad \frac{d^2i(t)}{dt^2} + \frac{1}{CL}i(t) = 0 \quad \text{for } t \geq 0$$

$$\frac{d^2i(t)}{dt^2} + 144i(t) = 0$$

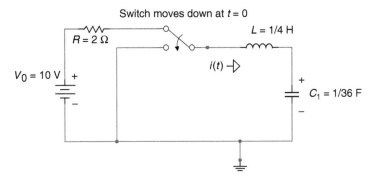

FIGURE 2.95
Network of Example 2.28.

with the boundary conditions given by

$$L\frac{di(t)}{dt}\Big|_{t=0} = v_C(0) = 10 \text{ V}$$

or

$$\frac{di(t)}{dt}\Big|_{t=0} = 40 \text{ V} \quad \text{and} \quad i(0) = 0 \text{ A}$$

$$\text{Note that } L\frac{di(t)}{dt} + \frac{1}{C}\int i(t) = 0 \quad \text{for } t \geq 0 \quad \text{satisfying KVL}$$

MATLAB Solution

```
>> current _ it = dsolve('D2y+144*y=0','y(0)=0,Dy(0)=40','t')

   current _ it =
               10/3*sin(12*t)
>> ezplot(it,[0 1])          % the plot is shown in Figure 2.96
>> xlabel('time t in sec')
>> ylabel('i(t) in amps')
>> title('i(t) vs t')
```

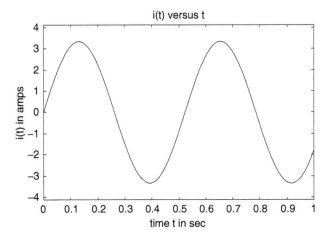

FIGURE 2.96
Plot of i(t) of Example 2.28.

Note that the network shown in Figure 2.95 is an ideal circuit (since there is no resistance, $R = 0$ in the LC loop for $t \geq 0$), and the solution clearly indicates that the current shows an oscillator behavior with $W = 12$ rad/s $= \frac{1}{\sqrt{LC}}$ and $T = \frac{\pi}{6}$ s.

Example 2.29

Steady-state conditions exist in the network shown in Figure 2.97 at $t = 0^-$, when the $V_1 = 120$ V source is connected to the RCL parallel circuit. At $t = 0^+$, the switch moves downward, and the source $V_1 = 120$ V and resistor $R = 10$ Ω are disconnected from the parallel (RLC) structure.

Analyze the transient response ($t > 0$) of the source-free parallel RLC circuit, for each of the following values of R, $R = 3, 9$, and 72 Ω.

1. Determine the analytical response $v_C(t)$ for each value of R, for $t \geq 0$
2. Create the script file *transient_RLC_parallel* that returns the MATLAB solutions of part 1 and its corresponding voltage plots
3. Compare the MATLAB solutions of part 2 with the analytical solutions of part 1

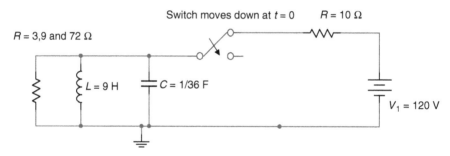

FIGURE 2.97
Network of Example 2.29.

ANALYTICAL Solution

From the circuit diagram of Figure 2.97 for $t \leq 0$, the initial conditions are $v_C(0) = 0$ and $i_L(0) = 120/10 = 12$ A, and

$$C \frac{dv_C(t)}{dt}\Big|_{t=0} = i_C(t = 0) = i_C(0) = i_L(0) + i_R(0)$$

then

$$\frac{dv_C(t)}{dt}\Big|_{t=0} = 36 i_C(0) = 36(12 + 0) = 432 \text{ V/s}$$

Recall that the node equation is

$$\frac{d^2 v_C(t)}{dt^2} + \frac{1}{RC}\frac{dv_C(t)}{dt} + \frac{1}{CL}v_C(t) = 0 \quad \text{for } t \geq 0$$

For $R = 3$ Ω, the resonant frequency is

$$w_0 = \frac{1}{\sqrt{LC}} = 2 \text{ rad/s}$$

The neper frequency is

$$\alpha = \frac{1}{2RC} = 6 \text{ Hz}$$

The complex frequencies are s_1 and s_2, given by

$$s_{1,2} = -\alpha \pm \sqrt{\alpha^2 - w_0^2}$$

then $0 > \sqrt{\alpha^2 - w_0^2}$ and is clearly the overdamped case.
Then the solution $v_C(t)$ is of the form $v_C(t) = A_1 e^{-s_1 t} + A_2 e^{-s_2 t}$
For $R = 9 \ \Omega$, the resonant frequency is

$$w_0 = \frac{1}{\sqrt{LC}} = 2 \text{ rad/s}$$

The neper frequency is

$$\alpha = \frac{1}{2RC} = 2 \text{ Hz}$$

Then s_1 and s_2 are negative and repeated frequencies, which are given by

$$s_{1,2} = -\alpha \pm \sqrt{\alpha^2 - w_0^2} = -\alpha$$

Since $0 = \sqrt{\alpha^2 - w_0^2}$, the case is critical damped.
Then the solution $v_C(t)$ is of the form $v_C(t) = A_1 e^{-\alpha t} + A_2 t e^{-\alpha t}$
For $R = 72 \ \Omega$, the resonant frequency is

$$w_0 = \frac{1}{\sqrt{LC}} = 2 \text{ rad/s}$$

The neper frequency is

$$\alpha = \frac{1}{2RC} = \frac{1}{4} \text{ Hz}$$

The complex frequencies s_1 and s_2 are given by $s_{1,2} = -\alpha \pm j\sqrt{w_0^2 - \alpha^2}$ (complex conjugate), and is clearly the underdamped case. Then the solution $v_C(t)$ is of the form

$$v(t) = e^{-\alpha t}[A_1 \cos(w_d t) + A_2 \sin(w_d t)]$$

where $w_d = \sqrt{w_0^2 - \alpha^2}$.

Let us use MATLAB to obtain and plot the solutions just presented, and compare them with the analytical results.

MATLAB Solution

```
% Script file : transient _ RLC _ parallel
% source free RLC parallel circuit analysis
disp('*************************************')
disp('*******R E S U L T S******************')
disp('Source free parallel RLC circuit')
disp('*************************************')
disp('Solution of diff. equation for R=3 ohms')
overdamped = dsolve('(1/36)*D2y+(1/3)*Dy+(1/9)*y=0','y(0) = 0','Dy(0)= 32','t')
figure(1)
ezplot(overdamped)
xlabel('time (sec)')
ylabel('voltage(volts)')
title('v(t) vs t (overdamped case; R=3 ohms)')
axis([0 3 0 40]);
disp('*************************************')
figure(2)
disp('Solution of diff. equation for R=9 ohms')
criticaldamped = dsolve('(1/36)*D2y+(1/9)*Dy+(1/9)*y=0','y(0)=0','Dy(0)=432','t')
simple _ crit=simple(criticaldamped);
ezplot(criticaldamped)
xlabel('time (sec)')
ylabel('voltage(volts)')
title('v(t) vs t (criticaldamped case; R=9 ohms)')
axis([0 4 0 90]);
disp('*************************************')
disp('Solution of diff. equation for R=72 ohms')
underdamped = dsolve('(1/36)*D2y+(1/72)*Dy+(1/9)*y=0','y(0)=0','Dy(0)=432','t')
figure(3)
ezplot(underdamped)
xlabel('time (sec)'); ylabel('voltage(volts)')
title('v(t) vs t (underdamped case; R=72 ohms)')
axis([0 6 -150 200]);
disp('*************************************')
```

The script file *transient_RLC_parallel* is executed, and the results and plots (Figures 2.98 through 2.100) are as follows:

```
>> transient _ RLC _ parallel
*************************************
*******R E S U L T S******************
Source free parallel RLC circuit
*************************************
Solution of diff. equation for R=3 ohms
overdamped =
  27*2^(1/2)*exp(2*(-3+2*2^(1/2))*t)-27*2^(1/2)*exp(-2*(3+2*2^(1/2))*t)
*************************************
Solution of diff. equation for R=9 ohms
criticaldamped =
                432*exp(-2*t)*t
*************************************
Solution of diff. equation for R=72 ohms
underdamped =
                576/7*7^(1/2)*exp(-1/4*t)*sin(3/4*7^(1/2)*t)
*************************************
```

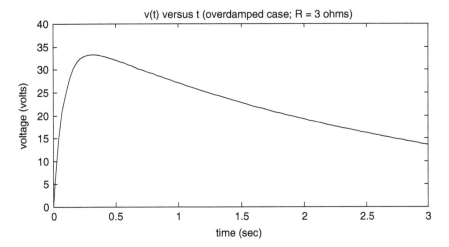

FIGURE 2.98
Transient overdamped plot ($t > 0$) of the source-free parallel RLC of Example 2.29.

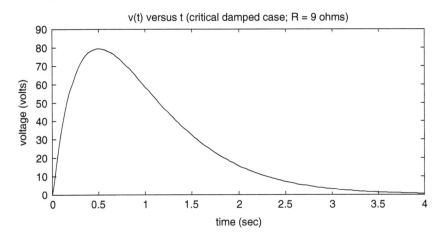

FIGURE 2.99
Transient critical damped plot ($t > 0$) of the source-free parallel RLC of Example 2.29.

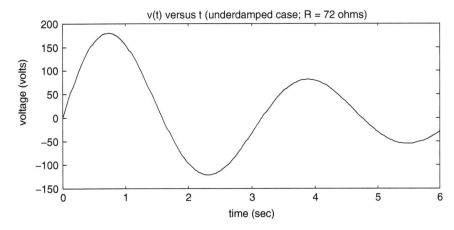

FIGURE 2.100
Transient underdamped plot ($t > 0$) of the source-free parallel RLC of Example 2.29.

Observe that by changing the value of R ($R = 3$, 9, and 72 Ω), the three solutions for the second-order differential equation—*over*, *critical*, and *underdamped*—are obtained. Observe that the analytical solutions completely agree with the MATLAB solutions.

Example 2.30

Steady-state conditions exist in the network shown in Figure 2.101, at $t = 0^-$, when the $V_0 = 90$ V source is connected to the RCL circuit. At $t = 0^+$, the switch opens (moves upward) and the source $V_0 = 90$ V is disconnected from the RLC structure.

Analyze the transient response ($t > 0$) of the source-free series RLC circuit for the following values of R, $R = 75$, 36, and 3 Ω.

1. Determine the analytical response $i_R(t)$ for each value of R, for $t \geq 0$
2. Create the script file *transient_RLC_series* that returns the MATLAB solutions of part 1, and its plots
3. Compare the MATLAB solutions of part 2 with the analytical solutions of part 1

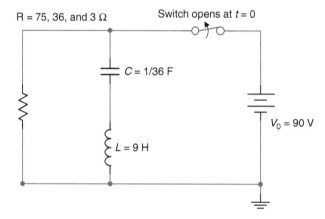

FIGURE 2.101
Network of Example 2.30.

ANALYTICAL Solution

For $t \leq 0^-$, the initial conditions are $v_C(0) = 90$ V and $i_L(0) = 0$ A.

$$L\frac{di(t)}{dt}\Big|_{t=0} + Ri(0) = v_C(0)$$

then

$$\frac{di(t)}{dt}\Big|_{t=0} = \frac{v_C(0) - Ri_L(t)}{L} = \frac{v_C(0)}{L} = \frac{90}{9} = 10 \text{ A/s}$$

Recall that the loop differential equation is

$$\frac{d^2i(t)}{dt^2} + \frac{R}{L}\frac{di(t)}{dt} + \frac{1}{CL}i(t) = 0 \quad \text{for } t \geq 0$$

For $R = 75\ \Omega$, the resonant frequency is

$$w_0 = \frac{1}{\sqrt{LC}} = 2\ \text{rad/s}$$

The neper frequency is

$$\alpha = \frac{R}{2L} = \frac{75}{18} = 4.1667\ \text{Hz}$$

The frequencies s_1 and s_2 are

$$s_{1,2} = -\alpha \pm \sqrt{\alpha^2 - w_0^2}$$

then

$$0 > \sqrt{\alpha^2 - w_0^2}$$

and the solution $i(t)$ is the overdamped case given by $i(t) = A_1 e^{-s_1 t} + A_2 e^{-s_2 t}$.
  For $R = 36\ \Omega$, the resonant frequency is

$$w_0 = \frac{1}{\sqrt{LC}} = 2\ \text{rad/s}$$

The neper frequency is

$$\alpha = \frac{R}{2L} = \frac{36}{18} = 2\ \text{Hz}$$

and s_1 and s_2 are repeated, given by

$$s_{1,2} = -\alpha \pm \sqrt{\alpha^2 - w_0^2} = -\alpha$$

then

$$0 = \sqrt{\alpha^2 - w_0^2}$$

and the solution $i(t)$ is the critical-damped case given by $i(t) = A_1 e^{-\alpha t} + A_2 t e^{-\alpha t}$.
  For $R = 3\ \Omega$, the resonant frequency is

$$w_0 = \frac{1}{\sqrt{LC}} = 2\ \text{rad/s}$$

The neper frequency is

$$\alpha = \frac{R}{2L} = \frac{3}{18} = \frac{1}{6} = 0.1667\ \text{Hz}$$

and the complex frequencies are s_1 and s_2 given by $s_{1,2} = -\alpha \pm j\sqrt{w_0^2 - \alpha^2}$, clearly the underdamped case.
  Then the solution $i(t)$ is of the form

$$i(t) = e^{-\alpha t}\ [A_1 \cos(w_d t) + A_2 \sin(w_d t)]$$

where $w_d = \sqrt{w_0^2 - \alpha^2}$.

```
MATLAB Solution
% Script file : transient _ RLC _ series
% source free RLC series circuit analysis
disp('****************************************')
disp('************R E S U L T S**************')
disp('*****Source free series RLC circuit***')
disp('****************************************')
disp('Solution of diff. equation for R=75 ohms')
overdamped = dsolve('D2y+(75/9)*Dy+4*y=0','y(0)=0', 'Dy(0)=10','t')

figure(1)
ezplot (overdamped,[0 5])
xlabel ('time (sec)')
ylabel ('current i(t) in amps')
title ('i(t) vs t (overdamped case; R=75 ohms)')
disp ('****************************************')

figure(2)
disp('Solution of diff. equation for R=36 ohms')
criticaldamped = dsolve('D2y+(36/9)*Dy+4*y=0','y(0)=0','Dy(0)=10','t')
ezplot(criticaldamped,[0 6])
xlabel('time (sec)'); ylabel('current i(t) in amps')
title('i(t) vs t (criticaldamped case; R=36 ohms)')
disp('**********************************************************')
disp('Solution of diff. equation for R=3 ohms')
underdamped = dsolve('D2y+(3/9)*Dy+4*y=0','y(0)=0','Dy(0)=10)

figure(3)
ezplot (underdamped)
xlabel ('time (sec)')
ylabel ('current i(t) in amps')
title ('i(t) vs t (underdamped case; R=3 ohms)')
disp('**********************************************************')
```

The script file *transient_RLC_series* is executed, and the results and plots (Figures 2.102 through 2.104) are as follows:

```
>> transient _ RLC _ series
****************************************
************R E S U L T S**************
          ***Source free series RLC circuit ***
**********************************************************************
Solution of diff. equation for R=75 ohms
overdamped =
          30/481*481^(1/2)*exp(1/6*(-25+481^(1/2))*t)- 30/481*481^(1/2)*exp(-
          1/6*(25+481^(1/2))*t)
**********************************************************************
Solution of diff. equation for R=36 ohms
criticaldamped =
          10*exp(-2*t)*t
**********************************************************************
Solution of diff. equation for R=3 ohms
underdamped =
          60/143*143^(1/2)*exp(-1/6*t)*sin(1/6*143^(1/2)*t)
**********************************************************************
```

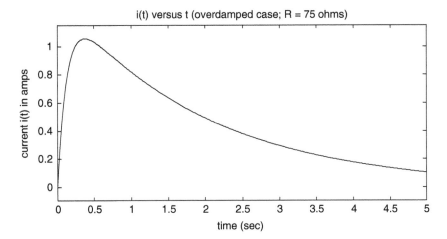

FIGURE 2.102
Transient overdamped plot ($t > 0$) of the source-free series RLC of Example 2.30.

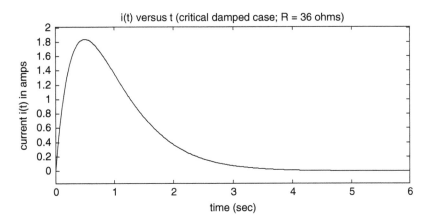

FIGURE 2.103
Transient critical damped plot ($t > 0$) of the source-free series RLC of Example 2.30.

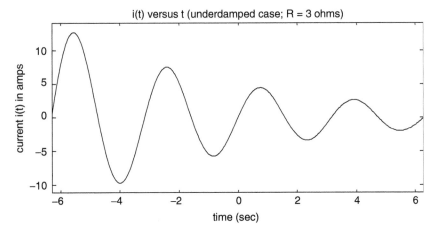

FIGURE 2.104
Transient underdamped plot ($t > 0$) of the source-free series RLC of Example 2.30.

Observe that by changing the value of R (R = 75, 36, and 3 Ω), the three solutions for the second-order differential equation—over, critical, and underdamped cases— are obtained. Also observe that the analytical solutions completely agree with the MATLAB solutions.

Example 2.31

Steady-state conditions exist in the circuit of Figure 2.105 for $t \leq 0$, while the voltage source $V_1 = 6$ V is connected to the network. At $t = 0^+$, the switch moves downward disconnecting the source while connecting R_1 to the rest of the circuit.

1. Obtain analytically the loop differential equation set and the initial conditions
2. Using the MATLAB symbolic solver, create the script file *transient_2loops* that returns the transient currents for each loop, and their respective plots, for $t \geq 0$
3. Also obtain *simplify* and *pretty* expressions for each of the transient loop currents of part 2

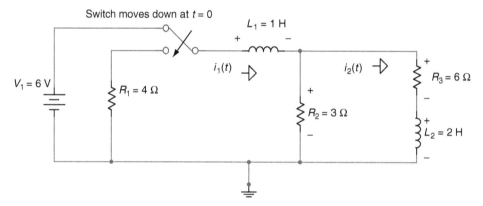

FIGURE 2.105
Network of Example 2.31.

ANALYTICAL Solution

The conditions at $t = 0$ are $i_1(0) = 3$ A and $i_2(0) = 1$ A. Applying KVL to the two-mesh network of Figure 2.105, for $t \geq 0$, results in the following set of simultaneous differential equations:

$$(R_1 + R_2)i_1(t) + L_1 \frac{di_1(t)}{di} - R_2 i_2(t) = 0$$

$$-R_2 i_1(t) + (R_3 + R_2)i_2(t) + L_2 \frac{di_2(t)}{di} = 0$$

Replacing the elements by their values yields the following set of differential equations:

$$7i_1(t) + \frac{di_1(t)}{di} - 3i_2(t) = 0$$

or

$$\frac{di_1(t)}{di} = -7i_1(t) + 3i_2(t)$$

and

$$-3i_1(t) + 9i_2(t) + 2\frac{di_2(t)}{di} = 0$$

or

$$2\frac{di_2(t)}{dt} = 3i_1(t) - 9i_2(t)$$

MATLAB Solution

```
% Script file : transient _ 2loops
% transient solutions for two loop network
disp('*************************************')
disp('*******R E S U L T S*****************')
disp('*************************************')
disp('Solution of the two diff. loop equations for t>0 ')
[y1 y2] = dsolve('Dy1=-7*y1+3*y2,2*Dy2=3*y1- 9*y2','y1(0)=3','y2(0)=2','t');
figure(1)
ezplot (y1,[0 3])
axis([0 1.5 0 3.5])
xlabel ('time (sec)')
ylabel ('current i1(t) in amps')
title ('i1(t) vs t ')

figure(2)
ezplot(y2,[0 3])
axis([0 1.5 0 2.3])
xlabel('time (sec)')
ylabel('current i2(t) in amps')
title('i2(t) vs t ')
disp('*************************************')
disp('The currents i1(t) and i2(t), in amps are :')
[y1 y2]'
disp('*************************************')
disp('The simplify solution for i1(t) is:')
simplify(y1)
disp('*************************************')
disp('The simplify solution for i2(t) is :')
simplify(y2)
disp('*************************************')
disp('The pretty solution for i1(t) is :')
pretty(y1)
disp('*************************************')
disp('The pretty solution for i2(t) is :')
pretty(y2)
disp('*************************************')
```

```
figure(3)
ezplot(y1)
hold on
ezplot(y2)
xlabel('time (sec)')
axis([0 1.0 0 3.5])
ylabel('currents [i1(t)&i2(t)] in amps')
title('i1(t) & i2(t) vs t ')
```

The script file *transient_2loops* is executed, and the results and plots (Figures 2.106 through 2.108) are as follows:

```
>> transient _ 2loops
```

```
*****************************************************
*******R  E  S  U  L  T  S***********************
*****************************************************
Solution of the two diff. loop equations for t>0
*****************************************************
The currents i1(t) and i2(t), in amps are :
ans =
[ conj(-9/194*97^(1/2)*exp(-1/4*(23+97^(1/2))*t)+9/194*97^(1/2)*exp(1/4*(-
23+97^(1/2))*t)+3/2*exp(1/4*(-23+97^(1/2))*t)+3/2*exp(-1/4*(23+97^(1/2))*t))]
[   conj(exp(1/4*(-23+97^(1/2))*t)-14/97*97^(1/2)*exp(-1/4*(23+97^
(1/2))*t)+14/97*97^(1/2)*exp(1/4*(-23+97^(1/2))*t)+exp(-1/4*(23+97^(1/2))*t))]
*************************************************************
The simplify solution for i1(t) is:
ans =
-9/194*97^(1/2)*exp(-1/4*(23+97^(1/2))*t)+9/194*97^(1/2)*exp(1/4*(-23+97^
(1/2))*t)+3/2*exp(1/4*(-23+97^(1/2))*t)+3/2*exp(-1/4*(23+97^(1/2))*t)
*****************************************************************
The simplify solution for i2(t) is :
ans =
exp(1/4*(-23+97^(1/2))*t)-14/97*97^(1/2)*exp(-1/4*(23+97^(1/2))*t)+14/97*97^
(1/2)*exp(1/4*(-23+97^(1/2))*t)+exp(-1/4*(23+97^(1/2))*t)
*****************************************************************
     The pretty solution for i1(t) is :
               1/2                      1/2
     - 9/194 97       exp(- 1/4 (23 + 97   ) t)
               1/2                1/2                          1/2
     + 9/194 97   exp(1/4 (-23 + 97   ) t) + 3/2 exp(1/4 (-23 + 97   ) t)
                        1/2
     + 3/2 exp(- 1/4 (23 + 97   ) t)
*****************************************************************
     The pretty solution for i2(t) is :
                    1/2              1/2                    1/2
     exp(1/4 (-23 + 97   ) t) -  14/97    exp(- 1/4 (23 + 97   ) t)
               1/2                1/2                        1/2
     + 14/19 97    exp(1/4 (-23 + 97   ) t) + exp(- 1/4 (23 + 97   ) t)
*****************************************************************
```

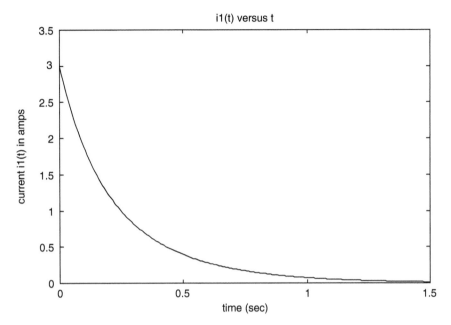

FIGURE 2.106
Plot of *i1(t)* of Example 2.31.

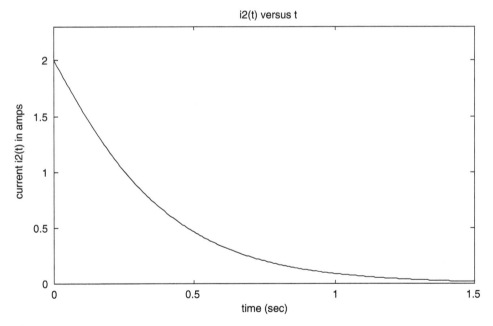

FIGURE 2.107
Plot of *i2(t)* of Example 2.31.

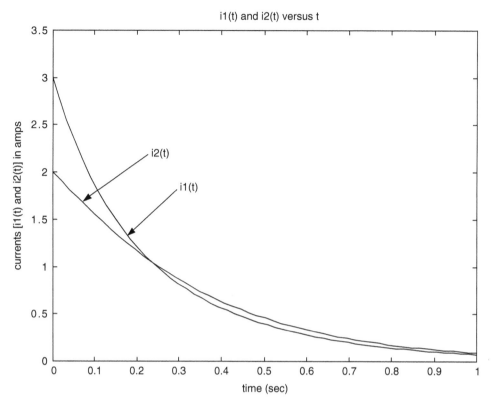

FIGURE 2.108
Plots of *i1(t)* and *i2(t)* of Example 2.31.

2.5 Application Problems

P.2.1   Create conversion tables and graphs for the first 21 integer values (from 0 to 20) for the following:

a. British thermal units to Joules

b. Joules to foot-pound force

c. Newton to pound-force

d. Inches to centimeter

e. Horsepower to watts

f. Foot to meter

g. Joules to calories

h. Horsepower to foot-pound force per second

i. Kilowatt per hour to Joules

j. Electrons to coulombs

The following relations may be of help:

- 1 Btu = 1055 J = 778.2 ft · lb$_f$
- 1 J = 0.737 ft · lb$_f$
- 1 N = 0.225 lb$_f$
- 1 in. = 2.54 cm
- 1 Hp = 746 W
- 1 ft = 0.3048 m
- 1 Btu = 252 Cal
- 1 Hp = 550 ft · lb$_f$/s = 33,000 ft · lb$_f$/min = 0.07067 Btu/s
- 1 kW/h = 3.6(106) J = 3.414 Btu/h
- 1 Cal = 3.088 ft · lb$_f$
- 1 C = 6.24 * 10^{18} electrons

P.2.2 Two electric charges given by Q_1 = +500 μC and Q_2 = −600 μC experience a force of 1500 N. Determine the distance separating the charges and the direction of the force (attraction or repulsion).

P.2.3 An electric current I consists of 25 * 1016 electrons/s. Express I in amperes (A), milliampere (mA), and microampere (μA).

P.2.4 An energy E of 650 J is required to move 12 C from point A to point B. Determine the potential difference between the points (A and B).

P.2.5 Obtain tables and plots of the force of attraction versus the separation (force versus distance) of two charges Q_1 = +1.500 μC and Q_2 = −600 μC, where the separation distance r is over the range 0.5 m ≤ r ≤ 5 m, in intervals of 0.25 m, where the variables are evaluated using the following units:

a. Newtons versus meters

b. Pound force versus inches

c. Newtons versus foot

d. Pound force versus centimeter

P.2.6 The current through a 2.7 kΩ resistor R is 3 mA. Determine the current direction and the polarity of the voltage V_R across the resistor.

P.2.7 A light bulb is rated 100 W at 120 V. Find the current through the bulb and calculate its resistance.

P.2.8 If the cost of electrical energy is $0.09 kW/h, what would be the cost of using a 100 W bulb during the following time periods?

a. One day

b. One week

c. One month

d. One year

P.2.9 Convert the MATLAB programs of Examples 2.2 and 2.3 into script files.

P.2.10 Determine the equivalent resistances of each of the networks shown in Figure 2.109 by hand and by using MATLAB.

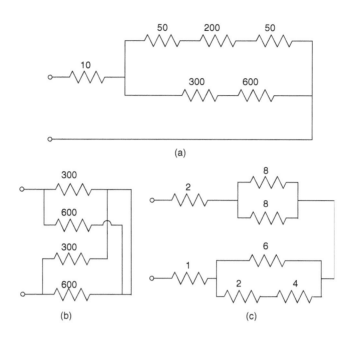

FIGURE 2.109
Network of P.2.10.

P.2.11   The potentiometer R_5 of the circuit diagram of Figure 2.110 varies over the range $0 \le R_5 \le 100\ \Omega$ in linear increments of 5 Ω. Write a MATLAB program that returns the following plots:

   a. R_5 versus I_1
   b. R_5 versus I_2
   c. R_5 versus I_3
   d. R_5 versus I_4
   e. R_5 versus I_5

P.2.12   Repeat P.2.11 for the following cases:

   a. R_5 versus V_{R1}
   b. R_5 versus V_{R3}
   c. V_{R1} versus V_{R5}
   d. V_{R3} versus I_5

P.2.13   For the circuit diagram shown in Figure 2.110, write a program that returns the following plots:

   a. R_5 versus power of R_3
   b. R_5 versus power of $(R_1||R_2)$
   c. R_5 versus power of $(R_4||R_5)$
   d. R_5 versus power of (V)

P.2.14   For the circuit shown in Figure 2.110, write and run a program that returns the following plots as R_5 varies over the range $0 \le R_5 \le 100\ \Omega$ in linear increments of 5 Ω:

   a. V_{R3} versus V_{R1}

b. I_5 versus I_3

c. I_1 versus I_5

P.2.15 The current source I, shown in the circuit of Figure 2.111, varies over the range $1\,\text{mA} \le I \le 0.5\,\text{A}$ in linear increments of 10 mA.

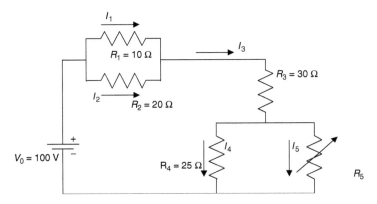

FIGURE 2.110
Network of P.2.11.

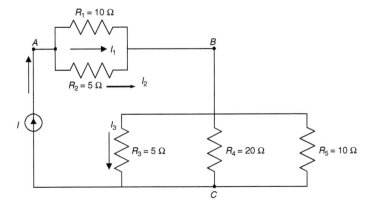

FIGURE 2.111
Network of P.2.15.

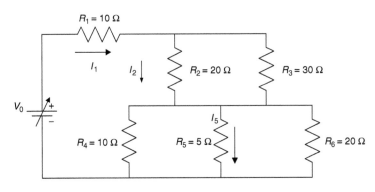

FIGURE 2.112
Network of P.2.16.

Write a program that returns the following plots:

a. I versus V_{AB}
b. I versus V_{BC}
c. I versus I_1
d. I versus I_3
e. I versus power $(R_1 \| R_2)$

P.2.16 The voltage source V_0 shown in Figure 2.112, varies from 10 to 100 V in steps (increments) of 5 V. Write a program that returns the following plots:

a. V_0 versus I_1
b. V_0 versus V_{R5}
c. V_0 versus I_2
d. V_0 versus I_5
e. V_0 versus power of R_2

P.2.17 The diagrams shown in Figure 2.113 represent the circuit equivalent Δ-to-Y transformations.

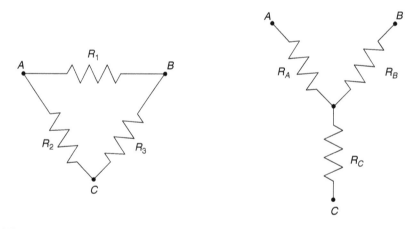

FIGURE 2.113
Network structures of P.2.17.

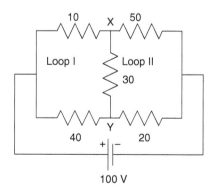

FIGURE 2.114
Network of P.2.17.

The Δ-to-Y transformation equations are

$$R_A = \frac{R_1 * R_2}{R_1 + R_2 + R_3}$$

$$R_B = \frac{R_1 * R_3}{R_1 + R_2 + R_3}$$

$$R_C = \frac{R_2 * R_3}{R_1 + R_2 + R_3}$$

Calculate the voltage V_{XY} in the circuit shown in Figure 2.114 by

a. Applying the Δ-to-Y transformation to loop I only
b. Applying the Δ-to-Y transformation to loop II only
c. Using loop equations (with no transformations)
d. Applying Thevenin's theorem (assuming that the load is $R = 30$)

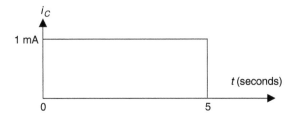

FIGURE 2.115
Current i_C of P.2.19.

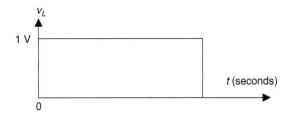

FIGURE 2.116
Voltage v_L of P.2.20.

e. Using node equations (with no transformations)
f. Applying Norton's theorem (assuming that the load is $R = 30$)
   Assume that all the resistors are given in ohms.

P.2.18 From the equations for R_A, R_B, and R_C defined in P.2.17, obtain expressions for R_1, R_2, and R_3 for the Y-to-Δ transformations by hand and by using MATLAB symbolic techniques.

P.2.19 The current $i_C(t)$ through a capacitor $C = 1$ μF is shown in Figure 2.115.
        Write a MATLAB program that returns the following plots:

a. The voltage $v_C(t)$ versus t

b. The energy $W_c(t)$ versus t over the range $0 \leq t \leq 10$ s

P.2.20 The voltage across the inductor $L = 1$ mH is shown in Figure 2.116. Write a MATLAB
        program that returns the following plots:

a. The current $i_L(t)$ versus t

b. The energy $w_L(t)$ versus t over the range $0 \leq t \leq 10$ s

P.2.21 The RC circuit shown in Figure 2.117 presents an initial voltage $v_C(0) = 0$ V at the
        instant the switch closes at $t = 0$.

Write a program that returns the following plots over the range $0 \leq t \leq 5$ s:

a. Current $i(t)$ versus t

b. Voltage $v_R(t)$ versus t

c. Voltage $v_C(t)$ versus t

d. Energy w_C in Joules stored in C versus t

e. Instantaneous power in watts dissipated by R versus t

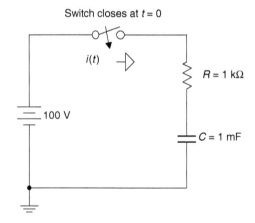

FIGURE 2.117
Network of P.2.21.

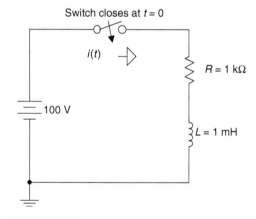

FIGURE 2.118
Network of P.2.22.

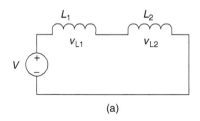

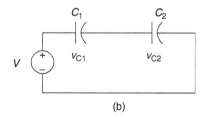

(a)   (b)

FIGURE 2.119
Networks of P.2.23.

P.2.22 The RL circuit in Figure 2.118 presents an initial current $i(0) = 0$ A at the instant the switch closes at $t = 0$, where it remains indefinitely. Write a program that returns the following plots over the range $0 \leq t \leq 10$ s:

a. [Current $i(t)$] versus t

b. Voltage $v_R(t)$ versus t

c. Voltage $v_L(t)$ versus t

d. [Energy w_L in Joules stored in L] versus t

e. [Power in watts dissipated by R] versus t

P.2.23 Determine V_{L2} and V_{C2} shown in the circuits of Figure 2.119 in a tablelike format; given $V = 100$ V, $L_2 = 10$ H, and $C_2 = 10$ F for the following cases:

a. L_1 varies from $L_2/2$ to $2L_2$ in linear steps of $L_2/10$.

b. C_1 varies from $C_2/2$ to $2C_2$ in linear steps of $C_2/10$.

P.2.24 For the circuit diagrams shown in Figure 2.120

a. Evaluate by hand the expressions for the current I_1

b. Plot I_1 as a function of L_2, if $L_1 = 5$ H and L_2 varies over the range $0 \leq L_2 \leq 15$ H in steps (increments) of 0.5 H

c. Repeat part b for the case where $C_1 = 5$ F and C_2 varies over the range $0 \leq C_2 \leq 15$ F in steps (increments) of 0.5 F

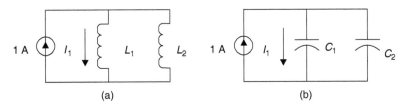

(a)   (b)

FIGURE 2.120
Networks of P.2.24.

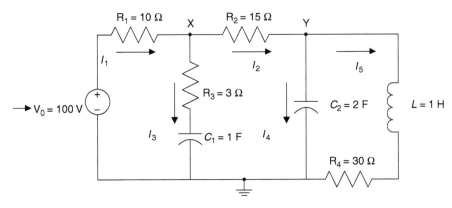

FIGURE 2.121
Network of P.2.25.

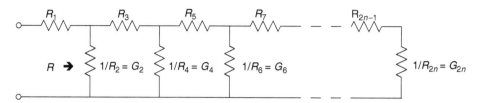

FIGURE 2.122
Lattice network of P.2.26.

P.2.25 For the circuit shown in Figure 2.121, evaluate
   a. $I_1, I_2, I_3, I_4,$ and I_5
   b. $V_X, V_Y,$ and V_{XY}
   c. Power of all the resistors in watts, horsepower, and foot-pound force per second (see P.2.1)
   d. The energy stored in $C_1, C_2,$ and L

P.2.26 The network structure shown in Figure 2.122 is called "lattice."
   Observe that the resistors, $R_1, R_3, \ldots, R_{2n-1}$ (ohms), placed horizontally are labeled with odd indexes, whereas the elements in the vertical positions are admittances $G_1, G_4, \ldots, G_{2n}$ (ohms^{-1} or siemans), which are labeled with even indexes.
   Verify the following expression:

$$R = R_1 + \dfrac{1}{G_2} + \dfrac{1}{R_3} + \dfrac{1}{G_4} + \dfrac{1}{R_5} + \cdots + \dfrac{1}{R_{2n-1}} + \dfrac{1}{G_{2n}}$$

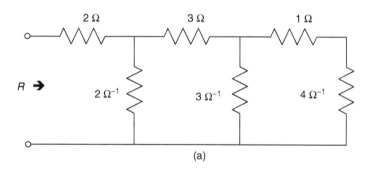

(a)

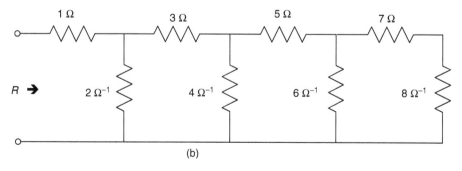

(b)

FIGURE 2.123
Network of P.2.27.

P.2.27 For the circuits shown in Figure 2.123 evaluate R by
 a. Writing a MATLAB program using the standard circuit analytical techniques
 b. Writing a MATLAB program that computes R by using the relation of P.2.26 given by

$$R = R_1 + \sum_{i=1}^{n} \frac{1}{G_{2i}} + \sum_{i=1}^{n} \frac{1}{R_{2i-1}} \quad \text{for } n = 3 \quad \text{and} \quad n = 4$$

P.2.28 Draw the possible circuit diagrams for the electrical network defined by the following set of mesh equations:
 a. $20I_1 - 5I_2 = 10$
   $-5I_1 + 10I_2 = -7$
 b. $10 = 6I_1 - 2I_2 - 3I_3$
   $-6 = -2I_1 + 6I_2 - 4I_3$
   $5 = -3I_1 - 4I_2 + 12I_3$
 c. Using MATLAB, solve each system (a and b) that consists of the set of loop equations

P.2.29 Verify that the equivalent resistance (between terminals a and a') of the circuit of Figure 2.124 is given by

$$R = \frac{R_1 * R_2(R_3 + R_4) + R_3 * R_4(R_1 + R_2) + R_5(R_1 + R_3)(R_2 + R_4)}{(R_1 + R_2)(R_3 + R_4) + (R_1 + R_2 + R_3 + R_4)R_5}$$

P.2.30 Evaluate the equivalent resistance between terminals a and a', of the circuit diagram shown in Figure 2.124, by applying the following techniques or procedures:
 a. Connect an arbitrary voltage source V across aa', solve for the current I, and then evaluate $Raa' = V/I$

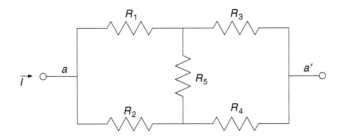

FIGURE 2.124
Network of P.2.29.

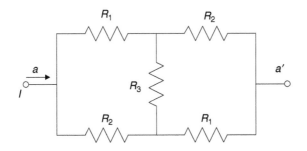

FIGURE 2.125
Network of P.2.31.

b. Direct evaluation of the expression for Raa' (P.2.29), where the element of Figure 2.124 have the following values: $R_1 = 10$, $R_2 = 20$, $R_3 = 25$, $R_4 = 30$, and $R_5 = 70$ (Ω)

c. Use Δ-to-Y transformation defined in P.2.17

P.2.31 The equivalent resistance between terminals aa' of the symmetric network shown in Figure 2.125 is given by

$$R = \frac{(R_1 + R_2)R_3 + 2R_1 * R_2}{R_1 + R_2 + 2R_3}$$

where $R_1 = 10\ \Omega$, $R_2 = 20\ \Omega$, and $R_3 = 30\ \Omega$.

Verify the preceding expression by

a. Connecting an arbitrary voltage source V across aa' (any value), solve for the current I, and then evaluate $Raa' = V/I$

b. Evaluating R by using the Δ-to-Y transformation defined in P.2.17

P.2.32 Draw possible circuit diagrams for the systems defined by the following set of node equations:

$$7V_1 - 3V_2 - 4V_3 = 10$$
$$-3V_1 + 6V_2 - 2V_3 = 4$$
$$-4V_1 - 2V_2 + 11V_3 = 20$$

Using MATLAB, solve for the voltages V_1, V_2, V_3, V_{12}, V_{13}, and V_{23}.

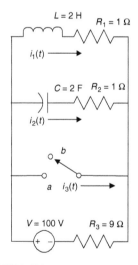

FIGURE 2.126
Network of P.2.33.

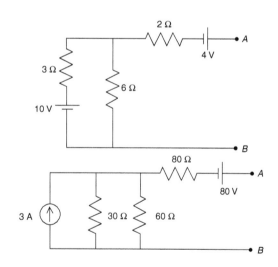

FIGURE 2.127
Network of P.2.34 and P.2.35.

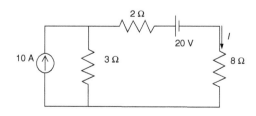

FIGURE 2.128
Network of P.2.36.

P.2.33 The switch shown in Figure 2.126 has been at (position) *a* for a very long time. At
$t = 0$, the switch is moved from position *a* to position *b*, where it remains indefinitely. Write a program that returns the following plots:

a. $i_1(t)$ versus t, $i_2(t)$ versus t, and $i_3(t)$ versus t

b. $v_L(t)$ versus t and $v_C(t)$ versus t over the range $0 < t < 6$ s

P.2.34 For the two circuits shown in Figure 2.127, determine the Thevenin's and Norton's
equivalent circuits as seen from terminals *A* and *B*.

P.2.35 Calculate, for each of the circuits shown in Figure 2.127, the value of the resistor that
connects across terminals *AB*, will dissipate maximum power.

P.2.36 For the circuit shown in Figure 2.128, calculate *I* by using

a. Node equations

b. Loop equations

c. Source transformations

d. Superposition principle

e. Thevenin's theorem

f. Norton's theorem

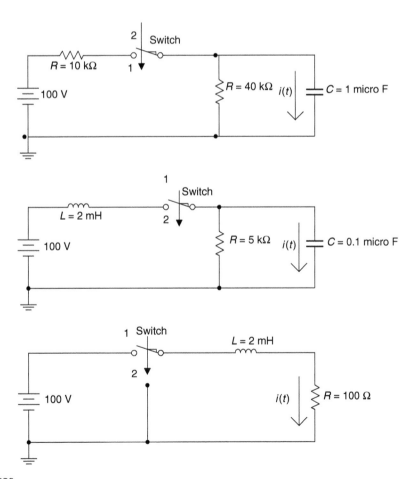

FIGURE 2.129
Networks of P.2.37.

P.2.37　For each of the circuits shown in Figure 2.129, evaluate and plot the expression for $i(t)$ for $0 < t \leq 5\tau$, if the switch is moved from position 1 to position 2 at $t = 0$ after having been in position 1 for a long time.

P.2.38　Choose appropriate values for R to obtain the three circuit responses: overdamped, critical, and underdamped. Write the differential equation, and use MATLAB to obtain the three second-order solutions, as well as their respective plots of the source-free series RLC circuit diagram shown in Figure 2.130, for $t \geq 0$.

P.2.39　Obtain and plot the transient response of the source-free RLC circuit shown in Figure 2.131—overdamped, critical, and underdamped—by choosing appropriate values for R (for $t > 0$).

P.2.40　Steady-state conditions exist for $t \leq 0$ in the circuit of Figure 2.132 (while the voltage source is connected to the circuit). At $t = 0^+$, the switch moves downward. Obtain the transient current responses and the respective plots for each of the loop currents.

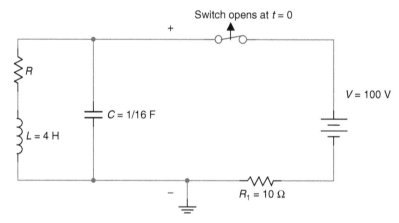

FIGURE 2.130
Network of P.2.38.

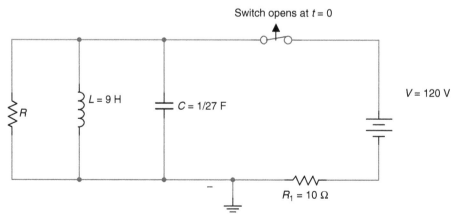

FIGURE 2.131
Network of P.2.39.

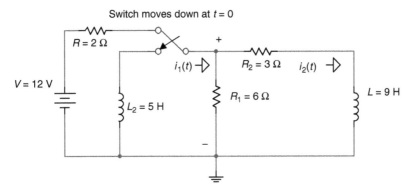

FIGURE 2.132
Network of P.2.40.

P.2.41 Steady-state conditions exist for $t \le 0$ in the circuit shown in Figure 2.133 (while the voltage source is connected to the circuit). At $t = 0^+$, the switch moves upward disconnecting the source. Obtain expressions and the respective time plots for each transient loop current.

P.2.42 Steady-state conditions exist for $t \le 0$ in the circuit of Figure 2.134 (while the voltage source is connected to the circuit). At $t = 0^+$, the switch moves downward

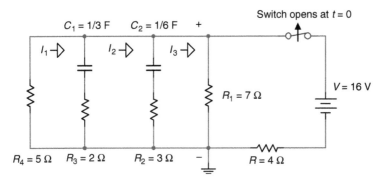

FIGURE 2.133
Network of P.2.41.

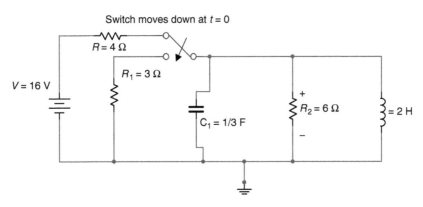

FIGURE 2.134
Network of P.2.42.

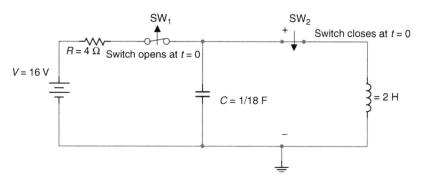

FIGURE 2.135
Network of P.2.43.

disconnecting the source. Obtain expressions for the transient voltage across C_1, the current $i_L(t)$, and their respective time plots.

P.2.43 Steady-state conditions exist for $t \leq 0$ in the circuit of Figure 2.135 (while the voltage source is connected to the circuit). At $t = 0^+$, the switch SW_1 which connects the voltage source opens, while the switch SW_2 closes. Obtain expressions for $v_C(t)$ and $i_L(t)$, and their respective time plots for $t \geq 0$.

P.2.44 Write a function file that returns the length L of a circular conductor with diameter d in meters (recall that $R = \rho L/A$), given R (ohms), ρ, and d.

P.2.45 Create a function file that returns the capacitance C (in Farads) and the strength of its electric field, given the area of the plates of the capacitor (in meters squared), the distance d between the plates (in meters), as well as the dielectric material of C.

P.2.46 Modify problem P.2.44 for the case when d is given in inches and L in foot.

P.2.47 Create a function file that returns a table and a plot of R versus d for a copper wire, where the diameter d varies over the range 0.01 in. $\leq d \leq$ 0.2 in. in steps (increments) of 0.02 in.

P.2.48 Convert the program of Example 2.4 (color code) into a function file.

P.2.49 Write a function file that requests the number and values of a set of resistors connected in series and outputs its equivalent resistance.

P.2.50 Repeat problem P.2.50 for the parallel case.

P.2.51 Write a function file that requests the value of R and n, and returns the series connections of the set of resistors with the following values:

$$R, 2R, 3R, \ldots, (n-1)R, nR$$

P.2.52 Repeat problem P.2.52 for the parallel case.

P.2.53 Write a function file that returns the Y connection, given the Δ.

P.2.54 Repeat problem P.2.55 for the Δ-Y transformation.

P.2.55 Develop a general function file that solves the set of n independent (loop or node) equations, for $2 \leq n \leq 10$.

3

Alternating Current Analysis

The task of Science is both to expand the range of our experience and to reduce it to order.

Niels Bohr

3.1 Introduction

DC is a network in which its currents and voltages have a fixed or constant magnitude and direction except when transients occur. AC networks, in contrast, are electrical networks where the currents and voltages are characterized by time-dependent alternating waveforms.

The majority of the AC signals in real life are sinusoids and they represent voltages or currents that in general are expressed instantaneously by the following equations:

$$v(t) = V_m \cos(\omega t + \theta) = V_m \sin\left(\omega t + \theta + \frac{\pi}{2}\right)$$

$$i(t) = I_m \sin(\omega t + \theta) = I_m \cos\left(\omega t + \theta - \frac{\pi}{2}\right)$$

The choice of sine or cosine-based calculations is a matter of taste, and the conversions are rather simple algebraic manipulations. However, it is imperative that the same choice be used throughout a problem.

The term alternating waveform, in general, refers to periodic waves of time-varying polarity. Therefore, alternating waveforms include a large family of waves such as square waves, saw-tooth waves, and triangular waves.

Most of the equipment and appliances used in homes, industries, and commercial and residential buildings operate with AC. The AC wave delivered to the consuming public by the utilities companies is the sinusoid and is the most common modern form of electric energy.

The three-phase AC is used in industries and commercial buildings, whereas the single phase is commonly used in homes and residences.

In general, AC principles and applications discussed in this book are used in power distribution, lighting, industrial systems, and consumer appliances and products.

The vast majority of AC problems deal with sinusoids, and because of it, mainly sinusoidal waves will be considered in this chapter. If the wave is not sinusoidal, it can always be approximated by sinusoids by means of a Fourier (trigonometric) series expansion (see Chapter 4 for additional information).

The general accepted universal AC wave model is, therefore, a sinusoid function, defined by three variables given as follows:

- V_m or I_m, the maximum or peak voltage or current
- ω the angular frequency
- θ the phase angle

Recall that the sinusoidal of the form $v(t) = V_m \cos(\omega t + \theta)$ is a periodic function with period $T = 2\pi/\omega$ therefore repeats indefinitely satisfying the time relation $f(tx) = f(tx + nT)$, for any integer value of n.

The analysis of AC circuits presented in this chapter follows mainly the model developed by Steinmetz,* which consists in reducing the AC time-dependent circuit into an equivalent phasor model.

This transformation consists basically in converting the sources and elements (impedances) into phasor representation, by using complex numbers and complex algebra (Chapter 6 of the book titled *Practical MATLAB® Basics for Engineers*) to evaluate and represent currents and voltages. The standard circuit equations, relations, and techniques developed for DC circuits are equally valid in AC, assuming that the interest is focused on the forced or steady-state response.

In most cases, the transient response decays rapidly to zero, although the force response persists indefinitely making the steady-state solution the solution of extreme practical importance.

3.2 Objectives

After completing this chapter the reader should be able to

- Express mathematically AC currents and voltages
- Express AC elements and sources in phasor format
- Understand the concepts of
  a. Instantaneous value
  b. Amplitude or peak value
  c. Peak to peak value
  d. Periodicity and period T
  e. Frequency in cycles/second (f) or radian/second (ω)
  f. Phase angle in radians and degrees (note that some of these concepts were introduced in early chapters, in particular Chapter 4 of the book titled *Practical MATLAB® Basics for Engineers* and in Chapter 1 of this book)

---

* Charles Proteus Steinmetz (1865–1923), a German–Austrian engineer, worked at the General Electric Corp (GE) and in the 1890s, he revolutionized the approach, used to analyze AC circuits, by reducing them to simple algebraic phasor equations, making the process easy and simple avoiding the complications of high level math. The method developed by Steinmetz was adapted by engineers and scientists around the world and is commonly referred as the *phasor transform method*.

- Sketch and plot sinusoidal wave forms
- State the standard analytic form to represent a sinusoidal wave
- Gain a working knowledge of the concepts of average, root-mean-square (RMS), or effective value
- Understand the concepts of leading and lagging
- Understand the relationship between current and voltage in a resistive and reactive network
- Recognize that the derivative of a sinusoid is directly proportional to its frequency (see Chapter 7 of the book titled *Practical MATLAB® Basics for Engineers*)
- Recognize that the integral of a sinusoid is inversely proportional to its frequency (see Chapter 7 of the book titled *Practical MATLAB® Basics for Engineers*)
- Calculate the currents, voltages, and reactances in a series and parallel AC circuits
- Learn and use complex numbers and notation to represent the electrical elements in a resistive/reactive (RLC)* AC system
- Recognize that the j operator is a 90° counterclockwise rotation angle, and the j represents a 90° clockwise rotation angle in the complex plane
- Understand that a phasor is a shorthand description or representation of a sinusoid wave in which the frequency is not present and it is assumed to be constant
- Solve simple series and parallel circuits using phasor techniques for currents, voltage, power, and energy
- Draw impedance diagrams
- Recognize that the voltage and current divider rule in a phasor network can be applied
- Recognize that the DC circuit theorems are equally valid in AC networks such as Thevenin, Norton, source transformation, and superposition
- Convert a Y-AC phasor network into an equivalent Δ-AC phasor network, and vice versa
- Write a set of loop and nodal equations using phasor notation
- State the maximum power theorems for the AC cases
- Know the meaning of the power factor (PF)
- Calculate the reactive power, apparent power, and effective power and construct the network power triangle
- State the conditions for the series and parallel resonant cases
- Know the meaning of bandwidth (BW), and quality factor (Q) of a resonant circuit
- Understand the concept of a three-phase (3 Φ) system
- Calculate currents and voltages in a 3 Φ system with a balanced and an unbalanced load
- Solve a set of simultaneous phasor equations (loop or node equations)
- Use many MATLAB® features in the analysis of AC circuits

---

* RLC stands for resistive, inductive, capacitive network.

3.3 Background

R.3.1   The general form of a sinusoid AC voltage is given by

$$v(t) = V_m \sin(\omega t \pm \theta_1)$$

or

$$v(t) = V_m \cos(\omega t \pm \theta_2)$$

where V_m is the maximum or peak value, given in volts, ω the angular velocity or frequency, given in radians/second, and θ_1 or θ_2 are the phase angles, expressed in radians or in degrees. (Conversions are sometimes required between radians and degrees depending on the nature of the problem.)

R.3.2   Sinusoidal waves are periodic functions, which means that the wave repeats itself indefinitely. Obviously, it can safely be stated that by knowing one cycle of a sinusoid wave the nature of the waveform for all and any time is then known.

Recall that one period or cycle of a sinusoid waveform is defined by the following formula:

$$T = \frac{2\pi}{\omega} \quad \text{(expressed in seconds)}$$

R.3.3   The frequency of a sinusoid is the reciprocal of its period, which means that if the value of the period T is known, then f is defined by

$$f = \frac{1}{T} \quad \text{(expressed in cycles per second (cps))}$$

where f is the frequency, given in Hertz or cps, and it is related to ω by $\omega = 2\pi f$, with units in radians per second.

R.3.4   A sinusoid AC signal can be expressed either by a sine or a cosine function because

$$\sin(\omega t + \theta) = \cos\left(\omega t + \theta - \frac{\pi}{2}\right)$$

or

$$\cos(\omega t + \theta) = \sin\left(\omega t + \theta + \frac{\pi}{2}\right)$$

Note that when changing a sine wave into a cosine wave or vice versa, it is necessary to change the phase angle by subtracting or adding $\pi/2$ (radians) or 90° (degrees).

R.3.5   Recall that sinusoidal waves are related to exponential functions by the Euler's identities and also be expressed by a Taylor's series as follows:

$$e^{j\theta} = \cos(\theta) + j\sin(\theta) \text{ (Euler's identity)}$$

where $j = \sqrt{-1}$ (Euler's identity) and

$$e^\theta = 1 + \theta + \frac{\theta^2}{2!} + \frac{\theta^3}{3!}, \cdots, \frac{\theta^n}{n!} \cdots \quad \text{(Taylor's series)}$$

$$\cos(\theta) = 1 - \frac{\theta^2}{2!} + \frac{\theta^4}{4!} - \frac{\theta^6}{6!} \cdots$$

$$\sin(\theta) = \theta - \frac{\theta^3}{3!} + \frac{\theta^5}{5!} - \frac{\theta^7}{7!} \cdots$$

Recall that $\cos(\theta) = real[e^{j\theta}]$ (from Chapter 5 of the book titled *Practical MATLAB® Basics for Engineers*), $\sin(\theta) = imaginary[e^{-j\theta}]$, and $|e^{j\theta}| = \sqrt{\cos^2\theta + \sin^2\theta} = 1$.

R.3.6 The average value of a given signal $x(t)$ is given by

$$X_{AVG} = \frac{1}{T}\int_0^T x(t)dt$$

assuming that $x(t)$ is a periodic wave with period T.

R.3.7 A DC voltmeter connected across an AC drop will indicate its average value.

R.3.8 The average value of a sinusoidal wave with period T is given by

$$X_{AVG} = \frac{1}{T}\int_0^T X_m \sin(\omega t)dt$$

which is also equal to

$$X_{AVG} = \frac{X_m}{2\pi}\int_0^{2\pi} \sin(\omega t)d(\omega t) = 0$$

R.3.9 The average value of a sinusoidal wave over 1/2 period (T) is given by

$$X_{AVG} = \frac{1}{T/2}\int_0^{T/2} X_m \sin(\omega t)dt$$

or by

$$X_{AVG} = \frac{X_m}{\pi}\int_0^\pi \sin(\omega t)d(\omega t)$$

which is equal to

$$X_{AVG} = \frac{2}{\pi}X_m = 0.64X_m$$

R.3.10 The RMS or effective value of the waveform $x(t)$ is given by

$$X_{RMS} = \sqrt{\frac{1}{T}\int_0^T x^2(t)\,dt}$$

R.3.11 If $x(t)$ is a sinusoidal wave, then the aforementioned relation can be evaluated and the result is given by

$$X_{RMS} = \frac{X_m}{\sqrt{2}} = 0.707\,X_m$$

Because of its importance this relation is verified as follows:
Let $x(t) = X_m \sin(\omega t)$ then

$$X_{RMS} = \sqrt{\frac{1}{T}\int_0^T [X_m \sin(\omega t)]^2\,dt}$$

$$X_{RMS} = \sqrt{\frac{1}{T}\int_0^T X_m^2\left[\frac{1}{2} - \frac{1}{2}\cos(2\omega t)\right]dt}$$

$$X_{RMS} = \sqrt{\frac{1}{T}\left[\int_0^T \frac{1}{2}X_m^2\,dt - \int_0^T \frac{1}{2}\cos(2\omega t)\,dt\right]}$$

Note that $\int_0^T (1/2)\cos(2\omega t)\,dt = 0$, then

$$X_{RMS} = \sqrt{\frac{1}{T}\int_0^T \frac{1}{2}X_m^2\,dt} = \sqrt{\frac{X_m^2}{2*T}t\Big|_0^T}$$

$$X_{RMS} = \sqrt{\frac{X_m^2 T}{2*T}} = \sqrt{\frac{X_m^2}{2}} = \frac{X_m}{\sqrt{2}} = X_m * 0.707$$

If a voltage (or current) consists of a DC and AC components, such as

$$v(t) = A + V_m \sin(\omega t)$$

then its V_{RMS} value is given by

$$V_{RMS} = \sqrt{\frac{1}{T}\int_0^T [A + V_m \sin(\omega t)]^2\,dt}$$

$$V_{RMS} = \sqrt{\frac{1}{T}\int_0^T [A^2 + AV_m \sin(\omega t) + V_m^2 \sin^2(\omega t)]\,dt}$$

$$V_{RMS} = \sqrt{\frac{1}{T}\left[\int_0^T A^2\,dt + \int_0^T AV_m \sin(\omega t)\,dt + \int_0^T V_m^2 \sin^2(\omega t)\,dt\right]}$$

Note that

$$\int_0^T AV_m \sin(\omega t)dt = 0, \quad \int_0^T A^2 dt = A^2 T$$

and

$$\int_0^T V_m^2 \sin^2(\omega t)dt = \int_0^T V_m^2 \left[\frac{1}{2} - \frac{1}{2}\cos(2\omega t)dt\right]$$

$$= \int_0^T \frac{V_m^2}{2} dt = \frac{V_m^2}{2} T$$

Then

$$V_{RMS} = \sqrt{\frac{1}{T}\left[A^2 T + \frac{V_m^2}{2} T\right]}$$

$$= \sqrt{A^2 + \frac{V_m^2}{2}} = \sqrt{A^2 + \left(\frac{V_m}{\sqrt{2}}\right)^2}$$

Note that

$$\frac{V_m}{\sqrt{2}} = V_m * 0.707 = V_{RMS-AC}$$

Then

$$V_{RMS} = \sqrt{A^2 + V_{RMS-AC}^2}, \quad \text{for } v(t) = A + V_m \sin(\omega t)$$

If

$$v(t) = A + V_{m1}\sin(\omega_1 t) + V_{m2}\sin(\omega_2 t)\cdots + V_{mn}\sin(\omega_n t)$$

then

$$V_{RMS} = \sqrt{A^2 + \sum_{k=1}^{n} \frac{V_{mk}^2}{2}} = \sqrt{A^2 + \sum_{k=1}^{n} V_{RMS-k}^2}$$

for k AC sources and one DC source (A).

R.3.12 The power dissipated by a resistor R, having a current $i(t)$ over an interval of time $[t_1, t_2]$, is given by

$$P = \frac{1}{t_2 - t_1} \int_{t_1}^{t_2} R i^2(t)\, dt$$

because $P = i(t)^2 R$, then the effective current is given by

$$I = \sqrt{\frac{1}{t_2 - t_1} \int_{t_1}^{t_2} i^2(t)\,dt}$$

R.3.13 AC measurements are usually given in RMS values. An AC current x given in RMS value, through a resistor R is equivalent to the DC current x through R, dissipating the same power (heat by R).

R.3.14 Let an AC signal be a sinusoidal function of the form $x(t) = X_m \sin(\omega t + \theta)$, then $x(t)$ can be expressed either in phasor form as $X = X_m \angle \theta$ using its peak value, or in RMS (also known as effective value) as $X = (X_m/\sqrt{2}) \angle \theta$. For example, let an RMS phasor current be given by $I = 8 \angle 30°$, then the current in the time domain representation is either $i(t) = 8\sqrt{2} \cos(\omega t + 30°)$, or $i(t) = 8\sqrt{2} \sin(\omega t + 30°)$.

As a second example, let a voltage $v(t) = 16 \sin(\omega t - 75°)$. Then $v(t)$ can be transformed into an effective (RMS) phasor voltage representation given by

$$V = \frac{16}{\sqrt{2}} \angle -75° = 8\sqrt{2} \angle -75°$$

R.3.15 The form factor ff of $x(t)$ is defined by $ff = X_{RMS}/X_{AVG}$. For a sinusoidal (current or voltage), the

$$ff = \frac{(X_m/\sqrt{2})}{(2X_m/\pi)} = \frac{\pi}{2\sqrt{2}} = 1.11$$

R.3.16 The crest factor (CF) of $x(t)$, also known as the peak factor or amplitude factor, is defined by

$$CF = \frac{X_m}{X_{RMS}}$$

For the case of a sinusoid (current or voltage) $CF = 1.41$.

R.3.17 Recall that the instantaneous power is defined by $p(t) = v(t)i(t)$.

R.3.18 Recall that the average power is defined by

$$P_{AVG} = \frac{1}{T} \int_{0}^{T} p(t)\,dt$$

assuming that $i(t)$ and $v(t)$ are the current through and voltage across a given load z, where $i(t)$ and $v(t)$ are periodic functions with the same period T (or frequency ω).

R.3.19 The PF is defined as

$$PF = \frac{P_{AVG}}{V_{RMS} \cdot I_{RMS}}$$

where

$$V_{RMS} = \sqrt{\frac{1}{T}\int_0^T v^2(t)\,dt}$$

and

$$I_{RMS} = \sqrt{\frac{1}{T}\int_0^T i^2(t)\,dt} \quad \text{and} \quad P_{AVG} = \frac{1}{T}\int_0^T p(t)\,dt$$

R.3.20 Let the current through and voltage across an arbitrary load be given by

$$i(t) = I_m \sin(\omega t)$$

and

$$v(t) = V_m \sin(\omega t + \theta) \quad \text{(an RL equivalent circuit since } v(t) \text{ leads } i(t) \text{ by } \theta)$$

Then the instantaneous power is given by

$$p(t) = i(t) \cdot v(t) = I_m V_m \sin(\omega t) \cdot \sin(\omega t + \theta)$$

Using trigonometric identities

$$p(t) = \frac{V_m I_m}{2}\cos(\theta) - \frac{V_m I_m}{2}\cos(\theta)\cos(2\omega t) + \frac{V_m I_m}{2}\sin(\theta)\sin(2\omega t)$$

where

$$\frac{V_m I_m}{2} = \frac{V_m}{\sqrt{2}}\frac{I_m}{\sqrt{2}} = V_{RMS} I_{RMS}.$$

Let $V_{RMS} I_{RMS} = A$
then, $p(t) = A\cos(\theta) - A\cos(\theta)\cos(2\omega t) + A\sin(\theta)\sin(2\omega t)$.

R.3.21 Let us explore the resistive case, where $\theta = 0°$. Then $p(t)$ of R.3.20 becomes

$$p(t) = A - A\cos(2\omega t)$$

And the average power, often referred as the real power, is given by

$$P_{AVG} = A = \frac{V_m I_m}{2} = V_{RMS} I_{RMS} \quad \text{(in watts)}$$

$$P_{AVG} = \text{real}(V_{RMS} I_{RMS}{}^*) \quad \text{(the character * denotes the complex conjugate of } I_{RMS})$$

The energy dissipated by the resistor R, in the form of heat over one full cycle, is given by

$$W_R = V_{RMS} I_{RMS} T \quad \text{(in joules)}$$

since $T = 1/f$, then

$$W_R = \frac{V_{RMS} I_{RMS}}{f} \quad \text{(in joules)}$$

R.3.22 The reactive power Q is defined as $Q = A \sin(\theta)$ with units given by volt-ampere-reactive (var)

Then,

$$Q = \left(\frac{V_m I_m}{2}\right)\sin(\theta)$$

or

$$Q = (V_{RMS} I_{RMS})\sin(\theta) = \text{imaginary}(V_{RMS} I_{RMS}{}^*)$$

where θ is the phase angle between V and I.

The reactive power Q can be either inductive or capacitive. Then,

$$Q_L = V_{RMS} I_{RMS} = I^2 X_L = V_{RMS}^2/X_L$$

$$Q_C = V_{RMS} I_{RMS} = I^2 X_C = V_{RMS}^2/X_C$$

and the respective energies are

$$W_L = \frac{V_{RMS} I_{RMS} T}{2\pi} = LI^2 \quad \text{(in joules)}$$

$$W_C = \frac{V_{RMS} I_{RMS} T}{2\pi} = CV^2 \quad \text{(in joules)}$$

R.3.23 The complex or apparent power S is defined as

$$S = V_{RMS} I_{RMS} \quad \text{with units in volt-ampere } (va)$$

$$S = I_{RMS}^2 Z = \frac{V_{RMS}^2}{Z} = P + jQ = abs(V_{RMS} \cdot I_{RMS}{}^*)$$

(Recall that * denotes the complex conjugate of I_{RMS}.)

R.3.24 The *PF* is defined by

$$PF = \cos(\theta) = P/S = R/Z$$

R.3.25 Observe that if the current through a resistance R is

$$i_R(t) = I_m \sin(\omega t)$$

Its voltage drop is then

$$v_R(t) = RI_m \sin(\omega t) = V_m \sin(\omega t)$$

where $V_m = RI_m$.

Observe that the phase angle θ between the current $i_R(t)$ and the voltage $v_R(t)$ in a resistive circuit is zero, a condition that is referred to as in phase.

R.3.26 Let the current through an inductor L be $i(t) = I_m \cos(\omega t)$, then its voltage is given by

$$v_L(t) = L\frac{di(t)}{dt} = L\frac{d}{dt}[I_m \cos(\omega t)]$$

$$v_L(t) = -\omega L I_m \sin \omega t$$

$$v_L(t) = \omega L I_m \cos\left(\omega t + \frac{\pi}{2}\right)$$

then

$$v_L(t) = V_m \cos\left(\omega t + \frac{\pi}{2}\right)$$

Clearly, if $v_L(t)$ is a sinusoidal wave, then $i_L(t)$ is also sinusoidal with the same frequency, but with a phase shift of $\pi/2$ rad.

Observe that the inductor voltage $v_L(t)$ leads its current $i_L(t)$, by an angle of $\pi/2$ rad.

R.3.27 From R.3.26, the following relations can be observed: $V_m = \omega L I_m$, then by Ohm's law, ωL is the inductive reactance or the impedance of the inductor L in ohms, expressed as $X_L(\omega) = j\omega L$, where j indicates a phase angle of $\pi/2$ rad. The inductive reactance opposes the flow of current, which results in the interchange of energy between the source and the magnetic field of the inductor.

R.3.28 In R.3.26, a current through the inductor was assumed and the voltage was then evaluated across the inductor L. The same result can be obtained by assuming a voltage across L, and solving for the current through $i_L(t)$ as illustrated as follows:

Let

$$v_L(t) = V_m \sin(\omega t)$$

then

$$i(t) = \frac{1}{L}\int v_L(t)dt = \frac{1}{L}\int V_m \sin(\omega t)dt$$

$$i(t) = \frac{V_m}{\omega L}(-\cos(\omega t)) + K \quad (\text{where } K \text{ is the initial current})$$

$$i(t) = \frac{V_m}{\omega L}\sin\left(\omega t - \frac{\pi}{2}\right) \quad (\text{assuming } K = 0 \text{ without any loss of generality})$$

Again, the reader can appreciate that by letting $X_L(\omega) = j\omega L$, then

$$I_L(\omega) = \frac{V_L(\omega)}{X_L(\omega)}$$

R.3.29 Let us assume that the voltage across the capacitor C is $v_C(t) = V_m \cos(\omega t)$ in volts, then its current is given by

$$i_C(t) = C\frac{dv_c(t)}{dt} = C\frac{d}{dt}[V_m \cos(\omega t)]$$

$$i_C(t) = -C\omega V_m \sin(\omega t) = \omega C V_m \cos\left(\omega t + \frac{\pi}{2}\right)$$

Let

$$i_C(t) = I_m \cos\left(\omega t + \frac{\pi}{2}\right)$$

then

$$I_m = \omega C V_m$$

or by Ohm's law

$$X_C(\omega) = \frac{1}{j\omega C} \quad \left(\text{where } j \text{ denotes a phase angle of } -\frac{\pi}{2}\right)$$

R.3.30 Let the voltage across C be $v_C(t) = V_m \cos(\omega t)$ (from R.3.29), then the current through C is given by $i_C(t) = I_m\cos(\omega t + (\pi/2))$. Clearly, the current leads the voltage by an angle of $\pi/2$ rad. Capacitive reactance X_C represents the opposition to the flow of charge, which results in the continuous interchange of energy between the source and the electric field of the capacitor.

R.3.31 Let the current be given by $i(t) = I_m \cos(\omega t)$, in the series RL circuit shown in Figure 3.1. Then,

$$v_R(t) = RI_m \cos(\omega t)$$

$$v_L(t) = -\omega L I_m \sin(\omega t) \quad \text{(from R.3.26) and applying KVL,}$$

$$v(t) = v_R(t) + v_L(t)$$

$$v(t) = RI_m \cos(\omega t) - \omega L I_m \sin(\omega t)$$

$$v(t) = I_m \sqrt{R^2 + (\omega L)^2} \cos(\omega t + \theta)$$

$$\theta = \tan^{-1}\left(\frac{\omega L}{R}\right) \quad \text{for } 0 \leq \theta \leq \frac{\pi}{2}$$

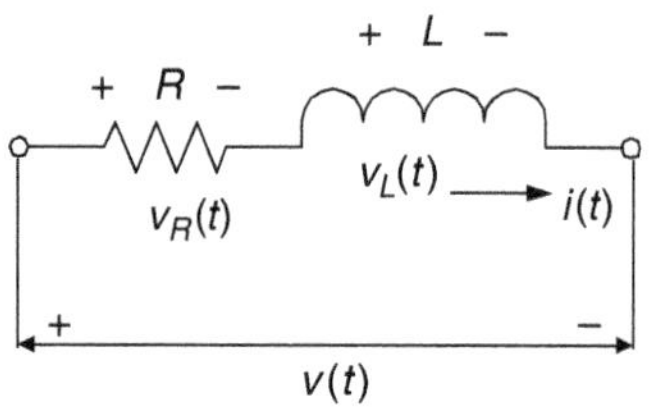

FIGURE 3.1
RL circuit diagram of R.3.31.

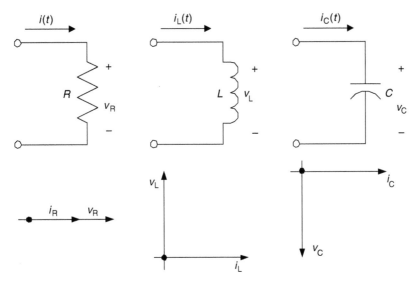

FIGURE 3.2
Phasor diagrams of R, L, and C.

or

$$v(t) = I_m \sqrt{R^2 + X_L^2} \cos(\omega t + \theta)$$

where $\theta = \tan^{-1}(X_L/R)$

Note that $|Z| = \sqrt{R^2 + X_L^2}$ (in ohms) represents the total impedance magnitude of the *RL* series circuit.

R.3.32 The behavior of the impedances of inductors and capacitors can be expressed in phasor form as

$$X_L(\omega) = j\omega L = |\omega L| \angle \left(+\frac{\pi}{2} \right)$$

$$X_C(\omega) = 1/j\omega C = |1/(\omega C)| \angle \left(-\frac{\pi}{2} \right)$$

The relation between the current and voltage for each of the three elements R, L, and C is expressed in terms of the phasor diagrams shown* in Figure 3.2.

R.3.33 The inverse of the impedance Z (G = 1/Z) is called the admittance G, with units $1/\Omega = \Omega^{-1}$ also referred to by the unit Siemen (*sie*).

R.3.34 For example, assume that the three elements R, L, and C are connected in series. Compute the equivalent impedance Z and admittance G for the following case:

$$R = 10 \, \Omega, \quad L = 0.02 \, \text{H}, \quad C = 20 \, \mu\text{F}, \quad \text{and} \quad \omega = 1000 \, \text{rad/s}$$

---

* Complex quantities are usually expressed in electrical engineering in polar form rather than in exponential form. The abbreviated complex representation is the phasor.

ANALYTICAL Solution

$$X_L = j\omega L = j*1000*0.02 = j20 \ \Omega$$

$$X_C = \frac{-j}{\omega C} = \frac{-j}{1000*20*10^{-6}} = -j50 \ \Omega$$

$$X = X_L + X_C = j20 - j50 = -j30 \ \Omega$$

Then, $Z = R + X = (20 - j30) \ \Omega$ and

$$G = \frac{1}{Z} = \frac{1}{20 - j30} = \frac{20}{20^2 + 30^2} + \frac{j30}{20^2 + 30^2} = \left(\frac{2}{130} + j\frac{3}{130}\right)\Omega^{-1} \quad (or \ sie)$$

R.3.35 For example, evaluate the voltage $v(t)$ across the series RL connection shown in Figure 3.3, assuming that the current through is given by $i(t) = I_m \sin(\omega t)$.

ANALYTICAL Solution

$$v(t) = \sqrt{R^2 + (\omega L)^2} \ I_m \sin(\omega t + \theta)$$

where $\theta = \tan^{-1}(\omega L/R)$.

The phase angle θ is positive since the voltage leads the current in an RL circuit (similar to R.3.31).

The phasor diagram is shown in Figure 3.4.

R.3.36 Let us explore now the parallel RL case. Compute the current $i(t)$ for the parallel RL circuit diagram shown in Figure 3.5, assuming that the applied voltage is given by $v(t) = V_m \cos(\omega t)$

ANALYTICAL Solution

The total admittance Y is given by

$$|Y| = \sqrt{G^2 + \left(\frac{1}{\omega L}\right)^2} \ \Omega^{-1}$$

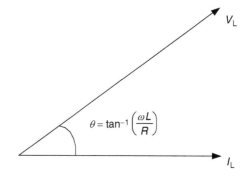

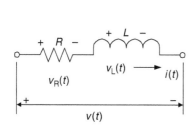

FIGURE 3.3
Circuit diagram of R.3.35.

FIGURE 3.4
Phasor diagram of R.3.35.

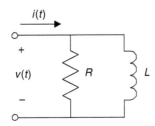

FIGURE 3.5
RL parallel circuit diagram of R.3.36.

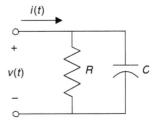

FIGURE 3.6
RC parallel circuit diagram of R.3.37.

where $G = 1/R$ then

$$i(t) = V_{m|Y|} \cos(\omega t + \theta)$$

where $\theta = \tan^{-1}(-R/\omega L)$.

Clearly, the voltage $v(t)$ across the parallel RL connection shown in Figure 3.5 leads the current $i(t)$, by a phase angle of θ rad.

R.3.37 Let us now explore the RC parallel case. Evaluate the current $i(t)$ of the parallel *RC* circuit, shown in Figure 3.6, assuming that the applied voltage is given by

$$v(t) = V_m \cos(\omega t)$$

ANALYTICAL Solution

Then the current $i(t)$ is given by $i(t) = V_{m|Y|}\cos(\omega t + \theta)$, where

$$|Y| = \sqrt{\left(\frac{1}{R}\right)^2 + (\omega C)^2} \quad \text{and} \quad \theta = \tan^{-1}(\omega CR)$$

Clearly the current $i(t)$ leads the voltage $v(t)$ by an angle θ.

R.3.38 Let us consider now a numerical example. Evaluate for the circuit diagram shown in Figure 3.7

a. The total admittance Y

b. V, I_R, and I_C

ANALYTICAL Solution

Part a

$$Y_T = \frac{1}{R} + \frac{1}{X_C} = \frac{1}{12.5} + \frac{1}{-j25} = \frac{1}{12.5} + \frac{j}{25}$$

$$Y_T = \sqrt{\left(\frac{1}{12.5}\right)^2 + \left(\frac{1}{25}\right)^2} \angle \tan^{-1}\left(\frac{12.5}{25}\right)$$

$$Y_T = 89 * 10^{-3} \angle 26.6°$$

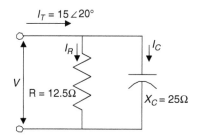

FIGURE 3.7
Network of R.3.38.

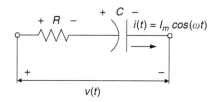

FIGURE 3.8
Network of R.3.39.

Part b

$$V = \frac{I_T}{Y_T} = \frac{15\angle 20°}{89 * 10^{-3}\angle 26.6°} = 168\angle -6.6°$$

$$I_R = \frac{V}{R} = \frac{168\angle -6.6°}{12.5\angle 0°} = 13.4\angle -6.6°$$

$$I_C = \frac{V}{X_C} = \frac{168\angle -6.6°}{25\angle -90°} = 6.72\angle 83.4°$$

R.3.39 Let us analyze now the RC series case. Evaluate the voltage $v(t)$ of the series RC circuit, shown in Figure 3.8, assuming that the current is given by

$$i(t) = I_m \cos(\omega t)$$

then

$$v(t) = I_m |Z| \cos(\omega t + \theta)$$

where

$$Z = R - \frac{j}{\omega C}$$

$$|Z| = \sqrt{R^2 + \left(\frac{1}{\omega C}\right)^2} \quad \text{and} \quad \theta = \tan^{-1}\left(-\frac{1}{\omega C R}\right) = -\tan^{-1}\left(\frac{1}{\omega C R}\right)$$

Clearly, in a series RC circuit the voltage lags the current by a phase angle bounded by $0 \le \theta \le (\pi/2)$.

R.3.40 Let us analyze the series LC circuit diagram shown in Figure 3.9, and assume that the current is given by

$$i(t) = I_m \cos(\omega t)$$

then the voltage

$$v(t) = v_L(t) + v_C(t) \quad \text{(KVL)}$$

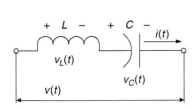

FIGURE 3.9
LC series circuit diagram of R.3.40.

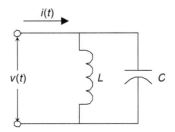

FIGURE 3.10
LC parallel circuit diagram of R.3.41.

$$v(t) = -I_m \omega L \sin(\omega t) + \frac{I_m}{\omega C} \sin(\omega t)$$

or

$$v(t) = XI_m \cos(\omega t + \theta)$$

where the total reactance is $X = \omega L - 1/(\omega C)$.

Observe that the reactance X can be positive or negative. If $X > 0$, then $v(t) = I_m X \cos(\omega t + \pi/2)$ (inductive equivalent), and if $X < 0$, then $v(t) = I_m X \cos(\omega t - \pi/2)$ (capacitive equivalent).

R.3.41   Let us turn our attention to the LC parallel case.

Evaluate the current $i(t)$ in the parallel *LC* circuit diagram shown in Figure 3.10, assuming that the applied voltage is $v(t) = V_m \cos(\omega t)$.

ANALYTICAL Solution

$$i(t) = BV_m \cos\left(\omega t + \frac{\pi}{2}\right) \quad \text{for } B > 0$$

where B represents the circuit admittance

$$i(t) = BV_m \cos\left(\omega t - \frac{\pi}{2}\right) \quad \text{for } B < 0$$

where $B = B_C - B_L$ (admittance), $B_C = \omega C$, and $B_L = 1/(\omega L)$.

R.3.42   Given a circuit where the current through and the voltage across are known. Then let us evaluate the equivalent impedance, and its phasor diagram representation for the following case:

$$i(t) = 2 \cos(60 \cdot 2 \cdot \pi \cdot t - 30°)$$

and

$$v(t) = 5 \cos(60 \cdot 2 \cdot \pi \cdot t + 45°)$$

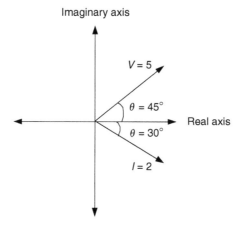

FIGURE 3.11
Phasor diagram of R.3.42.

Indicate also the type and its equivalent circuit in term of its simplest element representation.

ANALYTICAL Solution

The phasor diagram in which I and V are indicated is shown in Figure 3.11.
    The equivalent impedance can be calculated as indicated as follows:

$$Z = \frac{V}{I} = \frac{5\angle 45°}{2\angle -30°} = \frac{5}{2}\angle(45° + 30°) = 2.5\angle 75° \text{ (using peak values)}$$

its trigonometric form is $Z = 2.5\cos(75°) + j2.5\sin(75°)$ Ω.
    That means that the equivalent impedance is an RL circuit, with $R = 2.5\cos(75°)$ Ω and $L = \dfrac{25\sin(75°)}{120\pi}$ H.

R.3.43 Let us evaluate now the voltage $v(t)$ of a series RLC circuit, if the current $i(t)$ is given by $i(t) = I_m\cos(\omega t)$.

ANALYTICAL Solution

$$Z = R + j\left[\omega L - \frac{1}{\omega C}\right]$$

and

$$|Z| = \sqrt{R^2 + \left(\omega L - \frac{1}{\omega C}\right)^2}$$

The voltage across the series RLC is then given by

$$v(t) = I_m|Z|\cos(\omega t + \theta)$$

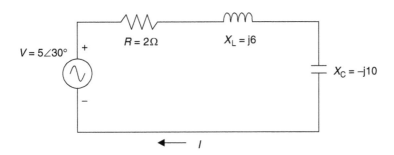

FIGURE 3.12
Network of R.3.44.

where

$$\theta = \tan^{-1}\left(\frac{\omega L - (1/\omega C)}{R}\right) \qquad \text{for } -\frac{\pi}{2} \le \theta \le +\frac{\pi}{2}$$

Clearly,

a. If $\omega L > 1/\omega C$, then θ is positive, meaning that the voltage leads the current and the equivalent circuit is RL.

b. If on the other hand, $\omega L < 1/\omega C$, then θ is negative, meaning that the voltage lags the current and the equivalent circuit is RC.

R.3.44 For example, evaluate for the circuit diagram shown in Figure 3.12 the following:

a. Z_T

b. V_R, V_L, V_C, and I as phasors

ANALYTICAL Solution

Part a

$$Z_T = 2 + j(6 - 10) = 2 - j4 \ \Omega$$

$$Z_T = \sqrt{2^2 + 4^2} \ \angle \tan^{-1}(-4/2) \ \Omega$$

$$Z_T = 4.46 \ \angle -63.4° \ \Omega$$

Part b

$$I = \frac{V}{Z_T} = \frac{5\angle 30°}{4.46\angle -63.4°} = 1.12 \ \angle(30° + 63.4°) \ \text{A}$$

$$I = 1.12 \ \angle 93.4° \ \text{A}$$

$$V_R = (1.12\angle 93.4°) * (2\angle 0°) = 2.24 \ \angle 93.4° \ \text{V}$$

$$V_L = (1.12\angle 93.4°) * (6\angle 90°) = 6.72\angle 183.4° \ \text{V}$$

$$V_C = (1.12\angle 93.4°) * (10\angle -90°) = 11.2\angle 3.4° \ \text{V}$$

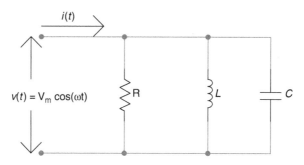

FIGURE 3.13
RLC parallel circuit diagram of R.3.45.

R.3.45 Let us evaluate now the current $i(t)$ of the parallel *RLC* circuit diagram shown in Figure 3.13, assuming that its voltage is $v(t) = V_m \cos(\omega t)V$.

ANALYTICAL Solution

Since $v(t) = I_m \cos(\omega t)$, then the current $i(t)$ can be determined by $i(t) = |Y|V_m \cos(\omega t + \theta)$, where

$$Y = \sqrt{\left(\frac{1}{R}\right)^2 + \left(\omega C - \frac{1}{\omega L}\right)^2}$$

and

$$\theta = \tan^{-1}\left(\frac{\omega C - (1/\omega L)}{1/R}\right)$$

R.3.46 Power analysis of electrical AC circuits is done by means of a right triangle called the power triangle, where the active power is given by $P = [I_{RMS}]^2 R$, the reactive power by $Q = V_{RMS} I_{RMS} \sin(\theta)$, and the apparent power by $S = V_{RMS} I_{RMS}^*$ (recall that I_{RMS}^* is the complex conjugate of I_{RMS}) (Figure 3.14). A summary of useful power relations are given as follows:

Active Power $P = I_{RMS} V_{RMS} \cos(\theta) = I_{RMS}^2 R = \dfrac{V_{RMS}^2}{R} = \text{real}\left(V_{RMS} I_{RMS}^*\right)$

Reactive Power $Q = I_{RMS} V_{RMS} \sin(\theta) = I_{RMS}^2 X = \dfrac{V_{X-RMS}^2}{X} = \text{imag}\left(V_{RMS} I_{RMS}^*\right)$

Apparent Power $S = P + jQ = \sqrt{P^2 + Q^2} \angle \tan^{-1} Q/P$

$$= I_{RMS}^2 Z = \dfrac{V_{Z-RMS}^2}{Z} = \text{abs}\left(V_{RMS} I_{RMS}^*\right)$$

Power Factor $PF = \cos(\theta) = \dfrac{R}{Z} = \dfrac{P}{S}$

R.3.47 The following example is used to illustrate the construction of the power triangle for the series circuit shown in Figure 3.15.

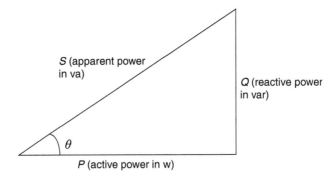

FIGURE 3.14
Power triangle.

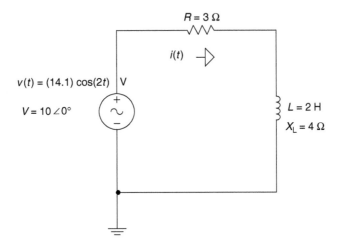

FIGURE 3.15
Network of R.3.47.

ANALYTICAL Solution

$$Z_T = R + jX_L = 3 + j4 = \sqrt{3^2 + 4^2} \angle \tan^{-1}\left(\frac{4}{3}\right)$$

$$Z_T = 5 \angle 53.13°$$

$$I = \frac{V}{Z_T} = \frac{10\angle 0°}{5\angle 53.13°} = 2\angle -53.13° \quad \text{(in RMS values)}$$

$$P \text{ (active power)} = I^2 \cdot R = 2^2 * 3 = 12 \text{ W}$$

$$Q_L \text{ (reactive power)} = I^2 * X_L = 2^2 * 4 = 16 \text{ var}$$

$$S \text{ (apparent power)} = P + jQ_L$$

$$S = 12 + j16 = \sqrt{12^2 + 16^2}\angle \tan^{-1}\left(\frac{16}{12}\right) va$$

$$S = 20 \angle 53.13° \ va$$

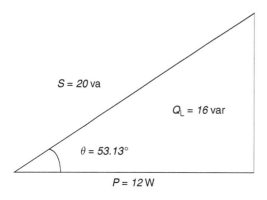

FIGURE 3.16
Power triangle of the circuit in Figure 3.15.

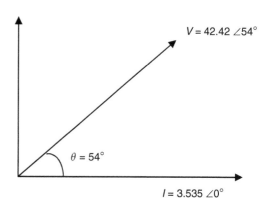

FIGURE 3.17
Phasor diagram of R.3.48.

or

$$S = V_{RMS}I_{RMS}{}^* = (10\angle 0°)(2\angle -53.13°) = 20\angle 53.13° \; va$$

The power triangle for the circuit diagram of Figure 3.15 is shown in Figure 3.16.

R.3.48 Evaluate the active power P, apparent power S, the reactive power Q, and the PF, if the impedance of a given circuit is $Z = 7.05 + j9.7 \; \Omega$ and the voltage across it is $V_m = 60$ V.

ANALYTICAL Solution

$$Z = 7.05 + j9.7 = \sqrt{7.05^2 + 9.7^2} \; \angle \tan^{-1}\left(\frac{9.7}{7.05}\right)\Omega$$

$$Z = 12\angle 54° \; \Omega$$

$$I_m = \frac{V_m}{|Z|} = \frac{60}{12} = 5 \; A$$

$$V_{RMS} = 0.707 * 60 = 42.42 \; V$$

$$I_{RMS} = 0.707 * 5 = 3.535 \; A$$

Taking the current I as reference, the phasor diagram that relates the current I to the voltage V is shown in Figure 3.17.

$$PF = \cos(54°) = 0.588$$

$$\sin(54°) = 0.809$$

then

$$S = 42.42 * 3.535 = 150 \; va$$

$$P = |S|\cos(\theta) = 150 * 0.588 = 88.1 \; W$$

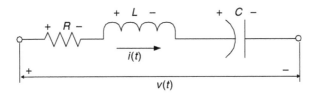

FIGURE 3.18
RLC series circuit diagram of R.3.49.

and

$$Q = |S|sin(\theta) = 150 * 0.809 = 121.2 \text{ var}$$

R.3.49 Let us revisit the *RLC* series circuit shown in Figure 3.18. The RLC circuit is said to be at resonance if the phase angle between the current *i(t)* and its voltage *v(t)* is zero. The network is then purely resistive (*R*) at this frequency and this particular frequency is referred to as the resonant frequency.

ANALYTICAL Solution

The total impedance of the series RLC circuit is given by

$$Z_T(\omega) = R + j\left[\omega L - \frac{1}{\omega C}\right]$$

and at resonant $\omega L = 1/\omega C$ or $\omega_R = 1/\sqrt{LC}$, and $fr = \omega_R/2\pi$ or $f_R = 1/(2\pi\sqrt{LC})$, where f_R is the resonant frequency, a phenomenon first observed by Thomson around 1853.

R.3.50 At the resonant frequency f_R, a series *RLC* circuit presents the following characteristics:

a. Z_T is purely resistive (phase angle between *I* and *V* is zero).

b. Z_T is at a minimum.

c. Current is at a maximum.

d. The effective power is at its maximum ($P = I^2R = V^2/R$).

R.3.51 The quality factor *Q* (do not be confused with reactive power) is a dimensionless ratio of the energy stored in the inductor to the average energy dissipated by the resistor, given by

$$Q = \frac{\text{inductive reactive power}}{\text{average power}}$$

The quality factor *Q* for an *RLC* series circuit is then given by

$$Q_S = \frac{\omega_R L}{R} = \frac{L}{2\sqrt{LC}} = \frac{X_L}{R}$$

and the dissipation factor *D* is defined as $D = 1/Q$ or $D = \frac{R}{\omega_R L}$ or $D = \omega_R RC$.

R.3.52 The BW and cutoff frequencies for the series resonant *RLC* circuit can be evaluated by the following equations:

$$\omega_1 = \omega_R\sqrt{1 + \left(\frac{1}{2Q_S}\right)^2} - \frac{\omega_R}{2Q_S} \text{ rad/s}$$

and

$$\omega_2 = \omega_R \sqrt{1 + \left(\frac{1}{2Q_S}\right)^2} + \frac{\omega_R}{2Q_S} \text{ rad/s}$$

if $Q_S > 10$ then $\omega_1 = \omega_R \left(1 - \frac{1}{2Q_S}\right)$ rad/s, and $\omega_2 = \omega_R \left(1 + \frac{1}{2Q_S}\right)$ rad/s

where the *BW* is defined as $BW = \omega_2 - \omega_1$ or by $BW = \omega_R/Q$. Recall that the BW represents the range of frequencies present in the output with a significant power content.

The frequency ω_R is the geometric mean of the two frequencies given by ω_1 and ω_2 (or f_1 and f_2) as

$$\omega_R = \sqrt{\omega_1 \cdot \omega_2}$$

or

$$f_R = \sqrt{f_1 \cdot f_2}$$

R.3.53 Let us analyze now the simple parallel *RLC* circuit, shown in Figure 3.19, at resonance.

The resonant frequency for the parallel case is still given by

$$f_R = \frac{1}{2\pi\sqrt{LC}}$$

The quality factor Q is given by

$$Q_P = \omega_R CR = \frac{R}{\omega_R L} = \frac{\omega_2 - \omega_1}{\omega_R}$$

and the BW cutoff frequencies are given by the following equations:

$$\omega_1 = \sqrt{\omega_R^2 - \left(\frac{1}{2RC}\right)^2} - \frac{1}{2RC} = \frac{1}{2C}\left[\frac{1}{R} - \sqrt{\frac{1}{R^2} + \frac{4C}{L}}\right]$$

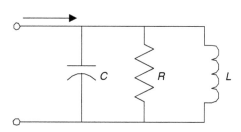

FIGURE 3.19
RLC parallel circuit diagram of R.3.53.

and

$$\omega_2 = \sqrt{\omega_R^2 + \left(\frac{1}{2RC}\right)^2} + \frac{1}{2RC} = \frac{1}{2C}\left[\frac{1}{R} + \sqrt{\frac{1}{R^2} + \frac{4C}{L}}\right]$$

The *BW* is then given by

$$BW = \omega_2 - \omega_1 = \frac{1}{2\pi}\left(\frac{1}{RC}\right) = \frac{\omega_R}{Q_P}$$

R.3.54 A parallel RLC circuit presents the following characteristics at resonance:
   a. Z_T is at a maximum.
   b. $Z_T = R$, since the angle between I and V is zero.
   c. Current is at a minimum.
   d. The effective power is at its minimum.

R.3.55 Practical resonant circuits are constructed by placing a capacitor and an inductor in parallel as shown in Figure 3.20, where R_1 and R_2 are the internal resistances of L and C, respectively. Recall that the condition at resonance is that the complex admittance Y is a real number, then

$$\frac{X_C}{R_2^2 + X_C^2} = \frac{X_L}{R_1^2 + X_L^2}$$

and the resonant frequency is given by

$$\omega_R = \frac{1}{\sqrt{LC}}\sqrt{\frac{R_1^2 - (L/C)}{R_2^2 - (L/C)}}$$

Since ω_R is real, $R_1^2 > L/C$ and $R_2^2 > L/C$.

R.3.56 Recall that the process used in the mesh or loop equations techniques was presented and discussed in Chapter 2, for the purely resistive DC case. The theory developed

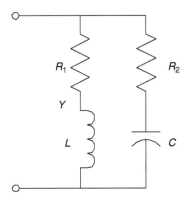

FIGURE 3.20
Network of R.3.55.

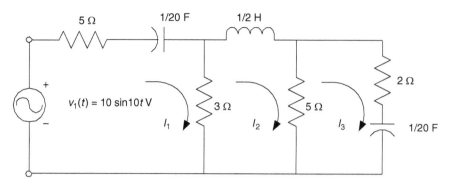

FIGURE 3.21
Network of R.3.57.

for DC circuits can easily be extended to include AC circuits. To use the mesh analysis technique it is required that all the forcing functions be sinusoidals (either sines or cosines) having the same frequency. The mesh equations are then expressed in terms of phasor currents, voltages, and impedances. The passive elements such as resistances, capacitances, and inductances can be combined using the standard complex algebra, presented in Chapter 6 of the book titled *Practical MATLAB® Basics for Engineers.*

R.3.57   The circuit diagram shown in Figure 3.21 is used to illustrate the general approach in the construction of the loop equations. First, we recognize the existence of three independent loops, then loop currents are assigned to each loop, thus we write the three independent loop equations and finally solve for each unknown loop current $(I_1, I_2, \text{ and } I_3)$, as illustrated as follows:

ANALYTICAL Solution

Observe that $\omega = 10$ rad/s (from the forcing function), the three loop currents are labeled assuming a clockwise direction, then all the network reactances are evaluated as indicated as follows:

$$X_C = \frac{1}{j\omega C} = \frac{1}{j(10)(1/20)} = -j2$$

$$X_L = j\omega L = j(10)\frac{1}{2} = j5$$

The circuit of Figure 3.21 is then redrawn with the network elements, replaced by their respective impedances and the source voltage, by a phasor, as shown in Figure 3.22. The three independent loop equations are then given by

$$(8 - j2)I_1 - 3I_2 + 0I_3 = 10$$

$$-3I_1 + (8 + j5)I_2 - 5I_3 = 0$$

$$+0I_1 - 5I_2 + (7 - j2)I_3 = 0$$

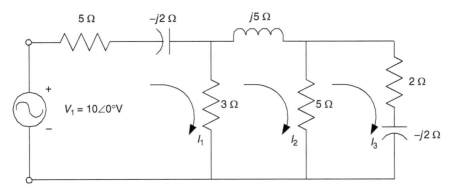

FIGURE 3.22
Phasor network of R.3.57.

The matrix loop equation is indicated as follows:

$$\begin{bmatrix} 8-j2 & -3 & 0 \\ -3 & 8+j5 & -5 \\ 0 & -5 & 7-j2 \end{bmatrix} \cdot \begin{bmatrix} I_1 \\ I_2 \\ I_3 \end{bmatrix} = \begin{bmatrix} 10 \\ 0 \\ 0 \end{bmatrix}$$

and the currents I_1, I_2, and I_3 are evaluated as indicated as follows:

$$I_1 = V_1\left(\frac{\det(I_1)}{\det(Z)}\right) \qquad I_2 = V_1\left(\frac{\det(I_2)}{\det(Z)}\right) \qquad I_3 = V_1\left(\frac{\det(I_3)}{\det(Z)}\right)$$

where

$$\det(Z) = \begin{bmatrix} 8-j2 & -3 & 0 \\ -3 & 8+j5 & -5 \\ 0 & -5 & 7-j2 \end{bmatrix} = 315\angle 16.2°$$

$$\det(I_1) = \begin{bmatrix} 8+j5 & -5 \\ -5 & 7-j2 \end{bmatrix} = 41.1\angle 24.9°$$

$$\det(I_2) = (-1)\begin{bmatrix} -3 & -5 \\ 0 & 7-j2 \end{bmatrix} = 21.8\angle -16°$$

$$\det(I_3) = \begin{bmatrix} -3 & 8+j5 \\ 0 & -5 \end{bmatrix} = 15\angle 0°$$

$$I_1 = 10\angle 0°\frac{41.1\angle 24.9°}{315\angle 16.2°} = 1.43\angle 8.7°\,\text{A}$$

$$I_2 = 10\angle 0°\frac{21.8\angle -16°}{315\angle 16.2°} = 0.693\angle -32.2°\,\text{A}$$

$$I_3 = 10\angle 0°\frac{15\angle 0°}{315\angle 16.2°} = 0.476\angle -16.2°\,\text{A}$$

The calculations for the currents I_1, I_2, and I_3 are verified using MATLAB as follows:

```
>>Z = [8 - 2j - 3 0; -3 8 + 5j - 5; 0 - 5 7 - 2j]
  Z =
      8.0000 - 2.0000i          -3.0000                  0
     -3.0000                8.0000 + 5.0000i          -5.0000
          0                     -5.0000          7.0000 - 2.0000i
>>V = [10;0;0]
  V =
      10
       0
       0
>>I = inv(z)*V
  I =
      1.4158 + 0.2159i
      0.5861 - 0.3682i
      0.4565 - 0.1326i
>>Current _ magnitude = abs(I)
  Current _ magnitude =
                      1.4322
                      0.6922
                      0.4754
>>Phase _ angle = angle(I)*180/pi
  Phase _ angle =
                      8.6689
                    -32.1402
                    -16.1948
```

Then the instantaneous currents are

$$i_1(t) = 1.43\sin(10t + 8.7°)\ A$$

$$i_2(t) = 0.693\sin(10t - 32.2°)\ A$$

$$i_3(t) = 0.476\sin(10t - 16.2°)\ A$$

R.3.58 The node equations technique presented in Chapter 2 for the case of the resistive DC networks can be extended to include AC elements (in phasor form), following the same steps outlined for the AC loop equations.

The following example illustrates the general approach for the circuit diagram shown in Figure 3.23. Note that this circuit has three nodes where one is grounded (reference). Then only two node equations are required, labeled E_1 and E_2.

ANALYTICAL Solution

Observe that the source frequency is $\omega = 10\ rad/s$. Then replacing all the elements by their respective admittances the following is obtained

$$R_1 = 1/10 \quad \text{and} \quad Y_1 = 10$$

$$R_2 = 1/3 \quad \text{and} \quad Y_2 = 3$$

$$X_L = j\omega L = j(10)\frac{1}{50} = j\frac{1}{5} \qquad Y_L = \frac{1}{j(1/5)} = -j5$$

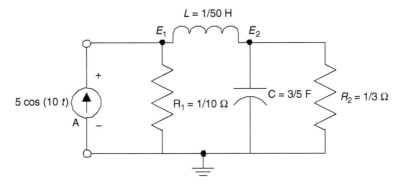

FIGURE 3.23
Network of R.3.58.

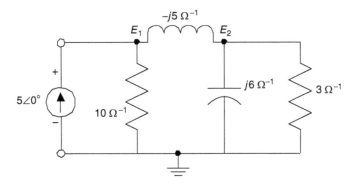

FIGURE 3.24
Phasor network of R.3.58.

and

$$X_C = \frac{1}{j\omega C} = \frac{1}{j(10)(3/5)} = \frac{1}{j6} \qquad Y_C = +j6$$

The circuit of Figure 3.23 is then redrawn with the elements replaced by its equivalent admittances and the current source by its phasor representation, as indicated in Figure 3.24.

The two node equations are

For node E_1, $5\angle 0° = (10 - j5)E_1 + j5E_2$

For node E_2, $0 = j5E_1 + (j + 3)E_2$

The corresponding matrix equation is shown as follows:

$$\begin{bmatrix} (10 - j5) & j5 \\ j5 & (j + 3) \end{bmatrix} \cdot \begin{bmatrix} E_1 \\ E_2 \end{bmatrix} = \begin{bmatrix} 5\angle 0° \\ 0 \end{bmatrix}$$

Solving for E_1 and E_2 by hand yields

$$E_1 = \frac{\begin{bmatrix} 5\angle 0° & j5 \\ 0 & j+3 \end{bmatrix}}{\begin{bmatrix} 10-j5 & j5 \\ j5 & j+3 \end{bmatrix}} = \frac{15+5j}{60-5j} = 0.2626\angle 23.1986°$$

$$E_2 = \frac{\begin{bmatrix} 10-j5 & 50\angle 0° \\ j5 & 0 \end{bmatrix}}{\begin{bmatrix} 10-j5 & j5 \\ j5 & j+3 \end{bmatrix}} = \frac{-25j}{60-5j} = 0.4152\angle -85.236$$

The voltages E_1 and E_2 are verified by using MATLAB as follows:

```
>>Y = [10 - 5j - 5j:5j j + 3]
  Y =
     10.0000 - 5.0000i        0 + 5.0000i
            0 + 5.0000i   3.0000 + 1.0000i
>>l = [5:0]
  l =
       5
       0
>>node _ voltage = inv(Y)*l
  node _ voltage =
                 0.2414 + 0.1034i
                 0.0345 - 0.4138i
>>magnitude _ voltages = abs(node _ voltage)
  magnitude _ voltages =
                     0.2626
                     0.4152
>>phase _ voltages = angle(node _ voltage)*180/pi
phase _ voltages =
                   23.1986
                  -85.2364
```

The instantaneous voltages are then given by

$$e_1(t) = 0.2626 \cos(10t + 23.1986°) \text{ V}$$

$$e_2(t) = 0.4152 \cos(10t - 85.2364°) \text{ V}$$

R.3.59 The Thevenin's and Norton's theorems developed for DC discussed in Chapter 2 can be extended to include the AC case. Recall that the Thevenin's theorem states that any linear network with a load connected to terminals aa' can be replaced by a voltage source called the Thevenin's voltage V_{TH} in series with an impedance Z_{TH}, where the V_{TH} is the open circuit voltage measured or calculated across terminals aa' (by removing the load), and Z_{TH} is the impedance across terminals aa' when all sources are set to zero (with no load).

Recall that the Norton's theorem states that any linear network can be replaced by a current source I_N, which is the short circuit current across aa', and the Thevenin's impedance Z_{TH}, connected in parallel.

R.3.60 Thevenin's as well as Norton's equivalent circuits can be evaluated if all the network sources have the same frequency, and they are expressed as either sines or cosines.

R.3.61 The example shown in Figure 3.25 uses the Thevenin's theorem to calculate the current I_L, assuming that the load is $Z_L = 4 - j4\ \Omega$.

ANALYTICAL Solution

Source transformation as presented in Chapter 2 can be extended to AC circuits as indicated as follows:

Note that the current source I, in parallel with the impedance Z_2, can be transformed into a voltage source in series with Z_2 as shown as follows (see R.3.62):

Therefore,

$$E_3 = I * Z_2 = (6 + j3)(8 - j4) = 60\ V$$

The circuit of Figure 3.25 is transformed into the circuit shown in Figure 3.26, where the load Z_L is removed, following Thevenin's theorem. Then,

$$I_1 = \frac{40 + 60}{16} = \frac{100}{16} = \frac{25}{4}A = 6.25\ A$$

Therefore, $V_{TH} = (8 - j4)6.25 + j30 - 60 = -10 + j5\ V$

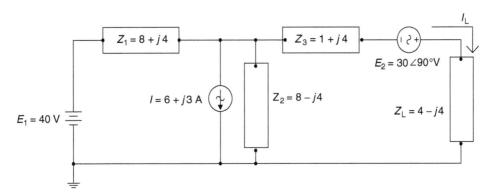

FIGURE 3.25
Network of R.3.61.

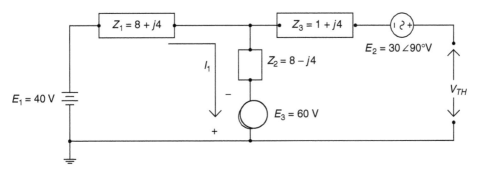

FIGURE 3.26
Thevenin's model of the circuit diagram of Figure 3.25.

and

$$Z_{TH} = 1 + j4 + \frac{(8 + j4)(8 - j4)}{8 + j4 + 8 - j4} = 6 + j4 \,\Omega$$

(by replacing the voltage sources by shorts).

The Thevenin's equivalent circuit is shown in Figure 3.27.

Then,

$$I_L = \frac{V_{TH}}{Z_L + Z_{TH}} = \frac{-10 + j5}{10} = -1 + j0.5 \text{ A}$$

R.3.62 Note that source transformation concept for the DC case discussed in Chapter 2 can be extended to include the AC. Note that source transformation was already employed in the example presented in R.3.61. Observe that a current source of $I = 6 + j3$ with a parallel impedance of $Z_2 = 8 - j4$ was converted into a voltage source of *60* V ($Z_2 * I$) in series with Z_2 (Figure 3.25).

R.3.63 The circuit shown in Figure 3.28 is used to illustrate Norton's theorem and source transformation in evaluating the current I_L through the load Z_L.

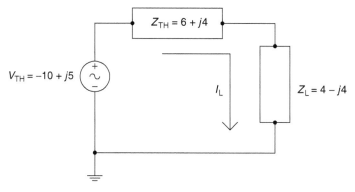

FIGURE 3.27
Thevenin's equivalent circuit of Figure 3.25.

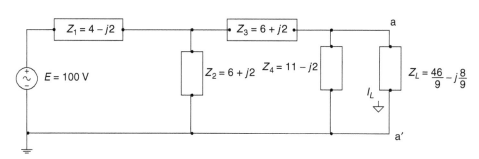

FIGURE 3.28
Network of R.3.63.

ANALYTICAL Solution

Replacing Z_L by a short (Norton's theorem) results in the circuit diagram shown in Figure 3.29.

Observe that Z_2 and Z_3 are connected in parallel, then

$$Z_5 = Z_2 \| Z_3 = Z_2/2 = 3 + j$$

The circuit of Figure 3.29 is further simplified and redrawn in Figure 3.30.

Then,

$$I = \frac{100}{7 - j} = 2(7 + j) \text{ A}$$

and

$$I_N = \frac{I}{2} = 7 + j \text{ A}$$

Then Z_{TH} can be evaluated from the circuit shown in Figure 3.31, where the voltage source of Figure 3.28 is replaced by a short circuit.

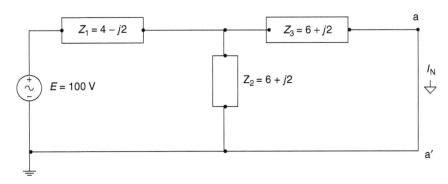

FIGURE 3.29
Norton's model of the circuit diagram of Figure 3.28.

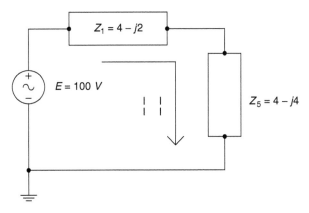

FIGURE 3.30
Simplified version of Figure 3.29.

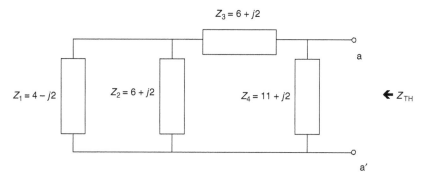

FIGURE 3.31
Z_{TH} model for the circuit diagram of Figure 3.28.

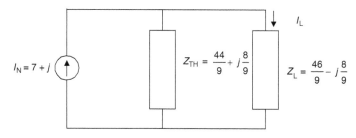

FIGURE 3.32
Norton's equivalent circuit of Figure 3.28.

Then,

$$Z_{TH} = [(Z_1 \| Z_2) + Z_3] \| Z_4 \, \Omega$$

$$Z_1 \| Z_2 + Z_3 = \frac{(6 + j2)(4 - j2)}{10} + 6 + j2 = 8.8 + j1.6 \, \Omega$$

and

$$Z_{TH} = \frac{(8.8 + j1.6) * (11 + j2)}{19.8 + j3.6} = \frac{44}{9} + j\frac{8}{9} \, \Omega$$

Finally, the Norton's equivalent circuit is shown in Figure 3.32, and the Thevenin's equivalent circuit is shown in Figure 3.33. The current I_L can then easily be evaluated from either circuit, as illustrated as follows:

$$I_L = \frac{((44/9) + j(8/9)) * (7 + j)}{(44/9) + j(8/9) + (46/9) - j(8/9)} \, \text{A} \quad \text{(from Figure 3.32)}$$

$$I_L = \frac{(300/9) + j(100/9)}{10} = \frac{30}{9} + j\frac{10}{9} \, \text{A} \quad \text{(from Figure 3.33)}$$

$$Z_{TH} = \frac{44}{9} + j\frac{8}{9} \, \Omega$$

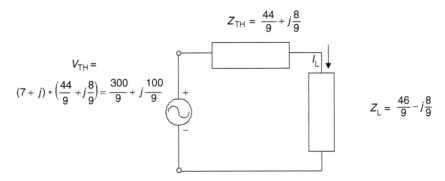

FIGURE 3.33
Thevenin's equivalent circuit of Figure 3.28.

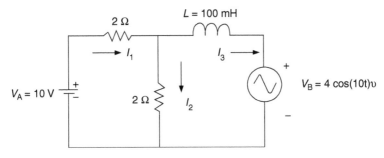

FIGURE 3.34
Network of R.3.65.

R.3.64 The superposition principle for the AC case follows closely to the DC case discussed in Chapter 2, that is, in a linear network containing n sources (not necessarily with the same frequencies) any network current or voltage is the algebraic sum of the responses or contributions due to each of the n sources acting separately by forcing all the remaining $n - 1$ sources to zero.

R.3.65 The circuit shown in Figure 3.34 uses the superposition principle to evaluate each current (I_1, I_2, and I_3). Observe that one source is DC and the other is AC (with $\omega = 10$ rad/s).

ANALYTICAL Solution

The circuit of Figure 3.34 is redrawn into the circuit shown in Figure 3.35, by setting $V_B = 0$ (short), and solving for all the currents contributed by the voltage source V_A.
Note that because $\omega = 0$, $X_L = j\omega L = 0$ (short), then the currents I_{A1}, I_{A2}, and I_{A3} are

$$I_{A1} = \frac{10}{2} = 5 \text{ A}$$

$$I_{A2} = 0 \text{ A}$$

$$I_{A3} = 5 \text{ A}$$

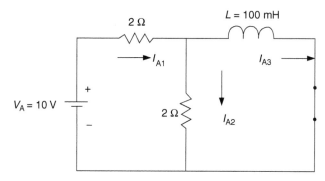

FIGURE 3.35
Network of Figure 3.34 with $V_B = 0$.

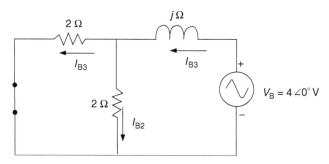

FIGURE 3.36
Network of Figure 3.34 with $V_A = 0$.

Now consider the AC source by setting the DC source to zero ($V_A = 0$), and transforming the source V_B into a phasor, the equivalent circuit is redrawn in Figure 3.36.

Solving for the loop currents I_{B1} and I_{B3} of Figure 3.36, using loop equations the following relations are obtained

$$I_{B3} = \frac{4 \angle 0°}{i + j} = \frac{4 \angle 0°}{\sqrt{2} \angle 45°} = 2\sqrt{2} \angle{-45°} \text{ A}$$

$$I_{B1} = I_{B2} = \frac{I_{B3}}{2}$$

Then

$$I_{B1} = \sqrt{2} \angle{-45°}$$

$$I_{B2} = I_{B1} = \sqrt{2} \angle{-45°}$$

Transforming the aforementioned phasor equations into the instantaneous currents results in

$$i_{B1}(t) = \sqrt{2} \cos(10t - 45°) \text{ A}$$

$$i_{B2}(t) = \sqrt{2} \cos(10t - 45°) \text{ A}$$

$$i_{B3}(t) = 2\sqrt{2} \cos(10t - 45°) \text{ A}$$

Then the currents of the circuit diagram shown in Figure 3.34 are evaluated by the algebraic addition of the individual contributions of each source (superposition principle), yielding the following:

$$i_1(t) = I_{A1} + i_{B1}(t)$$

$$i_1(t) = 5 - \sqrt{2}\cos(10t - 45°) \text{ A}$$

$$i_2(t) = I_{A2} + i_{B2}(t)$$

$$i_2(t) = \sqrt{2}\cos(10t - 45°) \text{ A}$$

$$i_3(t) = I_{A3} - i_{B3}(t)$$

$$i_3(t) = 5 - 2\sqrt{2}\cos(10t - 45°) \text{ A}$$

R.3.66 Let the Thevenin's impedance as seen across an arbitrary load Z_L be $Z_{TH} = R_{TH} + jX_{TH}$. Then maximum power is delivered to the load Z_L, when $Z_L = R_{TH} - jX_{TH}$ (note that Z_L is the complex conjugate of Z_{TH}), as illustrated in Figure 3.37.

Let us evaluate the power delivered (in watts) to the load Z_L.

The total impedance of the circuit is given by

$$Z_T = Z_{TH} + Z_L = 2R_{TH}$$

Then,

$$I = \frac{V_{TH}}{2R_{TH}} \qquad P_{RL} = I^2 R_{TH}$$

and

$$P_{ZL} = \left(\frac{V_{TH}}{2R_{TH}}\right)^2 R_{TH} = \frac{V_{TH}^2}{4R_{TH}} \text{ W}$$

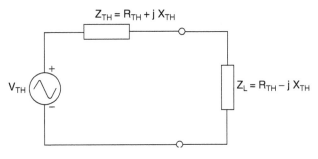

FIGURE 3.37
Condition for maximum power transfer to Z_L.

R.3.67 Let the load impedance of an arbitrary network be $Z_L = R_L + jX_L$, where R_L can be adjusted (variable), and let the load reactance X_L be fixed, where $X_L \neq X_{TH}$. Then maximum power delivered to the load Z_L occurs when R_L is (adjusted) to

$$R_L = \sqrt{R_{TH}^2 + (X_{TH} + X_L)^2}$$

and the power delivered to the load is then

$$P_{RL} = \frac{V_{TH}^2 R_X}{4}$$

where

$$R_X = \frac{R_{TH} + R_L}{2}$$

R.3.68 A Y impedance structure can be transformed into a Δ equivalent impedance structure (refer to Figure 3.38) by the following set of equations:

$$Z_1 = \frac{Z_A Z_B + Z_B Z_C + Z_C Z_A}{Z_C}$$

$$Z_2 = \frac{Z_A Z_B + Z_B Z_C + Z_C Z_A}{Z_B}$$

$$Z_3 = \frac{Z_A Z_B + Z_B Z_C + Z_C Z_A}{Z_A}$$

Observe that a Δ configuration is a structure consisting of three nodes (A, B, and C) and three elements Z_1, Z_2, and Z_3, where each node is the connection point of two elements. For example, node A is the connection point of Z_1 and Z_2.

A Y configuration is a four-node structure, where each of the Δ nodes (A, B, and C) is connected to one element Z_A, Z_B, and Z_C, and the fourth node (the center node) is the connection point of the three elements Z_A, Z_B, and Z_C.

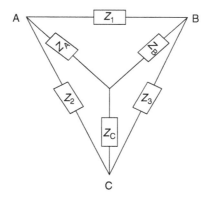

FIGURE 3.38
Y–Δ structures.

R.3.69 The transformation from Δ to Y is accomplished by the following set of equations:

$$Z_A = \frac{Z_1 Z_2}{Z_1 + Z_2 + Z_3}$$

$$Z_B = \frac{Z_1 Z_3}{Z_1 + Z_2 + Z_3}$$

$$Z_C = \frac{Z_2 Z_3}{Z_1 + Z_2 + Z_3}$$

Note that the Δ to Y and the Y to Δ transform equations for the AC case can be extended to the DC case, where the (Rs) resistances are replaced by (Zs) impedances.

R.3.70 Polyphase systems consist of two or more AC voltage sources with the same frequency, and with fixed differences in phase, connected in either a Δ or Y configuration.

R.3.71 A single-phase generator outputs a sinusoidal voltage for each rotation of its rotor. When more than one sinusoidal is generated by the system, then it is referred as a polyphase system. Multiple-phase voltages are in general more efficient to generate, transmit, and distribute electrical power than single phase, which results in copper reduction (thinner conductors) of up to 25%, and substantial saving in transmission lines and structures (transmission towers). The most common multiphase system is the three-phase system, denoted by 3 Φ.

R.3.72 A generator consists of a rotating shaft or rotor with coils around it that moves in a constant magnetic field. This arrangement results in an induced voltage in each coil. Due to the location of the windings as well as the number of turns, A 3 Φ system is a system that outputs three sinusoidal waves, *120°* apart in phase, with the same magnitudes. The generating frequency depends on the number of poles, as well as the angular speed of the rotor. The standard commercial and residential frequencies are 60, 50, and 400 Hz in the United States; Europe; and insulated, independent large-scale systems such as ships, aircrafts, and satellites, respectively.

R.3.73 The power distribution of a *3 Φ* system if implemented independently would need six wires as transmission lines, one per phase. It is far more efficient to interconnect the winding in a way as to reduce the number of transmission lines into two structures—Δ or Y (also known as star) configuration, reducing the number of transmission lines, and the resulting implementation cost.

R.3.74 A monophase generator is one that results in one or more outputs, but all of them with the same frequency and with the same phase.

R.3.75 A single-phase system that is often used in residential installations consists of three (secondary) wires and a transformer is illustrated in Figure 3.39.

R.3.76 Another popular connection consists of a two-phase (2 Φ) system with three wires connected to two loads Z_1 and Z_2 as shown in Figure 3.40, with its corresponding phasor diagram shown in Figure 3.41.
Where $I_N = I_1 - I_2$ and $E_{12} = E\sqrt{2} \angle{-45°}$.

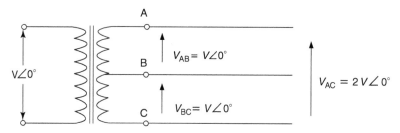

FIGURE 3.39
Single-phase system.

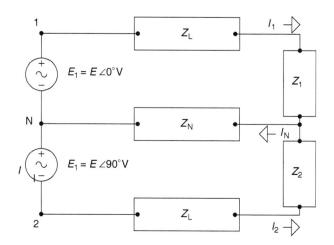

FIGURE 3.40
Two-phase system.

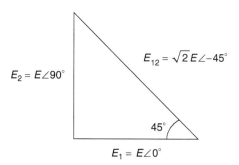

FIGURE 3.41
Phasor diagram of Figure 3.40.

R.3.77 $3\,\Phi$ Systems are the most common configurations used to generate and deliver electrical power. A $3\,\Phi$ system, with sequence ABC is shown in Figure 3.42, where the sequence label indicates the generation path.

R.3.78 The line voltages are the voltages V_{AN}, V_{BN}, and V_{CN}.

R.3.79 The system voltages are referred to as the voltage between any pair of line voltages such as V_{AB}, V_{BC}, and V_{CA}.

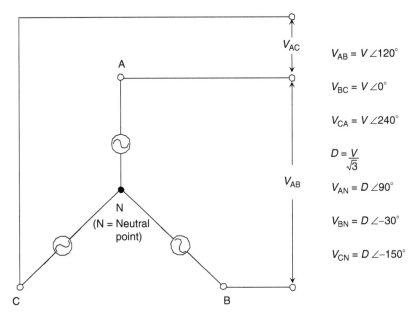

$V_{AB} = V \angle 120°$

$V_{BC} = V \angle 0°$

$V_{CA} = V \angle 240°$

$D = \dfrac{V}{\sqrt{3}}$

$V_{AN} = D \angle 90°$

$V_{BN} = D \angle -30°$

$V_{CN} = D \angle -150°$

FIGURE 3.42
3 Φ system with sequence ABC.

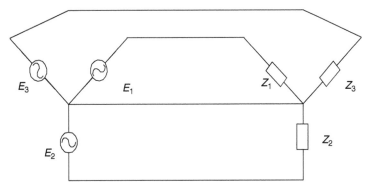

FIGURE 3.43
3 Φ–Y system connected to a Y load.

R.3.80 The system voltages, as well as the loads in a 3 Φ system can be connected in either a Δ or a Y configuration, as illustrated in Figures 3.43 through 3.46, where $E_1 + E_2 + E_3 = 0$.

R.3.81 A balance system consists of a system where the load impedances are $Z_1 = Z_2 = Z_3$ in a Y configuration, or when $Z_4 = Z_5 = Z_6$ in a Δ configuration; refer to Figures 3.42 through 3.46.

R.3.82 Standard circuit techniques such as loop or node equations can be used to analyze polyphase systems (to determine currents or voltages).

R.3.83 When a 3 Φ system is connected to an unbalanced Y load, the neutral nodal point referred by N is not at the neutral potential, and is labeled with an O. The change in voltage between the balanced and the unbalanced case is referred as V_{ON}, and is referred to as the neutral displacement voltage.

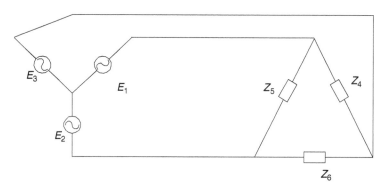

FIGURE 3.44
3 Φ–Y system connected to a Δ load.

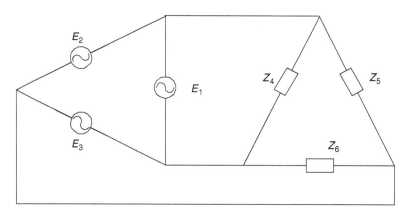

FIGURE 3.45
3 Φ–Δ system connected to a Δ load.

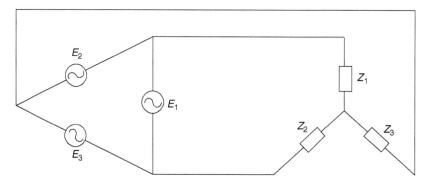

FIGURE 3.46
3 Φ–Δ system connected to a Y load.

R.3.84 The magnitude of the line currents in a balanced 3 Φ system, as well as the voltage drops across the loads are equal, and the phases are evenly spaced by 120°.

R.3.85 The magnitude of the voltages and currents in an unbalanced 3 Φ circuit can vary considerably across the load in which the line currents are not equal and the phases are no longer evenly spaced by 120°.

R.3.86 The example shown in Figure 3.47 illustrates the general approach employed to solve for all the currents (load and line) in a typical polyphase balanced network. The voltage specs are

$$V_{12} = 120 \angle 0°$$

$$V_{23} = 120 \angle -120°$$

$$V_{31} = 120 \angle 120° \quad \text{(in volts)}$$

The analysis shown in the following is for the load specs for the following structures:

a. A Δ impedance configuration shown in Figure 3.47 with

$$Z_{12} = Z_{23} = Z_{31} = j10 \ \Omega$$

b. A Y impedance configuration shown in Figure 3.48 with

$$Z_1 = Z_2 = Z_3 = j10 \ \Omega$$

ANALYTICAL Solution

Part a

Applying Ohm's law to the circuit shown in Figure 3.47, the following currents can be evaluated by

$$I_{12} = \frac{V_{12}}{Z_{12}} = \frac{120 \angle 0°}{j10} = \frac{120 \angle 0°}{10 \angle 90°} = 12 \angle -90° \ \text{A}$$

$$I_{23} = \frac{V_{23}}{Z_{23}} = \frac{120 \angle -120°}{10 \angle 90°} = 12 \angle(-120° - 90°) = 12 \angle -210° \ \text{A}$$

$$I_{23} = 12 \angle 150°$$

$$I_{31} = \frac{V_{31}}{Z_{31}} = \frac{120 \angle 120°}{10 \angle 90°} = 12 \angle(120° - 90°) = 12 \angle 30° \ \text{A}$$

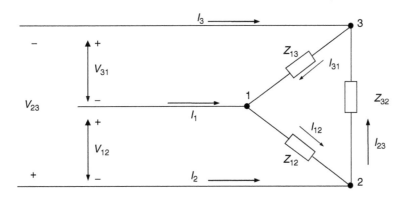

FIGURE 3.47
3 Φ system connected to a Δ load with $Z_{12} = Z_{23} = Z_{31} = j10$.

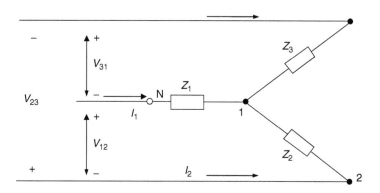

FIGURE 3.48
3 Φ system connected to a Y load with $Z_1 = Z_2 = Z_3 = 10j$.

Then, $I_1 = I_{12} - I_{31}$

$$I_1 = 12\angle-90° - 12\angle30° = 12\sqrt{3}\angle-120° \text{ A}$$

and $I_2 = I_{23} - I_{12}$

$$I_2 = 12\angle150° - 12\angle-90° = 12\sqrt{3}\angle120° \text{ A}$$

and $I_3 = I_{31} - I_{23} = 12\angle30° - 12\angle150° \text{ A}$

$$I_3 = 12\sqrt{3}\angle0° \text{ A}$$

Part b

The system for part b is shown in Figure 3.48.

Transforming the Y load into a Δ configuration, and making use of the symmetry of the system, the equivalent impedances are evaluated.

$$Z_{12} = \frac{Z_1 Z_2 + Z_2 Z_3 + Z_3 Z_1}{Z_3}$$

and since $Z_1 = Z_2 = Z_3 = j10$, then

$$Z_{12} = \frac{3Z_1^2}{Z_1} = 3Z_1 = 3(10j) = 30j \ \Omega$$

$$Z_{23} = \frac{Z_1 Z_2 + Z_2 Z_3 + Z_3 Z_1}{Z_1}$$

$$Z_{23} = 30j \ \Omega$$

$$Z_{31} = \frac{Z_1 Z_2 + Z_2 Z_3 + Z_3 Z_1}{Z_2}$$

and

$$Z_{31} = 30j \ \Omega$$

Then,

$$I_{12} = \frac{120\,\angle 0°}{30\,\angle 90°} = 4\angle -90°$$

$$I_{23} = \frac{120\,\angle -120°}{30\,\angle 90°} = 4\angle 150°$$

$$I_{31} = \frac{120\,\angle 120°}{30\,\angle 90°} = 4\angle 30°$$

Observe that the solution for part b is based in transforming the problem into one that is similar to part a, and the currents just evaluated (part b) are one-third of the currents of part a, because the loads become three times larger.

The line currents using KCL can be evaluated, and are indicated as follows:

$$I_1 = I_{12} - I_{31} + 4\sqrt{3}\,\angle -120° \text{ A}$$

$$I_2 = 4\sqrt{3}\,\angle 120° \text{ A}$$

$$I_3 = 4\sqrt{3}\,\angle 0° \text{ A}$$

The corresponding voltage drops are indicated as follows:

$$V_{1N} = I_1 * Z = 4\sqrt{3}\,\angle -120° * 10\,\angle 90° \text{ V}$$

$$V_{1N} = 40\sqrt{3}\,\angle -30° \text{ V}$$

$$V_{2N} = 40\sqrt{3}\,\angle -150° \text{ V}$$

$$V_{3N} = 40\sqrt{3}\,\angle 90° \text{ V}$$

3.4 Examples

Example 3.1

Create the script file *XL_XC* that returns the following plots:

1. *mag[$X_L(\omega)$]* versus ω
2. *mag[$X_C(\omega)$]* versus ω

for $L = 2$ H and $C = 1$ μF, over the frequency range $200 \le \omega \le 2000$ rad/s.

MATLAB Solution
```
% Script file: XL _ XC
L = 2;
```

```
C = 1e-6;
w = [200:50:2000];
XL = w*L;
XC = 1./(w*C);
plot(w,XL,'o',w, XC,'*')
title('[XL(w) and XC(w)] vs w')
xlabel('frequency in rad/sec'), ylabel('Magnitude in Ohms')
legend('XL','XC')
grid on
hold; plot(w,XL,w,XC)
```

The script file *XL_XC* is executed and the results are shown in Figure 3.49.

Example 3.2

Evaluate the impedance *Z* and admittance $Y = (1/Z)$ for the circuit shown in Figure 3.50 first by hand, and then by writing the script file *Z_Y*, for an angular frequency of $w = 1$ rad/s.

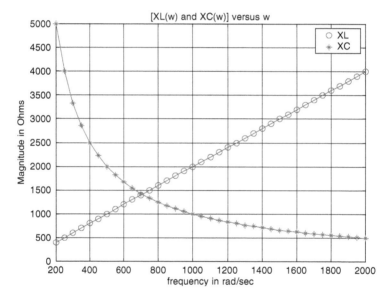

FIGURE 3.49
Plots of Example 3.1.

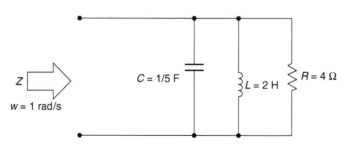

FIGURE 3.50
Network of Example 3.2.

ANALYTICAL Solution

$$X_C(\omega) = -j5$$

$$X_L(\omega) = j2$$

$$X(\omega) = X_C(\omega) \,/\!/\, X_L(\omega) = \frac{X_C(\omega)X_L(\omega)}{X_C(\omega) + X_L(\omega)} = \frac{-j5\,j2}{-j5 + j2} = \frac{10}{-j3} \quad (\Omega)$$

$$X(\omega) = j\frac{10}{3}$$

$$Z(\omega) = X(\omega) \,/\!/\, R = \frac{X(\omega)R}{X(\omega) + R} = \frac{j3.33 * 4}{j3.33 + 4} = 1.639 + j1.967 \ (\Omega)$$

and

$$Y(\omega) = \frac{1}{Z(\omega)}$$

$$Y(\omega) = \frac{j3.33 + 4}{j13.32} = 0.25 - j.3 \ (sie)$$

MATLAB Solution

```
% Script file: Z _ Y
W = 1;
C = 1/5;
L = 2;
R = 4;
XC = -j/(W*C);
XL = j*W*L;
X = XC*XL/(XC+XL);
Z = X*R/(X+R);
Y = 1/Z;
disp('^^^^^^^^^^^^^^^^^^^^^^^^^^^^^^^^^^^^^^^^^^^^^^^^^^^^^^^^^^^^^^^^^^^^^^^^^^')
disp('The impedance Z (in Ohms) and admittance (in Siemens) are given
by :');
Z,Y,
disp('^^^^^^^^^^^^^^^^^^^^^^^^^^^^^^^^^^^^^^^^^^^^^^^^^^^^^^^^^^^^^^^^^^^^^^^^^^')
```

```
^^^^^^^^^^^^^^^^^^^^^^^^^^^^^^^^^^^^^^^^^^^^^^^^^^^^^^^^^^^^^^^^^^^^^^^^^^^
The impedance Z (in Ohms) and admittance (in Siemens) are given by:
  Z =
      1.6393 + 1.9672
  Y =
      0.2500 - 0.3000i
^^^^^^^^^^^^^^^^^^^^^^^^^^^^^^^^^^^^^^^^^^^^^^^^^^^^^^^^^^^^^^
```

Example 3.3

Evaluate the impedance Z of the circuit of Figure 3.51 for $\omega = 10$ rad/s by hand, and by creating the script file *imp_Z*.

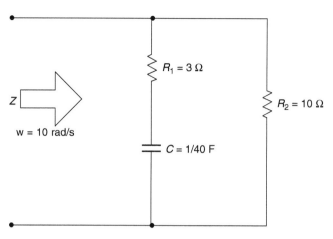

FIGURE 3.51
Network of Example 3.3.

ANALYTICAL Solution

$$X_C(\omega) = \frac{1}{j\omega C} = \frac{1}{j \cdot 10 \cdot (1/40)} = -j4\ \Omega$$

$$X = 3 - j4\ (\Omega)$$

Then, $Z = X//R_2$.

$$Z = \frac{X*10}{X+10} = 3.67\ \angle{-36°}\ \Omega$$

MATLAB Solution
```
% Script file: imp _ Z
W = 10;
R1 = 3;
R2 = 10;
C = 1/40;
XC = -j/(W*C);
X = R1+XC;
Z = X*R2/(X+R2);
Zmag = abs(Z);
Phase = angle(Z);
Phase _ deg = Phase*360/(2*pi);
disp('^^^^^^^^^^^^^^^^^^^^^^^^^^^^^^^^^^^^^^^^^^^^^^^^^^^^^^^^^^^^^^^^^^')
disp('Magnitude of Z (in Ohms) and phase (in degrees) is given by :');
Zmag
Phase _ deg
disp('^^^^^^^^^^^^^^^^^^^^^^^^^^^^^^^^^^^^^^^^^^^^^^^^^^^^^^^^^^^^^^^^^^')
```

The script file *imp_Z* is executed and the results are shown as follows.

```
>> imp _ Z
```

^^

```
Magnitude of Z (in Ohms) and phase (in degrees) is given by :
  Zmag =
          3.6761
  Phase _ deg =
              -36.0274
```

^^

Example 3.4

Evaluate by hand and by using MATLAB the voltage v(t) across the series RL circuit shown in Figure 3.52, if the current is $i(t) = 5\cos(100t)$ A.

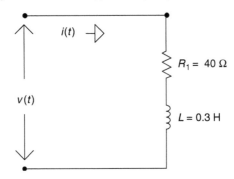

FIGURE 3.52
Network of Example 3.4.

ANALYTICAL Solution

$$I = 5\angle 0°$$

$$Z = 40 + j * 100 * 0.3 = 40 + j30$$

$$V = Z * I = (40 + j30) * 5\angle 0° = 250\angle(36.52°)$$

then

$$v(t) = 250\cos(100t + 36.87°)$$

MATLAB Solution
```
% Script file: vol _ RL
W=100;
Z=40+j*W*.3;
V=5*Z;
Vmax=abs(V);
Phase=angle(V);
Phasedegree=Phase*180/pi;
% Print results
disp('^^^^^^^^^^^^^^^^^^^^^^^^^^^^^^^^^^^^^^^^^^^^^^^^^^^^^^^^^^^^^^^')
disp('The peak value (in volts) and the phase (in degrees) of v(t)
  are=');
Vmax,
Phasedegree,
disp('^^^^^^^^^^^^^^^^^^^^^^^^^^^^^^^^^^^^^^^^^^^^^^^^^^^^^^^^^^^^^^^')
```

The script file *vol_RL* is executed and the results are as follows:

```
>> vol _ RL
^^^^^^^^^^^^^^^^^^^^^^^^^^^^^^^^^^^^^^^^^^^^^^^^^^^^^^^^^^^^^^^^^^^^^^^^^
The peak value (in volts) and the phase (in degrees) of v(t) are=
Vmax =
       250
Phasedegree =
              36.8699
^^^^^^^^^^^^^^^^^^^^^^^^^^^^^^^^^^^^^^^^^^^^^^^^^^^^^^^^^^^^^^^^^^^^^^^^^
```

The preceding result means that $v(t) = 250 \cos(100t + 36.8699°)$ V.

Example 3.5

Evaluate the current $i(t)$ for the circuit shown in Figure 3.53 by hand, and by using MATLAB.

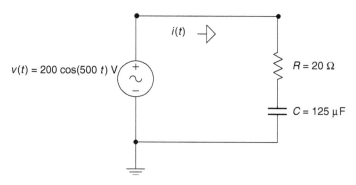

FIGURE 3.53
Network of Example 3.5.

ANALYTICAL Solution

$$V = 200 \angle 0°, \ Z = 20 - \frac{j}{125 * 10^{-6} * 500} = 20 - j16$$

then

$$I = \frac{V}{Z} = \frac{200}{20 - j16} = \frac{50}{5 - j4} = 7.8 \angle 38.6° \ \text{A}$$

Therefore, $i(t) = 7.8 \cos(500t + 38.6°)$ A.

MATLAB Solution
```
>> W = 500;
>> V = 200;
>> R = 20;
>> C = 125e-6
```

```
C =
        1.2500e-004
>> XC = -j/(W*C)

   XC =
        0 -16.0000I
>> Z = R+XC

   Z =
        20.0000 -16.0000I
>> Zmag = abs(Z);
>> phase = -angle(Z)*180/pi

   phase =
           38.6598
>> I = V/Zmag;
>> % Print results
>> disp('The peak value of the current (in amps) is ='),  disp(I);
>> disp('with a phase angle(in degrees) of ='), disp(phase)

   The peak value of the current (in amps) is =
                                              7.8087
   with a phase angle (in degrees) of =
                                      38.6598
```

The meaning of the aforementioned results is that the current is given by

$$i(t) = 7.8087 \cos(500t + 38.6598°) \text{ A}$$

Example 3.6

Create the script file *curr_volt* that returns the plots of $v(t)$ versus t and $i(t)$ versus t, over the range $0 \le t \le 0.06$ s, for the circuit shown in Figure 3.54, where $i(t) = 5 \sin(300t)$ A.

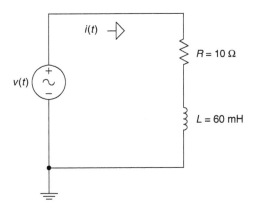

FIGURE 3.54
Network of Example 3.6.

MATLAB Solution
```
% Script file:  curr_volt
W = 300;
R = 19;
L = 60e-3;
Ang = W*L/R;
t = [0:.0001:.06];
V = (R^2+(W*L)^2)^.5*5*sin(W.*t+atan(Ang));
subplot(2,1,1)
I = 5*sin(W.*t);
plot (t,V), title(' v(t) vs. t')
grid on
ylabel('Amplitude in volts'), title('v(t) vs.t')
subplot(2,1,2)
plot(t,I)
grid on
title('i(t) vs. t'), ylabel(' Amplitude in amps')
xlabel('time in sec.')
```

The script file *curr_volt* is executed and the resulting plots are shown in Figure 3.55.

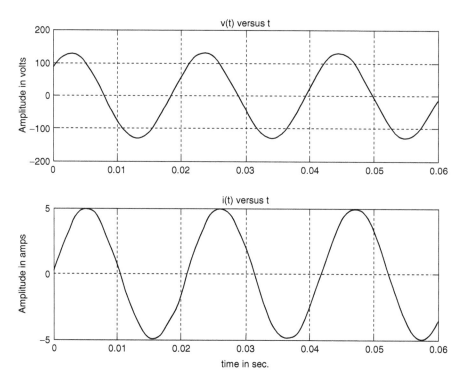

FIGURE 3.55
Plots of Example 3.6.

Example 3.7

Let the current in the RC circuit shown in Figure 3.56 be $i(t) = 5 \sin(200t)$ A. Create the script file *curr_vol_RC* that returns the plots of $v(t)$ versus t and $i(t)$ versus t, over the range $0 \le t \le 0.1\, s$.

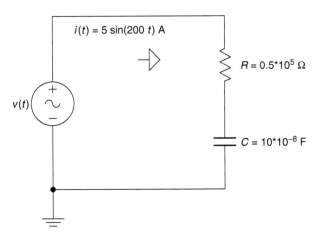

FIGURE 3.56
Network of Example 3.7.

MATLAB Solution
```
% Script file: curr _ vol _ RC
clf
W = 200;
R = 0.5e5;
C = 10e-8;
Ang = 1/(W*C*R);
t = 0:.001:.1;
V = (R^2+(1/W*C)^2)^.5*5*sin(W.*t-atan(Ang));
I =5*sin(W.*t);
subplot(2,1,1)
plot(t,V)
ylabel('Ampitude[v(t)]');
xlabel('time in seconds');
title('v(t)  vs. t ')
subplot(2,1,2)
plot(t,I)
title('i(t) vs. t  ')
xlabel('time in seconds');
ylabel('Amplitude[i(t)]')
```

The script file *curr_vol_RC* is executed and the returning plots are shown in Figure 3.57.

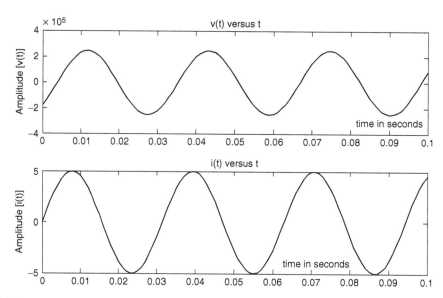

FIGURE 3.57
Plots of Example 3.7.

Example 3.8

Given the following periodic functions defined over one period (T):

a. $f_1(t) = 5\cos((2\pi/3)t)$, with $T = 3$ s
b. $f_2(t) = t$, for $0 \le t \le 1$, with $T = 1$ s
c. $f_3(t) = \begin{cases} 1 & 0 \le t \le 0.5 \\ 0 & 0.5 \le t \le 1 \end{cases}$ with $T = 1$ s

Create the script file *ave_rms_sym* that returns the following plots over the range $-1 \le t \le 3$:

a. $f_1(t)$ versus t
b. $f_2(t)$ versus t
c. $f_3(t)$ versus t

Also evaluate the average and the RMS (effective) value for each one of the preceding functions using

a. a symbolic approach by creating the script file *ave_rms_sym* (use *int*)
b. a numerical solution by creating the script file *ave_rms_num* (use *trapz*)

Recall that the average value of the arbitrary $f(t)$ is given by

$$F_{AVE} = \frac{1}{T}\int_0^T f(t)dt$$

and its RMS value is given by

$$F_{RMS} = \sqrt{\frac{1}{T}\int_0^T f(t)^2\,dt}$$

MATLAB Solution

```
% Script file : ave _ rms _ sym
% returns the functions f1,f2,& f3
% and computes the average and rms values of f1,f2 & f3
% using symbolic (int) technique
x = linspace(-3,3,100);
f1t = 5.*cos(2*pi./3.*x);
subplot(3,1,1)
plot(x,f1t);
title('Plots of f1(t),f2(t) & f3(t)');
ylabel('f1(t)'),axis([-3 3 -6 6]);
subplot(3,1,2)
a = linspace(0,1,25);
y = [a a a a];x=linspace(-1,3,100);
plot(x,y);ylabel('f2(t)');axis([-1 3 -.5 1.5]);
b = [ones(1,12) zeros(1,13)];bb = [b b b b];subplot(3,1,3)
plot(x,bb);ylabel('f3(t)');axis([-1 3 -.5 1.5]); xlabel('time');
                                                % Figure 3.58

syms t;
w = 2*pi/3;
T = 2*pi/w;
f1ave=1/T*int(5*cos(w*t),0,T);                  % part(a)
disp('****************************************************');
disp('***********Symbolic Results (using int )*********');
disp('****************************************************');
disp('The average value of f1(t) is =');
ans1=vpa(f1ave);disp(ans1);
f1rms=sqrt(1/T*int(25*cos(w*t)^2,0,T));         % part(b)
disp('The RMS value of f1(t) is =');
ans2=vpa(f1rms);disp(ans2);
T1=1;
w1=2*pi/T1;
Vave2 = 1/T1*int(t,0,T1);%part(b)
disp('The average value of f2(t) is =');
ans3 = vpa(Vave2); disp(ans3);
Vrms2 =sqrt(1/T1*int(t^2,0,T1));
disp('The RMS value of f2(t) is =');
ans4 = vpa(Vrms2); disp(ans4);
w2 = 2*pi/T1;
Vave3 =1/T1*int(sym('Heaviside(t)'),0,T1/2);    %part(c)
disp('The average value of f3(t) is =');
ans5 = vpa(Vave3);disp(ans5);
Vrms3 =sqrt(1/T1*int(sym('Heaviside(t)^2'),0,T1/2));
disp('The RMS value of f3(t) is =');
ans6 = vpa(Vrms3);disp(ans6);
disp('****************************************************');
```

The results of executing the script file *ave_rms_sym* are shown as follows:

```
>> ave_rms_sym
**************************************************************
************* Symbolic Results ( using int) **************
**************************************************************

The average value of f1(t) is =
                    0
```

```
The RMS value of f1(t) is =
                         3.5355339059327376220042218105243
The average value of f2(t) is =
                         .50000000000000000000000000000000
The RMS value of f2(t) is =
                         .57735026918962576450914878050196
The average value of f3(t) is =
                         .50000000000000000000000000000000
The RMS value of f3(t) is =
                         .70710678118654752440084436210485
*****************************************************************
```

```
% Script file: ave _ rmst _ num
% evaluates the average and rms values of f1,f2 and f3
% using numerical techniques (trapz)
t1= linspace(0,3,100);
y1= 5*cos(2*pi/3*t1);
avecos =1/3*trapz(t1,y1);
y1sq = y1.^2;
rmscos = sqrt(1/3*trapz(t1,y1sq));
t2 =linspace(0,1,100);
y2 = t2;
avesaw = trapz(t2,y2);
y2sq = y2.^2;
rmssaw = sqrt(trapz(t2,y2sq));
y3 = [ones(1,50) zeros(1,50)];
avesqr = trapz(t2,y3);
y3sq = y3.^2;
rmssqr = sqrt(trapz(t2,y3sq));
disp('********************************************************');
disp('  ********* Numerical Results (using trapz)***********');
disp('********************************************************');
disp('     ')
disp('The average value of f1(t) is =')
disp(avecos)
disp('The rms value of f1(t) is =')
disp(rmscos)
disp('     ')
disp('The average value of f2(t) is =')
disp(avesaw)
disp('The rms value of f2(t) is =')
disp(rmssaw)
disp('     ')
disp('The average value of f3(t) is =')
disp(avesqr)
disp('The rms value of f1(t) is =')
disp(rmssqr)
disp('**************************************************')
```

The script file *ave_rms_num* is executed and the results are as follows:

```
>> ave _ rms _ num
*****************************************************
******** Numerical Results (using trapz) *************
*****************************************************
```

```
The average value of f1(t) is =
                         -6.8464e-016
The rms value of f1(t) is =
                         3.5355
The average value of f2(t) is =
                         0.5000
The rms value of f2(t) is =
                         0.5774
The average value of f3(t) is =
                         0.5000
         The rms value of f1(t) is =
                             0.7071
```

****

Observe that the numerical results fully agree with the symbolic results.

Example 3.9

Create the script files *circ_Fig_359* and *circ_Fig_360* that return for the circuits shown in Figures 3.59 and 3.60, the following plots:

1. $|Z(\omega)|$ versus ω
2. $\angle Z(\omega)$ versus ω

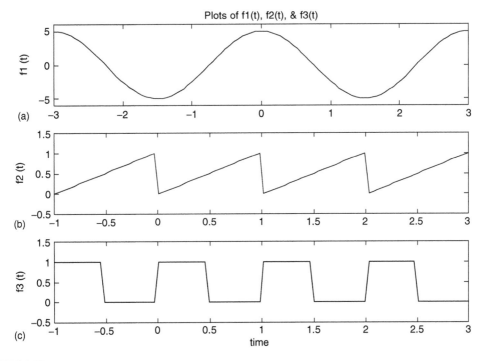

FIGURE 3.58
Plots of the periodic functions of Example 3.8 (a, b, c).

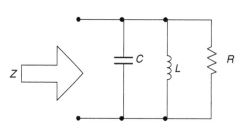

FIGURE 3.59
Network with $R = C = L = 1$, over the range $0 \leq \omega \leq 5$ rad/s of Example 3.9.

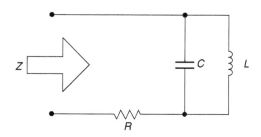

FIGURE 3.60
Network with $R = 1000\,\Omega$, $L = 30$ mH, and $C = 0.47\,\mu F$, over the range $7500 \leq \omega \leq 9500$ rad/s of Example 3.9.

ANALYTICAL Solution

For the circuit shown in Figure 3.59 (part a),

$$Z(\omega) = \frac{(1/C)(j\omega)}{(j\omega)^2 + (1/RC)(j\omega) + (1/LC)}$$

For $R = L = C = 1$,

$$Z(\omega) = \frac{(j\omega)}{(j\omega)^2 + (j\omega) + 1}$$

For the circuit shown in Figure 3.60 (part b),

$$Z(\omega) = \frac{R(j\omega)^2 + (1/C)(j\omega) + (R/LC)}{(j\omega)^2 + (1/LC)}$$

For $R = 1000\,\Omega$, $L = 30$ mH, and $C = 0.47\,\mu F$ (Figure 3.61),

$$Z(\omega) = \frac{1000 * (j\omega)^2 + \dfrac{1}{0.47 * 10^{-6}} * (j\omega) + \dfrac{1000}{30 * 10^{-3} * 0.47 * 10^{-6}}}{(j\omega)^2 + \dfrac{1}{30 * 10^{-3} * 0.47 * 10^{-6}}}$$

MATLAB Solution
```
% Script file: circ _ Fig _ 359
% R, L, and C in parallel
% R = C = L = 1.
echo on;
w = 0:0.1:5;
num = [0 0 1];
den = [1 1 1];
ZW = freqs (num, den, w);
subplot(2, 1, 1);
plot(w, abs(ZW));
grid on;
title('Magnitude and phase of Z(w)');
xlabel('w(rad/sec)');
ylabel('magnitude');
subplot(2, 1, 2);
plot(w, angle(ZW)*180/pi);              % Figure 3.61
xlabel('w(rad/sec)');
ylabel('phase in degrees');
grid on;
```

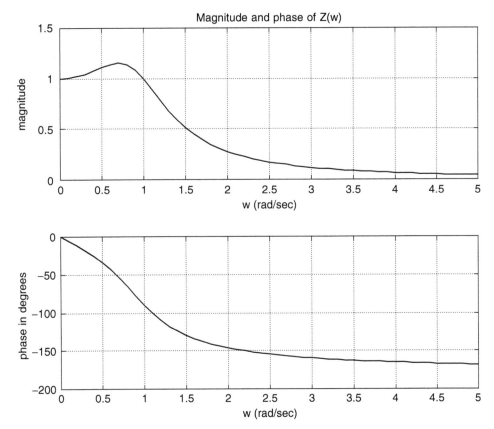

FIGURE 3.61
Plots for the circuit of Figure 3.59.

```
% Script file: circ _ Fig _ 360
echo on;
w = 7500:50:9500;
R = 1000;
L = 30e-3;
C = 0.47e-6;
num = [R 1/C R/(L*C)];
den = [1 0 1/(L*C)];
ZW = freqs(num, den, w);
subplot(2, 1, 1);
plot(w, abs(ZW));
grid on;
title('Magnitude and phase of Z(w)');
xlabel('w(rad/sec)');
ylabel('magnitude');
subplot(2, 1, 2);
plot(w, angle(ZW)*180/pi);
xlabel('w(rad/sec)');
ylabel('phase in degrees');
grid on;                          % Figure 3.62
```

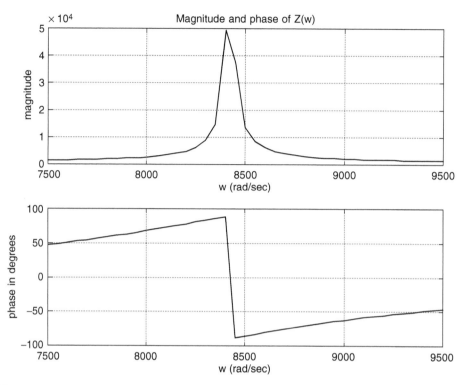

FIGURE 3.62
Plots for the circuit of Figure 3.60.

Example 3.10

Given the current and voltage shown in the circuit diagram of Figure 3.63, write a program that returns the following:

1. Plots of $i(t)$ versus t, $v(t)$ versus t, and $p(t)$ versus t over $0 \le t \le 0.01$
2. Evaluate using MATLAB
   a. The average power delivered to the load
   b. The RMS value of $i(t)$ and $v(t)$ by using numerical techniques and verify the results by $V_{RMS} = V_m * 0.707$ and $I_{RMS} = I_m * 0.707$
   c. The PF

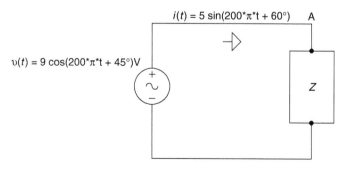

FIGURE 3.63
Network of Example 3.10.

MATLAB Solution

```
>> % Instantaneous power plot
>> T= 2*pi/(200*pi);
>> X= [0:.01:1];
>> t = X.*T;
>> IT = 5*cos(200*pi*t+pi/3);
>> VT = 9*cos(200*pi*t+pi/4);
>> PT = IT.*VT;
>> plot(t,PT,'o',t,VT,'*',t,IT,'+');        % Figure 3.64
>> legend('i(t)', 'v(t)', 'p(t)')
>> grid on
>> title('Instantaneous Power, Current, and Voltage')
>> xlabel('time in seconds'),ylabel(' Amplitudes of P, I, &V')
>> VTSQ=VT.^2;                              % v(t)^2
>> IntVTSQ = .01*T*(trapz(VTSQ))            % integral ofv(t)^2

   IntVTSQ =
              0.4050

>> ITSQ = (5*cos(200*pi*t+pi/3)).^2;        % i(t)^2
>> IntITSQ = .01*T*(trapz(ITSQ))            % integral of i(t)^2

   IntITSQ =
              0.1250

>> VRMS = sqrt((IntVTSQ)/T)                 % RMS value of voltage

   VRMS =
           6.3640

>> IRMS = sqrt((IntITSQ)/T)                 % RMS value of current.

   IRMS =
           3.5355

>>                                          % average power using trapz
>> PowerWave = .01*T*trapz(PT)/T

   PowerWave =
                21.7333

>>                                          % Power Factor numerical
>> PF = PowerWave/(VRMS*IRMS)

   PF =
         0.9659

>>                                          % verify results analytically
>> VRMSANA = 9*.707

   VRMSANA =
              6.3630

>> IRMSANA = 5*.707

   IRMSANA =
              3.5350
```

```
>> PowerWaveAna = (45/2)*cos(15*pi/180)

PowerWaveANA =
               21.7333

>>                                          % results are printed for
                                            comparisons
>> fprintf('Average Power, Numerical:%f\n',PowerWave)

        Average Power, Numerical: 21.733331

>> fprintf('Average Power, Analytical:%f\n',PowerWaveANA)

        Average Power, Analitical: 21.7333

>> fprintf('VRMS, Numerical:%f\n',VRMS)

        VRMS, Numerical: 6.3640

>> fprintf('VRMS, Analytical:%f\n',VRMSANA)

        VRMS, Analytical: 6.3630

>> fprintf('IRMS, Numerical:%f\n',IRMS)

        IRMS, Numerical: 3.5355

>> fprintf ('IRMS, Analytical:%f\n',IRMANA)

        IRMS, Analytical: 3.5350

>> fprintf ('PowerFactor, PF:%f\n',PF)

        PowerFactor,PF: 0.965926
```

Observe that the analytical results fully agree with the numerical results.

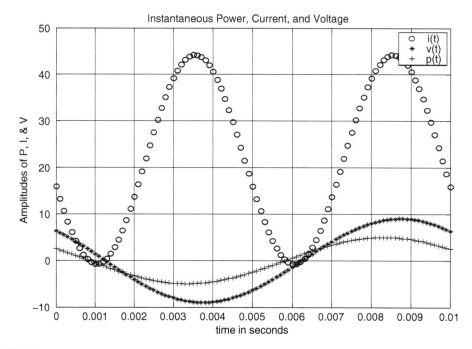

FIGURE 3.64
Plots of $i(t)$, $v(t)$, and $p(t)$ of Example 3.10.

Example 3.11

Write a MATLAB program that returns

1. The phasor diagram of the voltages $V_{(Applied)}$, V_L, V_C, and V_R of the circuit shown in Figure 3.65
2. The angle between the I and V
3. Verify KVL, that is, $V = V_L + V_C + V_R$

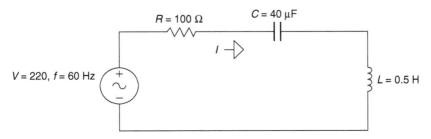

FIGURE 3.65
Network of Example 3.11.

MATLAB Solution

```
>> V = 220;
>> R = 100;
>> C = 40e-6;
>> L = .5;
>> Omega = 2*pi*60;
>> Z = R+j*((L*Omega)-1/(C*Omega));
>> Magnitude = abs(Z);
>> Phase = 180*angle(Z)/pi;
>> I = V/Z;
>> I = abs(I);
>> VR = I*R;
>> VL = j*L*Omega*I;
>> VC = -j*I/(C*Omega);
>> Vapplied = VR+VL+VC;              % the result should be 220 (KVL)
>> Check _ KVL = abs(VApplied)

   Check _ KVL =
                    220.0000

>> phase _ in _ deg = 180*angle(VApplied)/pi

   phase _ in _ deg =
                    50.7011

>>                                    % construction of the phasor diagram
>> L(1) = 0;
>> L(2) = VR;
>> L(3) = VR+VL;
>> L(4) = VR+VC;
>> Voltage = [0 VApplied];
>> axis('square')
>> plot(real(L), imag(L), real(Voltage), imag(Voltage))
>> grid on
>> xlabel('Real Axis'), ylabel('Imaginary Axis')
>> title('Phasor Diagram') (Figure 3.66)
```

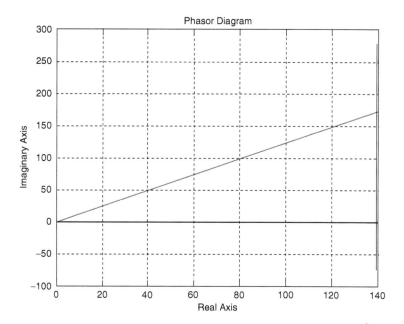

FIGURE 3.66
Phasor diagram of Example 3.11.

Example 3.12

Create the script file *phasor_time_plots* that returns the circuit diagram shown in Figure 3.67, the following plots:

1. The phasor diagram for V_S, V_C, and V_R
2. $v_S(t)$ versus t, $v_R(t)$ versus t, and $v_C(t)$ versus t

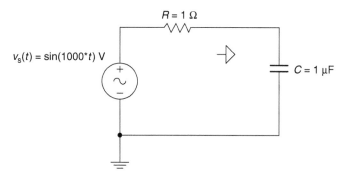

FIGURE 3.67
Network of Example 3.12.

MATLAB Solution
```
% Script file: phasor _ time _ plots
clf
VS = 1;
R = 1e3;
C = 1e-6;
```

```
F = 1000;
W = 2*pi*F;
ZC = -j/(W*C);
ZT = R+ZC;
I = VS/ZT;
VR = R*I;
VC = ZC*I;
subplot(2,1,1)
%  Phasors diagram.
VSS = [0 VS];
VVC = [0 VC];
VVR = [0 VR];
plot(real(VSS),imag(VSS),'*',real(VVC),imag(VVC),'+',real(VVR),imag(VVR),'o');
xlabel('Real'), ylabel('Imaginary'),
title('Phasor Diagram (Example 3.12)');    % Figure 3.68 (top)
grid on
hold ;                                      %hold plot
plot(real(VSS), imag(VSS), real(VVC),imag(VVC),real(VVR),imag(VVR));
legend('VS', 'VC', 'VR');
% Check the Phasor Diagram by applying Kirchhoffs Voltage Law.
disp('**********************************************************')
disp('****Check the Phasor Diagram results by applying *****')
disp('**********  Kirchhoffs Voltage Law  ******************')
Check _ Voltage = VR+VC
disp('**********************************************************')
disp(' (Note:Check _ V should be equal to VS=1.)')
T = 1/F;
t = [0:2*T/50:2*T]; % t for 2 periods
VS = sin(W.*t);
VRT = abs(VR)*sin(W.*t+angle(VR));
VCT = abs(VC)*sin(W.*t+angle(VC));
subplot (2,1,2)
plot (t,VS,'o',t,VRT,'*',t,VCT,'+')
xlabel ('time in seconds'),
ylabel ('Amplitude in Volts')
title('Voltages of the RC Circuit of Example 3.12');
                                            % Figure 3.68 (bottom)
grid on
legend('vs(t)','vr(t)','vc(t)')
```

The MATLAB script file *phasor_time_plots* is executed, and the results are shown as follows:

```
>> phasor _ time _ plots

Current plot held
*************************************************************
****Check the Phasor Diagram results by applying*****
************* Kirchhoffs Voltage Law *****************

   Check _ Voltage =
                     1

*********************************************************
```

(note that the correct result is verified by Check _ Voltage=1)

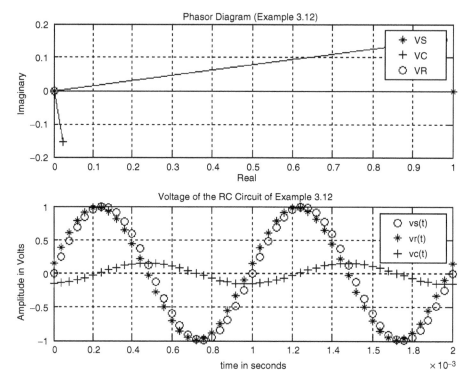

FIGURE 3.68
Plots of Example 3.12.

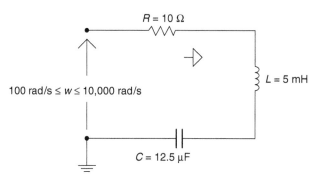

FIGURE 3.69
Network of Example 3.13.

Example 3.13

Create the script file *impedance_plots* that returns for the series RLC circuit shown in Figure 3.69, over the range $100 \leq \omega \leq 10,000$ rad/s, the following plots:

1. $|Z(\omega)|$ versus ω
2. $\angle Z(\omega)$ versus ω
3. Repeat parts 1 and 2 for the case where the three elements (R, L, and C) are connected in parallel

ANALYTICAL Solution (series case)

$$|Z(\omega)| = |R + j(X_L - X_C)|$$

$$|Z(\omega)| = \sqrt{R^2 + (X_L - X_C)^2}$$

The magnitude equation is given by

$$\left[|Z(\omega)| = \sqrt{R^2 - \left(\omega L - \frac{1}{\omega C} \right)^2} \right]$$

The phase equation is given by

$$\angle Z(\omega) = \arctan\left[\frac{X_L - X_C}{R} \right]$$

ANALYTICAL Solution (parallel case)

$$Y(\omega) = \frac{1}{R} + \frac{1}{j\omega L} + \frac{1}{1/j\omega C}$$

The magnitude equation is given by

$$|Y(\omega)| = \sqrt{\left(\frac{1}{R} \right)^2 + \left(\omega C - \frac{1}{\omega L} \right)^2}$$

The phase equation is given by

$$\angle Y(\omega) = \arctan\left[\frac{\left(\omega C - \frac{1}{\omega L} \right)}{\frac{1}{R}} \right]$$

MATLAB Solution

```
% Script file: impedance _ plots
clf
R = 10;
L = 5e-3;
C = 12.5e-6;
w = [100:100:10000];
Z = R+ j*(w*L-1./(C*w));                          % series case
subplot(2,2,1);
plot(w,abs(Z));
title('Mag. [Z(w)] vs w (series case)');
ylabel('Mag[Z(w)] in Ohms');
grid on;
subplot(2,2,2);
plot(w,angle(Z)*180/pi); grid on;
title('Phase[Z(w)] vs w (series case)');
ylabel('Phase angle in degrees');
```

```
XC = -j./(w*C);
XL = j*w*L;
ZLC = XC.*XL./(XC+XL);
ZRLC = R.*ZLC./(R+ZLC);                        % parallel case
subplot (2,2,3);
plot (w,abs(ZRLC));
title('Mag. [Z(w)] vs w (parallel case)');
xlabel ('w in rad/sec');
ylabel ('Mag[Z(w)] in Ohms');
grid on;
subplot(2,2,4);
plot(w,angle(ZRLC)*180/pi); grid on;
title('Phase[Z(w)] vs w (parallel case)');
xlabel('w in rad/sec');
ylabel('Phase angle in degrees');
```

The script file *impedance_plots* is executed, and the results are indicated in Figure 3.70.

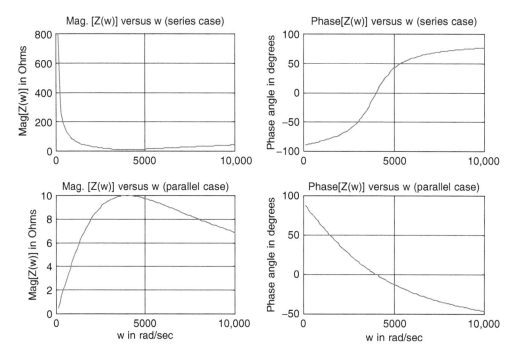

FIGURE 3.70
Plots of Example 3.13.

Example 3.14

Analyze the circuit diagram shown in Figure 3.71, and obtain by hand the system loop equations, as well as the matrix loop equation.

Create the script file *loops* that returns

1. The system matrices Z and V
2. The loop currents I_1, I_2, and I_3, in phasor form
3. The instantaneous loop currents $i_1(t)$, $i_2(t)$, and $i_3(t)$

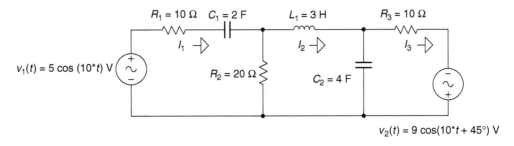

FIGURE 3.71
Network of Example 3.14.

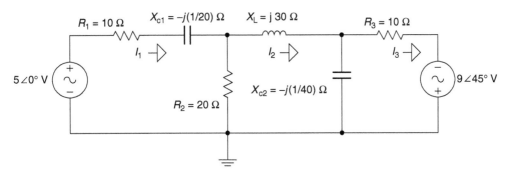

FIGURE 3.72
Phasor circuit diagram of Example 3.14.

ANALYTICAL Solution

The circuit of Figure 3.71 is transformed into the phasor circuit diagram shown in Figure 3.72.

The three loop equations are

$$\left[(10 + 20) - j\left(\frac{1}{20}\right)\right] * I_1 - 20 * I_2 + 0 I_3 = 50\angle 0°$$

$$-20 I_1 + \left[20 + j\left(30 - \frac{1}{40}\right)\right] * I_2 - \left(-j\frac{1}{40}\right) * I_3 = 0$$

$$0 I_1 - \left[-j\left(\frac{1}{40}\right)\right] * I_2 + \left(10 - j\frac{1}{40}\right) * I_3 = 9\angle 45°$$

The matrix loop equation is given by

$$\begin{bmatrix} (10 + 20) - j(1/20) & -20 & 0 \\ -20 & 20 + j(30 - (1/40)) & +j(1/40) \\ 0 & +j(1/40) & 10 - j(1/40) \end{bmatrix} * \begin{bmatrix} I_1 \\ I_2 \\ I_3 \end{bmatrix} = \begin{bmatrix} 5\angle 0° \\ 0 \\ 9\angle 45° \end{bmatrix}$$

Note that the aforementioned matrix equations can be expressed as $[Z] * [I] = [V]$, where

$$Z = \begin{bmatrix} (10+20) - j(1/20) & -20 & 0 \\ -20 & 20 + j(30 - (1/40)) & +j(1/40) \\ 0 & +j(1/40) & 10 - j(1/40) \end{bmatrix}$$

and

$$V = \begin{bmatrix} 5\angle 0° \\ 0 \\ 9\angle 45° \end{bmatrix}$$

The currents I_1, I_2, and I_3 can be evaluated by using the following MATLAB matrix relation

$$I = inv(Z) * [V]$$

MATLAB Solution
```
% Script file: loops
disp('*********************************')
disp('                System Matrices                ')
disp('*********************************')
disp('The impedance matrix is given by:')
Z = [30-j*1/20 -20 0;-20 20+j*(30-1/40) j*(1/40); 0 j*(1/40) 10-j*(1/40)]
A = 9*exp(j*45*pi/180);
disp('The voltage matrix is given by:')
V= [5; 0; A]
disp('*********************************')
disp('                Loop Currents                ')
disp('*********************************')
% Solve for loop currents I1, I2 and I3
I = inv(Z)*V;
% Solve for Magnitude and Phase Angle.
I1mag = abs(I(1));
I2mag = abs(I(2));
I3mag = abs(I(3));
I1ang = angle(I(1))*180/pi;
I2ang = angle(I(2))*180/pi;
I3ang = angle(I(3))*180/pi;
% Print currents
disp('^^^^^^^^^^^^^^^^^^^^^^^^^^^^^^^^^^^^^^^^^^^^^^^')
disp('                phasor domain                ')
disp('^^^^^^^^^^^^^^^^^^^^^^^^^^^^^^^^^^^^^^^^^^^^^^^')
fprintf('The magnitude of the Current I1 (in amps) is :%f\n and its
  phase angle in degrees is :%f\n',I1mag,I1ang)
fprintf('The magnitude of the current I2 (in amps) is :%f\n and its
  phase angle in degrees :%f\n',I2mag,I2ang)
fprintf('The magnitude of the current I3 (in amps)is :%f\n and its
  phase angle in degrees is :%f\n',I3mag,I3ang)
disp('^^^^^^^^^^^^^^^^^^^^^^^^^^^^^^^^^^^^^^^^^^^^^^^')
disp('                time domain                ')
disp('^^^^^^^^^^^^^^^^^^^^^^^^^^^^^^^^^^^^^^^^^^^^^^^')
```

```
fprintf('The current i1(t)=%fcos(10t %f)amps\n',I1mag,I1ang)
fprintf('The current i2(t)=%fcos(10t %f)amps\n',I2mag,I2ang)
fprintf('The current i3(t)=%fcos(10t+ %f)amps\n',I3mag,I3ang)
disp('*****************************************************')
```

The script file *loops* is executed and the results are indicated as follows:

```
>> loops

*****************************************
              System Matrices
*****************************************
  The impedance matrix is given by:
  Z =
     30.0000 - 0.0500i  -20.0000                  0
     -20.0000            20.0000 + 29.9750i  0 + 0.0250i
      0                   0 + 0.0250i        10.0000 - 0.0250i
  The voltage matrix is given by:
  V =
     5.0000
     0
     6.3640 + 6.3640i
*****************************************************
                   Loop Currents
*****************************************************
^^^^^^^^^^^^^^^^^^^^^^^^^^^^^^^^^^^^^^^^^^^^^^^^^^^^^^

                   phasor domain
^^^^^^^^^^^^^^^^^^^^^^^^^^^^^^^^^^^^^^^^^^^^^^^^^^^^^^

The magnitude of the Current I1 (in amps) is :0.195625
     and its phase angle in degrees is :-21.209060
The magnitude of the current I2 (in amps) is :0.109148
     and its phase angle in degrees:-77.628332
The magnitude of the current I3 (in amps)is :0.899767
     and its phase angle in degrees is:45.152608
^^^^^^^^^^^^^^^^^^^^^^^^^^^^^^^^^^^^^^^^^^^^^^^^^^^^^^

                   time domain
^^^^^^^^^^^^^^^^^^^^^^^^^^^^^^^^^^^^^^^^^^^^^^^^^^^^^^

The current i1(t) = 0.195625cos(10t -21.209060)amps
The current i2(t) = 0.109148cos(10t -77.628332)amps
The current i3(t) = 0.899767cos(10t+ 45.152608)amps
*****************************************************
```

Example 3.15

Analyze the circuit diagram shown in Figure 3.73, and obtain by hand the system node equations, as well as the matrix node equation (all elements are given as admittances).
   Create the script file *nodes* that returns

1. The system matrices Y and I
2. The node voltages V_1 and V_2 in phasor form
3. The instantaneous node voltages $v_1(t)$ and $v_2(t)$

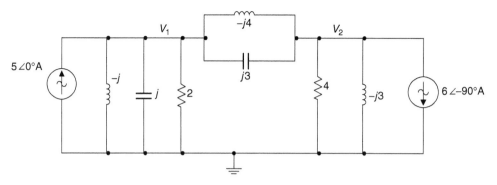

FIGURE 3.73
Network of Example 3.15.

ANALYTICAL Solution

The two node equations are

$$\text{for node } V_1: \quad (2 - j)V_1 + jV_2 = 5\angle 0°$$

$$\text{for node } V_2: \quad jV_1 + (4 - j4)V_2 = -6\angle -90°$$

The matrix node equation is given by

$$\begin{bmatrix} 2 - j & j \\ j & 4 - j4 \end{bmatrix} * \begin{bmatrix} V_1 \\ V_2 \end{bmatrix} = \begin{bmatrix} 5\angle 0° \\ -6\angle -90° \end{bmatrix}$$

The aforementioned matrix equation can be expressed as $[Y] * [V] = [I]$, where

$$Y = \begin{bmatrix} 2 - j & j \\ j & 4 - j4 \end{bmatrix}$$

and

$$I = \begin{bmatrix} 5\angle 0° \\ -6\angle -90° \end{bmatrix}$$

where the phasor current sources are converted to complex numbers as

$$5\angle 0° = 5 \quad \text{and} \quad -6\angle -90° = -(-j6) = j6$$

MATLAB Solution

```
% Script file: nodes
disp('*******************************************************')
disp('                    System Matrices                   ')
disp('*******************************************************')
disp('The admittance matrix Y is given by:')
Y = [2-j j;j 4-4*j]
disp('The current matrix I is given by:')
I = [5;j*6]
disp('*******************************************************')
disp('                    Node voltages                     ')
disp('*******************************************************')
```

```
% Solution for nodal voltages V1 and V2
V = inv(Y)*I;
% Solution for the magnitude and phase angle
V1mag = abs(V(1));
V2mag = abs(V(2));
V1ang = angle(V(1))*180/pi;
V2ang = angle(V(2))*180/pi;
% Print voltages
disp('^^^^^^^^^^^^^^^^^^^^^^^^^^^^^^^^^^^^^^^^^^^^^^^^^^^^^')
disp('                     phasor domain      ')
disp('^^^^^^^^^^^^^^^^^^^^^^^^^^^^^^^^^^^^^^^^^^^^^^^^^^^^^')
fprintf('The magnitude of the voltage V1 (in volts) is :%f\n and its
   phase angle in degrees is :%f\n',V1mag,V1ang)
fprintf('The magnitude of the voltage V2 (in voltss) is :%f\n and its
   phase angle in degrees is:%f\n',V2mag,V2ang)
disp('^^^^^^^^^^^^^^^^^^^^^^^^^^^^^^^^^^^^^^^^^^^^^^^^^^^^^')
disp('                      time domain      ')
disp('^^^^^^^^^^^^^^^^^^^^^^^^^^^^^^^^^^^^^^^^^^^^^^^^^^^^^')
fprintf('The voltage v1(t)=%fcos(10t+%f)voltss\n',V1mag,V1ang)
fprintf('The voltage v2(t)=%fcos(10t+%f)voltss\n',V2mag,V2ang)
disp('****************************************************')
```

The script file *nodes* is executed and the results are indicated as follows:

```
>> nodes

**********************************************
               System Matrices
**********************************************
The admittance matrix Y is given by:
 Y =
     2.0000 - 1.0000i       0 + 1.0000i
          0 + 1.0000i   4.0000 - 4.0000i
The current matrix I is given by:
 I =
     5.0000
          0 + 6.0000i
**********************************************
               Node voltages
**********************************************
^^^^^^^^^^^^^^^^^^^^^^^^^^^^^^^^^^^^^^^^^^^^^^^^^

               phasor domain
^^^^^^^^^^^^^^^^^^^^^^^^^^^^^^^^^^^^^^^^^^^^^^^^^

The magnitude of the voltage V1 (in volts) is: 2.523265
and its phase angle in degrees is : 29.811543
The magnitude of the voltage V2 (in volts) is: 0.709196
and its phase angle in degrees is:116.778840
^^^^^^^^^^^^^^^^^^^^^^^^^^^^^^^^^^^^^^^^^^^^^^^^^

                time domain
^^^^^^^^^^^^^^^^^^^^^^^^^^^^^^^^^^^^^^^^^^^^^^^^^

The voltage v1(t) = 2.523265 cos(10t+29.811543) volts
The voltage v2(t) = 0.709196 cos(10t+116.778840) volts
****************************************************
```

Example 3.16

The objective of this example is to verify the maximum power transfer theorem, when the load reactance varies over a given range.

Let the Thevenin's equivalent circuit of an electrical network be given by $V_{TH} = 5\angle 0°$ V and $Z_{TH} = 10 - j50$ Ω. Let the load impedance be given by $Z_L = 10 + jX_L$, where X_L varies over the range $0 \le X_L \le 150$ Ω, in steps of 1 Ω, as indicated in the circuit diagram of Figure 3.74.

Create the script file *max_power_AC* that returns the plot of P_{ZL} versus X_L.

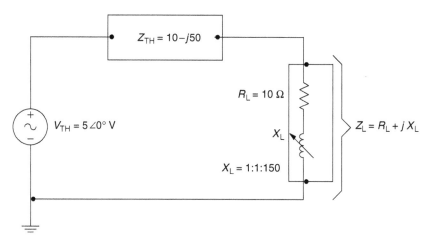

FIGURE 3.74
Network of Example 3.16

MATLAB Solution
```
% Script file: max_power_AC
% Max. power delivered to the load ZL = RL + jXL
% where XL varies from 0 to 100 Ohms
echo on
Xth = -50;
RL = 10;
Rth = 10;
Xl = 0:1:150;
Zth = Rth+50j;
Zl = RL+Xl.*j;
Zt = Zth+Zl;
magZt = abs(Zt);
magZl = abs(Zl);
Pload = (5./magZt).^2.*magZl;
plot(Xl,Pload);grid on;
xlabel('XL(Ohms)');
ylabel('Power deliver to ZL(Watts)');
title('Power ZL vs XL, for 0<XL<100')
text(41,.118,'XL=50j Ohms')
```

The script file *max_power_AC* is executed and the results are indicated in Figure 3.75.

Observe from Figure 3.75 that the maximum power delivered to the load Z_L occurs when $|X_L| = |X_{TH}|$.

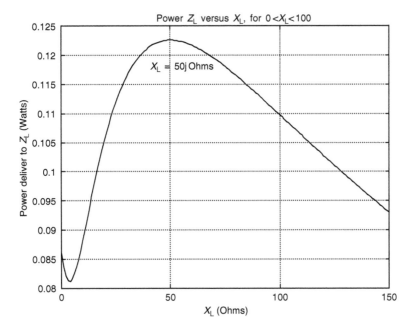

FIGURE 3.75
Plot of Z_L versus X_L of Example 3.16.

Example 3.17

The objective of this example is to verify the maximum power transfer theorem, when the load resistance varies over a given range.

Repeat Example 3.16 for the case where the load impedance is $Z_L = R_L + jX_L$, where $X_L = j30$, whereas R_L varies over the range $1 \le R_L \le 100\ \Omega$ in steps of 1 Ω, as indicated in Figure 3.76, with the Thevenin's equivalent circuit given by $V_{TH} = 5 \angle 0°$ V and $Z_{TH} = 10 - j50\ (\Omega)$.

Let us call this new file *max_power_AC_RL*.

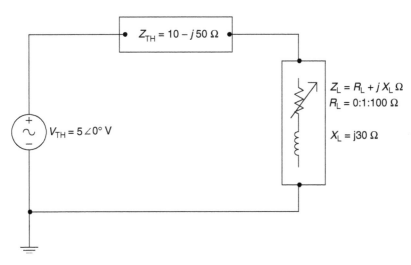

FIGURE 3.76
Network of Example 3.17.

MATLAB Solution

```
% Script file: max _ power _ AC _ RL
% Max. power delivered to ZL=RL+jXL
% where RL varies from 0 to 100 Ohms.
Xth = -50;
RL = 1:1:100;
Rth =10;
Xl = 30;
Zth = Rth+Xth*j;
Zl = RL+Xl*j;
Zt = Zth+Zl;
Pload = (RL.*25)./(abs(Zt).^2);
plot(RL,Pload);grid on;
xlabel('RL(Ohms)');
ylabel('Power deliver to ZL(Watts)');
title('Power ZL vs RL, for 0<RL<100')
text(18,.35,'RL=22.3607 Ohms')
text(13,.32,'RL=sqrt(Rth^2+(Xth+XL)^2)')
Rload _ cal=sqrt(Rth^2+(Xth+Xl)^2);
disp('*********************************');
disp('The calculated RL ,to deliver max.power is=');
disp(Rload _ cal);
disp('*********************************');
```

The script file *max_power_AC_RL* is executed, and the resulting plot is shown in Figure 3.77.

Observe that maximum power occurs at the theoretical value given by

$$R_L = \sqrt{R_{TH}^2 + (X_{TH} + X_L)^2} = 22.3607 \ \Omega$$

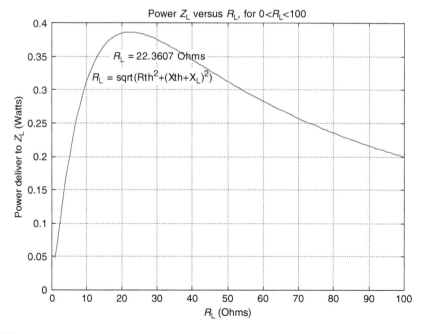

FIGURE 3.77
Plot of Z_L versus R_L of Example 3.17.

Example 3.18

Create the script file *I_versus V_RLC* that analyzes the effect of changing R of the series *RLC* circuit diagram shown in Figure 3.78, for the following values of R—540 Ω, 1.35 kΩ, and 2.7 kΩ, over the range $4 * 10^3 \le \omega \le 2\pi * 15 * 10^3$ rad/s, linearly spaced every $\Delta\omega = 500$ rad/s, and returns the following plots:

1. $Z(\omega)$ versus ω
2. $I(\omega)$ versus ω
3. $\angle Z(\omega)$ versus ω
4. $V_C(\omega)$ versus ω

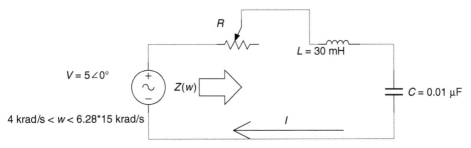

FIGURE 3.78
Network of Example 3.18.

MATLAB Solution
```
% Script file: I _ vsV _ RLC
% R = 540, L = 30mH, C =. 01uF
V = 5;
RI = 270;
R1 = RI*2;
C = .01e-6;
L = 30e-3;
W = [4000:500:2*pi*15e3];
XC = -1./(W.*C);
XL = W.*L;
Z1 = sqrt(R1^2+(XL+XC).^2);
Angle1 = atan((XL+XC)./R1);
I1 = V./Z1;
VC1 = (V*XC)./Z1;
% R = 1.35K, L = 30mH, C =. 01uF
R2 = RI*5;
Z2 = sqrt (R2^2+(XL+XC).^2);
Angle2 = atan((XL+XC)./R2);
I2 = V./Z2;
VC2 = (V*XC)./Z2;
% R = 2.7K, L = 30mH, C =. 01uF
R3 = RI*10;
Z3 = sqrt(R3^2+(XL+XC).^2);
Angle3=atan((XL+XC)./R3);
I3 = V./Z3;
VC3 = (V*XC)./Z3;
figure(1)
```

```
subplot(2,2,1)
plot(W,Z1,'*',W,Z2,'s',W,Z3,'o')
grid on
legend('R=540 Ohms', 'R=1.35KOhms', 'R=2.7KOhms');
title('[Z(w)] vs w')
ylabel('Impedance (Ohms)')
subplot(2,2,2)
plot(W,I1,'*',W,I2,'s',W,I3,'o')
title('[I(w)] vs w')
ylabel('Amplitude (amps)')
grid on
legend('I for 540', 'I for 1350', 'I for 2700');
subplot(2,2,3)
plot(W,Angle1,'*',W,Angle2,'s',W,Angle3,'o')
title(' [ang(Z(w)] vs w')
ylabel('Amplitude (radians)'); xlabel(' frequebcy w (rad/sec)');
legend('angle for 540', 'angle for 1350', 'angle for 2700');
grid on
subplot(2,2,4)
plot(W,VC1,'*',W,VC2,'s',W,VC3,'o')
grid on
title('[Vc(w)] vs w')
ylabel('Amplitude (volts)'); xlabel('frequency w (rad/sec)');
legend('Vc for 540', 'Vc for 1350', 'Vc for 2700');
```

The script file *I_vsV_RLC* is executed, and the resulting plots are shown in Figure 3.79.

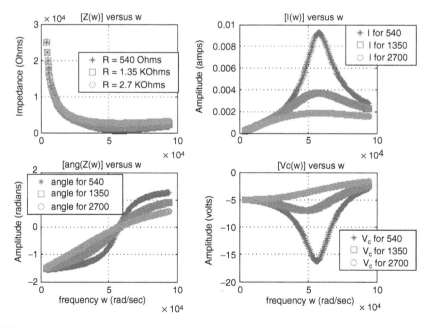

FIGURE 3.79
Plots of Example 3.18.

Example 3.19

Create the script file *RLC_parallel_analysis* that returns the following plots of the circuit shown in Figure 3.80, over the range $1000\ Hz \leq f \leq 20{,}000\ Hz$ in linear increments of $\Delta f = 250\ Hz$:

1. Figure(1)
    a. $|Z_T(\omega)|$ versus ω
    b. $\angle Z_T(\omega)$ versus ω
    c. $|Y_T(\omega)|$ versus ω
    d. $\angle Y_T(\omega)$ versus ω
2. Figure(2)
    a. $|I(\omega)|$ versus ω
    b. $\angle I(\omega)$ versus ω
    c. $|V_c(\omega)|$ versus ω
    d. $\angle V_c(\omega)$ versus ω
3. Figure(3)
    a. $|I_c(\omega)|$ versus ω
    b. $\angle I_c(\omega)$ versus ω
    c. $|I_L(\omega)|$ versus ω
    d. $\angle I_L(\omega)$ versus ω

where Z_T is the impedance seen by the source.

MATLAB Solution
```
% Script file: RLC _ parallel _ analysis
deg = 180/pi;
V= 5;
R1 = 47e3;
R2 = 270;
C = 0.01e-6;
L = 30e-3;
f = [1000:250:20e3];
w = 2*pi.*f;
XC =1./(j*w.*C);
```

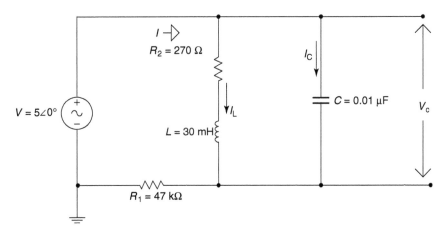

FIGURE 3.80
Network of Example 3.19.

```
XL = j*w.*L;ZS=R2+XL;
ZP = (ZS.*XC)./(ZS+XC);
ZT = R1+ZP;
YT =1./ZT;
I = V./ZT;
VC = (V.*ZP)./ZT;

figure(1)
subplot(2,2,1)
plot(w,abs(ZT))
grid on
set(gca,'fontsize',9)
title('abs[ZT(w)] vs. w','FontSize',9)
ylabel('Amplitude (Ohms)')
subplot(2,2,2)
plot(w,angle(ZT)*deg)
grid on
set(gca,'FontSize',9)
title('angle[ZT(w)] vs. w','FontSize',9)
ylabel('Angle (degrees)')
subplot(2,2,3)
plot(w, abs(YT))
grid on
set(gca,'FontSize',9)
title('abs(YT) vs. w','FontSize',9)
xlabel('frequency w (rad/sec)'), ylabel('Amplitude (Ohm^-1)')
subplot(2,2,4)
plot(w,angle(YT)*deg)
grid on
set(gca,'FontSize',9)
title('angle(YT) vs. w','FontSize',9)
xlabel('frequency in w (rad/sec)'), ylabel('Angle (degrees)')

figure(2)
subplot(2,2,1)
plot(w,abs(I))
grid on
set(gca,'fontsize',9)
title('abs[I(w)] vs. w','FontSize',9)
ylabel('Current (amps)')
subplot(2,2,2)
plot(w,angle(I)*deg)
grid on
set(gca,'FontSize',9)
title('angle[I(w)] vs. w','FontSize',9)
ylabel('angle (degrees)')
subplot(2,2,3)
plot(w, abs(VC))
grid on
set(gca,'FontSize',9)
title('abs(VC) vs. w','FontSize',9)
xlabel('frequency w (rad/sec)'), ylabel('Amplitude (volts)')
subplot(2,2,4)
plot(w,angle(VC)*deg)
grid on
set(gca,'FontSize',9)
```

```
title('angle(VC)  vs.  w','FontSize',9)
xlabel('frequency in w (rad/sec)'), ylabel('angle (degrees)')

figure(3)
IC = VC./XC; IL = VC./ZS;
subplot(2,2,1)
plot(w,abs(IC))
grid on
set(gca,'fontsize',9)
title('abs[IC(w)]  vs.  w','FontSize',9)
ylabel('Current (amps)')
subplot(2,2,2)
plot(w,angle(IC)*deg)
grid on
set(gca,'FontSize',9)
title('angle[IC(w)]  vs.  w','FontSize',9)
ylabel('angle (degrees)')
subplot(2,2,3)
plot(w, abs(IL))
grid on
set(gca,'FontSize',9)
title('abs(IL)  vs.  w','FontSize',9)
xlabel('frequency w (rad/sec)'), ylabel('Current (amps)')
subplot(2,2,4)
plot(w,angle(IL)*deg)
grid on
set(gca,'FontSize',9)
title('angle(IL)  vs.  w','FontSize',9)
xlabel('frequency in w (rad/sec)'), ylabel('angle (degrees)')
```

The script file *RLC_parallel_analysis* is executed and the results are shown in Figures 3.81 through 3.83.

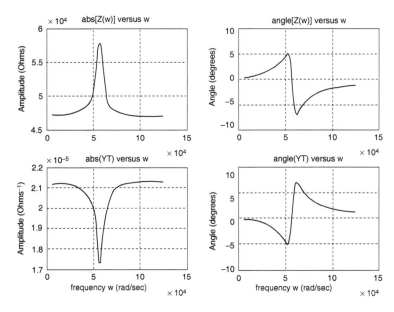

FIGURE 3.81
Plots of Example 3.19.

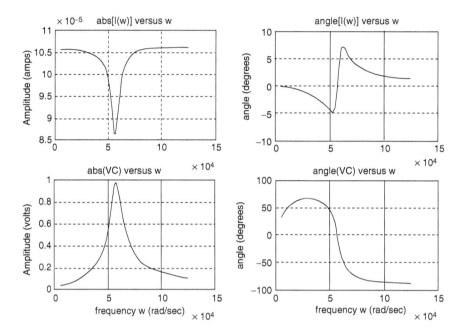

FIGURE 3.82
Plots of Example 3.19.

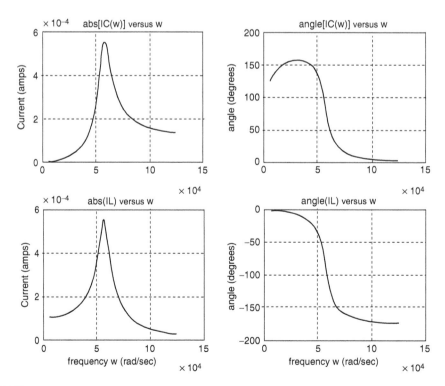

FIGURE 3.83
Plots of Example 3.19.

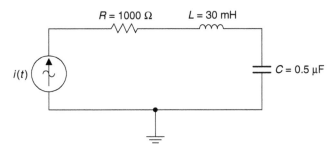

FIGURE 3.84
Network of Example 3.20.

Example 3.20

Let the current in the series RLC circuit diagram shown in Figure 3.84 be given by
$i(t) = t^3 + t^2 - 12t$ A.

Create the script file *sym_analysis_RLC* that returns the following plots using symbolic techniques:

1. $i(t)$ versus t
2. $i^2(t)$ versus t
3. $\int i(t)dt$ versus t
4. $(di(t))/dt$ versus t
5. $v_R(t)$ versus t, where $v_R(t) = R\,i(t)$
6. $v_L(t)$ versus t, where $v_L(t) = L\,di/dt$
7. $v_C(t)$ versus t, where $v_C(t) = 1/C \int i(t)dt$
8. $[v_L(t) + v_R(t)]$ versus t
9. $[v_C(t) + v_R(t)]$ versus t
10. $p_R(t)$ versus t, where $p_R(t) = i^2(t)R$
11. $p_L(t)$ versus t
12. $p_C(t)$ versus t

and the instantaneous expressions of $i(t)$, $V_R(t)$, $V_L(t)$, $V_C(t)$, $V_{RL}(t) = V_R(t) + V_L(t)$, $V_{RC}(t) = V_R(t) + V_C(t)$, $P_R(t)$, $P_L(t)$, and $P_C(t)$.

MATLAB Solution
```
% Script file : sym _ analysis _ RLC
% Analysis of an RLC series circuit
% where the current isiamps=t^3+t^2-12*t
% R=1000 Ohms, L=30 mH, and C=0.5 microF
echo off;
syms t;
iamps=t^3+t^2-12*t;

figure(1)
subplot(2,2,1)
ezplot(iamps)
title('i(t) vs.t');
ylabel('Amplitude (amps)');
grid on
isquare=iamps^2;
subplot(2,2,2)
ezplot(isquare)
```

```
title('[i(t)^2] vs. t'); ylabel ('Amplitude');
grid on
subplot(2,2,3)
inti=int(iamps);
ezplot(inti)
xlabel(' time ');ylabel('amplitide');
title('integral[i(t)] vs. t')
grid on
subplot(2,2,4)
diffi=diff(iamps);
ezplot(diffi)
xlabel(' time '); ylabel('Amplitude')
title('d[i(t)/dt] vs.t')
grid on

figure(2)
Vr=1000*iamps;
subplot(2,2,1)
ezplot(Vr)
title('vr(t) vs.t');ylabel('voltage (volts)')
grid on
subplot(2,2,2)
Vl=30e-3*diffi;
ezplot(Vl)
ylabel(' voltage (volts) ')
title('vl(t) vs.t')
grid on
subplot(2,2,3)
Vc=.5e-6*inti;
ezplot(Vc)
title('vc(t) vs. t')
xlabel(' time ')
ylabel('voltage (volts)')
grid on
subplot(2,2,4)
Vlr=Vl+Vr;
ezplot(Vlr)
title('[vr(t)+vl(t)] vs. t')
xlabel(' time ')
ylabel('voltage (volts)')
grid on

figure(3)
Vcr =Vc+Vr;
subplot(2,2,1)
ezplot(Vcr)
title('[vr(t)+vc(t)] vs. t')
ylabel('voltage (volts)')
grid on
subplot(2,2,2)
Pr=iamps*Vr;
ezplot(Pr)
title('pr(t) vs. t')
ylabel('power (watts)')
grid on
```

```
subplot(2,2,3)
Pl = iamps*Vl;
ezplot(Pl)
title('pl(t) vs t')
xlabel(' time ')
ylabel('power (watts)')
grid on
subplot(2,2,4)
Pc=iamps*Vc;
ezplot(Pc)
title('pc(t) vs. t')
xlabel(' time ')
ylabel('power (watts)')
grid on
disp('^^^^^^^^^^^^^^^^^^^^^^^^^^^^^^^^^^^^^^^^^^^^^^^^^^^^^^^^^^^^^')
disp('                        time domain results               ')
disp('^^^^^^^^^^^^^^^^^^^^^^^^^^^^^^^^^^^^^^^^^^^^^^^^^^^^^^^^^^^^^')
disp('The current i(t) (in amps) is given by: '), pretty(iamps)
disp('The voltage drop across R=1000 Ohms (in volts) is:'),pretty(Vr)
disp('The voltage drop across L=30 mH (in volts) is : '), pretty(Vl)
disp('The voltage drop across C=0.5 microF (in volts) is :'),pretty(Vc)
disp('The voltage drop across the RL (in volts) is :'),pretty(Vlr)
disp('The voltage drop across the CR (in volts) is : '),pretty(Vcr)
disp('The power _ R(t) (in watts) is :'),pretty(Pr)
fprintf('The power _ L(t) (in watts) is :'),pretty(Pl)
fprintf('The power _ C(t) (in watts is :'),pretty(Pc)
disp('^^^^^^^^^^^^^^^^^^^^^^^^^^^^^^^^^^^^^^^^^^^^^^^^^^^^^^^^^^^^^')
```

The script file *sym_analysis_RLC* is executed and the results are shown as follows (Figures 3.85 through 3.87):

```
>> sym _ analysis _ RLC
^^^^^^^^^^^^^^^^^^^^^^^^^^^^^^^^^^^^^^^^^^^^^^^^^^^^^^^^^^^^^
                    time domain results
^^^^^^^^^^^^^^^^^^^^^^^^^^^^^^^^^^^^^^^^^^^^^^^^^^^^^^^^^^^^^
The current i(t) (in amps) is given by:
                    3     2
                 t  +  t - 12 t
The voltage drop across R=1000 Ohms (in volts) is:
                 3          2
         1000 t  + 1000 t  - 12000 t
The voltage drop across L = 30 mH (in volts) is:
                    2
         9/100 t  + 3/50 t - 9/25
The voltage drop across C=0.5 microF (in volts) is:

    4722366482869645          4         4722366482869645   3
    --------------         t      +     --------------    t
  37778931862957161709568               2833419889721787 1282176


       14167099448608935      2
  -  -------------    ---    t
     4722366482869645213696
```

The voltage drop across the RL (in volts) is:

$$\frac{100009}{100} t^2 - \frac{599997}{50} t - 9/25 + 1000 t^3$$

The voltage drop across the CR (in volts) is :

$$\frac{4722366482869645}{37778931862957161709568} t^4 + \frac{28334198901940237765045645}{28334198897217871282176} t^3$$

$$+ \frac{4722366468702545765087065}{4722366482869645213696} t^2 - 12000 t$$

The power _ R(t) in watts is :

$$(t^3 + t^2 - 12 t) (1000 t^3 + 1000 t^2 - 12000 t)$$

The power _ L(t) (in watts) is:

$$(t^3 + t^2 - 12 t) (9/100 t^2 + 3/50 t - 9/25)$$

The power _ C(t) (in watts) is:

$$(t^3 + t^2 - 12 t) \left(\frac{4722366482869645}{37778931862957161709568} t^4 + \frac{4722366482869645}{28334198897217871282176} t^3 - \frac{14167099448608935}{4722366482869645213696} t^2 \right)$$

^^^

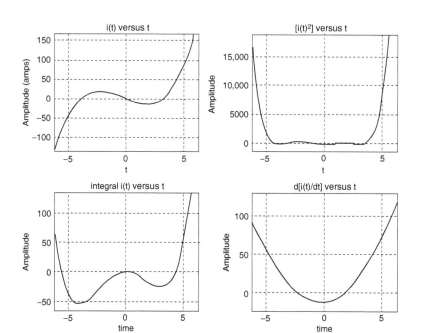

FIGURE 3.85
Plots of parts 1–4 of Example 3.20.

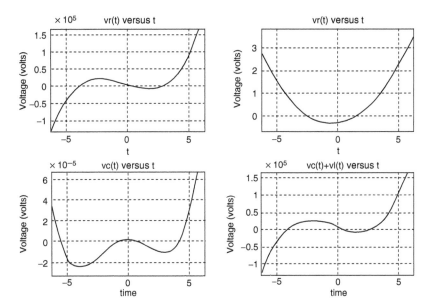

FIGURE 3.86
Plots of parts 5–8 of Example 3.20.

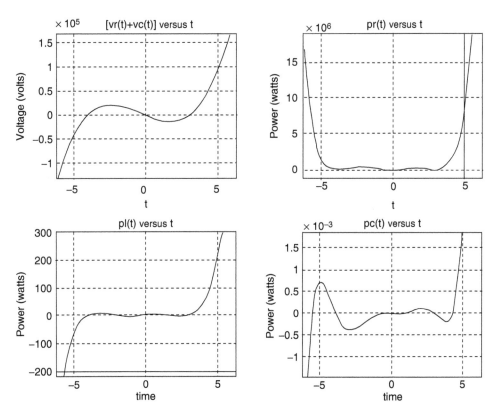

FIGURE 3.87
Plots of parts 9–12 of Example 3.20.

3.5 Application Problems

P.3.1   Which of the following relations hold
a. $3\angle135° = -3\cos(\omega t - 45°)$
b. $-2 + j2 = -2.82\cos(\omega t - 45°)$
c. $-3 - j4 = -5\cos(\omega t + 53°)$
d. $3\angle300° = 3\cos(\omega t - 60°)$
e. $4\angle-225° = -4\cos(\omega t - 45°)$

P.3.2   A sine wave $x(t)$ has a frequency of $f = 1000$ Hz, a peak value of 3.0, and a phase angle of $-45°$.
a. Determine its period T and angular frequency ω
b. Obtain the equation for $x(t)$
c. Sketch by hand $x(t)$ versus degrees, radians, and time
d. Write a program that returns the plots for part c over two complete cycles

P.3.3   Verify that the current is 53.8 A with a phase $\Phi = 57°30'$, when a 200 V RMS, 50 Hz is applied to a series RL circuit with $R = 2\ \Omega$ and $L = 0.01$ H.

P.3.4   Verify using MATLAB that $100\angle45 + 80\angle120° = 30.7 + j140$.

P.3.5   Verify that if $V = 100\angle45°$ and $Z = 20\angle60°$, then $I = 5\angle-15°$, by hand and by using MATLAB.

P.3.6   Verify that if $V = 50 + j75$ V and $I = 3 + j5$ A, then its power is $225 + j475$ W.

P.3.7   Show that when the frequency is 151 Hz and its impedance is $28.8\angle80°\ \Omega$, then the current is lagging the voltage by $80°$ if the circuit is a series RL, with $R = 5\ \Omega$ and $L = 30$ mH.

P.3.8   Let the impedance of a given circuit be $Z = 15 + 20j\ \Omega$, and the voltage across it be $v(t) = 100\cos(100t)$ V. Verify that the current through it is $i(t) = 4\cos(100t - 53°8')$ A.

P.3.9   For the problem P.3.8, verify that Z is equivalent to an inductor $L = .2$ H in series with a resistor $R = 15\ \Omega$.

P.3.10 For what frequency is the series RLC circuit of Figure 3.88 equivalent to the parallel RC circuit of Figure 3.89?

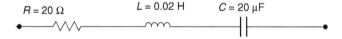

FIGURE 3.88
Series network diagram of P.3.10.

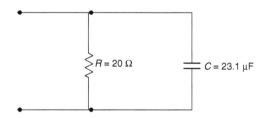

FIGURE 3.89
Parallel network diagram of P.3.10.

P.3.11 Let the impedance of a given circuit be given by $Z = 7.05 + j9.7\ \Omega$, and the peak sinusoidal voltage across be $V_m = 60$ V. Verify the following:

a. $Z = 12\ \angle 54°\ \Omega$

b. $I_m = 5$ A

c. $V_{eff} = 42.42$ V

d. $I_{eff} = 3.535$ A

e. $S = P_{AP} = 150\ va$

f. $P = 88.1$ W

g. $Q = P_{reac} = 121.2\ var$

h. $p(t) = V_m \cdot I_m\ [\cos(\theta)\sin(2\omega t) - (1/2)\sin(\theta)\sin(2\omega t)]$, where $\cos(\theta) = \cos(54°) = 0.588$ and $\sin(\theta) = \sin(54°) = 0.809$

i. Evaluate the maximum and minimum of $p(t)$ by solving the equation given by $\dfrac{dp(t)}{dt} = 0$

Check if the solutions occur at $\omega t = 27$ rad or 117 rad

$$\text{and if } p_{max} = 238\text{ W}$$

and

$$p_{min} = -61.7\text{ W}$$

P.3.12 A resistor of $50\ \Omega$ and an inductor of $L = 0.1$ H are connected in series where the applied voltage is sinusoidal with $110\ V_{RMS}$, and frequency $f = 60$ Hz. Verify the following:

a. $v(t) = 156\sin(2\pi\ 60t)$ V

b. $Z = 50 + j37.7\ \Omega$

c. $Z = 62.5\ \angle 37°\ \Omega$

d. $I = 2.5\ \angle{-37°}$ A

e. $i(t) = 2.5\sin(2\pi\ 60\ t - 37°)$ A

f. $I_{RMS} = 1.76$ A

g. $|Z| = 62.5\ \Omega$

P.3.13 A series RC circuit consisting of a resistor $R = 10$ KΩ and a capacitor $C = 10\ \mu$F, with an applied voltage given by sine wave of 10 V, with a frequency of 1 kHz. Verify the following:

a. $v(t) = 14.14\sin 6283t$ V

b. $Z = 10,000 - j15,900\ \Omega$

c. $Z = 18,800\ \angle{-58°}\ \Omega$

d. $I = 0.752\ \angle 58°$ mA

e. $i(t) = 0.752\sin(6283t + 58°)$ mA

f. $I_{RMS} = 0.532$ mA

g. $V_{R(peak)} = 7.52$ V

h. $V_{R(RMS)} = 5.32$ V

i. $V_{C(peak)} = 11.96$ V

j. $V_{C(RMS)} = 8.46$ V

P.3.14 Solve for the loop currents shown in the circuit diagram of Figure 3.90, and verify the following relations:

$$I_1 = 2 - j \text{ A}$$

$$I_2 = 0.5 - j1.5 \text{ A}$$

$$I_3 = -1 - j2 \text{ A}$$

where all the impedances are given in ohms.

P.3.15 Given the following loop system equations:

$$(2 + j2)I_1 - jI_2 = 5$$

$$-jI_1 + (1 - j)I_2 = j3$$

Draw its circuit diagram.

P.3.16 Solve for the loop currents shown in the circuit diagram of Figure 3.91 and verify if $I_1 = -j4$ A, $I_2 = 0.8$ A, and $I_3 = -1.6 - j4$ A.

P.3.17 For a given circuit, if the applied voltage is $v(t) = 200\sin(\omega t + 15°)$ V and the resulting current is $i(t) = 20\sin(\omega t - 30°)$ A, draw its power triangle.

P.3.18 Given the series RLC circuit of Figure 3.92, draw its power triangle.

P.3.19 Write a MATLAB program that returns the following plots:

a. $|Z(\omega)|$ versus ω

b. $\angle Z(\omega)$ versus ω

for circuit diagram of Figure 3.93.

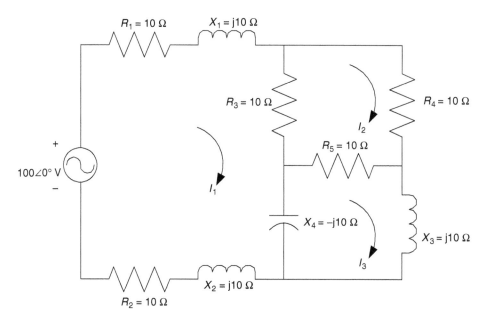

FIGURE 3.90
Network diagram of P.3.14.

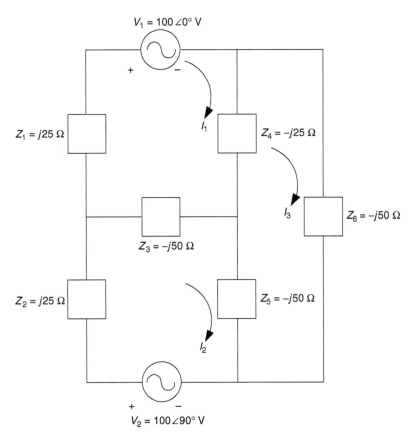

FIGURE 3.91
Network diagram of P.3.16.

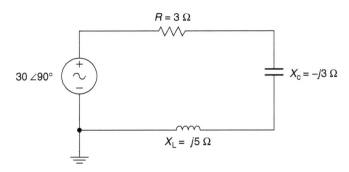

FIGURE 3.92
Network diagram of P.3.18.

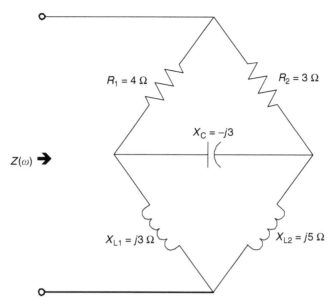

FIGURE 3.93
Network diagram of P.3.19.

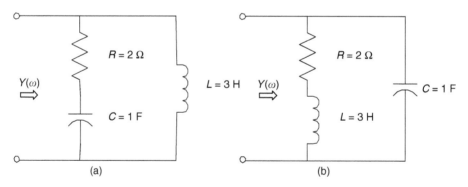

FIGURE 3.94
Network diagrams of P.3.20.

P.3.20 Write a MATLAB program for each circuit shown in Figure 3.94 that returns the following plots over 1 rad/s $\leq \omega \leq$ 150 K rad/s:

a. $[Y(\omega) = 1/Z(\omega)]$ versus ω

b. $\angle Y(\omega)$ versus ω

P.3.21 Write a program for the circuit diagram shown in Figure 3.95 that returns the voltage across $R_2 = 2\ \Omega$ by using

a. Node equations

b. Loop equations

c. Thevenin's theorem

d. Norton's theorem

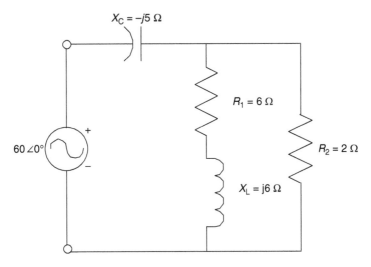

FIGURE 3.95
Network of P.3.21.

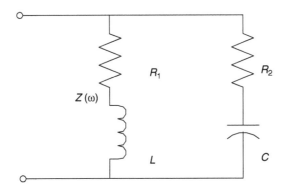

FIGURE 3.96
Network of P.3.22.

P.3.22 Verify that the resonant frequency for the circuit shown in Figure 3.96 is given by

$$\omega_R = \frac{1}{\sqrt{LC}} \sqrt{\frac{R_1^2 - (L/C)}{R_2^2 - (L/C)}}$$

P.3.23 Given the series RLC circuit consisting of $R = 5\ \Omega$, $L = 6$ mH, and $C = 10\ \mu$F, create a function file that returns the following plots:
   a. $|Z(\omega)|$ versus ω
   b. $\angle Z(\omega)$ versus ω
   c. $|I(\omega)|$ versus ω
   d. $\angle I(\omega)$ versus ω
   over the range $0.7\omega_R \leq \omega \leq 1.4\omega_R$, assuming that the applied voltage is 10 V RMS.

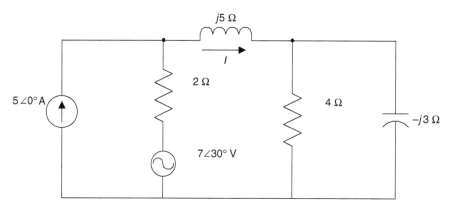

FIGURE 3.97
Network of P.3.24.

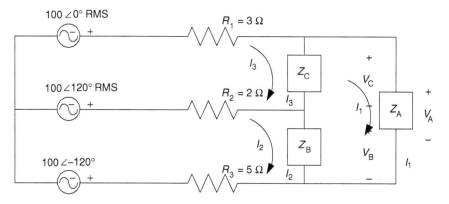

FIGURE 3.98
Unbalanced 3 Φ network of P.3.25.

P.3.24 Create a program that returns the current I (through the inductor) shown in the circuit diagram of Figure 3.97 by using

   a. Superposition

   b. Thevenin's theorem

   c. Norton's theorem

   d. Source transformation

P.3.25 For the unbalanced 3 Φ system shown in Figure 3.98, find the currents I_1, I_2, and I_3 and the voltages V_A, V_B, and V_C, where

$$Z_A = 2 + j3\ \Omega$$

$$Z_B = 4 - j6\ \Omega$$

$$Z_C = 12 + j8\ \Omega$$

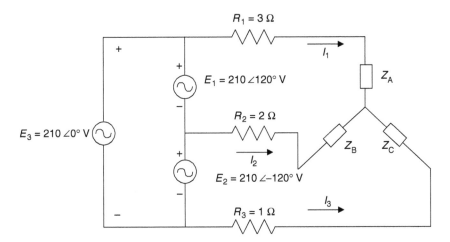

FIGURE 3.99
Unbalanced 3 Φ network of P.3.26.

P.3.26. Write a program that returns the line currents and the voltage drop across each impedance of the 3 Φ system shown in Figure 3.99, where

$$Z_A = 2 + j3 \ (\Omega)$$

$$Z_B = 4 - j6 \ (\Omega)$$

$$Z_C = 12 + j8 \ (\Omega)$$

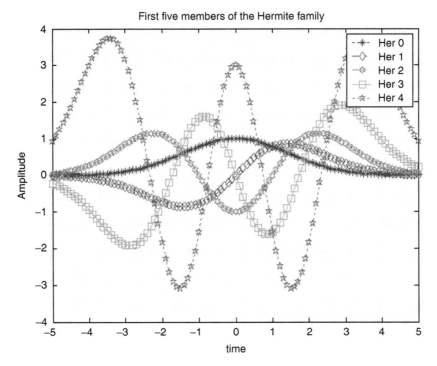

COLOR FIGURE 1.18
Plots of the Hermite family of R.1.49.

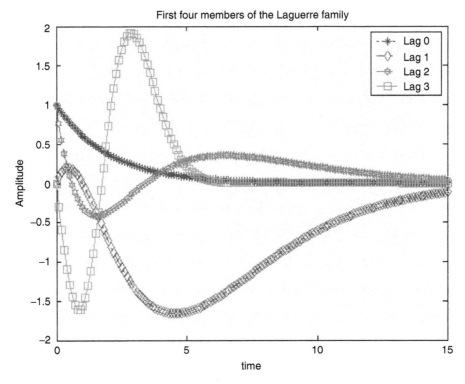

COLOR FIGURE 1.19
Plots of the Laguerre family of R.1.51.

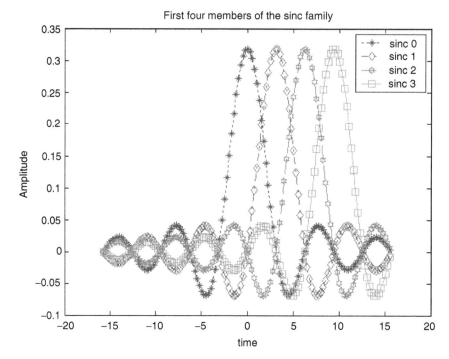

COLOR FIGURE 1.20
Plots of the *sinc* family of R.1.53.

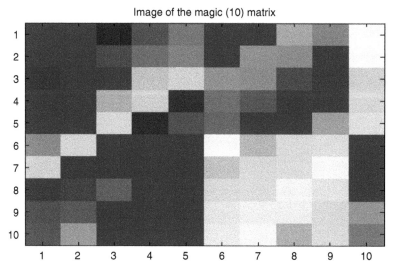

COLOR FIGURE 1.46
Color plot of the magic matrix of R.1.169.

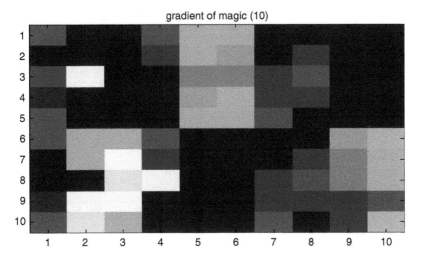

COLOR FIGURE 1.47
First color derivative plot of the magic matrix of R.1.169.

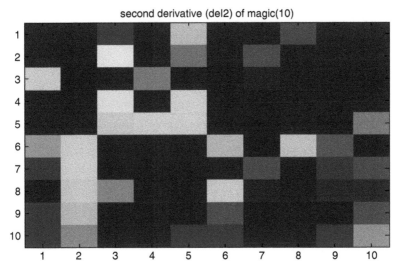

COLOR FIGURE 1.48
Second color derivative plot of the magic matrix of R.1.169.

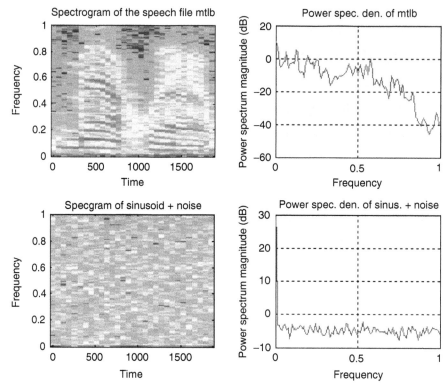

COLOR FIGURE 4.21
Plots of parts 5, 6, 7, and 8 of R.4.125.

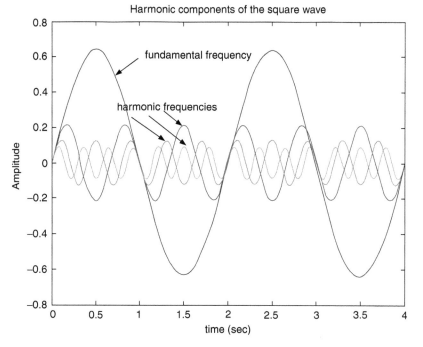

COLOR FIGURE 4.29
Plots of part b of Example 4.1.

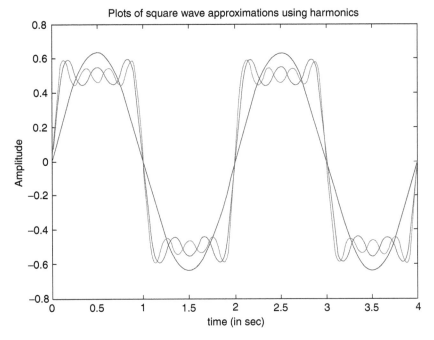

COLOR FIGURE 4.31
Plots of part d of Example 4.1.

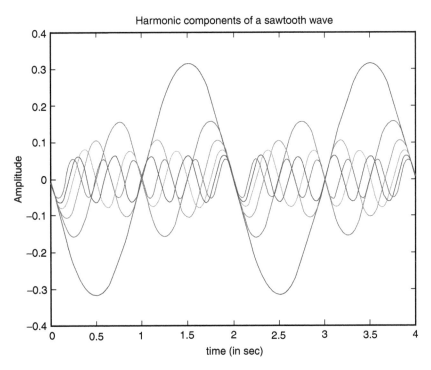

COLOR FIGURE 4.34
Plots of part b of Example 4.2.

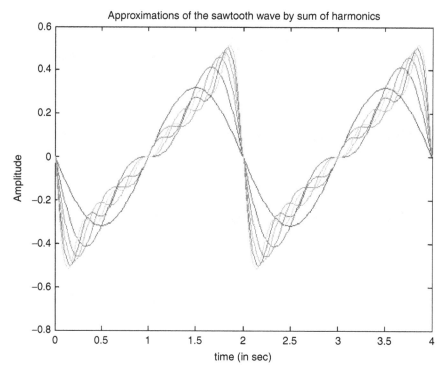

COLOR FIGURE 4.36
Plots of part d of Example 4.2.

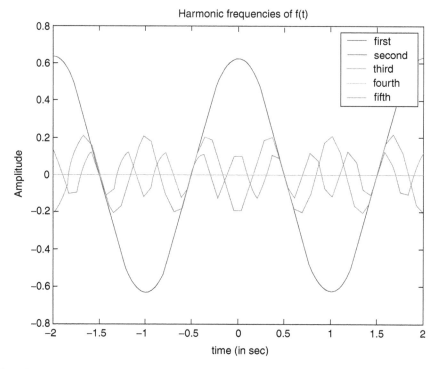

COLOR FIGURE 4.41
Plots of part b of Example 4.4.

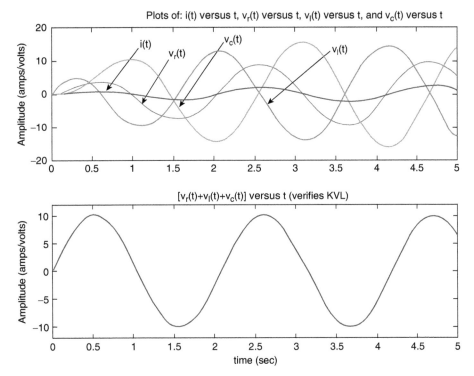

COLOR FIGURE 4.72
Plots of $i(t)$, $v_l(t)$, $v_r(t)$, and $v_c(t)$ on the same plot (top), and verification of KVL (bottom) of Example 4.16.

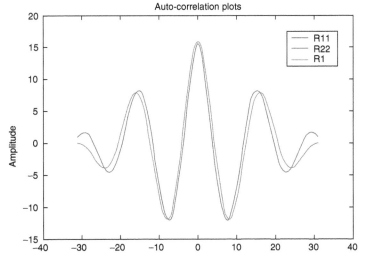

COLOR FIGURE 5.49
Autocorrelation plots of part c of Example 5.11.

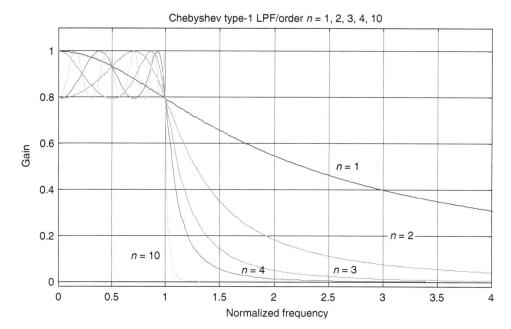

COLOR FIGURE 6.14
Magnitude plots of normalized analog Chebyshev type-1 LPFs of orders $n = 1, 2, 3, 4$, and 10.

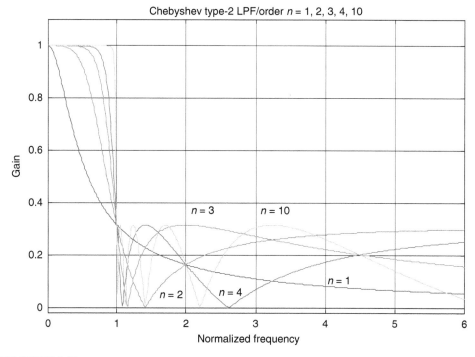

COLOR FIGURE 6.15
Magnitude plots of normalized analog Chebyshev type-2 LPFs of orders $n = 1, 2, 3, 4$, and 10.

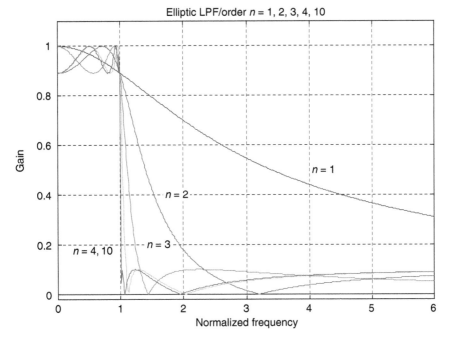

COLOR FIGURE 6.16
Magnitude plots of normalized analog elliptic LPFs of orders $n = 1, 2, 3, 4$, and 10.

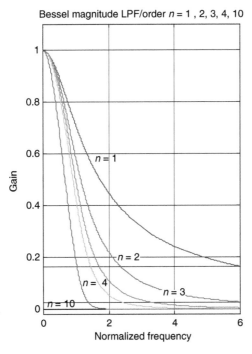

COLOR FIGURE 6.17
Magnitude plots of normalized analog Bessel LPFs of orders $n = 1, 2, 3, 4$, and 10.

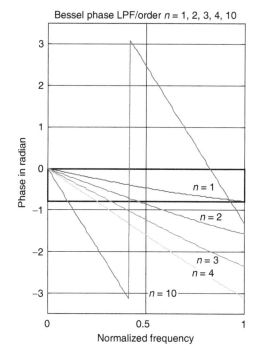

COLOR FIGURE 6.18
Phase plots of normalized analog Bessel LPFs of orders $n = 1, 2, 3, 4,$ and 10.

4

Fourier and Laplace

What we know is but a little thing;
what we are ignorant of is immense.

Pierre Simon de Laplace

4.1 Introduction

This chapter is devoted to the Fourier series (FS) and the Fourier and Laplace transforms named after the French mathematicians Jean Baptiste Joseph Fourier (1768–1830)* and Pierre Simon de Laplace (1749–1829).[†]

The FS expansion (presented in 1807), called "the great mathematical poem" by Lord Kelvin, deals with waves that are periodic in nature, such as sound, light, radio, thermal, ocean, pressure, or force waves.

Fourier (1822) while studying the problem associated with the flow, propagation, and conduction of heat, showed that any arbitrary wave can be expressed as a linear infinite series of sinusoidal (sines and cosines) functions of harmonic-related frequencies. The purpose of representing a function in terms of an FS is that the function can be analyzed in terms of its frequency components, making it much easier to understand and calculate its power and energy distribution content.

This model is, in many cases, the preferred form to analyze a function. The FS can be used to evaluate the amount of power concentrated at each harmonic; hence, the relative importance of each harmonic of a given wave. The theory developed by Fourier is used for solving many engineering problems in diverse areas such as communications, signal processing,

---

* Baron Jean Baptist Joseph Fourier—distinguished scientist, professor of mathematics and physicist, politician, and diplomat—served Napoleon and King Louis XVIII. He played an important role in Napoleon's expeditions, in particular the one to Egypt. He is credited (1807) with the idea that any arbitrary function, defined by different analytical expressions over adjacent segments of its range can be expressed by a single analytical expression.

† Pierre Simon Laplace (a count and years later a marquis)—distinguished scientist, mathematical astronomer, and mathematician, professor at the Paris École Militaire, a member of the Académic Royale des Sciences, astronomer, and an accomplished diplomat and politician—served Napoleon and King Louis XVIII governments, as president of the senate, and as a member of the Chamber of Peers. As a scientist he was second only to Newton; but as a person his qualities are mixed. Laplace was born at Beaumont-en-Auge, Normandy, on March 23, 1747. Facts of his life were destroyed by a fire in the year 1929, and lost during the bombardments of Caen, in WWII. Laplace adapted easily to the social and political changes of his time. He prospered financially, scientifically, socially, and politically. In 1784, Laplace was appointed examiner to the royal artillery, a lucrative and prominent post. There he had the good fortune to examine a young ambitious 16-year-old sublieutenant named Napoleon Bonaparte, who years later became the ruler of continental Europe. This relationship gave Laplace a number of opportunities that he took advantage of.

controls, heat and wave propagation, electronic power supplies, and signal generation, as well as nonengineering areas such as biology, physiology, economics, and music.

The illogical concept that a continuous varying function such as a sinusoidal could be used to approximate functions with square corners and discontinuities was first received by the best minds of the time, such as the members of the French Academy of Science, with skepticism. Probably this concept will also be received with skepticism by the reader, but Fourier was right.

The Fourier transform (FT) is an extension of the FS for the case of nonperiodic functions as well as the series; it represents the original function in a format that is, in many instances, easier to analyze and understand. The FT is a compact, clever, and symmetrical relation between a function and its Fourier complex expansion.

The transform of a function is referred to as its spectrum, and physically constitutes a model that is as good as, or better than the function itself. For example, an electrical, optical, or acoustical wave can be observed and studied using a spectrum analyzer in frequency, as well as an oscilloscope in time.

The Laplace transform (LT), a cousin of the FT, is introduced later in this chapter as a tool to solve linear, ordinary differential equations (DEs) (with initial conditions).

The main idea is to convert the DE such as the loop or node equations of an electric network from the time to the frequency domain (referred as the s-domain). This conversion transforms the DE into an algebraic equation. The algebraic frequency domain equation provides, in many situations, information and insight into a system, not evident in the time domain, such as stability and its pole/zero constellation, and is, in general, an easier system model to deal with.

Once a solution of the algebraic equation is found in the frequency domain (in terms of s), an inverse transform is required to convert the solution from s (frequency) into t (time domain).

For this purpose, a conversion table (Table 4.2) can be used, or if the symbolic MATLAB® toolbox is available, the commands *laplace* and *ilaplace* can be used for the evaluation of the direct and inverse transformations. The forerunner of the LT method was the operational calculus, created by Oliver Heaviside (1850–1925), which was used to solve transients in a circuit described by a set of DEs; a method not fully understood and accepted by the leading scientists and mathematicians of his time. Heaviside was a gifted practical engineer who had the physical insight to pick the correct solution from a number of possible solutions—a heuristic approach that lacked the mathematical rigor. Years after Heaviside's publications, his method was rigorously verified by men such as Bromwich, Giorgi, Carson, and others.

The verification came basically from Laplace's work in 1807. The Laplace method of solving DEs is particularly useful in circuit analysis since initial conditions are automatically incorporated into the equation, as sources, in the first step rather than the last.

The FTs as well as the LTs are sophisticated mathematical techniques that engineers and scientists have developed over the past 100 years, which are extensively employed in research and development. Many achievements, discoveries, major contributions, and applications in modern technology are based on the concepts developed by Fourier and Laplace in a variety of areas such as analysis and filtering of data, image and music processing, image and music enhancement, sound effects, system analysis, controls, and many other applications.

The Fourier and Laplace methods introduced over the past century a new point of view of system analysis and synthesis, new terminology and new concepts that evolved and over time became a part of the specialized vocabulary used by engineers and scientists to communicate with one another.

4.2 Objectives

After completing this chapter, the reader should be able to

- Mathematically state the trigonometric and exponential FS
- State that sine and cosine are the basic periodic (orthogonal) functions used to approximate an arbitrary function
- State that any arbitrary periodic function can be resolved into harmonic components
- Evaluate the coefficients of the FS expansion of any arbitrary signal
- Compute the FS coefficients of a square and sawtooth wave
- Improve the partial sums of the FS by the use of a window function
- State the FT and the inverse FT (IFT) equations
- Understand the importance of the Fourier analysis
- Determine the linear and power spectral distribution of arbitrary signals
- Demonstrate that any arbitrary time function can be represented in the frequency domain by two plots referred as the amplitude and phase spectrum plots
- Understand the physical implications of a periodic and nonperiodic signal in the frequency domain
- Evaluate the FT of periodic signals
- State the concept and conditions for amplitude and delay distortion
- Evaluate the percentage of harmonic and total distortion
- Define the concept and purpose of an equalizer circuit (to correct distortion)
- Use Parseval's theorem to calculate the power for periodic and nonperiodic time signals
- Understand the concept of energy and power signals
- Revisit the concepts of root mean square (RMS), power, and energy in the frequency domain
- State the integral definition of the direct and inverse LT (ILT)
- Use the LT pairs in the analysis of electrical circuits, systems, and signals
- Find the LT of simple functions by using tables, as well as the symbolic MATLAB toolbox
- Understand the purpose and power of the LT in the analysis of electrical circuits
- Express electrical circuit elements using the s-domain model (Laplace), where initial stored energy conditions are incorporated in the model (as sources)
- Understand the procedure by which integral differential and DEs can be solved by using the LT
- State and be able to apply the initial and final value theorems in the solution of electrical circuit problems
- Identify the effects of waveform symmetry in the computational steps involved in spectra analysis
- State the basic relationships (referred to as properties) between the time- and frequency domain

- Revisit the process of partial fraction expansion when evaluating the ILT
- Understand the differences between the FT and the LT
- Know when and how the FS, FT, and the LT are used
- Use MATLAB as a tool to solve circuit, signal, and system problems using Fourier and Laplace techniques

4.3 Background

R.4.1   Recall that a continuous time function $f(t)$ is periodic if it satisfies the following relation:

$$f(t) = f(t \pm nT), \quad \text{for } n = 0, 1, 2, 3, \ldots, \infty$$

where T is referred to as its period, which is the smallest possible value that satisfies the preceding relation.

R.4.2   Let $f(t)$ be a periodic function with period T, then the trigonometric FS is given by

$$f(t) = \frac{a_0}{2} + \sum_{n=1}^{\infty} [a_n \cos(nw_0t) + b_n \sin(nw_0t)]$$

where

$$a_n = \frac{2}{T} \int_0^T f(t)\cos(nw_0t)dt \quad \text{for } n \geq 0$$

$$b_n = \frac{2}{T} \int_0^T f(t)\sin(nw_0t)dt \quad \text{for } n \geq 1$$

where $w_0 = 2\pi/T$, for $n = 1, 2, 3, \ldots, \infty$.
    The cosine–sine series is the most popular way to define the FS and its coefficients, but as will be seen later in this chapter, it is not the most convenient.

R.4.3   An alternate form to evaluate the coefficients (a_ns and b_ns) is to evaluate the integrals with respect to w_0t, over the period defined by 2π radians. The equations used to evaluate the coefficients of the trigonometric FS are then given by

$$a_n = \frac{1}{\pi} \int_0^{2\pi} f(t)\cos(nw_0t)\, d(w_0t)$$

$$b_n = \frac{1}{\pi} \int_0^{2\pi} f(t)\sin(nw_0t)\, d(w_0t)$$

R.4.4   Observe that the limits of integration in the evaluation of the Fourier coefficients, in either of the preceding cases, with respect to t or w_0t, must be over one full period, but need not be from 0 to T, or from 0 to 2π, but may be from $-T$ to 0, $-T/2$ to $+T/2$, -2π to 0, or $-\pi$ to $+\pi$.

R.4.5 The FS converges uniformly to $f(t)$ at all the continuous points and converges to its mean value at the discontinuity locations.

R.4.6 The constant ω_0 is called the fundamental frequency, and all of its integer multiples $(n\omega_0)$ are referred to as harmonic frequencies. The fundamental frequency w_0 is the lowest sinusoidal frequency in the FS expansion. All other frequencies are integer multiples of the fundamental frequency.

R.4.7 The coefficient $[a_0/2]$ is referred to as the DC component of the series, and represents the average value of $f(t)$ over a period T.

R.4.8 The sufficient conditions that ensure the existence and convergence of the FS expansion for an arbitrary function $f(t)$ are referred as Dirichlet's conditions. These conditions are

a. $f(t)$ should present a finite number of discontinuities in any period

b. $f(t)$ should present a finite number of maxima and minima over any period T

c. $f(t)$ should be absolutely integrable over a period T, that is $\int_{-T/2}^{T/2} |f(t)| dt < k < \infty$, where k is a finite quantity

d. $f(t)$ must be single valued everywhere

All practical (electrical or mechanical) waveforms in nature satisfy the Dirichlet's conditions.

R.4.9 Let $f(t)$ be a periodic function, then it can be represented by a sum of either cosine or sine terms, indicated as follows:

$$f(t) = \frac{a_0}{2} + \sum_{n=1}^{+\infty} c_n \cos(n\omega_0 t + \theta n) = \frac{a_0}{2} + \sum_{n=1}^{\infty} c_n \sin\left(nw_0 t + \theta n + \frac{\pi}{2} \right)$$

where $c_n = \sqrt{a_n^2 + b_n^2}$ and $\theta_n = -\tan^{-1}(b_n/a_n)$.

Note that the FS is an expansion of $f(t)$, over the range $-\infty \le t \le +\infty$ (everywhere).

R.4.10 Let $f(t)$ be a periodic function, then it can be expressed in terms of an exponential FS as indicated in the following (by replacing the sinusoids of FS by the Euler's identities):

$$f(t) = \sum_{n=-\infty}^{n=+\infty} F_n e^{jnw_0 t}$$

where

$$F_n = \frac{1}{T} \int_0^T f(t) e^{-jnw_0 t} dt$$

R.4.11 The complex exponential coefficients F_n can be evaluated in terms of the trigonometric coefficients (a_ns and b_ns), by replacing the exponential $e^{-jw_0 t}$ with $\cos(nw_0 t) + j \sin(nw_0 t)$, and by equating the trigonometric FS with the exponential FS, obtaining the following relations:

$$a_n = 2 \, \text{real}\{F_n\}$$
$$b_n = -2 \, \text{imag}\{F_n\}$$

The Fourier coefficients a_n, b_n, and F_n represent the degree of similarity between $f(t)$ and each of the frequency components nw_0.

R.4.12 The relation between the exponential coefficients F_n and the trigonometric coefficients a_ns and b_ns are as follows:

$$F_0 = \frac{a_0}{2}$$

$$F_n = \frac{1}{2}(a_n - jb_n) \qquad F_{-n} = \frac{1}{2}(a_n + jb)$$

$$a_0 = F_n + F_{-n}$$

$$b_n = j(F_n - F_{-n})$$

$$c_n = \sqrt{(a_n^2 + b_n^2)} = 2|F_n|$$

R.4.13 Recall that the average power of a periodic function $f(t)$ may be evaluated in the time domain by

$$P_{ave} = \frac{1}{T}\int_0^T |f(t)|^2 dt$$

R.4.14 The time function $f(t)$ may be viewed as a current or a voltage that acts on a resistor of $1\,\Omega$ (normalized).

Recall that

$$\text{Power} = \frac{v(t)^2}{R} = v(t)^2 \text{ W}$$

or Power $= i^2(t)R = i^2(t)$W (assuming $R = 1\,\Omega$ without any loss of generality) and the average power is therefore given by

$$P_{ave} = \frac{1}{T}\int_0^T i(t)^2 dt = \frac{1}{T}\int_0^T v(t)^2 dt$$

R.4.15 Parseval's relation (also known as Parseval's theorem) states that if $f(t)$ is a real and periodic function, then the average power denoted by P_{ave} may be conveniently evaluated in the frequency domain by

$$P_{ave} = \frac{1}{T}\int_0^T |f(t)|^2 dt = \sum_{n=-\infty}^{\infty} |F_n|^2$$

Note that the evaluation of P_{ave} in the frequency domain is much easier than in the time domain, since integration is substituted by the summation. Observe that P_{ave} can be easily evaluated if the coefficients F_n are known by using MATLAB in the following way:

Let

$$F = [F_{-n}\ F_{-n+1}\ \cdots\ F_{-1}\ F_0\ F_1 \cdots\ F_n]$$

Then $P_{ave} = F * F'$.

R.4.16 For example, let us verify Parseval's theorem by means of the function $f(t) = 2\sin(100t)$, evaluating by hand the average power of $f(t)$ in

a. The time domain by using integration
b. The frequency domain by using addition

ANALYTICAL Solution

1. Time domain solution

$$w_0 = 100 \text{ rad/s} \qquad T = 2\pi/w_0 = [2\pi/100] \text{ s}$$

$$P_{ave} = \frac{1}{T}\int_0^T |f(t)|^2 \, dt = \frac{100}{2\pi} \int_0^{2\pi/100} \{2\sin(100t)\}^2 \, dt$$

$$P_{ave} = \frac{400}{2\pi} \int_0^{2\pi/100} \left\{ \frac{1}{2} - \frac{1}{2}\cos(200t) \right\} dt = \frac{400}{4\pi}\left(\frac{1}{2}\right) t \Big|_0^{2\pi/100}$$

$$P_{ave} = \left(\frac{400}{2\pi}\right)\left(\frac{1}{2}\right)\left(\frac{2\pi}{100}\right) = 2 \text{ W}$$

2. Frequency domain solution

$$f(t) = 2\sin(100t) = 2\left[\frac{e^{j100t} - e^{-j100t}}{2j}\right] \qquad \text{by Euler's identity}$$

$$2\sin(100t) = -je^{j100t} + je^{-j100t}$$

Then

$$F_0 = 0, \quad |F_1| = 1, \quad |F_{-1}| = 1$$

and all the coefficients

$$F_n = 0, \quad \text{for } n = 2, 3, \ldots, \infty$$

Then

$$P_{ave} = \sum_{n=-1}^{n=1} |F_n| = |F_1| + |F_{-1}| = 1 + 1 = 2 \text{ W}$$

R.4.17 Any periodic wave $f(t)$ can be approximated by cosine terms only; then

$$f(t) = \frac{a_0}{2} + \sum_{n=1}^{+\infty} c_n \cos(nw_0 t + \theta_n)$$

Then its RMS value can be evaluated by

$$F_{\text{RMS}} = \sqrt{\left(\frac{a_0}{2}\right)^2 + \left(\frac{c_1}{\sqrt{2}}\right)^2 + \left(\frac{c_2}{\sqrt{2}}\right)^2 + \cdots + \left(\frac{c_n}{\sqrt{2}}\right)^2}$$

Note that the RMS value can be easily evaluated using MATLAB as indicated in the following:

let

$$F = \begin{bmatrix} \sqrt{2}a_0/2 & c_1 & c_2 & \cdots & c_n \end{bmatrix} .* \left(\frac{1}{\sqrt{2}}\right)$$

Then the MATLAB command, Frms = norm(F), returns the RMS value of $f(t)$.

R.4.18 Harmonic distortion is an index that indicates the discrepancy between the series approximation of $f(t)$, and the actual waveform of $f(t)$. The percentage distortion due to a particular Fourier component (harmonic) is given by

$$\text{Percentage distortion for the n-harmonic component (PDn)} = \frac{c_n}{c_1} * 100\%$$

R.4.19 The percentage of total harmonic distortion (PTHD) is given by

$$\text{PTHD} = \sqrt{\left(\frac{c_2}{c_1}\right)^2 + \left(\frac{c_3}{c_1}\right)^2 + \cdots + \left(\frac{c_n}{c_1}\right)^2} * 100\%$$

R.4.20 Gibb's* phenomena states that if $f(t)$ presents some discontinuities, then the FS approximation at the discontinuities would show a significant amount of ripple, and the synthesized function converges to the average value at the point of the discontinuity, if a sufficiently large number of terms are employed in its series approximation. Recall that a discontinuity refers to any point of $f(t)$, whose amplitude abruptly changes from one value to another (step).

Josiah Willard Gibbs first published the preceding observation in 1899.

R.4.21 Recall that the word synthesis means that the sum of the parts constitutes the whole.

For the case of the Fourier analysis, the term synthesis means that the recombination of the terms of the FS, usually the first five or six terms, represent a good approximation of the original wave.

R.4.22 The coefficients F_0, F_1, F_{-1}, F_2, F_{-2}, ..., F_n of the exponential FS represent the magnitude of the fundamental and all the harmonic frequencies at 0, w_0, $-w_0$, $2w_0$, $-2w_0$, ..., $\pm nw_0$, respectively.

---

* Josiah Willard Gibbs (1839–1903), a physicist–chemist at Yale University, where he served honorarily for 10 years; he also served with a reduced salary at John Hopkins University.

A one-sided spectrum consists of only the positive frequencies $w \geq 0$, whereas the two-sided spectrum is over $-\infty \leq w \leq +\infty$. In this book, the term spectrum refers to the two-sided spectrum.

R.4.23 The plot of Fn versus nw_0 is called the line or discrete spectrum of $f(t)$.

Since Fn is, in general, a complex function, two plots are required to completely define its behavior. They are referred to as

a. The magnitude spectrum plot

b. The phase spectrum plot

R.4.24 If the line (spectrum) representing the magnitudes of the coefficients F_ns decrease rapidly, then the FS converges rapidly to $f(t)$, implying that the wave is continuous. If, on the contrary, $f(t)$ has discontinuities, then the F_ns slowly decrease, and $f(t)$ is referred to as having strong high harmonic components, implying that many terms are required for a good approximation of $f(t)$.

R.4.25 For example,

Let

$$f(t) = \sum_{n} Fn e^{jnw_0 t}$$

Then the generic sketches shown in Figures 4.1 and 4.2 represent typical plots of the following:

a. Line spectrum

b. Phase spectrum

c. Magnitude spectrum

d. Power spectrum, assuming that $T = 1$ s

ANALYTICAL Solution

The generic spectrum plots are shown in Figures 4.1 and 4.2.
Note that $F_0 = 3, F_1 = F-1 = 2, F_2 = F-2 = 0.5$ and $F_3 = F-3 = 1$.

R.4.26 Observe from the plot of Figure 4.1 that the amplitude spectrum is always symmetric, that is, $Fn = F - n$, whereas the phase spectrum is always asymmetric, given by $\theta n = -\theta - n$.

R.4.27 One of the features of the exponential FS coefficients is their symmetry regarding the variables t and nw_0. Symmetry considerations are of great computational advantage in evaluating the components of $f(t)$ as indicated in the following:

a. If $f(t) = f(-t)$, indicating that $f(t)$ is an even function with respect to t, then

$$f(t) = \sum_{n} a_n \cos(nw_0 t)$$

where

$$a_n = \frac{4}{T} \int_{0}^{T/2} f(t)\cos(nw_0 t)dt \quad \text{for } n \geq 1, \text{ and all the coefficients } b_n = 0$$

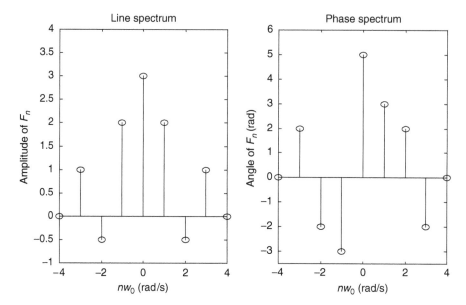

FIGURE 4.1
Plots of line spectrum.

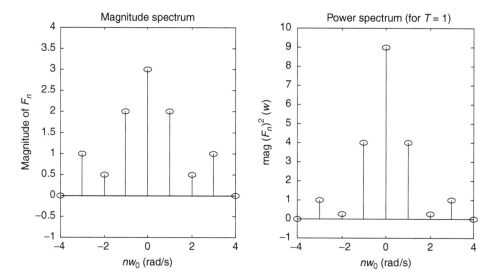

FIGURE 4.2
Plots of magnitude and power spectrum.

b. If $f(t) = -f(-t)$, indicating that $f(t)$ is an odd function with respect to t, then

$$f(t) = \sum_n b_n \sin(nw_0 t)$$

where

$$b_n = \frac{4}{T} \int_0^{T/2} f(t) \sin(nw_0 t) dt \quad \text{for } n \geq 1, \text{ and all the coefficients } a_n = 0$$

c. If $f(t) = -f(t \pm T/2)$, a condition referred to as half wave or rotational symmetry (with respect to t), then

$$a_0 = 0 \qquad a_n = b_n = 0 \quad \text{for } n \text{ even}$$

and

$$f(t) = \sum_{n=odd} [a_n \cos(nw_0 t) + b_n \sin(nw_0 t)]$$

where

$$a_n = \frac{4}{T} \int_0^{T/2} f(t) \cos(nw_0 t) dt$$

and

$$b_n = \frac{4}{T} \int_0^{T/2} f(t) \sin(nw_0 t) dt \quad \text{for } n = 1, \text{ and } a_n = 0 \text{ for all } ns$$

R.4.28 If $f(t)$ is a real and even function of t, then the coefficients F_ns are also real and even.

R.4.29 If $f(t)$ is a real and odd function of t, then the coefficients F_ns are imaginary and odd.

R.4.30 Recall that any arbitrary function $f(t)$ can be expressed as

$$f(t) = fe(t) + fo(t) \qquad \text{(see Chapter 1)}$$

where

$$fe(t) = 0.5[f(t) + f(-t)] \qquad \text{(even component of } f(t))$$

and

$$fo(t) = 0.5[f(t) - f(-t)] \qquad \text{(odd component of } f(t))$$

R.4.31 By decomposing a signal (or system) into its frequency components, the BW of the signal (or system) can be estimated, and the relative contributions and importance of each frequency, or range of frequencies can then be estimated. Recall that the BW represents a range of frequencies that can pass through a device, system, or communication channel without significant attenuation.

R.4.32 The exponential FS of an arbitrary signal $f(t)$ states that $f(t)$ can be decomposed into components of the form

$$Fne^{-jnw_0 t} \quad \text{for } n = 0, \pm 1, \pm 2, \ldots, \pm \infty$$

Then by applying each individual component as an input to a given system (assuming the system is linear), and by evaluating the output of each individual component, it is possible to predict the output of the composite input by adding all the individual outputs (superposition principle).

Recall that the superposition principle states that the sum of the inputs applied to a linear system returns as its output, the sum of the individual outputs.

R.4.33 Practical considerations dictate that the FS representation of an arbitrary function $f(t)$ cannot have an infinite number of terms and must be truncated at some point, and will, therefore, consist of a finite number of terms. This truncation process creates an error that can be best evaluated by what is referred to as the mean square error (MSE), defined as follows:

$$\text{MSE} = \frac{1}{T}\int_0^T [\text{error}(t)]^2\,dt$$

where

$$\text{error}(t) = f(t) - \sum_{n=-m}^{n=m} F_n e^{-jw_0 nt} \quad \text{for a finite } m$$

then

$$\text{MSE} = \frac{1}{T}\int_0^T \left[f(t) - \sum_{n=-m}^{n=m} F_n e^{-jw_0 nt} \right]^2 dt$$

R.4.34 The process of truncation, or approximation of $f(t)$ by a partial sum is equivalent to passing $f(t)$ through an ideal low pass filter, and eliminating the high frequencies. If $f(t)$ presents discontinuities, then high frequencies are present with a considerable amount of power, and the elimination of these frequencies would create distortion and oscillations.

The resulting truncated waveform can be improved if a *window* (see Chapter 1) is used with the objective of gradually reducing the high-frequency components.

R.4.35 Observe the existence of negative frequency components in the exponential FS expansion. These frequencies have no physical meaning at this point, and represent a convenient mathematical model of representing the series. The physical implications will become evident when analyzing, for example, the process of modulation.

R.4.36 Let $f(t)$ be now a nonperiodic function of t. Then by using the FT equations given below, $f(t)$ can be transformed from the time domain to the frequency domain, and vice versa. These equations are also referred as the inverse and direct FT, given by

$$f(t) = \frac{1}{2\pi}\int_{-\infty}^{+\infty} F(\omega)e^{jwt}\,dt$$

and

$$F(w) = \int_{-\infty}^{+\infty} f(t)e^{-jwt}\,dt$$

Observe the notation used. The same letter is used for $f(t)$ and its transform $F(w)$, where the lowercase denotes a time function and the uppercase is used to define their transforms (Fourier and later in this chapter Laplace). Thus, the transform of $v(t)$ would be $V(w)$ for a voltage, and the transform of $i(t)$ would be $I(w)$ for a current.

Note also that the FT can be considered a limiting case of an FS, as the period T is extended to infinity.

R.4.37 The notation used to indicate the (direct) FT is given by

$$\Im[f(t)] = F(w)$$

whereas

$$\Im^{-1}[F(w)] = f(t)$$

denotes the inverse FT of $F(w)$.

R.4.38 The FT of $f(t)$ exists, if the following condition is satisfied:

$$\int\limits_{-\infty}^{+\infty} \left| f(t)e^{-jwt} \right| dt < k < \infty$$

where k is a finite value constant.

R.4.39 The existence of the FT of $f(t)$ denoted by $F(w)$ is guaranteed if the Dirichlet's conditions are satisfied. The Dirichlet's conditions state (similar to the FS case)

a. $f(t)$ may have a finite number of maxima and minima and a countable number of finite discontinuities within a given time interval

b. $f(t)$ must be absolutely integrable, that is, $\int_{-\infty}^{+\infty} |f(t)| dt < \infty$

Note that, strictly speaking, a periodic function does not have a transform, but if

$$\int\limits_{-T/2}^{+T/2} |f(t)| dt < \infty$$

then in the limit, as T approaches infinity, the FT exists. The preceding signals are referred to as power signals and, therefore, satisfy the relation given by

$$\lim_{T \to \infty} \left\{ \frac{1}{T} \int\limits_{-T/2}^{+T/2} f(t)^2 dt \right\} < \infty$$

R.4.40 On the contrary, if

$$\int\limits_{-T/2}^{+T/2} |f(t)| dt < \infty$$

then $f(t)$ is referred to as a finite energy signal. The FT exists for finite energy signals and its evaluation is an exercise in calculus.

R.4.41   Let $f(t)$ be a real function of t, then

$$F(w) = F^*(w)$$

(Recall that the character * denotes the *complex conjugate of*).

R.4.42   Since $F(w)$ is in general a complex function, then two equations that are translated into plots are required to fully specify $F(w)$. They are referred as

a. The magnitude spectrum

b. The phase spectrum

R.4.43   The magnitude spectrum is an even function of w, that is, $F(w) = F(-w)$; whereas the phase spectrum is an odd function of w, that is, $\theta(w) = -\theta(-w)$. Recall that those conditions are similar to the case of a periodic function, in which the FS was employed. Recall that in the series case

$$Fn = F - n$$

whereas

$$\theta(nw_0) = -\theta(-nw_0)$$

R.4.44   Table 4.1 summarizes some of the most frequently used time/frequency function transforms, where

$$f(t) = \frac{1}{2\pi} \int\limits_{-\infty}^{+\infty} F(w)e^{jwt}dw \Leftrightarrow F(w) = \int\limits_{-\infty}^{+\infty} f(t)e^{-jwt}dt$$

R.4.45   The functions $f(t)$ and $F(w)$ constitute an FT pair, indicated by the following notation:

$$f(t) \leftrightarrow F(w) \quad \text{or} \quad f(t) \Leftrightarrow F(w)$$

R.4.46   The FT is one of the most common forms of describing the frequency domain characteristics of a signal or system. Note that the FT can also be used for periodic signals (Table 4.1, transform No. 17), instead of the more traditional FS expansion.

R.4.47   Note that the FT of a periodic function $f(t)$ is given by

$$\Im\{f(t)\} = \sum_{n=-\infty}^{+\infty} 2\pi F_n \delta(w - nw_0)$$

where

$$F_n = \frac{1}{T} \int\limits_{0}^{T} f(t)e^{-jw_0nt}dt$$

Observe that the difference between the FS and the FT in the case of a periodic function is given by 2π, a constant that equally affects all the frequencies, and therefore, with no effect on the relative importance of the frequencies that constitute the spectrum.

TABLE 4.1

Time Frequency Transformations

Signal/Time Domain	Transform/Frequency Domain
$A\delta(t)$	A
A	$2\pi\delta(w)$
$A[u(t + a/2) - u(t - a/2)]$	$\dfrac{Aa}{(wa/2)}\sin(wa/2)$
$\dfrac{Aa}{(\tau a/2)}\sin(\tau a/2)$	$2\pi A[u(w + a/2) - u(w - a/2)]$
$Ae^{jw_{au}t}$	$2\pi A\delta(w - w_0)$
$A\cos(w_0 t)$	$\pi A[\delta(w - w_0) + \delta(w + w_0)]$
$A\sin(w_0 t)$	$-j\pi A[\delta(w - w_0) + \delta(w + w_0)]$
$A\mathrm{sgn}(t)$	$2A/jw$
$Au(t)$	$\pi A\delta(w) + (A/jw)$
$Ae^{-at} \cdot u(t),\ a > 0$	$A/(jw + a)$
$Ate^{-at} \cdot u(t)$	$1/(jw + a)^2$
$Ate^{-at} \cdot u(-t),\ a > 0$	$A/(-jw + a)$
$Ae^{-a|t|}$	$2Aa/(w^2 + a^2)$
$Ae^{-b|t|}\cos(w_0 t)u(t),\ b > 0$	$A\dfrac{b + jw}{b^2 + w_0^2 - w^2 + 2jwb}$
$Ae^{-b|t|}\sin(w_0 t)u(t),\ b > 0$	$A\dfrac{w_0}{b^2 + w_0^2 - w^2 + 2jbw}$
$A|t|$	$-2A/w^2$
$\displaystyle\sum_{n=-\infty}^{+\infty} F_n e^{jnw_0 t}$	$2\pi\displaystyle\sum_{n=-\infty}^{+\infty} Fn\delta(w - nw_0)$
$\displaystyle\sum_{k=-\infty}^{+\infty} \delta(t - kT)$	$\dfrac{2\pi}{T}\displaystyle\sum_{k=-\infty}^{+\infty} \delta\left(w - \dfrac{2\pi k}{T}\right)$

R.4.48 It is useful to know the relations and the effects that operations in the time domain have in the frequency domain, and vice versa, to be able to get an insight of the transformation process, and in many cases to avoid the integral definition of the transform.

Let the FT pair time/frequency be represented using the short notation

$$f(t) \leftrightarrow F(w)$$

Then the most important properties that relate the two domains are summarized as follows:

a. Linearity

Let

$$a_1 f_1(t) \leftrightarrow a_1 F_1(w) \qquad \text{and} \qquad a_2 f_2(t) \leftrightarrow a_2 F_2(w)$$

Then $a_1 f_1(t) + a_2 f_2(t) \leftrightarrow a_1 F_1(w) + a_2 F_2(w)$, where a_1 and a_2 are arbitrary (may be complex) constants, and the addition is complex.

b. Time scaling

$$f(at) \leftrightarrow \frac{1}{a} F\left(\frac{w}{a}\right)$$

where a is a real nonzero constant. If $a < 1$, then $f(t)$ is expanded in time, whereas $F(w)$ is compressed in frequency by a similar factor.

Note that if $a > 1$, then the function $f(at)$ is compressed in time, whereas $F(w)$ is expanded in frequency by a similar factor.

c. Duality or symmetry

$$F(t) \leftrightarrow 2\pi f(-w)$$

obtained by the following subtitution $t \to w$ and $w \to t$.

Observe that this property can be used to double the table of transform pairs, by simply interchanging the time and frequency variables.

d. Time shifting

$$f(t - t_0) \leftrightarrow F(w)e^{-jt_0 w}$$

where a time shift translates as a linear phase shift in frequency.

e. Frequency shifting

$$f(t)e^{jw_0 t} \leftrightarrow F(w - w_0)$$

This property is also referred to as the modulation property. It is a fundamental property in communication theory, and is used to prove the modulation theorem that states

$$f(t) \cdot \cos w_0 t \leftrightarrow \frac{1}{2}\{F(w + w_0) + F(w - w_0)\}$$

Observe that multiplying an arbitrary time function $f(t)$ by $\cos(w_0 t)$ shifts the spectrum of $f(t)$, so that half the original spectrum is centered at w_0, and the other half is centered at $-w_0$. The signal $f(t)$ is referred to as the modulating signal, whereas $\cos(w_0 t)$ is referred as the carrier signal (in practical applications the frequency w_0 must be much higher than the highest frequency of $f(t)$).

f. Time differentiation

$$\frac{df(t)}{dt} \leftrightarrow jwF(w)$$

or the more general relation given by

$$\frac{d^n f(t)}{dt^n} \leftrightarrow (jw)^n F(w)$$

g. Frequency differentiation

$$(-jt)f(t) \leftrightarrow \frac{dF(w)}{dw}$$

or the more general relation given by

$$(-jt)^n f(t) \leftrightarrow \frac{d^n F(w)}{dw^n}$$

h. Time integration

$$\int_{-\infty}^{t} f(\lambda)d\lambda \leftrightarrow \left(\frac{1}{jw}\right) F(w) + \pi F(0)\delta(w)$$

Note that differentiation of $f(t)$ in the time domain has the effect of multiplying $F(w)$ by jw in the frequency domain, similarly integration of $f(t)$ in the time domain has the effect of dividing $F(w)$ by jw in the frequency domain.

i. Frequency integration

$$\frac{1}{-jt} f(t) \leftrightarrow \int_{-\infty}^{w} F(\lambda)d\lambda$$

j. Multiplication in time

$$f_1(t) \cdot f_2(t) \leftrightarrow \frac{1}{2\pi} \int_{-\infty}^{+\infty} F_1(\lambda) \cdot F_2(w - \lambda)d\lambda = \frac{1}{2\pi} [F_1(w) \otimes F_2(w)]$$

This property states that multiplying two time domain functions given by $f_1(t)$ and $f_2(t)$ in the time domain has the effect of evaluating the convolution of their spectrums $F_1(w)$ with $F_2(w)$ in the frequency domain times $\frac{1}{2\pi}$. This property is used extensively in linear controls and communication system analysis.

k. Convolution in time

$$f_1(t) \otimes f_2(t) = \int_{-\infty}^{+\infty} f_1(\lambda) \cdot f_2(t - \lambda)d\lambda \leftrightarrow [F_1(w)F_2(w)]$$

This property states that the convolution of the time functions $f_1(t)$ with $f_2(t)$ (in the time domain) has the effect of multiplying their spectrums $F_1(w)$ with $F_2(w)$ (in the frequency domain). Observe that the convolution integral (indicated by the character $\otimes$) is in general a process not easy to evaluate, and that is precisely the reason why the transform is used just to avoid it. Note that the convolution process in one domain is translated into a product in the other domain.

l. Cross correlation in time

$$\int_{-\infty}^{+\infty} f_1(\lambda)f_2(t + \lambda)d\lambda \leftrightarrow F_1^*(w)F_2(w), = F_1(-w)F_2(w)$$

The cross correlation of two signals $f_1(t)$ with $f_2(t)$ provides an indication of the degree of similarity between them. If $f_1(t) = f_2(t)$, then the cross correlation becomes an autocorrelation. Note the similarity between the correlation and convolution process.

m. Time reversal

$$f(-t) \leftrightarrow F(-w)$$

R.4.49 Recall that the integral equation

$$\int_{-\infty}^{+\infty} f_1(\lambda) \cdot f_2(t - \lambda)d\lambda$$

is called the convolution integral of $f_1(t)$ with $f_2(t)$, and is indicated using the notation

$$f_1(t) \otimes f_2(t)$$

Recall also that the convolution integral is a powerful way of defining the input–output relations of a time invariant linear system (see Chapter 1).

R.4.50 From the transform equation pair, and their properties, the following can be observed: if $f(t)$ is a real and an even function of t, then $F(w)$ is also a real and an even function of w.

R.4.51 If $f(t)$ is a real and an odd function of t, then $F(w)$ is an imaginary function of w.

R.4.52 If $f(t)$ is a periodic function in the time domain, then $F(w)$ is a discrete function in the frequency domain.

R.4.53 If $f(t)$ is a periodic function in the time domain, then $F(w)$ is a nonperiodic function in the frequency domain.

R.4.54 If $f(t)$ is a continuous and nonperiodic function in the time domain, then $F(w)$ is a nonperiodic and continuous function in the frequency domain.

R.4.55 Let the input to a given system be $x(t)$, and its impulse system response be $h(t)$, as indicated in Figure 4.3. Recall then that the output $y(t)$ of the system is given by

$$y(t) = x(t) \otimes h(t) = \int_{-\infty}^{+\infty} x(\lambda)h(\lambda - t)d\lambda$$

R.4.56 Let $x(t)$ be the input to a given system (as indicated in Figure 4.3), and let

$$x(t) \leftrightarrow X(w) \quad \text{and} \quad h(t) \leftrightarrow H(w)$$

Then the general block box system diagram given in the time domain indicated in Figure 4.3 can be transformed to the equivalent block box system diagram in the frequency domain indicated in Figure 4.4.

Recall that the convolution integral in the time domain is transformed into a product of its spectrums in the frequency domain (R.4.48, property k).

FIGURE 4.3
System expressed in the time domain.

FIGURE 4.4
System expressed in the frequency domain.

R.4.57 Let the input to a system be $x(t) = \delta(t)$, since $X(w) = \Im\{\delta(t)\} = 1$, then

$$Y(w) = H(w)$$

where $h(t) \leftrightarrow H(w)$.
    Recall that $h(t)$ is referred to as the impulse system response, for obvious reasons.

R.4.58 Recall some useful convolution properties (from Chapter 1)
    a. $f_1(t) \otimes f_2(t) = f_2(t) \otimes f_1(t)$
    b. $f_1(t) \otimes [f_2(t) + f_3(t)] = f_1(t) \otimes f_2(t) + f_1(t) \otimes f_3(t)$

R.4.59 As a consequence of the convolution properties of R.4.58, the block box system transformation diagrams shown in Figure 4.5 can be obtained.
    Note that the system input and the system equation can be interchanged without affecting its output. Also note that there is no distinction between a signal or system, and the analytical tools developed for signals can be applied equally well for systems.

R.4.60 Recall that a system can be thought of as a method or algorithm of processing or changing an input signal $x(t)$ (or $x(n)$) into an output signal $y(t)$ (or $y(n)$). If the system is linear then it can be characterized by $h(t)$ or $H(w)$, its system impulse response in time, or its FT in frequency. The system transfer function $H(w)$ thus modifies, or filters the spectrum of the input $X(w)$. The objective of the system transfer function is to change the relative importance of the frequencies contained in the input signal, both in amplitude and phase. $H(w)$ is also called the gain function since it weights the various frequencies' components of the input $x(t)$ to generate its output $y(t)$.

R.4.61 Let $x(t)$ be the input to a linear system, then its output $y(t)$ is said to be distortionless if it is of the form $y(t) = kx(t - a)$, where k and a are constants, referred as the gain and delay, respectively.

R.4.62 The distortionless time–frequency relation is then given by

$$kx(t - a) \rightarrow KX(w)e^{-jwa}$$

    See R.4.48, property d. Therefore, for distortionless transmission the system returns a constant gain K, and a linear phase shift of the form $-aw$, when its input

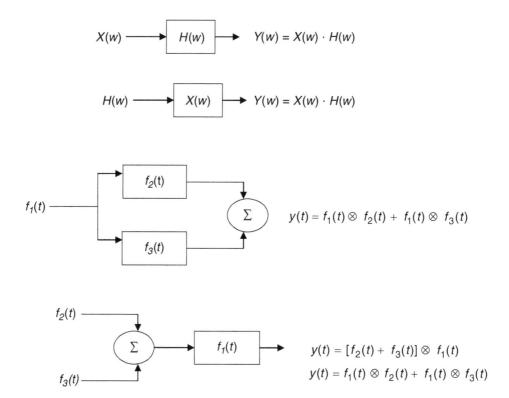

FIGURE 4.5
Block box system representations

has a gain of k, where k and K are constants, and in general $K \neq k$ (in passive systems $k > K$).

R.4.63 Strictly speaking the process of distortion occurs if the output $y(t)$ of a system is not its input scaled by a constant, and its phase is not linear with respect to w.

R.4.64 If the phase shift associated with its output is not linear with respect to w, the various input frequencies will appear at the output with different time delays, creating what is referred to as delay distortion. Most electronic devices introduce both amplitude and delay distortion to same degree. The amount of distortion that can be tolerated by a device depends on the application.

R.4.65 Waveform distortion can be compensated or reduced by a network connected to the system output called an equalizer. For example, if $Y(w) = X(w)H(w)$, and the system or filter $H(w)$ introduces distortion, then by placing an equalizing network, defined by the transfer function $H_e(w) = 1/H(w)$, the input signal $x(t)$ can be recovered distortionless. Meaning that the poles of $H_e(w)$ should fall exactly at the zero locations of $H(w)$, and the zeros of $H_e(w)$ should fall at the locations of the poles of $H(w)$, such that $H(w) \cdot H_e(w) = K$, where K is a constant. Of course, perfect pole–zero cancellations may be difficult to implement, and only approximations may be achievable.

R.4.66 In summary, note that the spectrum of a time signal is a valuable tool when studying the effects of processing it, such as sampling, modulation, and transmission of that signal through a linear system.

R.4.67 Parseval's theorem, used to evaluate the energy of a periodic function or signal, can be extended to include the nonperiodic function $f(t)$, given by

$$\int\limits_{-\infty}^{+\infty} f(t)^2 dt = \frac{1}{2\pi} \int\limits_{-\infty}^{+\infty} |F(w)^2| dw$$

The proof of Parseval's theorem is beyond the scope of this book, but its relationship is verified by numerical examples in the following sections.

R.4.68 Let $H(w) = Y(w)/X(w)$, then the MATLAB function *[Real_H, Imag_H, w] = nyquist (Y, X)* returns the real and imaginary part of the transfer function, where $Y(w)$ and $X(w)$ are its system output and input in vector form that represent the coefficients of the polynomials arranged in descending powers of jw. The range of w, as well as, the number of points used are chosen automatically by MATLAB.

   If no output argument is specified then MATLAB returns the Nyquist* polar plot consisting of the plot of *[Real_H]* versus *[Imag_H]*, referred as the *nyquist* plot.

   The frequency w defines its output range, is an optional variable, and may be included in the function indicated by

$$[Real\_H, Imag\_H] = nyquist(Y, X, w)$$

The nyquist command can be used with the system transfer function (Y and X) or by specifying the systems zeros, poles, and gain k. The range w may be supplied by the user in the form of a frequency vector (in radian/second), or it may be specified by its upper and lower limit by *wmax* and *wmin* indicated as follows:

$$[Real\_H, Imag\_H] = nyquist(Y, X, (wmax\ wmin))$$

Recall from Chapter 7 of the book entitled *Practical MATLAB® Basics for Engineers* that the log–log plot of *mag[H(w)]* versus w, and semilog *phase[H(w)]* versus w, is another popular way to represent $H(w)$ by engineers, referred as the Bode[†] plots.

R.4.69 For example, let

$$H(s) = \frac{s+5}{s^3 + 3s^2 + 4s + 5}$$

where $s = jw$. Create the script file *nyquist_bode* that returns the following (Figures 4.6 and 4.7):

a. The *nyquist* plot (over the range $w = -1:0.4:10$, and without w)

b. *Bode* plots (over the range $w = 0.1:0.1:5$, and without w)

c. *Nyquist* default points

d. *Bode* default points

---

[*] Harry Nyquist (1889–1976), Swedish electrical engineer employed by the Bell Telephone Laboratories, made important contributions concerning system stability. He is also credited with formulating the sampling conditions by which a continuous band-limited signal can be converted into a discrete sequence. The sampling rate used in the analog-digital conversion is referred as the Nyquist rate or Nyquist frequency.

[†] Hendrik Wade Bode (1905–1984), an engineer employed by Bell Telephone Laboratories, and later in his life a faculty member at Harvard University is credited with being the first in using logarithmic scales to represent the system gain H(jw) in db versus frequency.

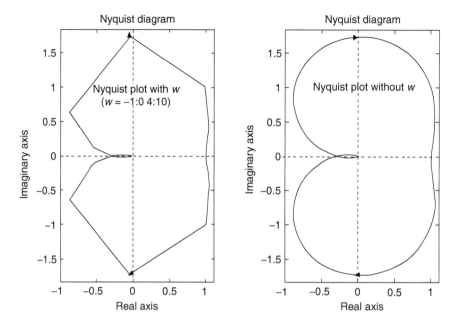

FIGURE 4.6
Nyquist plots of R.4.69.

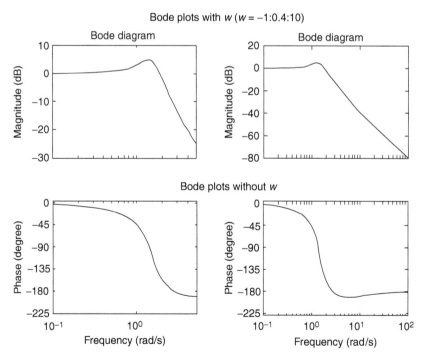

FIGURE 4.7
Bode plots of R.4.69.

MATLAB Solution
```
% Script file: nyquist _ bode
Y=[1 5]; X=[1 3 4 5];
[real _ H, imag _ H,w1] = nyquist(Y,X);
[mag, phase,w2] = bode(Y,X);
disp('^^^^^^^^^^^^^^^^^^^^^^^^^^^^^^^^^^^^^^')
disp(' Bode default points ')
disp(' w mag [H(w)] angle [H(w)] ')
disp('^^^^^^^^^^^^^^^^^^^^^^^^^^^^^^^^^^^^^^')
[w2 mag phase]
disp('*********************************')
disp('^^^^^^^^^^^^^^^^^^^^^^^^^^^^^^^^^^^^^^')
disp(' Nyquist default points ')
disp(' w real [H(w)] imag [H(w)] ')
disp('^^^^^^^^^^^^^^^^^^^^^^^^^^^^^^^^^^^^^^')
[w1 real _ H imag _ H]
disp('*********************************')

figure(1)                        % nyquist plots
w=-1:0.4:10;
subplot(1,2,1)
nyquist(Y,X,w)
subplot(1,2,2)
nyquist(Y,X)

figure(2)
ww=0.1:0.1:5;                    % bode plots
subplot(1,2,1)
bode(Y,X,ww)
subplot(1,2,2)
bode(Y,X)
```

The script file *nyquist_bode* is executed and the results are indicated as follows:

```
>> nyquist _ bode
^^^^^^^^^^^^^^^^^^^^^^^^^^^^^^^^^^^^^^^^
 Bode default points
 w mag[H(w)] angle[H(w)]
^^^^^^^^^^^^^^^^^^^^^^^^^^^^^^^^^^^^^^^^

ans =
  0.1000 1.0030 -3.4442
  0.1184 1.0042 -4.0793
  0.1401 1.0059 -4.8330
  0.1658 1.0083 -5.7285

  .........................................
  1.4810 1.6782 -104.0615
  1.5850 1.5145 -119.4969
  1.7139 1.2510 -135.5458

  .........................................
  60.3237 0.0003 -181.8893
  71.3935 0.0002 -181.5988
  84.4947 0.0001 -181.3524
  100.0000 0.0001 -181.1436
  *********************************
```

```
^^^^^^^^^^^^^^^^^^^^^^^^^^^^^^^^^^^^^^^^^^^
Nyquist default points
w real[H(w)] imag[H(w)]
^^^^^^^^^^^^^^^^^^^^^^^^^^^^^^^^^^^^^^^^

ans =
 1.0e+003 *
      0    0.0010         0
 0.0001    0.0010   -0.0001
 0.0001    0.0010   -0.0001
 0.0001    0.0010   -0.0001
.........................................
 0.0010    0.0010   -0.0011
 0.0011    0.0009   -0.0012
.........................................
 0.0014   -0.0003   -0.0017
 0.0015   -0.0004   -0.0016
.........................................
 0.0031   -0.0002    0.0000
 0.0036   -0.0001    0.0000
.........................................
 0.0202   -0.0000    0.0000
.........................................
 0.5356   -0.0000    0.0000
 0.6261   -0.0000    0.0000
 0.7318   -0.0000    0.0000
 0.8555   -0.0000    0.0000
 1.0000   -0.0000    0.0000

***********************************
```

R.4.70 The MATLAB command $F = fourier\ (f, x)$ returns the FT of f (denoted by $f(t) \rightarrow F(w)$), where f is given as a symbolic object.

R.4.71 For example, use MATLAB to obtain the FT of the following time functions:

a. $f1(t) = \cos(3t)$

b. $f2(t) = \cos(3t + \pi/4)$, and factor $F2(w)$

c. $f3(t) = \sin(3t)$

d. $f4(t) = 1/t$

e. $f5(t) = e^{-t}u(t)$, and pretty $F5(w)$

f. $f6(t) = t \cdot e^{-t}u(t)$

g. $f7(t) = e^{-|t|}$

h. $f8(t) = Pul_2(t) = u(t + 1) - u(t - 1)$

i. $f9(t) = u(t)$

MATLAB Solution[[TI1]]
```
>> syms t w
>> F1 = fourier(cos(3*t))                          % part (a)
```

```
    F1 =
        pi*Dirac(w-3)+pi*Dirac(w+3)

>> F2 = fourier(cos(3*t+pi/4))                                    % part (b)

    F2 =
        1/2*2^(1/2)*pi*Dirac(w-3)+1/2*i*2^(1/2)*pi*Dirac(w-
        3)+1/2*2^(1/2)*pi*Dirac(w+3)1/2*i*2^(1/2)*pi*Dirac(w+3)

>> factor(F2)

    ans =
        1/2*2^(1/2)*pi*(Dirac(w-3)+i*Dirac(w-3)+Dirac(w+3)- i*Dirac(w+3))

>> F3 = fourier(sin(3*t))                                          % part (c)

    F3 =
        -i*pi*Dirac(w-3)+i*pi*Dirac(w+3)

>> F4 = fourier(1/t)                                              % part (d)

    F4 =
        i*pi*(Heaviside(-w)-Heaviside(w))

>> F5 = fourier(exp(-t)*sym('Heaviside(t)'),t,w)                   % part (e)

    F5 =
        1/(1+i*w)

>> pretty(F5)

                  1
               -------
               1 + i w

>> F6 = fourier(t*exp(-t)*sym('Heaviside(t)'),t,w)                % part (f)

    F6 =
        1/(1+i*w)^2

>> F7 = fourier(exp(-abs(t)),t,w) % part (g)

    F7 =
        2/(1+w^2)

>> F8 = fourier(sym('Heaviside(t+1)')
        -sym('Heaviside(t- 1)'),t,w)                              % part (h)
```

```
    F8 =
        exp(i*w)*(pi*Dirac(w)-i/w)-exp(-i*w)*(pi*Dirac(w)-i/w)

>> F9 = fourier(sym('Heaviside(t)'),t,w)                    % part (i)

    F9 =
        pi*Dirac(w)-i/w
```

R.4.72 The MATLAB function $f = ifourier(F)$, where F is a symbolic object with an independent variable w, returns the inverse FT of F, which is a function of x, denoted by $F(w) \rightarrow f(x)$. The MATLAB command $f = ifourier(F, w, t)$ returns the function $f(t)$, given the symbolic function $F(w)$.

R.4.73 For example, use MATLAB to obtain the inverse FTs of the following functions:

a. $F1 = pi * Dirac(w − 3) + pi * Dirac(w + 3)$, default version

b. $F2 = 1/2 * 2^{(1/2)} * pi * (Dirac(w − 3) + i * Dirac(w − 3) + Dirac(w + 3) − i * Dirac(w + 3))$ $(w \rightarrow t)$, default and factor

c. $F3 = −i * pi * Dirac(w − 3) + i * pi * Dirac(w + 3)$, $(w \rightarrow t)$

d. $F4 = i * pi * (Heaviside(−w) − Heaviside(w))$; $(w \rightarrow t)$

e. $F5 = 1/(1 + i * w)$, $(w \rightarrow t)$

f. $F6 = 1/(1 + i * w)^2$, $(w \rightarrow t)$

g. $F7 = 2/(1 + w^2)$, $(w \rightarrow t)$

h. $F8 = exp(i * w) * (pi * Dirac(w) − i/w) − exp(−i * w) * (pi * Dirac(w) − i/w)$, $(w \rightarrow t)$ and simplify

i. $F9 = pi * Dirac(w) − i/w$, $(w \rightarrow t)$ and simplify.

MATLAB Solution
```
>> sym t w
>> F1 = pi*Dirac(w-3)+pi*Dirac(w+3);
>> f1 = ifourier(F1)

    f1 =
        1/2*exp(3*i*x)+1/2*exp(-3*i*x)

>> F2 =1/2*2^(1/2)*pi*(Dirac(w-3)+i*Dirac(w-3)+Dirac(w+3)- i*Dirac(w+3));
>> f2 = ifourier(F2,t,w)

    f2 =
        1/2*2^(1/2)*pi*(Dirac(w-3)+i*Dirac(w-3)+Dirac(w+3)-i*Dirac(w+3))
        *Dirac(w)

>> factor(f2)                       % observe that f2 is a function of x

    ans =
        -1/4*2^(1/2)*(-exp(3*i*x)-i*exp(3*i*x)-exp(- 3*i*x)+i*exp(-3*i*x))

>> F3 = -i*pi*Dirac(w-3)+i*pi*Dirac(w+3);
>> f3 = ifourier(F3,w,t)

    f3 =
        1/2*i*(-exp(3*i*t)+exp(-3*i*t))
```

```
>> F4 = i*pi*(Heaviside(-w)-Heaviside(w));
>> f4 = ifourier(F4,w,t)

    f4 =
          -i*pi*(-Heaviside(-t)+Heaviside(t))*Dirac(t)

>> F5 = 1/(1+i*w);
>> f5 = ifourier(F5,w,t)

    f5 =
          1/(1+i*t)*Dirac(t)

>> F6 = 1/(1+i*w)^2;
>> f6 = ifourier(F6,w,t)

    f6 =
          1/(1+i*t)^2*Dirac(t)

>> F7 = 2/(1+w^2);
>> f7 = ifourier(F7,w,t)

    f7 =
          2/(1+t^2)*Dirac(t)

>> F8 =exp(i*w)*(pi*Dirac(w)-i/w)-exp(-i*w)*(pi*Dirac(w)-i/w);
>> f8 = ifourier(F8,w,t)

    f8 =
          -(pi*Dirac(t)*t-i)*(-exp(i*t)+exp(-i*t))*Dirac(t)/t

>> simplify(f8)

    ans =
           2*Dirac(t)

>> F9 = pi*Dirac(w)-i/w F9;
>> f9 = ifourier(F9,w,t)

    f9 =
          (pi*Dirac(t)*t-i)*Dirac(t)/t

>> simplify(f9)

    ans =
          -i*Dirac(t)/t
```

R.4.74 Another popular transform, employed in system analysis, as well as in the solution of initial value DEs is the LT.

Let a real signal (or function) be given by $f(t)$. Then the LT of $f(t)$ is given by

$$F(s) = \int\limits_{-\infty}^{+\infty} f(t)e^{-st}dt$$

R.4.75 The LT can be viewed as a generalized case of the FT.
        Recall that the FT was defined by

$$F(w) = \int_{-\infty}^{+\infty} f(t)e^{-jwt}dt$$

Observe that in the LT, $f(t)$ is expressed in terms of e^{-st}, whereas the FT represents $f(t)$ in terms of e^{-jwt}. The FT is then a special case of the LT in which $s = jw$. In general, the Laplace variable s is complex, defined as $s = \sigma + jw$, where σ and w are real.

R.4.76 $F(s)$ is referred as the direct LT of $f(t)$, denoted by

$$F(s) = £[f(t)]$$

R.4.77 The process of obtaining the time function $f(t)$ from the LT $F(s)$ is referred to as the ILT. Assuming its existence, ILT is denoted by

$$f(t) = £^{-1}[F(s)]$$

Mathematically, the variable s represents a complex frequency; however, it is not necessary to pursue this interpretation to make use of the transformation.

R.4.78 The sufficient conditions for the existence of the LT are that $f(t)$ must be sectionally continuous in every finite interval $0 \le t \le M$, and of exponential order a for $t > M$, then the LT $F(s)$ as well as the ILT exist and are unique over the range $s > a$. Uniqueness will always be assumed unless otherwise stated (Lerch's theorem).

R.4.79 The region of convergence (ROC) of the LT is the region where the transform exists, and is unique. Recall that causal signals are defined by $f(t) = 0$, for $t < 0$ and causal systems are defined by $h(t) = 0$, for $t < 0$. In either case the ROC is in the right half of the complex plane, and the transform used is called, for obvious reasons, the unilateral LT, given as follows:

$$£[f(t)] = F(s) = \int_{0}^{\infty} f(t)e^{-st}ds$$

R.4.80 As mentioned earlier, the notation used to define $f(t)$ from its ILT is given by

$$f(t) = £^{-1}[F(s)]$$

        where

$$f(t) = \frac{1}{2\pi j}\int_{\sigma-jw}^{\sigma+jw} F(s)e^{st}ds$$

R.4.81 Table 4.2 summarizes some of the standard time–frequency (assuming causality) LT pairs.

TABLE 4.2

Transform Pairs

Signal/System—$f(t)$	Transform—$F(s)$
$\delta(t)$	1
$u(t)$	$1/s$
$tu(t)$	$1/s^2$
$t^n u(t)$	$n!/s^{n+1}$
$e^{-at}u(t)$	$1/(s+a)$
$te^{-at}u(t)$	$1/(s+a)^2$
$\cos(\omega_a t)u(t)$	$s/(s+\omega_0^2)$
$\sin(\omega_a t)u(t)$	$\omega_0/(s^2-\omega_0^2)$
$e^{-at}\cos(\omega_a t)u(t)$	$(s+a)/((s+a)^2+\omega_0^2)$
$e^{-at}\sin(\omega_a t)u(t)$	$(s+\omega_0)/((s+a)^2+\omega_0^2)$
$\cosh(\omega_0 t)u(t)$	$s/(s^2-\omega_0^2)$
$\sinh(\omega_0 t)u(t)$	$\omega_0/(s^2+\omega_0^2)$
$\dfrac{1}{(n-1)!}t^{n-1}e^{at}u(t)$	$\dfrac{1}{(s-a)^n}$
$\dfrac{1}{a-b}(e^{at}-e^{bt})u(t)$	$\dfrac{1}{(s-a)(s-b)}$
$(1-e^{at})u(t)$	$\dfrac{-a}{s(s-a)}$
$[1-\cos(w_0 t)]u(t)$	$\dfrac{w_0{}^2}{s(s^2+a^2)}$
$[\sin(w_0 t+\theta)]u(t)$	$\dfrac{s\sin(\theta)+w_0\cos(\theta)}{s^2+w_0{}^2}$
$[\cos(w_0 t+\theta)]u(t)$	$\dfrac{s\cos(\theta)-w_0\sin(\theta)}{s^2+w_0{}^2}$

R.4.82 The most important time (t)–frequency (s) properties and relations of the unilateral LT are given as follows (similar to the FT properties):

a. Linearity

Let

$$\pounds[f_1(t)]=F_1(s) \quad\text{and}\quad \pounds[f_2(t)]=F_2(s)$$

then

$$a_1 f_1(t)+a_2 f_2(t) \longleftrightarrow a_1 F_1(s)+a_2 F_2(s)$$

b. Time scaling

$$f(at)\leftrightarrow \frac{1}{a}F\left(\frac{s}{a}\right)$$

c. Time shifting

$$f(t-t_0)\leftrightarrow F(s)e^{-st_0}$$

d. s-Shifting

$$f(t)e^{s_0 t}\leftrightarrow F(s-s_0)$$

e. Differentiation in freq(s)

$$-t \cdot f(t) \leftrightarrow \frac{dF(s)}{ds}$$

or a more general relation is given by

$$(-1)^n t^n f(t) \leftrightarrow \frac{d^n}{ds^n}[F(s)]$$

f. Differentiation in t

$$\frac{df(t)}{dt} \leftrightarrow sF(s) - f(0^-)$$

The second derivative is given by

$$\frac{d}{dt}\left[\frac{df(t)}{dt}\right] = \frac{d^2}{dt^2}[f(t)] \leftrightarrow s[sF(s) - f(0^+)] - f'(0^+)$$

$$s^2F(s) - sf(0^+) - f'(0^+)$$

The preceding process can be repeated for higher-order derivatives.

g. Integration in s

$$\frac{f(t)}{t} \leftrightarrow \int_0^\infty F(s)ds$$

h. Periodic functions $f(t)$, with period $T > 0$

$$f(t) \leftrightarrow \frac{\int_0^T e^{-st} f(t)dt}{(1 - e^{-sT})}$$

i. Integration in t

$$\int_0^\tau f(\lambda)d\lambda \leftrightarrow \frac{1}{s}F(s) + \frac{f(0)}{s}$$

j. Convolution in t or product in s

$$f_1(t) \otimes f_2(t) = \int_{-\infty}^\infty f_1(\lambda)f_2(t - \lambda)d\lambda \leftrightarrow [F_1(s)F_2(s)]$$

k. Product in t or convolution in s

$$f_1(t)f_2(t) \leftrightarrow \frac{1}{2\pi j}[F_1(s) \otimes F_2(s)]$$

R.4.83 The initial value $(t = 0)$ and the final value $(t = \infty)$ of a given function $f(t)$ can be evaluated from its LT $F(s)$, by the relations stated as follows, known as the initial and final value theorems, respectively.

a. The initial value theorem is given by

$$f(0^+) = \lim_{s \to \infty} [s \cdot F(s)]$$

(test that all the poles of $sF(s)$ are real and negative)

b. The final value theorem states that

$$f(\infty) = \lim_{s \to 0} [s \cdot F(s)]$$

(test if all the poles are on the left half of the s-plane)

R.4.84 The initial and final value theorems permit to calculate $f(0^+)$ and $f(\infty)$, if one exists, directly from the transform $F(s)$, without the need of inverting the transform.

R.4.85 For example, the initial and final value for the case of $f(t) = u(t)$ are evaluated as follows (recall that $\mathcal{L}[u(t)] = 1/s$), then

a. $u(t = 0^+) = \text{limit}_{s \to \infty} \ [sF(s)] = \text{limit}_{s \to \infty} [s(1/s)] = 1$

b. $u(t = \infty) = \text{limit}_{s \to 0} \ [sF(s)] = \text{limit}_{s \to 0} [s(1/s)] = 1$

The results just obtained confirm what is already known, that is, $u(t = 0^+) = u(t = \infty) = 1$.

R.4.86 The final value theorem evaluates $f(\infty)$, if all the singularities are in the left half of the s-plane. A simple pole (of $F(s)$) at the origin is permitted, but all the remaining poles must be in the left half of the s-plane, and the degree of the denominator must be greater or equal to the degree of the numerator of $F(s)$. If any of the preceding conditions are not met, $f(t)$ becomes unbounded as t approaches infinity, or physically $f(t)$ would sustain nondecaying oscillations.

R.4.87 The transfer function, or system function, is the equation that defines the dynamic properties of a linear system given by

$$H(s) = \frac{\text{Laplace transform of the output}}{\text{Laplace transform of the input}}$$

where $H(s)$ is in general a rational function, given as the ratio of two polynomials $Y(s)$ (its output) and $X(s)$ (its system input).

R.4.88 Recall from previous chapters that the values of s that make $H(s)$ go to zero and infinity are called the system zeros and poles, respectively. Hence the transfer function can be completely defined in terms of its poles, zeros, and a constant multiplier referred to as gain.

R.4.89 The LT of any function, including $H(s)$ must be accompanied by its ROC; only then the corresponding time function $h(t)$ is unique. If $H(s)$ is a one-sided transform then the ROC is not needed, and defines a unique inverse.

For obvious reasons, the ROC must exclude the system's poles.

R.4.90 A causal system can be defined in the frequency domain as the one with the ROC located to the right of the pole having the largest real part.

R.4.91 Similarly, a noncausal system can be defined as the one with the ROC located to the left of the pole having the smallest real part.

R.4.92 Mix systems are the ones that have poles in the left-half plane (causal), as well as, poles in the right-half plane (noncausal). The ROC then lies between the pole having the smallest real part, of the causal subsystem, and the pole having the smallest real part of the noncausal subsystem.

R.4.93 The stability of a system is given by the location of its system's poles. Stable causal systems must have all the poles to the left of the imaginary axis, and the axis itself may be considered part of the ROC.

R.4.94 The MATLAB function *pzmap (Y, X)* returns the plot of the poles and zeros specified by the coefficients of the polynomials of Y and X, the numerator and denominator of $H(s)$, expressed as vectors arranged in descending powers of s. It is customary to indicate the location of the system's poles by *x*s, and the system's zeros by *o*s, on the s-plane (complex plane).

The command *pzmap* uses those characters and returns the pole and zero constellation.

R.4.95 For example, create the script file *map_pz* that returns the constellation plot of the poles and zeros of the transfer function $H(s)$ given by

$$H(s) = \frac{5s^2 + 5s - 30}{s^3 + 3s^2 + 4s + 2}$$

their respective values (poles and zeros), and indicate the ROC (Figure 4.8).

MATLAB Solution
```
% Script file: map _ pz
num = [5 5 -30];
den = [1 3 4 2];
pzmap(num,den);
disp('****************************************************************')
disp('The zeroes and poles of H(s) are:')
disp('****************************************************************')
zeroes = roots(num)
poles = roots(den)
real _ s = min(real(poles));
disp('****************************************************************')
disp('The ROC lie in the region given by : real part greater than ')
roc
disp('****************************************************************')
```

The script file *map_pz* is executed and the results are indicated as follows:

```
>> map _ pz

**************************************************
The zeroes and poles of H(s) are:
**************************************************
zeroes =
        -3
         2
```

```
poles =
          -1.0000 + 1.0000i
          -1.0000 - 1.0000i
          -1.0000
**************************************************
The ROC lie in the region given by: real part greater than
  real_s =
              -1.0000
**************************************************
```

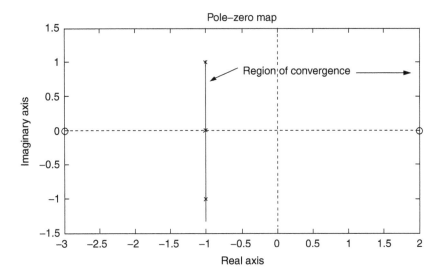

FIGURE 4.8
Pole/zero plot of R.4.95.

R.4.96 Recall that the linear time invariant (LTI) system transfer function $H(s)$ is a rational function, given as the ratio of two polynomials in s. These rational functions can be expressed in terms of a partial fraction expansion, a format that can be used in the evaluation of the ILT, by using the transformation Table 4.2.

Recall from Chapter 7 of the book entitled *Practical MATLAB® Basics for Engineers*, that the MATLAB function $[r, p, k] = residue(num, den)$ returns the coefficients (residues) r, the poles p, and the stand-alone term k of the partial fraction expansion given by the ratio of the $num(Y)$ and $den(X)$ polynomials expressed as vectors consisting of its coefficients arranged in descending powers of s.

Recall also that the poles can be distinct real, repeated, and complex, where complex poles always occur as conjugate pairs. In the evaluation process of the LT of a real function of t, distinct real poles are rather easy to deal with (see Chapter 7 of the book entitled *Practical MATLAB® Basics for Engineers*).

When repeated poles are present in the form $(s + a)^n$, the partial fraction expansion must include the following terms: $b_1/(s + a), b_2/(s + a)^2, \ldots, b_n/(s + a)^n$.

R.4.97 For example, let

$$H(s) = \frac{5s^4 + 7s^2 + 3s^2 + 5s - 30}{s^4 + 4s^3 + 7s^2 + 6s + 2}$$

be a system transfer function.

Verify that multiple poles are present in *H(s)*, and evaluate the MATLAB partial fraction expansion of *H(s)*.

MATLAB Solution
```
>> num = [5 7 3 5 -30];
>> den = [1 4 7 6 2];
>> [r,p,k] = residue(num,den)


r =                        % PFE coefficients
    -6.5000 -20.5000I
    -6.5000 +20.5000i
    0.0000
    -34.0000


p =                        % poles
    -1.0000 + 1.0000i
    -1.0000 - 1.0000i
    -1.0000
    -1.0000


k =                        % stand alone term
    5
```

The partial fraction coefficients given by the column vector *r* are then matched with the corresponding poles given by the column vector *p*, obtaining the following expansion:

$$H(s) = 5 + \frac{-6,5 - 20.5i}{s+1-i} + \frac{-6.5 + 20.5i}{s+1+i} + \frac{0}{s+1} + \frac{-34}{(s+1)^2}$$

Observe that *H(s)* has repeated poles at *s* = −1, then the PFE consists of two terms, a linear and a quadratic as a consequence of the repeated pole, as well as, two other terms as a consequence of the pair of complex poles, and the stand-alone term *k*.

R.4.98 Four examples of the evaluations of the direct and inverse LTs by hand calculations, using Table 4.2, are provided as follows, to gain practice and insight into the process.

a. Example (#1)

Let

$$f(t) = 5u(t) + 2e^{-3t}u(t) + 10\sin(3t)\,u(t) + 5e^{-2t}\cos(7t)$$

Find *F(s)*.

ANALYTICAL Solution

(From Table 4.2)

$$F(s) = £[f(t)] = £[5 + 2e^{-3t}\,u(t) + 10\sin(3t)\,u(t) + 5e^{-2t}\cos(7t)]$$

$$F(s) = £[5] + 2£[e^{-3t}\,u(t)] + 10£[\sin(3t)\,u(t)] + 5£[e^{-2t}\cos(7t)]$$

then

$$F(s) = \frac{5}{s} + \frac{2}{s+3} + \frac{10(3)}{s^2 + 3^2} + \frac{5(s+2)}{(s+2)^2 + 7^2}$$

b. Example (#2)

Let

$$F(s) = \frac{10}{s} + \frac{20}{s^2} + \frac{30}{s+3}$$

Find $f(t)$.

ANALYTICAL Solution

(From Table 4.2)

$$f(t) = \mathcal{L}^{-1}[F(s)] = f(t) = \mathcal{L}^{-1}\left\{\frac{10}{s} + \frac{20}{s^2} + \frac{30}{s+3}\right\}$$

then

$$f(t) = 10 + 20t\,u(t) + 30e^{-3t}\,u(t)$$

c. Example (#3)

Let

$$F(s) = \frac{8s + 30}{s^2 + 25}$$

Find $f(t)$.

ANALYTICAL Solution

(From Table 4.2)

$$F(s) = \frac{8s}{s^2 + 5^2} + 6\left[\frac{5}{s^2 + 5^2}\right]$$

then

$$f(t) = \mathcal{L}^{-1}[F(s)] = 8\cos(5t)\,u(t) + 6\sin(5t)\,u(t)$$

d. Example (#4)

Let

$$Y(s) = \frac{4s + 10}{s(s+1)(s+2)}$$

Find $y(t)$.

ANALYTICAL Solution

$$Y(s) = \frac{A}{s} + \frac{B}{s+1} + \frac{C}{s+2}$$

where $y(t) = £^{-1}[Y(s)]$. The process of evaluating the coefficients A, B, and C by hand is illustrated as follows:

$$A = \frac{(4s + 10) \cdot (s)}{s(s - 1)(s + 2)}\Big|_{s=0} = \frac{10}{(-1)(-2)} = 5$$

$$B = \frac{(4s + 10)(s + 1)}{s(s + 1)(s + 2)}\Big|_{s=-1} = -6$$

$$C = \frac{(4s + 10)(s + 2)}{s(s + 1)(s + 2)}\Big|_{s=-2} = 1$$

then

$$Y(s) = \frac{5}{s} - \frac{6}{s + 1} + \frac{1}{s + 2}$$

and

$$y(t) = £^{-1}[Y(s)] = 5u(t) - 6e^{-t}u(t) + e^{-2t}u(t)$$

R.4.99 Let us use the concepts developed by the Laplace technique, in the analysis of electrical networks. Recall that $v_R(t) = R\,i(t)$ (Ohm's law), and its Laplace transform is given by $£[v_R(t)] = £[R \cdot i(t)] = R\,£[i(t)]$, then

$$V_R(s) = R * I(s)$$

Ohm's law holds in the frequency domain, and the impedance $Z(s)$ (Ω) is defined by

$$Z(s) = \frac{V(s)}{I(s)}$$

The time–frequency domain relation for a pure resistor R is illustrated in Figure 4.9.

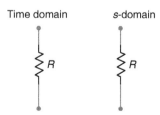

FIGURE 4.9
Time–frequency domain relation for R.

R.4.100 The voltage across a capacitor C, denoted by $v_C(t)$ is expressed in the time domain and its equivalent in the frequency domain as follows by (using FT properties):

$$v_C(t) = \frac{1}{C} \int_{-\infty}^{t} i_c(\lambda) d\lambda \;\leftrightarrow\; V_c(s) = \frac{I(s)}{sC} + \frac{V_C(0)}{s}$$

Recall that $V_C(0)$ denotes the initial voltage $v_C(t)$ at $t = 0$.

R.4.101 For example, let us analyze the case of a capacitor C of 2 F, charged with an initial voltage of +5 V. Then its frequency domain representation, using the LT is given by

$$V_C(s) = \frac{I(s)}{2s} + \frac{5}{s}$$

R.4.102 Note that if $V_C(s) = \dfrac{I(s)}{sC}$, then the impedance of the capacitor C in the frequency domain is given by

$$X_C(s) = \frac{1}{sC} \; \Omega$$

R.4.103 The equivalent circuit models of a capacitor in the time and frequency domain are shown in Figure 4.10 using either a voltage source in series with the impedance $X_C(s) = 1/(sC)$, or by source transformation, a current source in parallel with $X_C(s)$, assuming that its initial voltage is $v_C(0) = V_0$ V.

R.4.104 The voltage across an inductor L, denoted by $v_L(t)$, is expressed as follows in the time and frequency domain by

$$v_L(t) = L\frac{di(t)}{dt} \longleftrightarrow V_L(s) = sLI(s) - LI_0$$

Recall that $i(0) = I_0$ denotes the initial current through L at $t = 0$.

R.4.105 Note that if $V_L(s) = sLI(s)$, then the impedance of the inductor L in the frequency domain is given by

$$X_L(s) = sL \; \Omega$$

R.4.106 The equivalent circuit model of an inductor L in the time and frequency domains are shown in Figure 4.11, using either a voltage source in series with the impedance

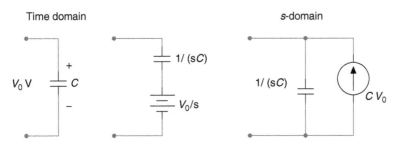

FIGURE 4.10
Time–frequency domain relation for C.

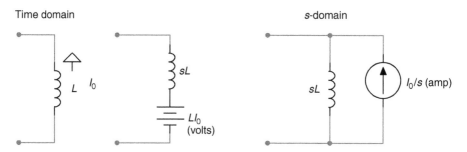

FIGURE 4.11
Time–frequency domain relation for L.

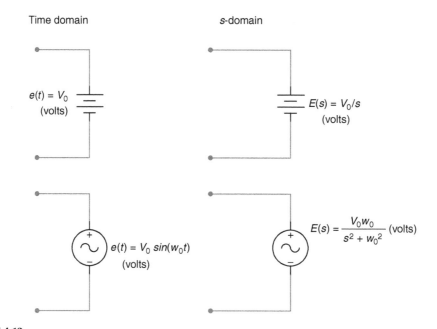

FIGURE 4.12
Time–frequency domain relation for DC and AC voltage sources.

$X_L(s) = sL$, or by source transformation, a current source in parallel with $X_L(s)$, assuming that the initial inductor current is I_0 (amp).

R.4.107 The time–frequency domain LTs of DC (constant) and AC (sinusoidal) voltage sources are shown in the circuit diagrams of Figure 4.12 (assuming that $e(t) = 0$ (volt) for $t \leq 0$).

R.4.108 The time–frequency domain LTs of DC (constant) and AC (sinusoidal) current sources are shown in the circuit diagram of Figure 4.13 (assuming that $i(t) = 0$ (amp) for $t \leq 0$).

R.4.109 A number of illustrative problems are presented as follows, where the LT is used in the analysis of electrical networks. For example, analyze the RC circuit shown in Figure 4.14, where the switch (sw) has been open for a long time and closes at $t = 0$, where it remains for $t > 0$. Find an expression for $i(t)$, using the LT.

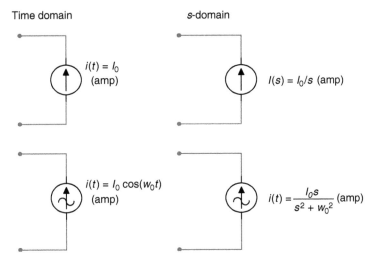

Time domain   s-domain

$i(t) = I_0$ (amp)

$I(s) = I_0/s$ (amp)

$i(t) = I_0 \cos(w_0 t)$ (amp)

$i(t) = \dfrac{I_0 s}{s^2 + w_0^2}$ (amp)

FIGURE 4.13
Time–frequency domain relation for DC and AC current sources.

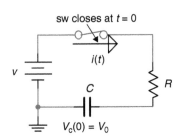

FIGURE 4.14
Network of R.4.109.

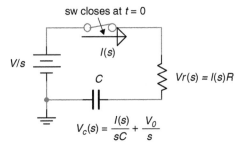

FIGURE 4.15
Frequency domain representation of Figure 4.14.

Note that these type of problems referred as transients were presented, discussed, and solved in Chapter 2, using DEs (time domain).

ANALYTICAL Solution

The first step in the solution process is to redraw the circuit diagram shown in Figure 4.14 in the frequency domain (using the LTs) illustrated in Figure 4.15.

Observe that the voltage source V in the time domain is transformed to V/s in the frequency domain, due to the step characteristics of the switch (sw). Its transform is given by $Vu(t) \rightarrow V/s$. Also observe that the initial voltage of the capacitor also acts as a step function in time, therefore, its transform is given by

$$V_0 u(t) \rightarrow \frac{V_0}{s}$$

Then, applying Kirchhoff's voltage law to the loop of the circuit diagram, shown in Figure 4.15, and performing some algebraic manipulation the current $I(s)$ can be solved, and an expression for $i(t)$ (ILT) can be obtained, as indicated in the following:

$$\frac{V}{s} = RI(s) + \frac{I(s)}{sC} + \frac{V_0}{s}$$

$$I(s)R + \frac{I(s)}{sC} = \frac{V}{s} - \frac{V_0}{s}$$

$$I(s)\left[R + \frac{1}{sC}\right] = \left[\frac{V - V_0}{s}\right]$$

$$I(s) = \left[\frac{V - V_0}{R}\right] \cdot \left[\frac{1}{s + (1/RC)}\right]$$

and the current $i(t)$ is therefore given by

$$i(t) = \pounds^{-1}\left[I(s)\right] = \left[\frac{V - V_0}{R}\right]e^{-t/RC}u(t)$$

R.4.110 Now consider the simple RL circuit shown in Figure 4.16, and let us solve for $i(t)$, assuming that the initial inductor's current is $I_L(0) = I_0 = 10$ (amp), in a counter-clockwise direction.

ANALYTICAL Solution

The time domain circuit diagram of Figure 4.16 is transformed using the LT into the frequency domain circuit diagram shown in Figure 4.17.
Then the analytical solution leading to $i(t)$ is obtained by the following steps:

a. Write the resulting equation by applying Kirchhoff's voltage law around the loop of the circuit shown in Figure 4.17
b. Solve for $I(s)$ (using partial fractions expansion if necessary)
c. Obtain $i(t)$ by taking the ILT of $I(s)$

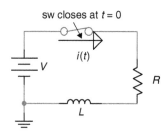

FIGURE 4.16
Network of R.4.110.

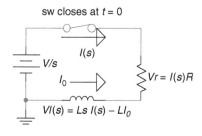

FIGURE 4.17
Frequency domain representation of the network of Figure 4.16.

The process is illustrated as follows:

$$\frac{V}{s} = RI(s) + LsI(s) - LI_0$$

$$I(s) \cdot [R + Ls] = \frac{V}{s} + LI_0$$

$$I(s) = \frac{V}{s}\left[\frac{1}{R + Ls}\right] + LI_0\left[\frac{1}{R + Ls}\right]$$

$$I(s) = \frac{V}{L}\left[\frac{1}{s(s + R/L)}\right] + I_0\left[\frac{1}{s + R/L}\right]$$

$$I(s) = \frac{A}{s} + \frac{B}{s + R/L} + \frac{I_0}{s + R/L}$$

Solving for the constants A and B, and substituting in $I(s)$, the following expression is obtained:

$$I(s) = \frac{V/R}{s} - \frac{V/R}{s + R/L} + \frac{I_0}{s + R/L}$$

Finally, transforming $I(s)$ from the frequency domain back to the time domain, the following solution for $i(t)$ is obtained:

$$i(t) = \pounds^{-1}[I(s)] = \frac{V}{R}(1 - e^{-Rt/L})u(t) + I_0 e^{-Rt/L}u(t)$$

R.4.111 As mentioned earlier, one of the most powerful applications of the LT is the solution of integrodifferential equations illustrated by the transient analysis of RC and RL circuits of R.4.109 and R.4.110. These concepts are extended to include loop or node equations.

The steps involved are summarized as follows:

a. Write the integrodifferential set of equations (loop or node equations) for a given circuit.

b. Transform the integrodifferential equations of part a using Laplace into an algebraic set of equations, in which the initial conditions are automatically inserted.

c. The algebraic set of equations of part b are then solved for either the currents $\{I(s)\}$ (loop equations) or voltages $\{V(s)\}$ (node equations).

d. Finally, the time solution is obtained by taking the ILT of the expressions obtained in part c.

R.4.112 The example shown in the circuit diagram of Figure 4.18, is solved for the current $i(t)$, is used to illustrate the steps followed in the solution of an integrodifferential system, in which each step is labeled according to R.4.111.

ANALYTICAL Solution

Step a

$$v(t) = v_L(t) + v_R(t) \quad \text{for } t \geq 0 \text{ (KVL)}$$

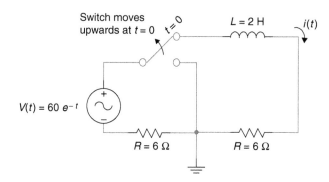

FIGURE 4.18
Network of R.4.112.

$$60e^{-t}u(t) = 2\frac{di(t)}{dt} + 6i(t)$$

Step b

$$\frac{60}{s+1} = 2sI(s) + 6I(s) \quad \text{(Taking the LT)}$$

observe that the initial current $i_L(0) = 0$, then

$$\frac{60}{s+1} = I(s)[2s+6]$$

Step c

$$I(s) = \frac{60}{(s+1)(2s+6)} = \frac{60}{2(s+1)(s+3)}$$

$$I(s) = \frac{30}{(s+1)(s+3)} = \frac{A}{s+1} + \frac{B}{s+3} \quad \text{(by partial fractions expansion)}$$

$$A = \frac{30}{s+3}\bigg|_{s=-1} = \frac{30}{2} = 15$$

$$A = \frac{30}{s+1}\bigg|_{s=-3} = \frac{30}{-2} = -15$$

then

$$I(s) = \frac{15}{s+1} + \frac{-15}{s+3} \quad \text{taking the ILT}$$

Step d

$$i(t) = 15e^{-t}u(t) - 15e^{-3t}u(t)$$

R.4.113 To gain additional experience, five examples of DEs are presented and solved manually, using the LT technique.

The following examples are solved for $y(t)$ for each of the given DEs.

a. Example (#1)

Given the DE

$$2\frac{dy(t)}{dt} - 4y(t) = 0 \quad \text{with IC (initial condition)} \, y(0) = 20$$

ANALYTICAL Solution

Taking the LT of the given equation results in

$$2[sY(s) - 20] - 4Y(s) = 0$$

$$sY(s) - 20 = \frac{4Y(s)}{2} = 2Y(s)$$

then

$$sY(s) - 2Y(s) - 20 = 0$$

$$Y(s)(s - 2) - 20 = 0$$

then

$$Y(s) = \frac{20}{s - 2}$$

taking the ILT

$$y(t) = 20e^{-2t}u(t)$$

b. Example (#2)

Let the initial value of DE be $\dfrac{d^2y(t)}{dt^2} = -4y(t)$

with the IC given by

$$y(0) = 0 \quad \text{and} \quad \frac{dy(t)}{dt}\bigg|_{t=0} = 8$$

ANALYTICAL Solution

Applying the LT to the preceding equation

$$s^2Y(s) - sy(0) - \frac{dy(t)}{dt}\bigg|_{t=0} = 4Y(s)$$

$$s^2Y(s) - s(0) - 8 = 4Y(s)$$

$$Y(s)[s^2 - 4] = 8$$

$$Y(s) = \frac{8}{s^2 - 4} = 4\frac{2}{s^2 - 2^2}$$

and using Table 1.2

$$y(t) = £^{-1}[Y(s)]$$

$$y(t) = 4 \sinh(2t)u(t)$$

c. Example (#3)

　Let

$$\frac{d^2y(t)}{dt^2} = -9y(t)$$

　with the IC given by

$$y(0) = 0 \quad \text{and} \quad \left.\frac{dy(t)}{dt}\right|_{t=0} = 15$$

ANALYTICAL Solution

Taking the LT of the given equation

$$s^2Y(s) - sy(0) - \left.\frac{dy(t)}{dt}\right|_{t=0} = -9Y(s)$$

$$s^2Y(s) - sy(0) - 15 = -9Y(s)$$

$$Y(s)(s^2 + 9) = 15$$

$$Y(s) = \frac{15}{s^2 + 9} = 5\left[\frac{3}{s^2 + 3^2}\right]$$

then

$$y(t) = £^{-1}[Y(s)]$$

$$y(t) = 5\sin(3t)u(t)$$

d. Example (#4)

　Let

$$2\frac{dy(t)}{dt} + 3y(t) = 30$$

　with the IC given by $y(0) = 0$.

ANALYTICAL Solution

Taking the LT of the given equation

$$2\left[sY(s) - y(0)\right] + 3Y(s) = \frac{30}{s}$$

$$Y(s)[2s + 3] = \frac{30}{s}$$

then

$$Y(s)(2s + 3) = \frac{30}{s}$$

$$Y(s) = \frac{30}{s(2s + 3)} = \frac{15}{s(s + 1.5)} = \frac{A}{s} + \frac{B}{s + 1.5}$$

by partial fraction expansion

$$Y(s) = \frac{10}{s} - \frac{10}{s + 1.5}$$

then

$$y(t) = \mathcal{L}^{-1}[Y(s)]$$

and

$$y(t) = 10u(t) - 10e^{-1.5t}u(t)$$

e. Example (#5)

Let

$$\frac{dy(t)}{dt} + 4y(t) = 0$$

with $y(0) = 10$.

ANALYTICAL Solution

Taking the LT of the preceding equation

$$sY(s) - 10 + 4Y(s) = 0$$

$$Y(s)(s + 4) = 10, \qquad Y(s) = \frac{10}{s + 4}$$

and

$$y(t) = \mathcal{L}^{-1}[Y(s)]$$

then

$$y(t) = 10e^{-4t}u(t)$$

R.4.114 The MATLAB function $F = laplace(f)$ returns the LT of f, denoted by $f(t) \rightarrow F(s)$, where f is a symbolic object with independent variable t.

R.4.115 The following examples illustrate the procedure used to obtain the LT of the standard time functions (signals or systems) using the symbolic MATLAB toolbox.

Obtain the LT of the following time-dependent functions, and compare the results with the transforms of Table 4.2.

a. $f1(t) = exp(t)$

b. $f2(t) = t\,exp(t)$

c. $f3(t) = cos(t)$

d. $f4(t) = sin(t)$

e. $f5(t) = t^7$

f. $f6(t) = u(t)$

g. $f7(t) = pul(t/2)$

MATLAB Solution
```
>> syms s t x ;
>> F1 = laplace(exp(t))                    % part (a)

F1 =
     1/(s-1)

>> F2 = laplace(t*exp(t))                   % part (b)

F2 =
     1/(s-1)^2

>> F3 = laplace(cos(t))                     % part (c)

F3 =
     s/(s^2+1)

>> F4 = laplace(sin(t))                     % part (d)

F4 =
     1/(s^2+1)

>> F5 = laplace(t^7)                        % part (e)

F5 =
     5040/s^8

>> F6 = laplace(sym('Heaviside(t)'))        % part (f)

F6 =
     1/s

>> F7 = laplace(sym('Heaviside(t+1)')
   -sym('Heaviside(t-1)'))                  % part(g)

F7 =
     1/s-exp(-s)/s
```

Note that the transforms obtained using the symbolic MATLAB toolbox fully agree with the transforms of Table 4.2.

R.4.116 The MATLAB function $f = ilaplace(F)$, where F is a symbolic expression with independent variable s, returns the ILT of F, denoted by f which is a function of t, denoted by $F(s) \rightarrow f(t)$.

R.4.117 For example, using the symbolic MATLAB toolbox obtain the ILT of the following
frequency-dependent functions (*F1* through *F7* of R.4.115):

a. $F1 = 1/(s - 1)$

b. $F2 = 1/(s - 1)^2$

c. $F3 = s/(s^2 + 1)$

d. $F4 = 1/(s^2 + 1)$

e. $F5 = 5040/s^8$

f. $F6 = 1/s$

g. $F7 = (1/s) - e^{-s}/s$

MATLAB Solution
```
>> syms s t
>> F1=1/(s-1)
>> f1 = ilaplace(F1)

   f1 =
            exp(t)
>> F2 = 1/(s-1)^2;
>> f2 = ilaplace(F2)

   f2 =
            t*exp(t)
>> F3 = s /(s^2+1);
>> f3 = ilaplace(F3)

   f3 =
            cos(t)
>> F4 =1/(s^2+1);
>> f4 = ilaplace(F4)

   f4 =
            sin(t)
>> F5 = 5040/s^8;
>> f5 = ilaplace(F5)

   f5 =
            t^7
>> F6 = 1/s;
>> f6 = ilaplace(F6)

   f6 =
            1
>> F7 = 1/s-e^-s/s;
>> f7 = ilaplace(F7)

   f7 =
            1-Heaviside(t-1)
```

R.4.118 Let us now illustrate the power of MATLAB by solving again the four Example
problems of R.4.98, using MATLAB, given below by the script file *solve_DE*.

Recall that the examples of R.4.98 were solved by hand. Compare the results, as
well as the labor involved in each process.

MATLAB Solution
```
% Script file: solve _ DE
syms s t ft Fsa ftb Fsb ftc Fsc yt Y
```

```
disp('******************************')
disp('****Solutions using Matlab****')
disp('******************************')
disp('(a) Example(#1), the time function f(t) is:')
ft = 5+2*exp(-3*t)+10*sin(3*t)+5*exp(-2*t)*cos(7*t);
pretty(ft)
disp ('its Laplace transform F(s) is given by:')
Fsa = laplace(ft);
pretty(Fsa)
disp('******************************')
disp('(b) Example(#2), the frequency function F(s) is:')
Fsb = 10/s+20/s^2+30/(s+3);
pretty(Fsb)
disp('its inverse Laplace transform is, f(t):')
ftb=ilaplace(Fsb);
pretty(ftb)
disp('******************************')
disp('(c)Example(#3), the frequency function F(s) is:')
Fsc = (8*s+30)/(s^2+25);
pretty(Fsc)
disp('its inverse Laplace transform, f(t) is:')
ftc = ilaplace(Fsc);
pretty(ftc)
disp('******************************')
disp('(d) Example(#4), the frequency function Y(s) is:')
Y = (4*s+10)/(s*(s+1)*(s+2));
pretty(Y)
disp('its inverse Laplace transform, y(t) is given by:')
yt = ilaplace(Y);
pretty(yt)
disp('******************************')
```

The script file *solve_DE* is executed and the results are as follows:

```
>> solve _ DE


**************************************
****** Solutions using Matlab ******
**************************************
(a) Example(#1), the time function f(t) is:
    5 + 2 exp(-3 t) + 10 sin(3 t) + 5 exp(-2 t) cos(7 t)
    its Laplace transform F(s) is given by :

             2        30         s + 2
    5/s + ----- + ------ + 5 -------------
          s + 3      2                 2
                   s + 9      (s + 2) + 49
**************************************

(b) Example(#2), the frequency function F(s) is:
    10     20      30
    ---- + ---- + -----
     s      2s     s + 3
    its inverse Laplace transform is, f(t):
    10 + 20 t + 30 exp(-3 t)
**************************************
```

(c) **Example (#3)**, the frequency function F(s) is:

```
8 s + 30
--------
s² + 25
```

its inverse Laplace transform, f(t) is:

```
8 cos(5 t) + 6 sin(5 t)
```
****

(d) **Example(#4)**, the frequency function Y(s) is:

```
4 s + 10
-----------------
s (s + 1) (s + 2)
```

its inverse Laplace transform, y(t) is given by:

```
5 - 6 exp(-t) + exp(-2 t)
```
****

Note that the results obtained fully agree with the results obtained in R.4.98, with a significant reduction of mathematical insight and labor.

R.4.119 Let $f_T(t)$ be a periodic function (with a period $T \neq 0$). Then its LT is given by

$$\mathcal{L}\big[f_T(t)\big] = \frac{1}{1 - e^{-Ts}} F(s)$$

where

$$F(s) = \int_0^T e^{st} f_T(t) dt$$

Observe that this frequency–time relation pair was presented in R.4.82 (#*h*), using a slightly different notation. It is important to understand that the LT can be equally useful in the analysis of periodic, as well as nonperiodic functions, as was the case of the FT.

R.4.120 A way to evaluate the ILT of a given transfer function H(s) = *num/den* is by evaluating its impulse response h(t). Recall that the MATLAB function *impulse (num,den,t)* returns the system impulse response, where *num* and *den* are the numerator and denominator polynomials of H(s) entered as row vectors arranged in descending powers of *s*, and *t* defines the time range of interest.

R.4.121 For example,

$$\text{Let } F(s) = \frac{3s + 25}{s^2 + 3s + 2} = \frac{num(s)}{den(s)}$$

then the program that returns the expression for $f(t)$ ($f(t) \rightarrow F(s)$), and the plot $f(t)$ versus t, evaluated by numerical and symbolic techniques, as well as the plot $F(s)$ versus s, over the ranges: $0 \leq t \leq 10$, and $0 \leq s \leq 10$, is given as follows by the script file F_f.

MATLAB Solution
```
% Script file: F _ f
num = [0 3 25];
den = [1 3 2];
t = 0:.1:10;
f = impulse (num, den, t);
subplot(3,1,1)
```

```
plot (t,f)
title (' plot of [ ILT of F(s)] vs. t (numerical solution)')
ylabel ('Amplitude'); xlabel ('time t');

disp('*************RESULTS************')
syms s t;
disp('F(s) is given by:')
F = (3*s+25)/(s^2+3*s+2);
pretty(F)
disp('The ILT of F(s)is given by:')
FILT = ilaplace(F);
pretty (FILT)
subplot (3,1,2)
ezplot (FILT,[0 10])
title (' plot of  f(t) vs. t (symbolic solution) ')
xlabel ('time t'); ylabel('Amplitude')
subplot(3,1,3)
ezplot (F,[0 10])
title (' plot of F(s) vs.s (symbolic) ')
xlabel ('s (frequency)')
ylabel('Amplitude')
disp('********************************')
```

The script file *F_f* is executed and the results are indicated as follows.

```
>> F _ f
```

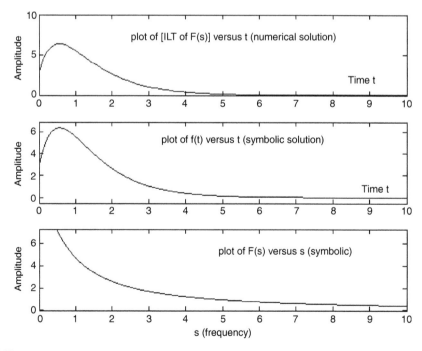

FIGURE 4.19
Plots of R.4.121.

```
***********RESULTS*************
F(s) is given by:

      3 s + 25
     ------------
     s² + 3 s + 2

The ILT of F(s) is given by:

    -19 exp(-2 t) + 22 exp(-t)
**********************************
```

R.4.122 A plot that relates to the amplitude of a given function *f(t)*, its frequency *w*, and its time *t*, is called a spectrogram. In its simplest version, the MATLAB function *specgram(f)* returns the plots of the spectrogram of the function *f(t)* specified by the row vector *f*, with a default sampling frequency of $F_s = 1/Ts = 2$.

R.4.123 In its simplest version, the MATLAB function *psd(f)* returns the power spectrum density of the function *f(t)* defined by the vector *f*.

R.4.124 Recall that MATLAB provides with a speech file named *mtlb.mat* that consists of 4001 samples, sampled at 1418 Hz. Recall also that this file (*mtlb.mat*) is often used for testing purposes (Chapter 1).

R.4.125 The following example illustrates how the MATLAB functions *specgram* and *psd* can be used on an arbitrary file (voice, audio, video, or data), or on a data array.

For example, create the script file *specpsd* that returns the following plots (Figures 4.20 and 4.21):

1. The speech file *mtlb.mat* versus *t*
2. *y(t)* = 3cos(2π*t*/2000), sampled with a rate of *T* = 1, for *t* = 0, 1, 2, 3, 4,..., 4000 (returning 4001 samples) versus *t*
3. A random noisy signal defined by noise = 2 * rand(1:4001) versus *t*
4. *ynoise* = *y(t)* + *noise*, over the first 1000 samples.
5. The spectrogram of the file *mtlb*
6. *psd* of the file mtlb
7. The spectrogram of the sequence *ynoise*
8. *psd* of the sequence *ynoise*

MATLAB Solution
```
% Script file: specpsd
% load speech file mtlb
load mtlb;
x = 1:4001;                      % time sequence
figure(1)
subplot(2,2,1);
plot(x,mtlb);
axis([0 4000 -4 4]);
ylabel('Amplitude ');
title('[Speech file mtlb] vs. t')
noise= rand(1,4001)*2;           % noise sequence;
y = 3*cos(2*pi*x/2000);
```

```
ynoise = 3*cos(2*pi*x/2000)+noise;
subplot(2,2,2)
plot(x,y);title('Sinusoid test funct ');
ylabel('Amplitude')
axis([0 4000 -4 4]);
subplot(2,2,3)
plot(x,noise);title('noise vs. t')
ylabel('Amplitude')
axis([0 100 -1 3]);
xlabel('time');
subplot(2,2,4)
plot(x,ynoise)
xlabel('time');title('[sinusoid+noise] vs. t');
ylabel('Amplitude')
axis([0 1000 -6 6]);
figure(2)
subplot(2,2,1)
specgram(mtlb);title('spectrogram of the speech file mtlb')
subplot(2,2,2)
psd(mtlb);title('power spec. den.of mtlb')
subplot(2,2,3)
specgram(ynoise);
title('specgram of sinusoid+noise')
subplot(2,2,4)
psd(ynoise);
title('power spec. den. of sinus.+noise')
```

The script file *specpsol* is executed and the resulting plots are shown in Figures 4.20 and 4.21.

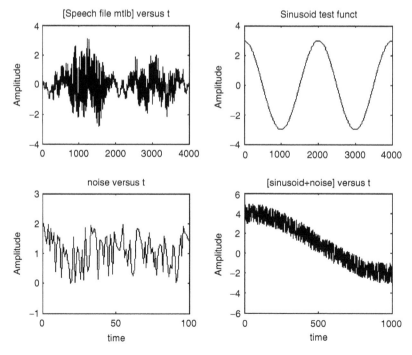

FIGURE 4.20
Plots of parts 1, 2, 3, and 4 of R.4.125.

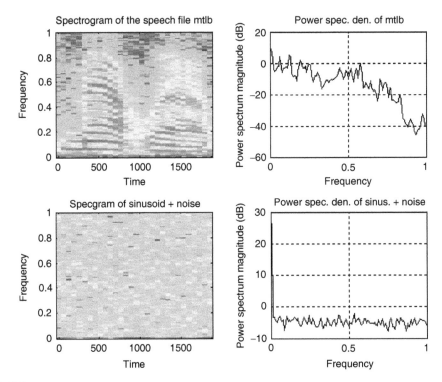

FIGURE 4.21
(See color insert following page 374.) Plots of parts 5, 6, 7, and 8 of R.4.125.

R.4.126 The MATLAB function *spectrum(x)* returns the script file *specpsd* is executed and the resulting plots are shown in Figures 4.20 and 4.21 plot of *psd(f)* with the default given by *length(f)* = 256 or less.

R.4.127 The MATLAB function *spectrum(x,y)* estimates the power density of the sequences *x* and *y*, and returns the spectral analysis for *x* and *y* in a tablelike format, where the first five columns are denoted by P_{xx}, P_{yy}, P_{xy}, T_{xy}, and C_{xy}, where

P_{xx} denotes the *psd* of *x*

P_{yy} represents the *psd* of *y*

P_{xy} represents the cross-spectral density

T_{xy} represents the complex transfer function that relates input *x* to the output *y*

C_{xy} denotes the coherence relation between the system input *x* and its output *y*

The remaining three columns provide confidence ranges.

R.4.128 The coherent relation C_{xy} indicates by means of the coefficients the degree of similarity between the two given sequences *x* and *y*, over a normalized range of frequencies. The range of the coefficients of similarity are defined between 0 and 1, where 0 indicates no similarity and 1 indicates a high degree of similarity between *x* and *y*.

R.4.129 The MATLAB function *spectrum(x,y)*, with no explicit output, returns these five plots: P_{xx}, P_{yy}, *magnitude* $[T_{xy}]$, *phase*$[T_{xy}]$, and C_{xy}.

R.4.130 The MATLAB function *cohere(x,y)* returns the plot of the coefficients of similarity between *x* and *y*, indicating how well the input *x* corresponds to the output *y*, over a normalized frequency range.

R.4.131 The MATLAB function *specplot(p)* uses the output of the function *p=spectrum(x,y)* (R.4.126) to return the plots of P_{xx}, P_{yy}, *magnitude[T_{xy}]*, *phase[T_{xy}]*, and C_{xy}.

R.4.132 The script file *specgrph* illustrates the use of the function *spectrum*, with input arguments *ynoise* and *mtlb*, defined in R.4.124 and R.4.125.

MATLAB Solution
```
% Script file: specgrph
% load speech file mtlb
% echo on;
load mtlb;
x =1:4001;                                    % time sequence
noise = rand(1,4001)*2;                       % noise sequence
y = 3*cos(2*pi*x/2000);
ynoise = 3*cos(2*pi*x/2000)+noise;
disp('*******************************************')
disp('*******************************************')
disp('Press Enter to get the plots of:')
disp('Pyy,mag(Txy),angle(Txy),Cxy')
disp('*******************************************')
disp('*******************************************')
spectrum(ynoise,mtlb)
```

The script file *specgrph* is executed and the resulting plots are shown in Figures 4.22 through 4.26.

R.4.133 It is often required to analyze a given system transfer function *H(s)*, in terms of its circuit elements, given by impedances. Review of simple network structures, expressed in terms of their impedance (in ohms), denoted by *Z(s)* are shown in Figure 4.27, with the corresponding equations. These simple building blocks when interconnected are used to build complex networks.

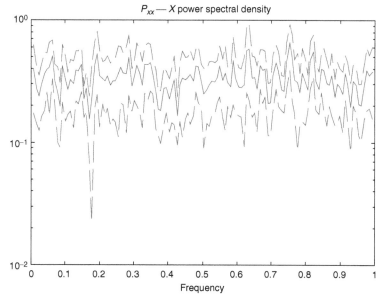

FIGURE 4.22
Plot of *psd* of *ynoise* of R.4.132.

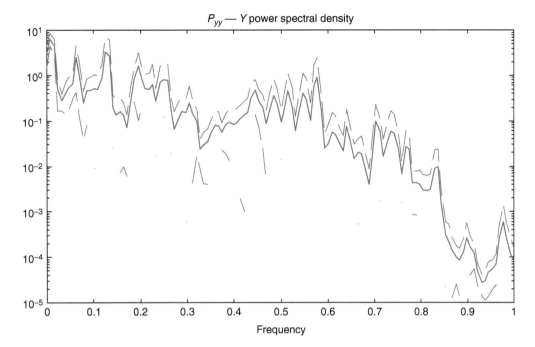

FIGURE 4.23
Plot of *psd* of *mtlb* of R.4.132.

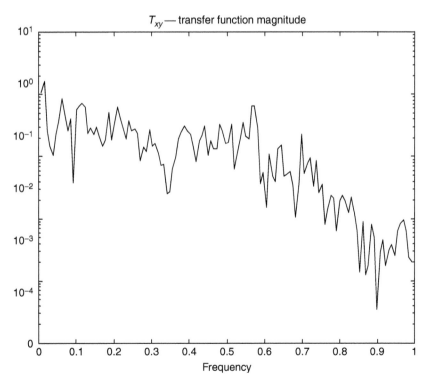

FIGURE 4.24
Plot of magnitude of T_{xy} of R.4.132.

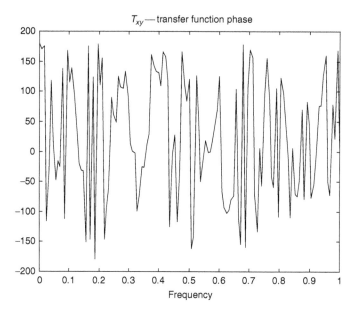

FIGURE 4.25
Plot of phase of T_{xy} where x = *ynoise* and y = *mtlb* of R.4.132.

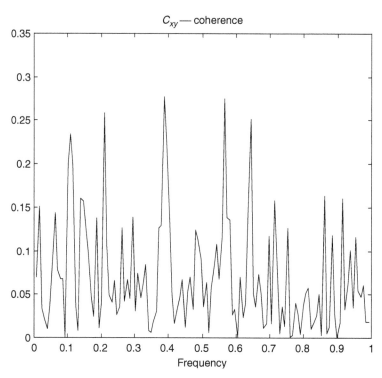

FIGURE 4.26
Plot of C_{xy} coherent relation between x and y of R.4.132.

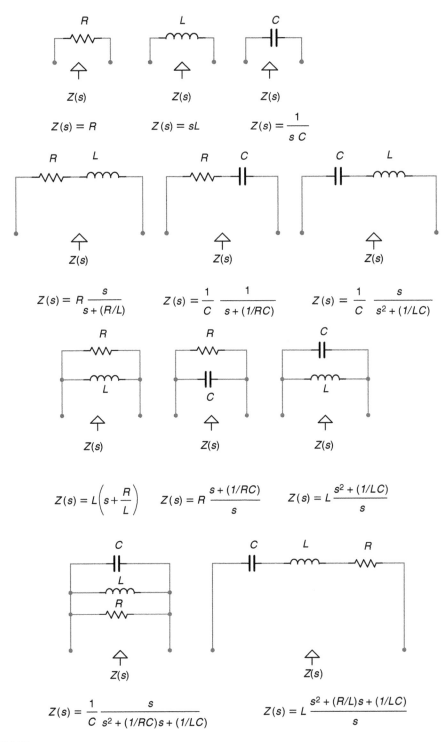

FIGURE 4.27
Standard impedances $Z(s)$ of simple network structures.

R.4.134 To conclude, let us discuss the reason for having defined and used two very similar transforms, the FTs and LTs in this chapter.

Observe that both transforms basically state the same mathematical relation. Both are obtained by integrating over time $f(t)$ multiplied by an exponential raised to a negative time complex frequency. Recall that for Laplace's case $f(t)$ is multiplied by e^{-st} (where s is complex), and the reason is that the convergence of the Fourier integral is enhanced, meaning that the LT may exist for many functions for which there is no FT. This is one of the main advantages of the LT, the ability to transform functions which are not otherwise transformable by means of Fourier.

In addition, Laplace incorporates the initial conditions in the evaluation of its transform and provides a more complete system model representation, in which information in the frequency domain are directly related to the time domain, as is the case of the initial and final value theorems.

4.4 Examples

Example 4.1

Let the FS expansion of a periodic square wave be given by

$$f(t) = \frac{2}{\pi} \sum_{n=\text{odd}}^{7} \frac{1}{n}\sin(nw_0 t)$$

where the period is $T = 2$ s, and the fundamental frequency is $w_0 = 2\pi/T = \pi$.

Create the script file *square_Fourier* that returns the following plots, over the range $0 \le t \le 4$:

a. The first four nonzero Fourier components of $f(t)$ versus t (for $n = 1, 3, 5, 7$) on separate plots

b. The components of part a on one plot (indicating the phase, frequency, and amplitude relations)

c. Successive partial approximation sums of the FS expansion on separate plots

d. Successive partial sums of the FS expansion on the same plot

e. The *errors(t)* versus t for each one of the approximations (sums) of part c

MATLAB Solution
```
% Script file: square _ Fourier
% Harmonic Analysis of a square wave by a Fourier Series
% approximation, fundamental frequency wo=pi
echo off;
T = 2;
w0 = 2*pi/T;
t = 0:.01:2*T;
% Harmonics are generated below
Harmonic _ 1 = 2/pi*sin(w0.*t);
Harmonic _ 3 = 2/(3*pi)*sin(3*w0.*t);
Harmonic _ 5 = 2/(5*pi)*sin(5*w0.*t);
Harmonic _ 7 = 2/(7*pi)*sin(7*w0.*t);
```

```
figure(1);                                    % Figure 4.28, part(a)
subplot(2,2,1);
plot(t,Harmonic _ 1);
title('fundamental freq. wo=pi');
ylabel('Amplitude');
subplot(2,2,2);
plot(t,Harmonic _ 3);
title('3rd. harmonic, freq.w=3*wo');
ylabel('Amplitude');
subplot(2,2,3);
plot(t,Harmonic _ 5);
title('5th. harmonic, freq.w=5*wo');
ylabel('Amplitude');
xlabel(' time (in sec)');
subplot(2,2,4);
plot(t,Harmonic _ 7);
title('7th. harmonic, freq.w=7*wo');
ylabel('Amplitude'); axis([0 4 −0.15 0.15])
xlabel(' time (in sec)');

figure(2);                                    % Figure 4.29, part (b)
plot(t,Harmonic _ 1,t,Harmonic _ 3,t,Harmonic _ 5,t,Harmonic _ 7);
title('Harmonic components of the square wave');
ylabel('Amplitude');
xlabel('time (sec) ');

figure(3);                                    % Figure 4.30, part(c)
subplot(2,2,1)
plot(t,Harmonic _ 1);
title('fundamental frequency (square)');
ylabel('Amplitude');
subplot(2,2,2);
Har _ 13 = Harmonic _ 1+ Harmonic _ 3;
plot(t,Har _ 13);
title('fund.+ 3rd. harmonic ');
ylabel('Amplitude');
subplot(2,2,3);
Har _ 135 = Harmonic _ 1+ Harmonic _ 3+ Harmonic _ 5;
plot(t,Har _ 135);
title('fund.+3rd.+ 5th harmonic ');
ylabel('Amplitude');
xlabel('time (in sec)');
subplot(2,2,4);
Har _ 1357 = Harmonic _ 1+ Harmonic _ 3+Harmonic _ 5+Harmonic _ 7;
plot(t,Har _ 1357);
title('fund.+ 3rd.+ 5th.+ 7th. harmonic ');
ylabel('Amplitude');
xlabel('time (in sec)');

figure(4);                                    % Figure 4.31, part (d)
plot(t,Harmonic _ 1,t,Har _ 135,t,Har _ 1357);
title('Plots of square wave approximations using harmonics ');
xlabel('time (in sec)');
ylabel('Amplitude');
```

```
figure(5);                                    % Figure 4.32, part (e)
sqr = [.5*ones(1,101) -.5*ones(1,100) .5*ones(1,100) -.5*ones(1,100)];
error1= sqr-Harmonic _ 1;
error3 = sqr-Har _ 13;
error5 = sqr-Har _ 135;
error7 = sqr-Har _ 1357;
subplot(2,2,1)
plot(t,sqr,t,error1);
title('error1(t)=square wave-harmonics:1')
ylabel('Amplitude');axis([0 4 -0.6 0.6]);
subplot(2,2,2)
plot(t,sqr,t,error3);
title('error3(t)=square wave-harmonics:1,3')
ylabel('Amplitude');axis([0 4 -0.6 0.6]);
subplot(2,2,3)
plot(t,sqr,t,error5);
title('error5(t)=square wave-harmonics:1,3.5')
xlabel('time (in sec)');axis([0 4 -0.6 0.6]);
ylabel('Amplitude');
subplot(2,2,4)
plot(t,sqr,t,error7);axis([0 4 -0.6 0.6]);
title('error7(t)=square wave-harmonics:1,3,5,7')
xlabel('time (in sec)');
ylabel('Amplitude');
```

The script file *square_Fourier* is executed and the results are shown in Figures 4.28 through 4.32.

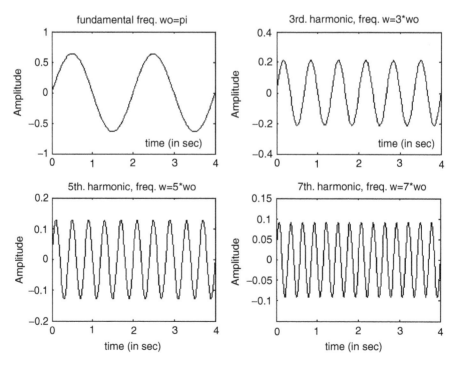

FIGURE 4.28
Plots of part a of Example 4.1.

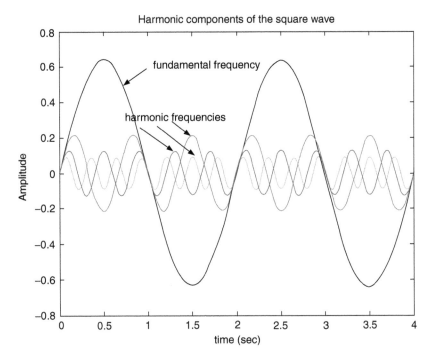

FIGURE 4.29
(See color insert following page 374.) Plots of part b of Example 4.1.

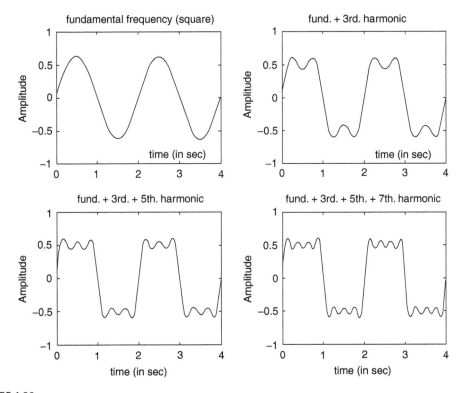

FIGURE 4.30
Plots of part c of Example 4.1.

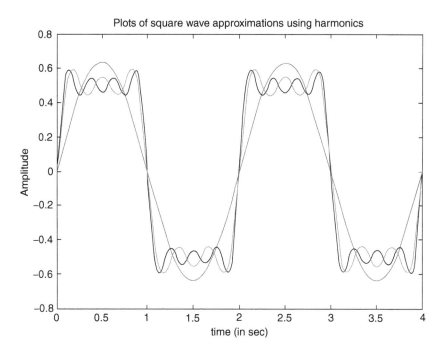

FIGURE 4.31
(See color insert following page 374.) Plots of part d of Example 4.1.

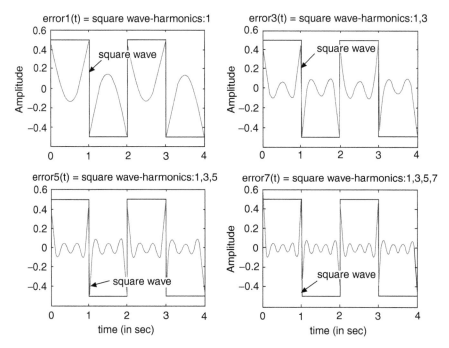

FIGURE 4.32
Plots of part e of Example 4.1.

Observe that at the discontinuity points or corners (at $t = 0, 1, 2, 3...$), an overshoot is clearly present. The overshoot is about 9%, even if the number of terms of the approximation is increased to infinity. This overshoot was first observed by Josiah Willard Gibbs (1839–1903), and is referred to as the Gibbs phenomenon, discussed and presented, in this chapter background.

Observe also that the amplitudes of the different harmonics decrease as the frequencies increase.

Example 4.2

Let the FS expansion for a periodic sawtooth wave be given by

$$f(t) = \frac{-1}{\pi} \sum_{n=1}^{6} \frac{1}{n} \sin(n w_0 t)$$

with period $T = 2$ s, and a fundamental frequency given by $w_0 = 2\pi/T = \pi$.

Create the script *filesawtooth_fourier* that returns the following plots over the range $0 \le t \le 4$:

a. The first six nonzero Fourier components of $f(t)$ versus t (for $n = 1, 2, 3, 4, 5, 6$) on separate plots
b. The components of part a on one plot
c. Successive partial sums of the FS expansion on separate plots
d. Successive partial approximation sums of the FS expansion on the same plot
e. The *[error(t)]* versus t, for each of the FS approximations (sums) of part c

MATLAB Solution

```
% Script file: sawtooth _ fourier
% Harmonic Analysis for a sawtooth wave by a Fourier Series
% approximation, with fundamental frequency wo = pi
echo off;
T = 2;
w0 = 2*pi/T;
t = 0:.01:2*T;
%Harmonics are created below
Harmonic _ 1 = -1/pi*sin(w0.*t);
Harmonic _ 2 = -1/(2*pi)*sin(2*w0.*t);
Harmonic _ 3 = -1/(3*pi)*sin(3*w0.*t);
Harmonic _ 4 = -1/(4*pi)*sin(4*w0.*t);
Harmonic _ 5 = -1/(5*pi)*sin(5*w0.*t);
Harmonic _ 6 = -1/(6*pi)*sin(6*w0.*t);

figure(1);                    % Figure 4.33; part(a)
subplot(3,2,1);
plot(t,Harmonic _ 1);
title('fundamental freq. wo=\pi');
ylabel('Amplitude');
subplot(3,2,2);
plot (t,Harmonic _ 2);
title ('2d. harmonic, freq.w=2\pi');
ylabel ('Amplitude');
subplot (3,2,3);
plot (t,Harmonic _ 3);
```

```
title ('3rd. harmonic, freq.w =3\pi');
ylabel ('Amplitude');
subplot (3,2,4);
plot (t,Harmonic _ 4);
title ('4th. harmonic, freq. w = 4\pi');
ylabel ('Amplitude');
subplot (3,2,5);
plot (t,Harmonic _ 5);
title('5th. Harmonic, freq. w =5\pi');
ylabel('Amplitude');
xlabel('time in sec.')
subplot (3,2,6);
plot (t,Harmonic _ 6);
title ('6th. Harmonic, freq. w = 6\pi');
ylabel ('Amplitude');
xlabel ('time in sec.')

figure(2);                              % Figure 4.34, part (b)
plot(t,Harmonic _ 1,t,Harmonic _ 2,t,Harmonic _ 3,t,Harmonic _ 4);
hold on;
plot(t,Harmonic _ 5,t,Harmonic _ 6);
title('Harmonic components of a sawtooth wave');
ylabel('Amplitude');
xlabel('time (in sec)');

figure(3);                              % Figure 4.35; part (c)
subplot(3,2,1)
plot(t,Harmonic _ 1);
title('fundamental components of the sawtooth');
ylabel('Amplitude');
subplot(3,2,2);
Har _ 12=Harmonic _ 1+Harmonic _ 2;
plot(t,Har _ 12);
title('Fund.+2nd.Harmonic of the sawtooth');
ylabel('Amplitude');
subplot(3,2,3)
Har _ 123=Harmonic _ 1+Harmonic _ 2+Harmonic _ 3;
plot(t,Har _ 123);
title('Fund.+2nd.+3rd Harmonics');
ylabel('Amplitude');
subplot(3,2,4)
Har _ 1234=Harmonic _ 1+Harmonic _ 2+Harmonic _ 3+Harmonic _ 4;
plot(t,Har _ 1234);
title('Fund.+2nd.+3rd.+4th.Harmonics ');
ylabel('Amplitude');
subplot(3,2,5);
Har _ 12345=Har _ 1234+Harmonic _ 5;
plot(t,Har _ 12345)
title('Fund.+2nd.+3rd.+4th.+5th.Harmonics ');
axis([0 2 -.55 .55])
ylabel('Amplitude');
xlabel('time (in sec)');
subplot(3,2,6);
Har _ 123456=Har _ 12345+Harmonic _ 6;
```

```
plot(t,Har _ 123456)
axis([0 2 -.6 .6])
title('Fund.+2nd.+3rd.+4th.+5th.+6th.Harm. ');
ylabel('Amplitude');
xlabel('time (in sec)');

figure(4);                              % Figure 4.36; part (d)
plot(t,Harmonic _ 1,t,Har _ 12,t,Har _ 123,t,Har _ 1234,t,Har _ 12345,t,
                                             Har _ 123456);
title('Approximations of the sawtooth wave by sum of harmonics');
ylabel('Amplitude');
xlabel('time (in sec)');

figure(5)                               % Figure 4.37; part(e)
x =0:.01:T;y =0:.01:T-.01;
trian0 =.5.*x-.5;trian1 =.5.*y-.5;
triag = [trian0 trian1];
error1 = triag-Harmonic _ 1;
error2 = triag-Har _ 12;
error3 = triag-Har _ 123;
error4 = triag-Har _ 1234;
error5 = triag-Har _ 12345;
error6 = triag-Har _ 123456;
subplot(3,2,1)
plot(t,error1)
title('error1(t) = sawtooth wave-harmonics 1')
ylabel('Amplitude')
subplot(3,2,2)
plot(t,error2)
title('error2(t) = sawtooth wave-harmonics 1,2')
ylabel('Amplitude')
subplot(3,2,3)
plot(t,error3)
title('error3(t) = sawtooth wave-harmonics 1,2,3')
ylabel('Amplitude')
subplot(3,2,4)
plot(t,error4)
title('error4(t) = sawtooth wave-harmonics 1,2,3,4')
ylabel('Amplitude')
subplot(3,2,5)
plot(t,error1)
title('error5(t) = sawtooth wave-harmonics 1,2,3,4,5')
ylabel('Amplitude')
xlabel('time (in sec)');
subplot(3,2,6)
plot(t,error6)
title('error6(t)=sawtooth wave-harmonics 1,2,3,4,5,6')
ylabel('Amplitude')
xlabel('time (in sec)');
```

The script file *sawtooth_fourier* is executed and the results are shown in Figures 4.33 through 4.37.

```
>> sawtooth _ fourier
```

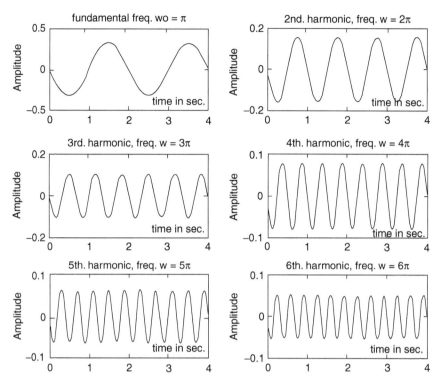

FIGURE 4.33
Plots of part a of Example 4.2.

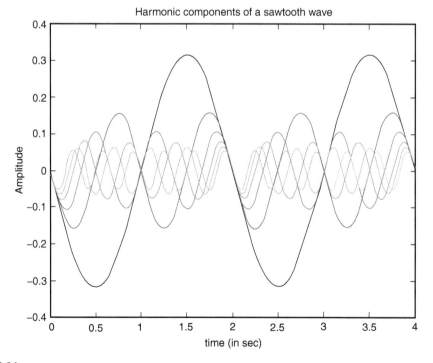

FIGURE 4.34
(See color insert following page 374.) Plots of part b of Example 4.2.

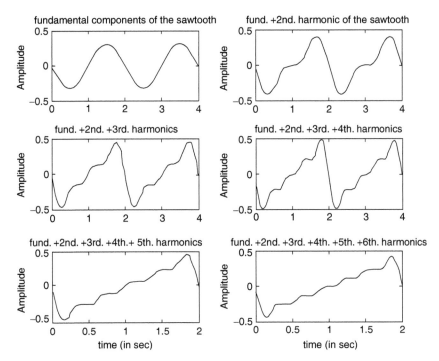

FIGURE 4.35
Plots of part c of Example 4.2.

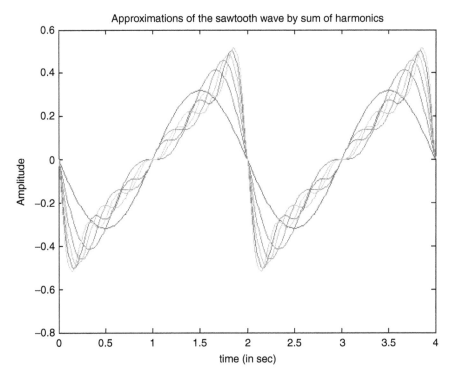

FIGURE 4.36
(See color insert following page 374.) Plots of part d of Example 4.2.

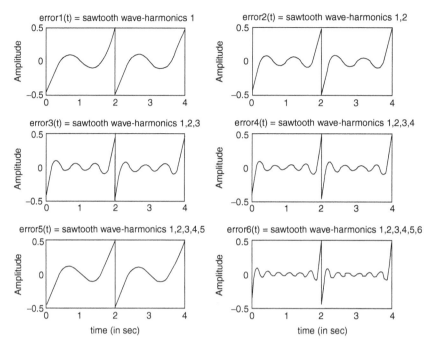

FIGURE 4.37
Plots of part e of Example 4.2.

Example 4.3

Let the FS expansion for a periodic sawtooth wave be given by

$$f(t) = \frac{-1}{\pi} \sum_{n=1}^{6} \frac{1}{n} \sin(nw_0 t) \quad \text{(same as in Example 4.2)}$$

with period $T = 2$ s, and fundamental frequency given by $w_0 = 2\pi/T = \pi$.

Let $f(t)$ represent the voltage drop across a resistor $R = 5\,\Omega$. Create the script file *Fourier_applic* that returns the following:

1. The first 10 FS coefficients
2. Its RMS value
3. The percentage of the total harmonic distortion
4. The average power dissipated by $R = 5\,\Omega$
5. The power concentrated in the first 10 harmonics
6. The plot of the percentage of total power (dissipated) versus its harmonic frequencies
7. Check the % of total power dissipated in the first 10 harmonics.

MATLAB Solution
```
% Script file: Fourier _ applic
% Calculations of the Fourier coefficient
for n = 1:10;
  c(n)  = -1/(n*pi);
  p(n)  = c(n).^2;
end
C = c;
% RMS calculations
```

```
F _ rms = norm(c./sqrt(2));
% Pave calculations
Pave = (F _ rms^2)/5;
% Perc. of total harmonic distortion calculations
a = c(1);c(1)=[];
PTHD = (norm(c)/a)*100;m=1:10;
disp('**********************************************')
disp('***************R  E  S  U  L  T  S*****************')
disp('**********************************************')
disp('The Fourier Series coefficients are:')
disp(' Harmonics coef. values')
[m' C']
fprintf('The rms value (in volts) of f(t) is: %8.3f\n',F _ rms)
fprintf('The Pave of f(t) is: %8.3f watt\n',Pave)
fprintf('The percentage of total harmonic distortion is: %8.3f\n',PTHD)
% calculations of % of total power in each harmonic
P = p.*100./sum(p);stem(m,P)
xlabel ('harmonics in rad/sec')
ylabel ('% of total power in each harmonic')
title('[% of total power in each harmonic] vs harmonics')
axis([-1 10 0 80]);grid on
disp('**********************************************')
disp('****************Check results*****************')
disp('**********************************************')
disp('The total % power in the first 10 harmonics is=')
sum(P)
disp(' percent ')
disp('**********************************************')
```

The script file *Fourier_applic* is executed and the results are indicated in the following (Figure 4.38):

```
>> Fourier _ applic
**************************************************
***************R  E  S  U  L  T  S ******************
**************************************************
The Fourier Series coefficients are:
Harmonics coef. values
  ans =
        1.0000  -0.3183
        2.0000  -0.1592
        3.0000  -0.1061
        4.0000  -0.0796
        5.0000  -0.0637
        6.0000  -0.0531
        7.0000  -0.0455
        8.0000  -0.0398
        9.0000  -0.0354
       10.0000  -0.0318
The rms value (in volts) of f(t) is : 0.280
The Pave of f(t) is: 0.016 watt
The percentage of total harmonic distortion is: 74.146
**************************************************
```

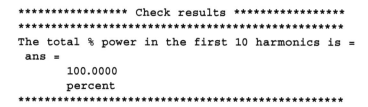

```
**************** Check results *****************
***********************************************
The total % power in the first 10 harmonics is =
 ans =
       100.0000
       percent
***********************************************
```

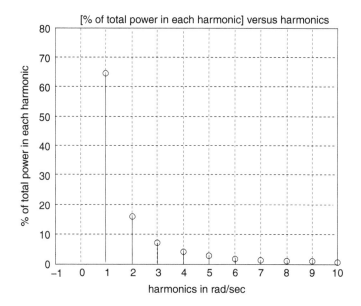

FIGURE 4.38
Plot of power concentration in harmonics of Example 4.3.

Example 4.4

Given the function $f(t)$ shown in Figure 4.39, defined over the range $-1 \leq t \leq 1$,

$$f(t) = \begin{cases} 0.5 & -1 < t < -0.5 \\ 1.5 & -0.5 \leq t \leq 0.5 \\ 0.5 & 0.5 < t \leq 1 \end{cases}$$

1. Evaluate by hand the coefficients of the exponential FS of $f(t)$.
2. Create the script file *Fourier_series* that returns the following:
   a. Plots of up to the fifth harmonic of $f(t)$ versus t (including the DC)
   b. Plots on one graph showing the components of $f(t)$ of part 2a
   c. Plots showing the partial sums of the Fourier components of $f(t)$ up to the sixth component
   d. The line spectrum plot of $f(t)$
   e. The power spectrum plot of $f(t)$
   f. Verify Parseval's theorem. Recall that $P_{ave} = (1/T)\int_0^T f(t)^2 dt = \sum_{n=0}^{6} |F_n|^2$, and determine the error
   g. Evaluate the percentage error for part f when using time versus frequency

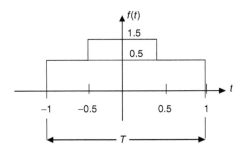

FIGURE 4.39
Plot of $f(t)$ of Example 4.4.

ANALYTICAL Solution

Part 1

$$f(t) = \sum_{-\infty}^{+\infty} F_n e^{-jnw_0 t}$$

$$F_n = \frac{1}{2} \int_{-0.5}^{+0.5} (1)e^{-jnw_0 t}\, dt; \quad T = 2 \text{ then } w_0 = \frac{2\pi}{2} = \pi$$

$$F_n = \frac{1}{2} \int_{-0.5}^{+0.5} (1)e^{-jnw_0 t}\, dt = \frac{1}{2jn\pi} e^{jnw_0 t}\Big|_{-0.5}^{+0.5}$$

$$F_n = \frac{1}{2} \frac{\sin(n\pi/2)}{(n\pi/2)}$$

Therefore,

$$f(t) = \frac{1}{2} + \frac{1}{2} \sum_{n=1}^{+\infty} \frac{\sin(n\pi/2)}{(n\pi/2)}$$

MATLAB Solution

```
% Script file: Fourier _ series
n = -5:1:5;                          % creates harmonics
% add eps to avoid division by zero
Fn = 0.5*sin(n.*pi/2)./((n.*pi/2)+eps);   % creates the coefficients Fn
t =linspace (-2,2,50);               % limits from -1 to 1,one
                                       period
F1=Fn(5)*exp(j*pi.*t)+Fn(7)*exp (-j*pi.*t);   % first component
F2=Fn(4)*exp(j*2*pi.*t)+Fn(8)*exp (-j*2*pi.*t);
                                     % second component
F3=Fn(3)*exp(j*3*pi.*t)+Fn(9)*exp (-j*3*pi.*t);
                                     % third component
F4=Fn(2)*exp(j*4*pi.*t)+Fn(10)*exp (-j*4*pi.*t);
                                     % fourth component
F5=Fn(1)*exp(j*5*pi.*t)+Fn(11)*exp (-j*5*pi.*t);
                                     % fifth component

figure(1);                           % see Figure 4.40; part (2a)
F0 = .5.*ones(1,50);
subplot (3,2,1);plot (t,F0);
title ('DC component'); ylabel ('Amplitude (F0)')
subplot (3,2,2);plot (t,F1);
```

```
title ('Fundamental frequency'); ylabel ('Amplitude of F1')
subplot (3,2,3); plot (t,F2);
title ('Second harmonic'); ylabel ('Amplitude of F2')
subplot (3,2,4); plot (t,F3);
title ('Third harmonic'); ylabel ('Amplitude of F3');
subplot (3,2,5); plot (t,F4);
title ('Fourth harmonic'); ylabel ('Amplitude of F4'); xlabel ('time (in sec)')
subplot (3,2,6); plot (t,F5);
title ('Fifth harmonic'); ylabel ('Amplitude ofF5'); xlabel ('time (in sec)')

figure(2);                             % see Figure 4.41, part (2b)
plot(t,F1,t,F2,t,F3,t,F4,t,F5);
title('Harmonic frequencies of f(t)')
ylabel ('Amplitude');
xlabel ('time (in sec)') ;
legend ('first','second','third','fourth','fifth')

figure(3);                             % Figure 4.42, part (2c)
F0 =1;
App1 =F0+F1;
App2=App1+F2+F3;
App3=App2+F4;
App4=App3+F5;
subplot(2,2,1);
plot(t,App1)
title('Partial sum of first 2 components (F0+F1)')
ylabel('Amplitude');
subplot(2,2,2);plot(t,App2);
title('Partial sum of first 3 harmonics + DC')
ylabel('Amplitude')
subplot(2,2,3);plot(t',App3);
title('Partial sum of first 4 harmonics + DC')
xlabel('time(in sec)');ylabel('Amplitude');
subplot(2,2,4);plot(t',App4);
title('Partial sum of first 5 harmonics + DC')
xlabel('time (in sec)');
ylabel('Amplitude')

figure(4)                              % Figure 4.43, parts (2d) & (2e)
Fn(6) =.5;
subplot (2,1,1);stem(n.*pi,Fn);
grid on
title('line spectrum');ylabel('Amplitude of Fn')
subplot(2,1,2);stem(n.*pi,Fn.^2);
grid on
title('power spectrum');xlabel('frequency');
ylabel('Magnitude of Fn^2');
disp('********************************');
disp(' Verification of Parsevals theorem');
disp('********************************');
Fnsquare=Fn.^2;
Pave=sum(Fnsquare);                    % parts (2f)
disp(' ');
disp(' ');
disp('The average power using summations of Fn(coefficients) is:')
disp(Pave);
```

```
disp(' ');
syms x;
intft = 1/2*int('Heaviside(x+.5)',-.5,.5);
Pavesym = vpa(intft);
disp(' ');
disp('*******************************');
disp(' ');
disp('The average power using integration in time is:');
vpa(intft)
disp(' ');
disp('*******************************');
disp(' ');
disp('The percentage error is :'); % parts (2g)
error = (Pavesym-Pave)*100/Pavesym;disp(error)
disp(' ');disp('*******************************');
```

The script file *Fourier_series* is executed and the results are as follows:

```
>> Fourier _ series
**************************************************
Verification of Parseval's theorem
**************************************************
The average power using summation of Fn (coefficients) is:
  0.4833
**************************************************
The average power using integration in time is:
  ans =
        .50000000000000000000000000000000
**************************************************
The percentage error is :
  3.3472238873502480061006281175660
**************************************************
```

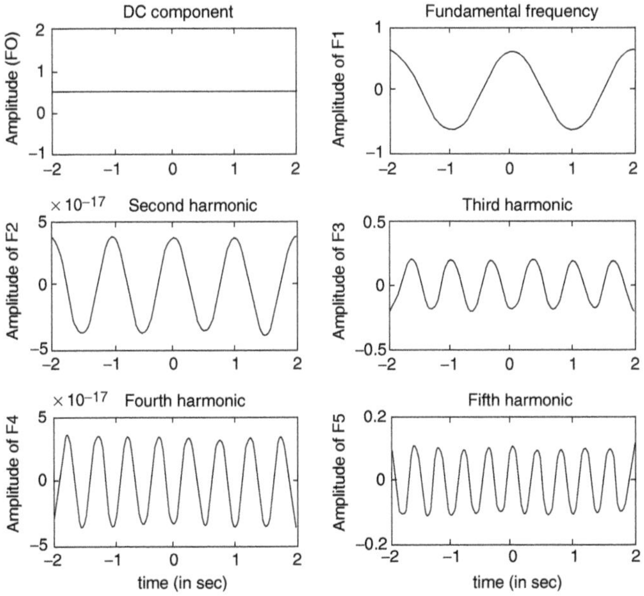

FIGURE 4.40
Plots of part a of Example 4.4.

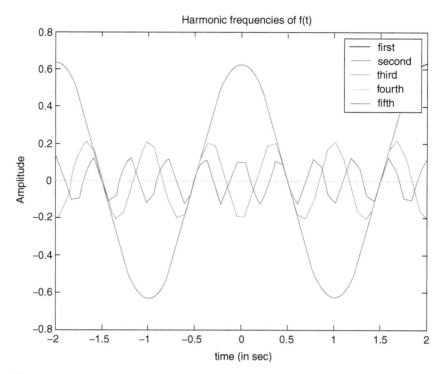

FIGURE 4.41
(See color insert following page 374.) Plots of part b of Example 4.4.

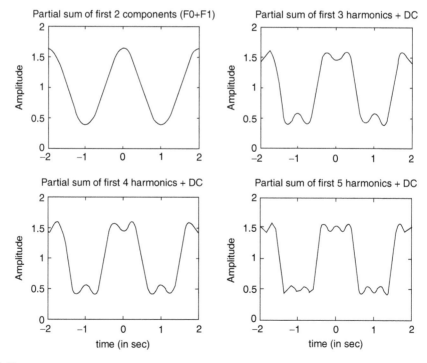

FIGURE 4.42
Plots of part c of Example 4.4.

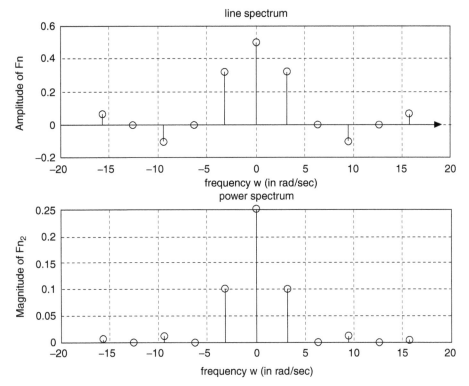

FIGURE 4.43
Plots of parts d and e of Example 4.4.

Note that the function $f(t)$ in this example is similar to the function analyzed in Example 4.1. The differences are a DC shift in magnitude by 1 and a time shift by 0.5 s.

Observe that simple shifts change the symmetry conditions of the function $f(t)$, creating new harmonics.

Example 4.5

Create the script file *Fourier_coeff* that returns the first five coefficients of the exponential FS (F_1, F_2, F_3, F_4, and F_5), of the function of Example 4.1, given by

$$f(t) = \frac{2}{\pi} \sum_{n=\text{odd}}^{7} \frac{1}{n} \sin(n w_0 t)$$

by using

1. Symbolic techniques
2. Numerical techniques
3. Compare the results of part 1 with part 2

MATLAB Solution
```
% Script file : Fourier _ coeff
T = 2;w0 = 2*pi/T; nn =1:5;
```

```
syms coesquare expon t ;
for k =1:5
expon = exp(-j*k*w0*t);
coesquare = .5*int(expon,-.5,.5);
Fn = .5*sin(k.*pi/2)./((k.*pi/2)+eps);
fn(k) = vpa(Fn);
f(k) = vpa(coesquare);
end
disp('************************************************************');
disp('Coefficients Fn of the exponential Fourier Series');
disp('Harmonics Sym. Coeff. ');
disp('************************************************************');
results1 = [nn' f'];
disp(results1)
disp('************************************************************');
disp('Harmonics Num. Coeff. ');
disp('************************************************************');
results2 = [nn' fn'];
disp(results2)
disp('************************************************************');
```

The script file *Fourier_coeff* is executed and the results are as follows:

```
>> Fourier _ coeff

**************************************************************************
Coefficients Fn of the exponential Fourier Series
Harmonics Sym. Coeff.
**************************************************************************
[1, .31830988618379067153776752674503]
[2, 0]
[3, -.10610329539459689051258917558168]
[4, 0]
[5, .63661977236758134307553505349006e-1]
**************************************************************************


Harmonics Num. Coeff.
**************************************************************************
[1, .31830988618379063570529297066969]
[2, .19490859162596877471531966954150e-16]
[3, -.10610329539459689707214806730917]
[4, -.19490859162596877471531966954150e-16]
[5, .63661977236758135467731278822612e-1]
```

■■

Note that the symbolic results fully agree with the numerical results. Also note that the magnitude of the error is of the order 10^{-16}.

Example 4.6

Create the script file *square_time_frq* that explores the effects in the frequency domain of changing the τ (tau) of the rectangular periodic wave function shown in Figure 4.44 in the time domain, by plotting their respective spectrums, for the following values of $\tau =$ 1, 0.5, 0.25, and 0.125. Discuss the results obtained.

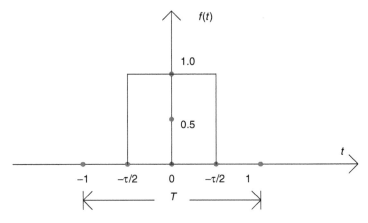

FIGURE 4.44
Plots of *f(t)* of Example 4.6.

ANALYTICAL Solution

The coefficients of the exponential FS for *f(t)* are given by

$$F_n = \frac{\tau}{T}\ \frac{\sin(n\pi/2)}{(n\pi/2)}$$

where $T = 2$ and

$$w_0 = \frac{2\pi}{2} = \pi \quad \text{for } n \geq 0$$

MATLAB Solution

```
% Script file: square _ time _ frq
echo off;
n =-10*pi:pi:10*pi;
tau1 =1;T=2;Tx1=tau1/T;
Fn1 = Tx1*sin(n*pi*Tx1)./(n*pi*Tx1+eps);
Fn1(11)=Tx1;

subplot(2,2,1);stem(n,Fn1);
title(' spectrum of square wave with tau=1');axis([-30 30 -.6 .6]);
ylabel('Amplitude'); xlabel('w= n*wo (rad per sec)');
tau2=0.5;T=2;Tx2=tau2/T;
Fn2=Tx2*sin(n*pi*Tx2)./(n*pi*Tx2+eps);
Fn2(11)=Tx2;
subplot(2,2,2);stem(n,Fn2);axis([-30 30 -.3 .3]);
title(' spectrum of square wave with tau=0.5');
ylabel('Amplitude'); xlabel('w= n*wo (rad per sec)');
tau3 = 0.25;T=2;Tx3=tau3/T; Fn3=Tx3*sin(n*pi*Tx3)./(n*pi*Tx3+eps);
Fn3(11) = Tx3;
subplot(2,2,3);stem(n,Fn3);axis([-30 30 -.15 .15]);
title(' spectrum of square wave with tau=0.25');
xlabel('w= n*wo (rad per sec)');ylabel('Amplitude');
tau4 = 0.125;T = 2;Tx4 = tau4/T; Fn4 =Tx4*sin(n*pi*Tx4)./(n*pi*Tx4+eps);
Fn4(11) = Tx4;
subplot(2,2,4);stem(n,Fn4);axis([-30 30 -.1 .1]);
title(' spectrum of square wave with tau=0.125');
xlabel('w= n*wo (rad per sec)');ylabel('Amplitude');
```

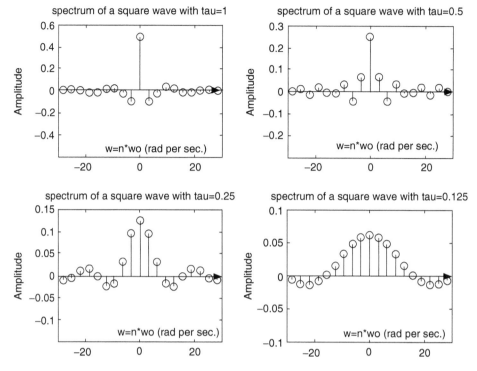

FIGURE 4.45
Spectrum plots of Example 4.6.

The script file *square_time_frq* is executed and the spectrums of the pulses for $\tau = 1, 0.5,$ 0.25, and 0.125 are shown in Figure 4.45.

Note that contracting $f(t)$ in the time domain, from $\tau = 1$ s to $\tau = 0.125$ s (by 1/8) produces an expansion in the frequency domain by a factor of 8, with the magnitude scaled by 1/8, verifying the property $f(at) \leftrightarrow (1/a)F(w/a)$.

For example, for $\tau = 1$ s, the estimated bandwidth is 1 rad/s, and the magnitude at $w = 0$ is 0.5. For $\tau = 0.125$ s, the estimated bandwidth is 8 rad/s, and the magnitude at $w = 0$ is 0.06 (which fully agrees with the theoretical value $0.5/8 = 0.06$).

Example 4.7

Let us analyze the periodic triangular function $f(t)$ shown in Figure 4.46, defined over a period of $T = 1$, given by the following trigonometric FS:

$$f(t) = \frac{1}{\pi} \sum_{n=1}^{\infty} \frac{1}{n} \sin(nw_0 t)$$

where $T = 1$ and

$$w_0 = \frac{2\pi}{T} = 2\pi$$

The equation for $f(t)$ is given by $f(t) = -t + 1$, over the range $0 \leq t \leq 1$ (one period).

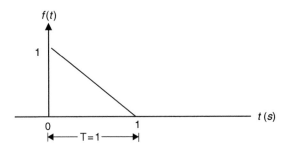

FIGURE 4.46
Plot of *f(t)* of Example 4.7.

Create the script file *Fourier_approx* that returns the following:

1. The plots of the approximation for *f(t)* using the first 26, 11, and 4 terms of the trigonometric FS, over the range $-3 \leq t \leq +3$.
2. For each of the approximation plots of part 1, evaluate the values of *f(t)* at $t = 0$ and 0.1.
3. Plot the four-term trigonometric FS approximation over the range $1 \leq t \leq 2$, with and without the help of a *Hamming* window, and observe the effects of the windowing process.

MATLAB Solution
```
%Script file : Fourier _ approx
echo off ;
t =-3:0.1:3;
x25=0;
for n=1:1:25;
xn25=1/(pi*n)*sin(2*pi*n*t);

x25=x25+xn25;                          % sum of 25 terms
end;
ft26=0.5+x25; % add the DC

figure(1)
subplot(3,1,1);plot(t,ft26)
title('26 term approximation ')
ylabel('Amplitude')
x10=0;
for n=1:1:10;
    xn10=1/(pi*n)*sin(2*pi*n*t);
    x10=x10+xn10;                      % sum of 10 terms
end;
x3=0;
for n=1:1:3;
xn3 =1/(pi*n)*sin(2*pi*n*t);
x3 = x3+xn3;                           % sum of 3 terms
end;
ft4 = 0.5+x3;                          % add DC
ft11 = 0.5+x10;
subplot(3,1,2);plot(t,ft11);
ylabel('Amplitude');
title('11 term approximation ')
subplot(3,1,3);
plot(t,ft4);ylabel('Amplitude')
```

```
xlabel('time (in sec)')
title('4 term approximation ')
minft26 = min(ft26); maxft26 = max(ft26);
minft11 = min(ft11); maxft11 = max(ft11);
minft4 = min(ft4);maxft4 = max(ft4)
A=[minft26 maxft26];
disp('*****************************************')
disp(' *********R E S U L T S ****************')
disp('**********************************************************');
disp ('The magnitude of f(t) using 26 terms at t = 0, and t = 0.1
are:');
disp (A);
B=[minft11 maxft11];C=[minft4 maxft4];
disp ('The magnitude of f(t) using 11 terms at t = 0, and t = 0.1
are:');
disp (B)
disp ('The magnitude of f(t) using 4 terms at t = 0 , and t = 0.1
are:');
disp (C);
disp('**********************************************************');
win=hamming(21)

figure(2)
subplot(2,1,1); tt=1:.1:2;
Ft4=ft4(21:41); F4=ft4(31:41)
plot(tt,F4)
title('Four term approximation over one cycle');
ylabel('Amplitude');
subplot(2,1,2)
ft4 _ win = Ft4.*win';ttt =1:.1:2, f4 = ft4 _ win(11:21)
plot(ttt,f4)
title('Four term hamm. window approximation over one cycle');
ylabel('Amplitude');
xlabel('time (in sec)')
```

Back at the command window, the script file *Fourier_ approx* is executed and the results are as follows:

```
>> Fourier _ approx

*****************************************
*********R E S U L T S ******************
**********************************************************
The magnitude of f(t) using 26 terms at t=0 , and t=0.1 are:
     0.0806 0.9194

The magnitude of f(t) using 11 terms at t=0 , and t=0.1 are:
     0.1469 0.8531

The magnitude of f(t) using 4 terms at t=0 , and t=0.1 are:
     0.0606 0.9394
**********************************************************
```

Observe that the approximation using four terms is better than the approximation using 26 or 11 terms evaluated at $t = 0$ and 0.1; but of course not over a larger range (Figure 4.47).

Figure 4.48 shows the fourier trigonometric series expansion improvement over one cycle with and without *windowing* the partial sum. Observe that the windowing process removes some ripples from the partial sum of the triangular waveform, and the result is a smoother wave.

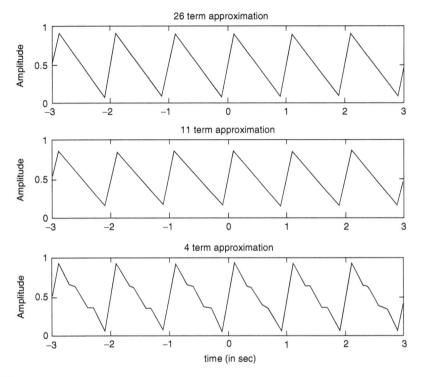

FIGURE 4.47
Approximation plots of *f(t)* of Example 4.7.

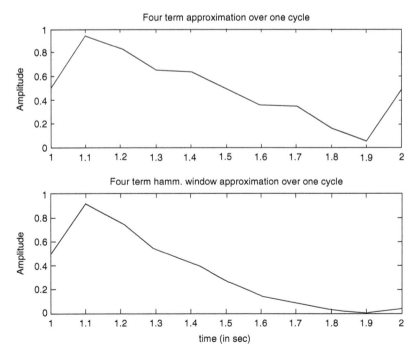

FIGURE 4.48
Approximation plots of *f(t)* using four terms of Example 4.7.

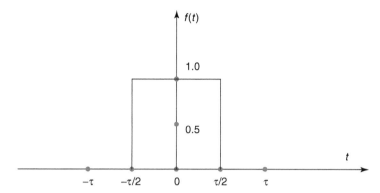

FIGURE 4.49
Plot of *f(t)* of Example 4.8.

Example 4.8

1. Evaluate by hand the FT (non-periodic) of the rectangular pulse *f(t)* shown in Figure 4.49 for the following cases $\tau = 1, 5$, and 10 using the transform pairs provided in Table 4.1.

2. Create the script file *square_FT_num* that returns the plots of the FT of part 1 using a numerical approach.

3. Create the script file *square_FT_sym* that returns the magnitude spectrum expressions as well as their plots of the FT of part 1 for different *τ*s using the MATLAB symbolic toolbox. Discuss and compare the results obtained.

ANALYTICAL Solution

Part 1

The expression of the FT of *f(t)* is

$$f(t) \leftrightarrow \frac{\tau \sin(w\tau/2)}{w\tau/2} \quad \text{(from Table 4.1)}$$

where *τ* takes the value of 1, 5, and 10.

MATLAB Solution
```
% Script file: square _ FT _ num
% This file returns the plot of the pulse spectrums for the tau's=
                                                    1,5, and 10
echo off;
w =-10*pi:0.1:10*pi;
F1 = ones(size(w));
F10 = ones(size(w));
tau1=1;tau5 =5;tau10 = 10;
F1= sin(w*tau1/2)./(w*tau1/2);
F5 = 5*sin(w*tau5/2)./(w*tau5/2);
```

```
F10 =10*sin(w*tau10/2)./(w*tau10/2);
subplot(3,1,1)
plot(w,F1);
ylabel('Amplitude');
title('Spectrum of pulse with tau=1');
axis([-10 10 -1 2]);
subplot(3,1,2);plot(w,F5);
ylabel('Amplitude');
title('Spectrum of pulse with tau=5');
axis([-7 8 -4 6]);
subplot(3,1,3);plot(w,F10);
xlabel('frequency w (in rad/sec)');
ylabel('Amplitude');
title('Spectrum of pulse with tau=10');
axis([-10 10 -4 11]);
```

The script file *square_FT_num* is executed and the results are as follows (Figure 4.50):

```
>> square _ FT _ num
```

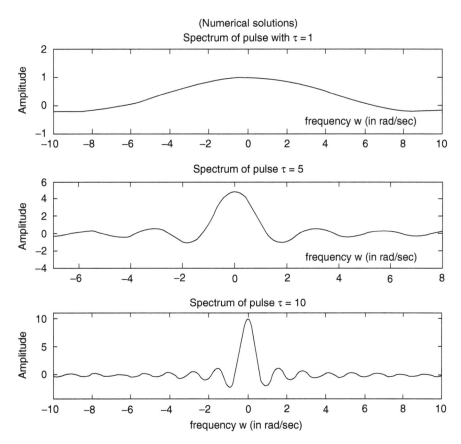

FIGURE 4.50
Numerical plots of part 2 of Example 4.8.

```
% Script file: square _ FT _ sym
% This file returns the expressions and plots for the pulse
% spectrums for tau's= 1,5, and10
syms t w
Fw1 = fourier(sym('Heaviside(t+1/2)')- sym('Heaviside(t- 1/2)'),t,w);
Fw5 = fourier(sym('Heaviside(t+2.5)')- sym('Heaviside(t- 2.5)'),t,w);
Fw10 = fourier(sym('Heaviside(t+5)')- sym('Heaviside(t-5)'),t,w);
disp('********************************************')
disp(' **********R  E  S  U  L  T  S  ***********')
disp('********************************************');
disp('The FT of the rect pulse with tau =1, 5 and 10 are given by:');
Fw1sim = simplify(Fw1)
Fw5sim = simplify(Fw5)
Fw10sim = simplify(Fw10)
disp('*****************************************');

figure(1)
subplot(3,1,1)
ezplot(Fw1sim,[-10 10])
ylabel(' Amplitude')
title(' Spectrum of the pulse for tau=1.0')
subplot(3,1,2)
ezplot(Fw5sim,[-10 10])
axis([-10 10 -4 6])
ylabel(' Amplitude')
title(' Spectrum of the pulse for tau= 5')
subplot(3,1,3)
ezplot(Fw10sim,[-10 10])
axis([-4 4 -6 11])
xlabel('frequency w (in rad/sec'),
ylabel(' Amplitude')
title(' Spectrum of the pulse for tau=10')
```

The script file *square_FT_sym* is executed and the results are as follows (Figure 4.51):

```
>> square _ FT _ sym

*****************************************************************
************************R  E  S  U  L  T  S  ********************
*****************************************************************
The FT of the rect pulse with tau =1, 5 and 10 are given by :
        Fw1sim =
                2*sin(1/2*w)/w

        Fw5sim =
                2.*sin(2.5000000000000000000000000000000*w)/w

        Fw10sim =
                2*sin(5*w)/w
*****************************************************************
```

Observe that the spectrum expressions obtained using the MATLAB symbolic toolbox shown in Figure 4.51 fully agree with the transform of Table 4.1 and the spectrum plots of part 2 shown in Figure 4.52.

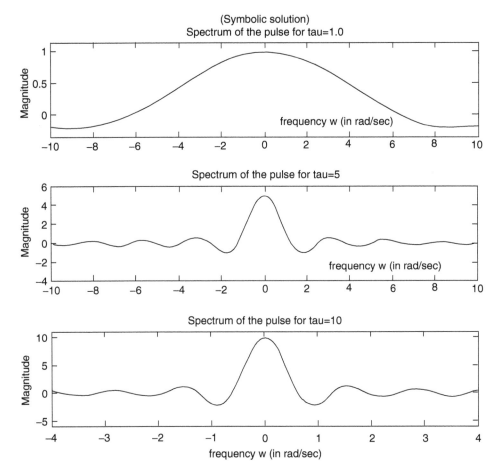

FIGURE 4.51
Symbolic plots of part 3 of Example 4.8.

Example 4.9

Let $f(t) = e^{-2t}u(t)$, evaluate by hand

1. The FT of $f(t) \leftrightarrow F(w)$, using Table 4.1
2. The LT of $f(t) \leftrightarrow F(s)$, using Table 4.2
3. The analytic expressions of $|F(w)|$ and $|F(s)|$ (magnitude), and $angle[F(w)]$ and $angle[F(s)]$ (phase)
4. Create the script file *Fourier_Laplace* that returns the following plots:
   a $f(t)$ versus t, over $-2 \le t \le 2$
   b $[f(t)]^2$ versus t (power, assuming $R = 1\ \Omega$ and $f(t)$ is either a current or a voltage)
5. Evaluate using MATLAB
   a. Area of $[f(t)] = \int_0^{+\infty} |f(t)| dt$ (test the existence of the FT)
   b. Energy of $[f(t)] = \int_0^{+\infty} |f(t)|^2 dt$
6. Evaluate the Fourier and Laplace transforms. $F(w)$ and $F(s)$, using the MATLAB symbolic toolbox
7. The symbolic and numerical plots of $|F(w)|$ versus w and $|F(s)|$ versus s

ANALYTICAL Solution

Part 1. From Table 4.1, the FT of $f(t) = e^{-2t}u(t) \leftrightarrow 1/(jw + 2)$ (transform # 10).
Part 2. From Table 4.2, the LT of $f(t) = e^{-2t}u(t) \leftrightarrow 1/(s + 2)$ (transform # 5).
Part 3. The magnitude equations for the Fourier and LTs are

$$|F(w)| = \frac{1}{\sqrt{(w^2 + 2^2)}} \qquad |F(s)| = \frac{1}{\sqrt{(s^2 + 2^2)}}$$

The phase equations for the Fourier and LTs are

$$Phase[F(w)] = -tan^{-1}(w/2) \ (Fourier)$$

$$Phase[F(s)] = -tan^{-1}(s/2) \ (Laplace)$$

MATLAB Solution

```
% Script file: Fourier _ Laplace
t = -2:0.05:2;
ft = exp(-2.*t).*stepfun(t,0);

figure(1)
subplot(2,1,1)
plot(t,ft)
xlabel('time (in sec)');ylabel('Amplitude');
axis([-2 2 -.5 1.25]);
title('exp(-2*t)*u(t) vs. t ')
subplot(2,1,2)
ftsquare = ft.^2; plot(t,ftsquare)
xlabel('time (in sec)');ylabel('Amplitude');
axis([-2 2 -.5 1.25]);
title('[exp(-2*t)*u(t)]^2 vs t')
% evaluates integral of exp(-2*t) from 0 to inf.
symint = int('(exp(-2*t)*Heaviside(t))',0,inf);
symintsq = int('(exp(-4*t)*Heaviside(t))',0,inf);
disp('****************************')
disp('******** R E S U L T S *********')
disp('****************************')
disp('The int(exp(-2*t)*u(t)) from t=0 to inf is =')
disp(symint)
vpa(symint)
disp('****************************')
disp('****************************')
disp('The int(exp(-2*t)^2*u(t)) from t=0 to inf is =')
disp(symintsq)
vpa(symintsq)
disp('****************************')
w =-6:0.1:6;
Fw =1./sqrt(4+w.^2);                    % mag [e^(-2*t) u(t)] and
phase =-atan(w/2);                      % phase from table of F.T

figure(2)
subplot(2,2,1);plot(w,Fw)
xlabel('frequency w');ylabel('Magnitude of F(w)')
title('Magnitude spectrum/Fourier');
```

```
axis([-5 5 0 0.6])
subplot(2,2,2);plot(w,phase*180/pi)
xlabel('frequency w');ylabel('Angle in degrees')
title('Phase spectrum/Fourier');
a = [1];                              % F(s) = 1/(s+2)
b = [1 2];
H = freqs(a,b,w);
mag = abs(H);ang = angle(H);
degrees = ang*180/pi;
subplot(2,2,3);plot(w,mag)
xlabel('frequency s');ylabel('Magnitude of F(s)')
title('Magnitude spectrum/Laplace');
axis([-5 5 0.1 0.6])
subplot(2,2,4);plot(w,degrees)
xlabel('frequency s');ylabel('Angle in degrees')
title('Phase spectrum/Laplace');

figure(3)
% Fourier transform using the symbolic toolbox
syms v;
disp('******************************')
disp('The sym Fourier transform ')
disp('of(exp(-2*t)*u(t)) is=')
FT= fourier(exp(-2*v)*sym('Heaviside(v)'))
disp('******************************')
% Laplace transform using symbolic toolbox
disp('******************************')
disp('The sym Laplace transform ')
disp('of(exp(-2*t)*u(t)) is=')
LT=laplace(exp(-2*v)*sym('Heaviside(v)'))
disp('******************************')
subplot(2,1,1);
ezplot(abs(FT));
title('sym Fourier transf of(exp(-2*t)*u(t))')
ylabel('Magnitude of F(w)');xlabel('frequency w');
subplot(2,1,2);
ezplot(abs(LT));
title('sym Laplace transf of(exp(-2*t)*u(t))')
ylabel('Magnitude of F(s)');
xlabel('frequency s');axis([-4 0 0 45])
```

The script file *Fourier_Laplace* is executed and the results are as follows (Figures 4.52 through 4.54):

```
>> Fourier _ Laplace

*****************************************
*********** R E S U L T S ************
*****************************************
The int(exp(-2*t)*u(t)) from t=0 to inf is =
                                    1/2


ans =
      .50000000000000000000000000000000
```

```
******************************************
******************************************
The int(exp(-2*t)^2*u(t)) from t=0 to inf is =
                                              1/4


ans =
        .25000000000000000000000000000000
******************************************

******************************************
The sym Fourier transform
of(exp(-2*t)*u(t)) is=
   FT =
        1/(2+i*w)
******************************************

******************************************
The sym Laplace transform
of(exp(-2*t)*u(t) is=
   LT =
        1/(s+2)
******************************************
```

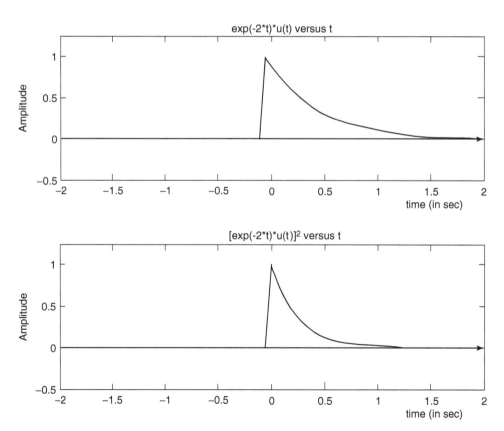

FIGURE 4.52
Symbolic plots of part 4 of Example 4.9.

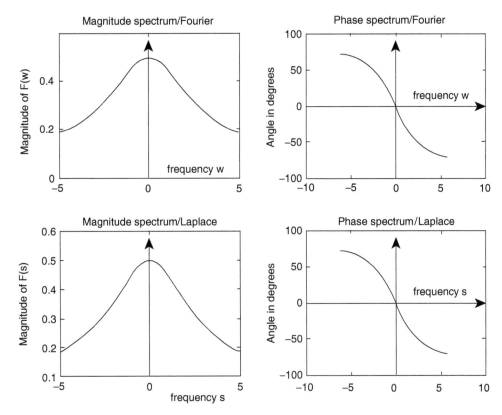

FIGURE 4.53
Numerical plots of *F(w)* and *F(s)* of part 7 of Example 4.9.

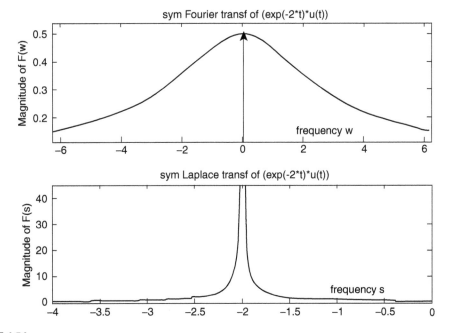

FIGURE 4.54
Symbolic plots of *F(w)* and *F(s)* of part 7 of Example 4.9.

FIGURE 4.55
Block box system diagram of Example 4.10.

Example 4.10

Analyze by hand and using MATLAB (include pretty) the block box system diagram shown in Figure 4.55, and obtain

a. Analytical expressions of
  i. $X(w)$
  ii. $H(w)$
  iii. $Y(w)$
  iv. $y(t)$
b. Create the script file *syst_anal* that returns the following plots:
  i. $X(w)$ versus w (numerical)
  ii. $H(w)$ versus w (numerical)
  iii. $Y(w)$ versus w (numerical)
  iv. $|H(w)|$ versus w, and verify that its spectrum consists of two impulses, located at $w = \pm 3$ rad/s
  v. $x(t)$ versus t, $h(t)$ versus t, and $y(t)$ versus t
  vi. $X(w)$ versus w (symbolic)
  vii. $Y(w)$ versus w (symbolic)

ANALYTICAL Solution

Part a: From Table 4.1, the following transforms are obtained:

$$\Im[x(t) = \text{pul}_\pi(t)] = X(w) = \frac{2\pi\sin(w\pi)}{(w\pi)}$$

$$\Im[h(t) = e^{-3|t|}] = H(w) = \frac{6}{w^2 + 9}$$

$$Y(w) = X(w) \cdot H(w) = \frac{12\pi\sin(w\pi)}{(w\pi) \cdot (w^2 + 9)}$$

$$y(t) = x(t) \otimes h(t)$$

where

$$y(t) = \int_{-\infty}^{+\infty} x(\tau)\, h(t - \tau)\, d\tau \quad \text{(convolution of } x(t) \text{ with } h(t))$$

MATLAB Solution
[part(b)]

```
% Script file: syst _ anal
echo off;
w =-2*pi:0.01:2*pi;
w =w+eps;
```

```
xw =2*pi*sin(w*pi)./(w*pi);
% FT of x(t) from Table 4.1,transf.#3
% H(w) = 6/(w^2+9);
% FT of h(t) from Table 4.1,transf.#13
a = [6];
b = [1 0 9];
Hw = freqs(a,b,w);
magxw = abs(xw);
magHw = abs(Hw);
yw = magxw.*magHw;

figure(1)
subplot(3,1,1);plot(w,magxw);
ylabel('Magnitude');axis([-6 6 0 8]);
title(' X(w) vs.w');xlabel('frequency w(rad/sec)');
subplot(3,1,2);plot(w,magHw);axis([-6 6 0 8]);
ylabel('Magnitude');
title(' H(w) vs.w');xlabel('frequency w(rad/sec)');
magyw=magxw.*magHw;
subplot(3,1,3)
plot(w,magyw);axis([-6 6 0 5]);
ylabel('Magnitude');
title(' Y(w) vs.w');xlabel('frequency w(rad/sec)');

figure(2)
plot(w,magHw);
title(' H(w) vs. w (showing impulses)');
ylabel('Magnitude');xlabel('fequency w (rad/sec)')
magyw=magxw.*magHw;
%time domain analysis

figure(3)
t=-5:.1:5;
xt=[zeros(1,20) ones(1,61) zeros(1,20)];
ht=exp(-3.*abs(t));
hut=[zeros(1,50) ones(1,51)];htt=ht.*hut;
conxh=conv(xt,ht);
tt=-10:.1:10;
subplot(3,1,1)
plot(t,xt);axis([-5 5 0 1.5])
title('x(t) vs. t');
ylabel('Amplitude');xlabel('time (in sec)')
subplot(3,1,2)
plot(t,ht);
title('h(t) vs. t');
axis([-2 2 0 1.5])
ylabel('Amplitude');xlabel('time (in sec)');
subplot(3,1,3)
plot(tt,conxh);
title('y(t) vs. t');
ylabel('Amplitude');
xlabel('time (in sec)')

% Symbolic evaluations
syms x;
FTx = fourier(sym('Heaviside(x-pi)')-sym('Heaviside(x+pi)'));
```

```
FTxsym =simplify(FTx) ; % X(w)
FTh = fourier(exp(- 3*x)*sym('Heaviside(x)')+
exp(3*x)*sym('Heaviside(-x)'));
FTy = FTx*FTh;
FTysym=simplify(FTy) ; %Y(w)
Disp('*****************************************************')
disp('******* S y m b o l i c **** R e s u l t s*********')
disp(' ')
disp('The Fourier Transf. of x(t) is:')
disp(FTxsym)
pretty(FTxsym)
disp('The Fourier Transf. of h(t) is:')
disp(FTh)
pretty(FTh)
disp('The Fourier Transf. of y(t) is:')
disp(FTysym)
pretty(FTysym)
disp('*****************************************************')

figure(4)
subplot(2,1,1)
ezplot(abs(FTxsym));
title('Symbolic plot of the Fourier Transf. of X(t) vs. w ')
ylabel('Magnitude');xlabel(' frequency (in rad/sec)');axis([-6 6 0 6])
subplot(2,1,2)
ezplot(abs(FTysym),[-6 6]);title('Symbolic plot of the Fourier Transf. of Y(t)
vs. w ')
ylabel('Magnitude');xlabel('frequency w (in rad/sec)');
axis([-6 6 0 5]);
```

The script file *syst_anal* is executed and the results are as follows (Figures 4.56 through 4.59):

```
>> syst _ anal

*****************************************************
******* S y m b o l i c **** R e s u l t s*********

The Fourier Transf. of x(t) is:
  -2*sin(pi*w)/w
   sin(pi w)
-2 ---------
     w

The Fourier Transf. of h(t) is:
  1/(3+i*w)+1/(3-i*w)
    1          1
 -------  +  --------
 3 + i w    3 - i w

The Fourier Transf. of y(t) is:
  12*sin(pi*w)/w/(3+i*w)/(-3+i*w)
          sin(pi w)
 12 ----------------------
    w (3 + i w) (-3 + i w)
*****************************************************
```

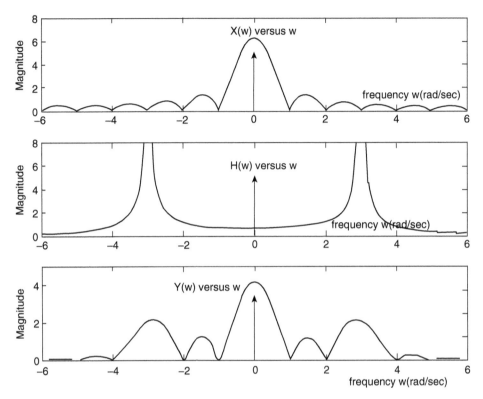

FIGURE 4.56
Plots of $X(w)$, $H(w)$, and $Y(w)$ of Example 4.10.

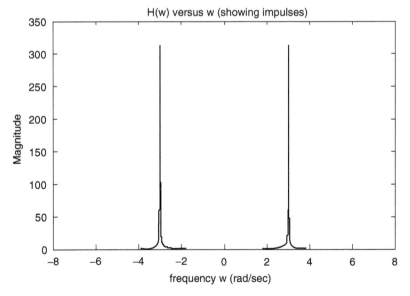

FIGURE 4.57
Plot of $H(w)$ of Example 4.10.

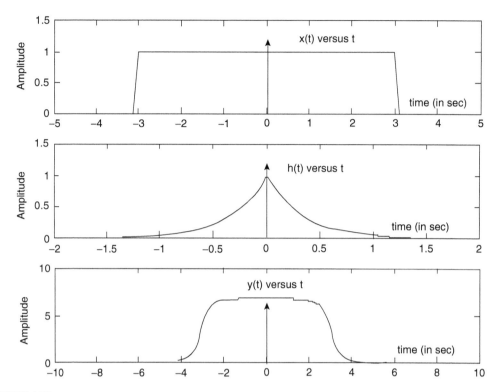

FIGURE 4.58
Plots of *x(t)*, *h(t)*, and *y(t)* of Example 4.10.

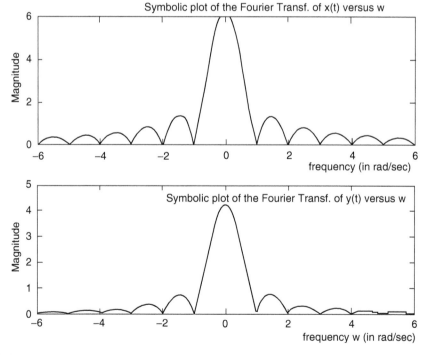

FIGURE 4.59
Symbolic plots of *X(w)* and *Y(w)* of Example 4.10.

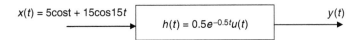

FIGURE 4.60
Block box system diagram of Example 4.11.

Example 4.11

Let us analyze the block system diagram shown in Figure 4.60, by creating the script file *analysis* that returns the following plots using numerical techniques:

1. $x(t)$ versus t and $|X(w)|$ versus w
2. $h(t)$ versus t, $|H(w)|$ versus w (magnitude), and phase of $[H(w)]$ versus w
3. $y(t)$ versus t, $|Y(w)|$ versus w, and phase $[Y(w)]$ versus w

MATLAB Solution
```
% Script file: analysis
t = -5:0.01:5;
xt = 5*cos(t)+15*cos(15*t);
subplot(3,2,1),plot(t,xt)
title(' x(t) vs. t') ,ylabel(' Amplitude')
w =-30:1:30;
xw =pi*[zeros(1,15) 15 zeros(1,13) 5 0 5 zeros(1,13) 15 zeros(1,15)];
                                        % spectrum of x(t)
%from Table 4.1 transform 6
subplot(3,2,2),stem(w,xw)
title('X(w) vs. w ');ylabel(' Magnitude');
axis([-30 30 -5 60]);
to = 0;
ut = stepfun(t,to);
h = 0.5*exp(-0.5*t);
ht = h.*ut;
subplot (3,2,3);
plot (t,ht);
title (' h(t) vs. t');ylabel('Amplitude')
axis([-3 5 -.5 1]);
% FT of h(t) is 0.5/(s+0.5), s=jw;
% from Table 4.1
a = [0.5];
b = [1 0.5];
Hw = freqs(a,b,w);
magHw = abs(Hw);
phase = angle(Hw);
subplot(3,2,4),plotyy(w,magHw,w,phase)
title('Magnitude & phase of H(w)');
% Convolution of x(t) and h(t)
yt = conv(xt,ht);
lengyt = length(yt)-1;
n = [0:1:lengyt]-[5*ones(1,lengyt+1)];
subplot(3,2,5),plot(n/100,yt/100)
```

```
xlabel('time'),title(' y(t) vs. t');
ylabel('Amplitude')
yw = magHw.*xw;
axis([5 15 -5 5])
subplot(3,2,6),plotyy(w,yw,w,phase);
xlabel('frequency w (in rad/sec)'), title('Magnitude & phase of Y(w)');
```

The script file *analysis* is executed and the results are as follows (Figure 4.61):

TIME DOMAIN FREQUENCY DOMAIN
****

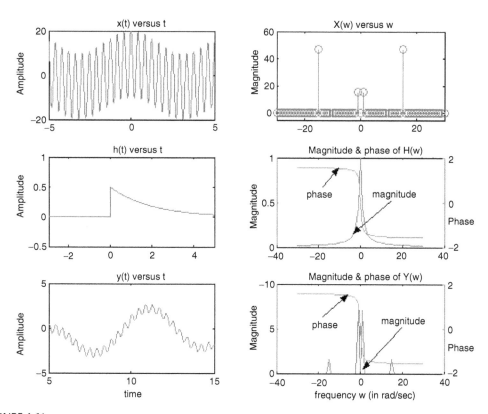

FIGURE 4.61
Time and frequency plots of the signals of the block box system of Example 4.11.

Example 4.12

Given the circuit shown in Figure 4.62,

1. Determine by hand the system transfer function $H(s) = V_0(s)/V_i(s)$
2. Obtain an expression for $v_0(t)$, from part 1, where $v_i(t) = u(t)$
3. Use MATLAB to obtain plots of $v_0(t)$ versus t, using an analytical and symbolic approach

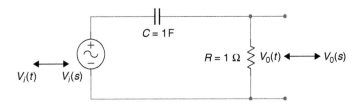

FIGURE 4.62
Circuit diagram of Example 4.12.

ANALYTICAL Solution

Part 1

$$H(s) = \frac{V_0(s)}{V_i(s)} = \frac{R}{R + (1/sC)} = \frac{s}{s+1}$$

Part 2

$$V_i(s) = \frac{1}{s} \qquad V_0(s) = \frac{1}{s+1} \qquad v_0(t) = \pounds^{-1}[V_0(s)] = e^{-t}u(t) \quad \text{(from Table 4.2)}$$

Part 3

MATLAB Solution
```
% Script file :analysis _ RC
t =-1:0.1:8;
t0 = 0;
ut =stepfun(t,t0);
vo =exp(-t);
vot =vo.*ut;
subplot(2,1,1);
plot(t,vot,t,vot,'o');
title('vo(t) vs. t,numerical solution');
xlabel('time(sec)');
ylabel(' Amplitude (volts)');
axis([-1 5 -.5 1.3]);
syms s x
Hs = s/(s+1);
Vis =1/s;
Vos = Hs*Vis;
vot = ilaplace(Vos);
subplot(2,1,2);
ezplot(vot);title('vo(t) vs. t, symbolic solution');
ylabel(' Amplitude (volts)');
xlabel('time(sec)');
axis([0 5 -.5 1.3]);
```

The script file *analysis_RC* is executed and the results are shown in Figure 4.63.

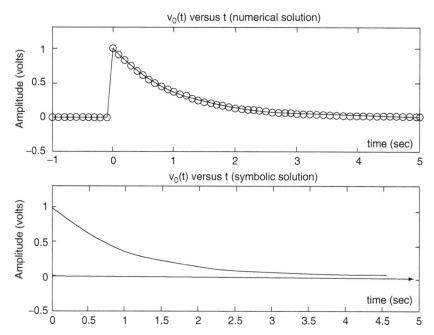

FIGURE 4.63
Numerical and symbolic plots of $v_0(t)$ of Example 4.12.

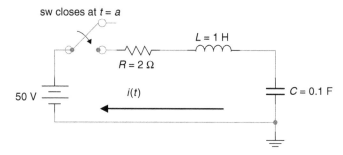

FIGURE 4.64
Circuit diagram of Example 4.13.

Example 4.13

The switch *sw* shown in the circuit diagram of Figure 4.64 has been opened for a long time. At $t = a = 0$, the *sw* closes and remains closed for $t \geq 0$. Assume that all the initial conditions are zero. Determine by hand an expression for $i(t)$. Create the script file *analysis_RLC* that returns $i(t)$ using the LT, and obtain plots of $i(t)$ versus t for $t \geq 0$, using a numerical and symbolic approach.

Also obtain the symbolic expression fot $i(t)$, for $t \geq 0$, using Laplace for the case when the *sw* closes at $t = a = 1$.

ANALYTICAL Solution

$$Z(s) = R + sL + \frac{1}{sC} \quad \text{(total impedence)}$$

$$Y(s) = \frac{1}{Z(s)} \qquad Y(s) = \frac{1}{R + sL + (1/sC)} = \frac{s}{s^2 + 2s + 10} = \frac{s}{(s+1)^2 + 3^2}$$

$$I(s) = \frac{V(s)}{Z(s)} = V(s)Y(s)$$

Since

$$V(s) = \pounds[50\,u(t)] = \frac{50}{s}$$

then

$$I(s) = \frac{50}{s} \cdot \frac{s}{(s+1)^2 + 3^2} = \frac{50}{3}\frac{3}{(s+1)^2 + 3^2}$$

$$i(t) = \pounds^{-1}[(I(s)] = \frac{50}{3} \cdot e^{-t}\sin(3t) \cdot u(t)$$

MATLAB Solution
```
% Script file : analysis _ RLC
t =-1:0.1:8;t0=0;
u = stepfun(t,t0);
it = 50/3*exp(-t).*sin(3*t).*u;
subplot(2,1,1)
plot(t,it);
title('i(t) vs t (numeric solution)')
xlabel('time(sec)')
ylabel('Amplitude (amps)')
axis([0 6 -4 11])
syms s zs
zs = 2+s+(1/(s*.1));
vs = 50/s;
ILs = vs/zs;
IL = ilaplace(ILs);
disp('*********************************************')
disp(' **** symbolic result (sw closes at t=0)*****')
disp('The current i(t) (in amps) is :')
simIL = simplify(IL);
disp(simIL)
subplot(2,1,2)
ezplot(simIL)
title('i(t) vs. t (symbolic solution)')
axis([0 6 -4 11])
xlabel('time(sec)')
ylabel('Amplitude (amps)')
t0=1;
IL = ilaplace((ILs)*exp(-s)); %sw closes at t=1
disp('******* *symbolic result(sw closes at t=1)  *****')
disp('The current i(t) (in amps) is given by :')
simIL = simplify(IL);
disp(simIL)
disp('***************************************')
```

The script file *analysis_RLC* is executed and the results are as follows (Figure 4.65):

```
>> analysis _ RLC

**********************************************************
**** symbolic result (sw closes at t =0)***********
     The current i(t) (in amps) is given by :
        25/3*i*(-exp((-1+3*i)*t)+exp((-1-3*i)*t))
********** symbolic result (sw closes at t =1)*****
     The current i(t) (in amps) is :
        50/3*Heaviside(t-1)*exp(-t+1)*sin(3*t-3)
***************************************
```

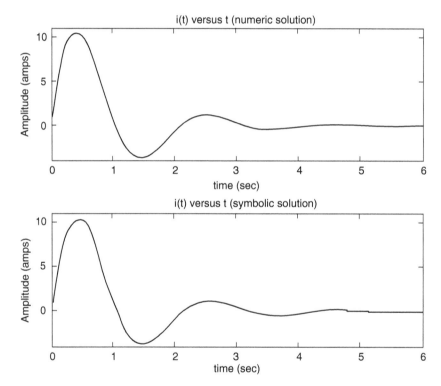

FIGURE 4.65
Numerical and symbolic plot of *i(t)* of Example 4.13.

Example 4.14

Create the script file *repeat_RLC* that returns the current *i(t)* of Example 4.13, shown in Figure 4.64, for $t \geq 0$, assuming that the *sw* closes at $t = 0$, for the following voltage source:

1. A unit impulse (use the *impulse* function)
2. A unit step (use the *step* function)

ANALYTICAL Solution

Recall that

$$I(s) = V(s) \cdot \frac{s}{s^2 + 2s + 10}$$

where the applied voltage is either $\delta(t)$ or $u(t)$, with a LT of 1 or $1/s$ (Table 4.2).

MATLAB Solution

```
% Script file : repeat _ RLC
num = [0 1 0];
den = [1 2 10];
t = 0:0.1:8*pi;
impres= impulse(num,den,t);
stepres= step(num,den,t);
subplot(2,1,1)
plot(t,impres);title('Impulse response ');
ylabel('magnitude');axis([0 5 -1 1]);
subplot(2,1,2)
plot(t,stepres);title('Step response ');
xlabel('time(sec)');ylabel('magnitude');
axis([0 5 -.1 .3])
```

The script file *repeat_RLC* is executed and the results are shown in Figure 4.66.

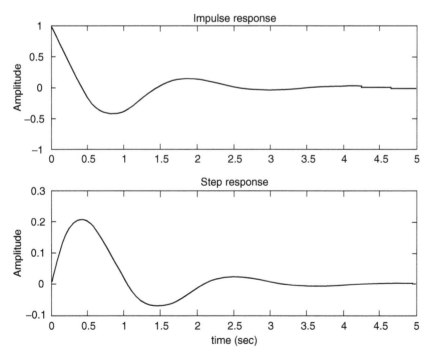

FIGURE 4.66
Impulse and step plots of Example 4.14.

Example 4.15

The switch in the circuit diagram of Figure 4.67 is closed for a long time. At $t = 0$, the switch is opened and remains opened for $t \geq 0$. Analyze by hand the given circuit using the LT and obtain analytical expressions for

$$v(t),\ i_r(t),\ i_c(t),\ i_l(t) \quad \text{for } t \geq 0$$

Create the script MATLAB file *RLC_parallel* that returns the plots of the analytical results as well as the symbolic (Laplace) plots of $v(t)$, $i_r(t)$, $i_c(t)$, $i_l(t)$, over the range $-2 \leq t \leq 5$, and observe that both solutions are identical.

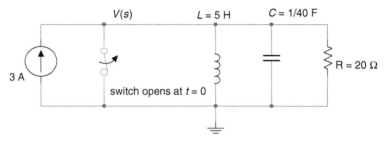

FIGURE 4.67
Circuit diagram of Example 4.15.

ANALYTICAL Solution

The nodal equation yields

$$\frac{V(s)}{20} + \frac{V(s)}{5s} + \frac{sV(s)}{40} = \frac{3}{s} \quad \text{for } t \geq 0$$

then

$$V(s)\left(\frac{1}{20} + \frac{1}{5s} + \frac{s}{40}\right) = \frac{3}{s}$$

$$V(s)\left(\frac{2s + 8 + s^2}{40s}\right) = \frac{3}{s}$$

$$V(s)(s^2 + 2s + 8) = 120, \text{ solving for voltage } V(s)$$

$$V(s) = \frac{120}{s^2 + 2s + 8} = \frac{120}{\sqrt{7}} \cdot \frac{\sqrt{7}}{(s+1)^2 + \sqrt{7}^2}$$

Then from Table 4.2

$$v(t) = \frac{120}{\sqrt{7}} \cdot e^{-t} \sin(\sqrt{7}t) \cdot u(t)$$

and therefore,

$$i_R(t) = \frac{v(t)}{R} = \frac{6}{\sqrt{7}} e^{-t} \sin(\sqrt{7}t) \cdot u(t)$$

$$i_C(t) = C\frac{dv(t)}{dt} = \frac{1}{40}\frac{d}{dt}\left[\frac{120}{\sqrt{7}} \cdot e^{-t} \sin(\sqrt{7}t)u(t)\right]$$

$$i_C(t) = \frac{3}{\sqrt{7}}\left[\sqrt{7}e^{-t}\cos(\sqrt{7}t) - \sin(\sqrt{7}t)e^{-t}\right] = \frac{3}{\sqrt{7}}e^{-t}\left[\sqrt{7}\cos(\sqrt{7}t) - \sin(\sqrt{7}t)\right]u(t)$$

$$i_L(t) = 3 - i_R(t) - i_C(t)$$

$$i_L(t) = 3 - \frac{6}{\sqrt{7}}e^{-t} \sin(\sqrt{7}t) u(t) - \frac{3}{\sqrt{7}}e^{-t}\left[\sqrt{7}\cos(\sqrt{7}t) - \sin(\sqrt{7}t)\right]u(t)$$

MATLAB Solution

```
% Script file: RLC _ parallel
t =-2:0.1:5;
t0 =0;
u = stepfun(t,t0);
vt =120/sqrt(7).*exp(-t).*sin(sqrt(7).*t).*u ;

figure(1)
subplot(2,2,1);
plot(t,vt);title(' v(t) vs.t (analytical)');
ylabel('Amplitude (volts)')
ir = 6/sqrt(7)*exp(-t).*sin(sqrt(7).*t).*u;
subplot(2,2,2);
plot(t,ir);title(' ir(t) vs.t (analytical)');
ylabel(' Amplitude (amps)')
ic=3/sqrt(7).*exp(-t).*(sqrt(7).*cos(sqrt(7).*t)- sin(sqrt(7).*t)).*u;
subplot(2,2,3);
plot(t,ic);title(' ic(t) vs.t (analytical)');
ylabel('Amplitude (amps)');xlabel('time(sec)');
subplot(2,2,4);
il = 3*ones(1,71)-ir-ic ;
plot(t,il);title('il(t) vs.t (analytical)');
ylabel('Amplitude (amps)') ;xlabel('time (sec)');

figure(2)
subplot(2,2,1)
num = [0 0 120];
den = [1 2 8];
vti = impulse(num,den,t);vti=[zeros(1,20) vti'];
plot(t,vti);title(' v(t) vs. t (sym. Laplace)');
```

```
ylabel('Amplitude (volts)')
subplot(2,2,2)
numr = [0 0 6];%IR=6/s^2+s+8;
denr = [1 2 8];
ir = impulse(numr,denr,t);ir = [zeros(1,20) ir'];
plot(t,ir);title(' ir(t) vs. t (sym. Laplace)');
ylabel('Amplitude (amps)')
subplot(2,2,3)
numc =[0 3 0];%IC=3s/(s^2+2s+8);
denc = [1 2 8];
ic = impulse(numc,denc,t);ic=[zeros(1,20) ic'];
plot(t,ic);title(' ic(t) vs. t (sym. Laplace)');
ylabel('Amplitude (amps)');xlabel('time(sec)');
subplot(2,2,4)
numl = [0 0 0 120];
denl = [5 10 40 0];%IL(s)=120/(5s^3+10s^2+40s);
il =impulse(numl,denl,t);il=[3*ones(1,20) il'];
plot(t,il);
title(' il(t) vs. t (sym. Laplace)');
ylabel('Amplitude (amps)') ;xlabel('time (sec)');
```

The script file *RLC_parallel* is executed and the results are shown in Figures 4.68 and 4.69.

Note that the analytical solutions fully agree with the symbolic solutions.

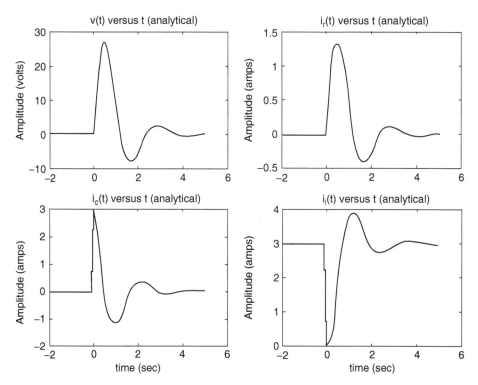

FIGURE 4.68
Plots of the analytical solutions of Example 4.15.

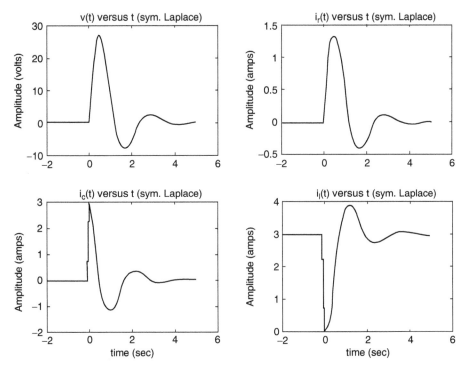

FIGURE 4.69
Plots of the symbolic solutions of Example 4.15.

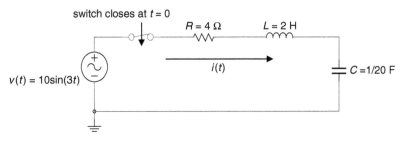

FIGURE 4.70
Circuit diagram of Example 4.16.

Example 4.16

The switch of the circuit of Figure 4.70 closes at $t = 0$, assuming that the circuit is initially relaxed, obtain analytical expressions of $i(t)$, $v_l(t)$, $v_r(t)$, and $v_c(t)$ for $t \geq 0$.

Create the script MATLAB file *RLC_series* that returns the analytical and symbolic (Laplace) plots of $i(t)$, $v_l(t)$, $v_r(t)$, and $v_c(t)$, over the range $0 \leq t \leq 10$ s. Also verify KVL given by

$$v(t) = v_l(t) + v_r(t) + v_c(t) \quad \text{over the given time range}$$

ANALYTICAL Solution

The loop equation of the preceding circuit in the *s domain* (KVL) is given by

$$\frac{10(3)}{s^2 + 3^2} = 4I(s) + 2sI(s) + \frac{20}{s}I(s)$$

$$\frac{30}{s^2 + 9} = \frac{I(s)}{s}\left[2s^2 + 4s + 20\right]$$

Then solving for *I*(*s*)

$$I(s) = \frac{15s}{s^4 + 2s^3 + 19s^2 + 18s + 90}$$

and the voltage drop across each element are

$$V_R(s) = \frac{60s}{s^4 + 2s^3 + 19s^2 + 18s + 90}$$

$$V_L(s) = \frac{30s^2}{s^4 + 2s^3 + 19s^2 + 18s + 90}$$

$$V_C(s) = \frac{300}{s^4 + 2s^3 + 19s^2 + 18s + 90}$$

The time functions $i(t)$, $v_R(t)$, $v_l(t)$, and $v_c(t)$ can be obtained by evaluating the inverse LTs ($£^{-1}$) on each of the frequency functions $I(s)$, $V_R(s)$, $V_l(s)$, and $V_c(s)$.

MATLAB Solution
```
% Script file : RLC _ series
% plot of i(t) vs t
t = 0:.1:10;
Inum = [0 0 0 15 0];
Iden = [1 2 19 18 90];

figure(1)
subplot(2,2,1)
it = impulse(Inum,Iden,t);
plot(t,it);title('i(t) vs t');ylabel('Amplitude (amps)')
%plot of VR(t) vs t
VRnum=[0 0 0 60 0];
VRden=Iden;
subplot(2,2,2)
vrt = impulse(VRnum,VRden,t);
plot(t,vrt);title('vr(t) vs t');ylabel('Amplitude (volts)')
% plot of VL(t) vs t
VLnum=[0 0 30 0 0];
VLden=Iden;
vlt=impulse(VLnum,VLden,t);
subplot(2,2,3)
```

```
plot(t,vlt);title('vl(t) vs t');ylabel('Amplitude (volts)');
xlabel('time (sec)')
%plot of VC(t) vs t
VCnum = [0 0 0 0 300];
VCden = [1 2 19 18 90];
vct = impulse(VCnum,VCden,t);
subplot(2,2,4)
plot(t,vct);title('vc(t) vs t');ylabel('Amplitude (volts)');
xlabel('time (sec)')

figure(2)
subplot(2,1,1)
plot(t,it,t,vrt,t,vlt,t,vct)
title('Plots of: i(t) vs.t, vr(t) vs.t, vl(t)vs.t, and vc(t) vs.t')
axis([0,5,-20,20])
ylabel('Amplitude (amps/volts)')
subplot(2,1,2)
vsourse = vrt+vlt+vct;
plot(t,vsourse);
title('[vr(t)+vl(t)+vc(t)] vs. t (verifies KVL)')
xlabel('time (sec)');
ylabel(' Amplitude (volts)')
axis([0 5 -12 12])
```

The script file *RLC_series* is executed and the results are shown in Figures 4.71 and 4.72.

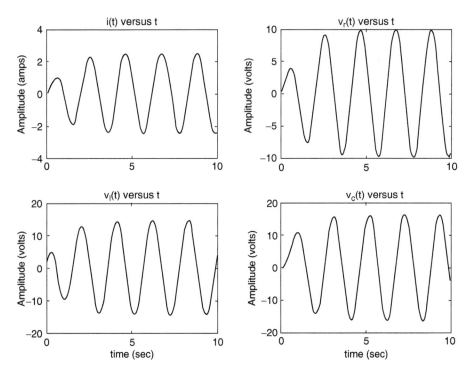

FIGURE 4.71
Plots of *i(t)*, $v_l(t)$, $v_r(t)$, and $v_c(t)$ of Example 4.16.

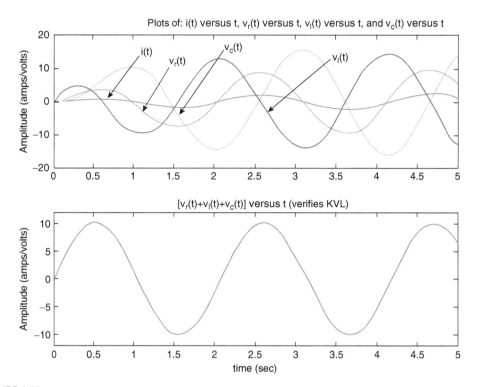

FIGURE 4.72
(See color insert following page 374.) Plots of *i(t)*, $v_l(t)$, $v_r(t)$, and $v_c(t)$ on the same plot (top), and verification of KVL (bottom) of Example 4.16.

Example 4.17

Let *f(t)* be the sum of two cosine waves with frequencies $f_1 = 50$ Hz and $f_2 = 200$ Hz, and unit amplitudes.

Create the script file *powers* that returns over the range $0 \le t \le 15$ s.

1. The average power of *f(t)* using
   a. The *norm* function
   b. The MATLAB functions *sum* and *abs*
2. The *psd* (power spectrum density) of *f(t)*, using 1024 points, over the normalized range $0 \le w \le 1$

MATLAB Solution
```
% Script file: powers
% echo off;
t = 0:0.01:15; % time sequence
x = 3*cos(2*pi*50.*t);
y = cos(2*pi*200.*t); ft=x+y;
```

```
power1 = norm(ft).^2/length(t);
power2 = sum(abs(ft).^2)/length(t);
disp('*******************************************')
disp(' ********** R E S U L T S**********')
disp('*******************************************')
disp('The power of f(t)(in watts) using the norm function is =')
disp(power1)
disp('The power of f(t) (in watts) using the sum/abs functions is =')
disp(power2)
disp('*******************************************')
Ps = spectrum(ft,1024);
specplot(Ps)
title('Power Spectral Density (PSD)')
```

The above script file *powers* is executed and the results are as follows (Figure 4.73):

```
>> powers

***************************************************************
*********************R E S U L T S ********************
***************************************************************
The power of f(t) (in watts) using the norm function is =
                                           10.0040

The power of f(t) (in watts) using the sum/abs functions is =
                                           10.0040
```

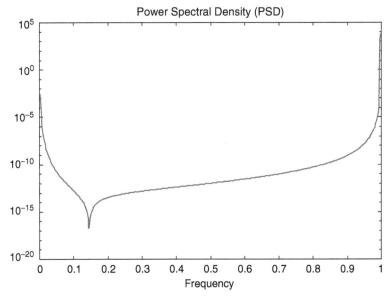

FIGURE 4.73
PSD plot of Example 4.17.

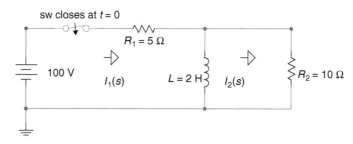

FIGURE 4.74
Circuit diagram of Example 4.18.

Example 4.18

Analyze the circuit diagram shown in Figure 4.74, after the switch "sw" closes at $t = 0$, assuming that the initial current through the inductor L is zero.

1. Write the set of loop equations in the frequency domain using Laplace
2. Write the loop matrix equation in the frequency domain
3. Evaluate the loop currents $i_1(t)$ and $i_2(t)$, by obtaining the ILT of the solutions of the matrix equation of part b
4. Evaluate the loop initial currents $i_1(0)$ and $i_2(0)$, using the initial value theorem
5. Evaluate the loop final currents $i_1(\infty)$ and $i_2(\infty)$, using the final value theorem
6. Obtain expressions for $V_{R1}(t)$, $V_{R2}(t)$, and $V_L(t)$ for $t \geq 0$
7. Obtain plots of the loop currents $i_1(t)$ versus t, $i_2(t)$ versus t, and $i_2(t)$ versus t
8. Obtain plots of the voltages: $V_L(t)$ versus t, $V_{R1}(t)$ versus t, and $V_{R2}(t)$ versus t
9. Discuss the results obtained

The MATLAB solution is given in the following text by the script file *loop_laplace_eqs*.

ANALYTICAL Solution

Part 1

The two loop equations for $t \geq 0$, assuming that the loop currents (directions) $I_1(s)$ and $I_2(s)$ are indicated in Figure 4.74, are given by

$$(5 + 2s)\, I_1(s) - 2s\, I_2(s) = 100/s$$

$$-2s\, I_1(s) + (10 + 2s)\, I_2(s) = 0$$

The resulting matrix equation is given by

$$\begin{bmatrix} 100/s \\ 0 \end{bmatrix} = \begin{bmatrix} 5 + 2s & -2s \\ -2s & 10 + 2s \end{bmatrix} \begin{bmatrix} I_1(s) \\ I_2(s) \end{bmatrix}$$

MATLAB Solution

```
% Script file: loop _ laplace _ eqs
syms s Zs Is Vs it y
```

```
Zs = [5+2*s -2*s;-2*s 10+2*s];
Vs = [100/s;0];
Is = inv(Zs)*Vs;
it = ilaplace(Is);VLs=(Is(1)-Is(2))/(2*s);
disp('*****************************************')
disp('*********** R E S U L T S ************')
disp('*********** C U R R E N T S *************')
disp('The loop currents i1(t) and i2(t) (in amps) are: ')
disp('i1(t)='),pretty(it(1))
disp('i2(t)='),pretty(it(2))
disp('*****************************************')
disp('The currents i1(t=0) and i2(t=0) (in amps)')
disp('using the initial value theorem are:')
i1 _ 0 = limit(s*Is(1),s,inf)
i2 _ 0 = limit(s*Is(2),s,inf)
disp('*****************************************')
disp('The current i1(t=inf) and i2(t=inf) (in amps)')
disp(' using the final value theorem are :')
i1 _ inf =limit(s*Is(1),s,0)
i2 _ inf =limit(s*Is(2),s,0)
disp('*********** V O L T A G E S*************** ')
disp('The voltage across the inductor L (in volts) is given by:')
VL _ t=ilaplace(VLs)
disp('The voltage across the resistor R1 (in volts) is given by:')
VR1 _ t=5*it(1)
disp('The voltage across the resistor R2 (in volts) is given by:')
VR2 _ t=10*it(2)
disp('*****************************************')

figure(1)
subplot (3,1,1)
ezplot(it(1))
title('i1(t) vs t');xlabel('time (sec)')
ylabel(' Amplitude (amps)');axis([0 2 0 21]);
subplot(3,1,2)
ezplot(it(2))
title('i2(t) vs t');axis([0 2 0 8]);
xlabel('time (sec)');ylabel(' Amplitude (amps)')
subplot(3,1,3)
ezplot(it(1)-it(2))
title('iL(t) vs. t');axis([0 2 0 8]);
xlabel('time (sec)');ylabel(' Amplitude (amps)')
axis([0 2.5 0 21])

figure(2)
subplot (3,1,1)
ezplot(VR2 _ t)
title('VL(t) vs t');xlabel('time (in sec)')
ylabel(' Amplitude (in volts)');axis([0 3 0 110]);
subplot(3,1,2)
ezplot(VR1 _ t)
title('VR1(t) vs t'); axis([0 3 0 110]);
xlabel('time (in sec)');ylabel(' Amplitude (in volts)')
```

```
subplot(3,1,3)
ezplot(VR2 _ t)
title('VR2(t) vs. t');axis([0 3 0 110]);
xlabel('time (in sec)');ylabel(' Amplitude (in volts)')
```

The script file *loop_laplace_eqs* is executed and the results are as follows (Figures 4.75 and 4.76):

```
>> loop _ laplace _ eqs


*******************************************
*********** R E S U L T S *************
*********** C U R R E N T S **************
The loop currents i1(t) and i2(t) (in amps) are:
i1(t)=
      - 40/3 exp(- 5/3 t) + 20
i2(t)=
      20/3 exp(- 5/3 t)
*******************************************
The currents i1(t=0) and i2(t=0) (in amps)
using the initial value theorem are:
i1 _ 0 =
        20/3
i2 _ 0 =
        20/3
*******************************************
The current i1(t=inf) and i2(t=inf) (in amps)
 using the final value theorem are :
i1 _ inf =
          20
i2 _ inf =
          0


*********** V O L T A G E S****************
The voltage across the inductor L (in volts) is given by:

VL _ t =
       6*exp(-5/3*t)+10*t-6

The voltage across the resistor R1 (in volts) is given by:

VR1 _ t =
         -200/3*exp(-5/3*t)+100

The voltage across the resistor R2 (in volts) is given by:

VR2 _ t =
         200/3*exp(-5/3*t)
```

Note that at $t = 0$, the inductor acts as an open circuit, then $i_1(0) = i_2(0) = 100$ V/15 Ω = 20/3 A; and at $t = \infty$, the inductor becomes a short circuit, then $i_1(\infty) = 100$ V/5 Ω = 20 A, and $i_2(\infty) = 0$ A. Values that completely agree with the results obtained using the MATLAB symbolic toolbox.

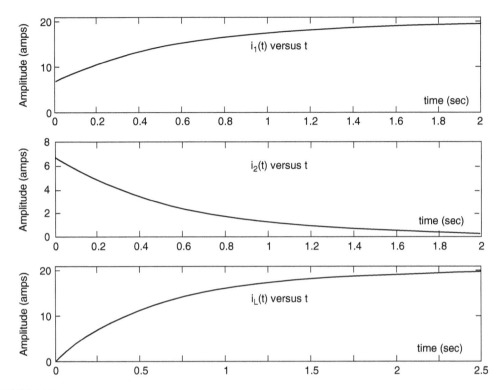

FIGURE 4.75
Current plots of Example 4.18.

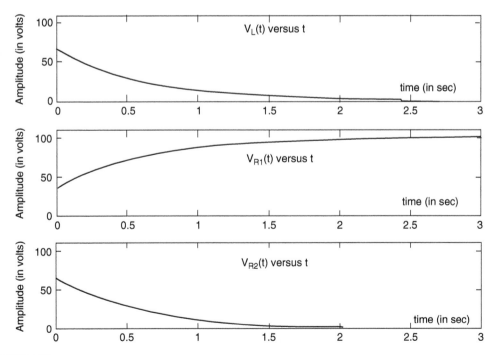

FIGURE 4.76
Voltage plots of Example 4.18.

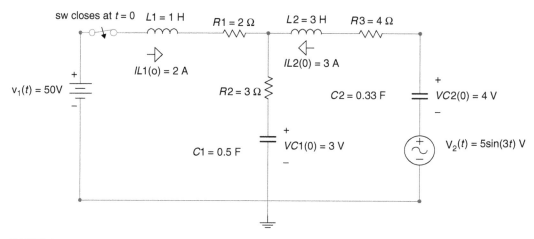

FIGURE 4.77
Circuit diagram of Example 4.19.

Example 4.19

The switch in the circuit diagram of Figure 4.77 closes at $t = 0$, where it remains for $t > 0$.
   The storing energy elements such as the (two) capacitors and the (two) inductors are initially charged as indicated in Figure 4.77 (for $t = 0$).

   1. Redraw the equivalent circuit in the s-domain
   2. Obtain in the s-domain the loop equations
   3. Obtain the matrix loop equation of part 2
   4. Use MATLAB and obtain expressions for $I_1(s)$ and $I_2(s)$
   5. Obtain the system's poles
   6. Use MATLAB and verify the initial and final value loop currents by applying the initial and final value theorems
   7. Use MATLAB and obtain plots of $I_1(s)$ versus s and $I_2(s)$ versus s
   8. Discuss the results

The MATLAB solution is given as follows by e script file *IC_loop_diff_eqs*.

ANALYTICAL Solution

Part 1
   The time domain circuit shown in Figure 4.77 is transformed into the equivalent s-domain circuit shown in Figure 4.78, where the initial conditions are converted into sources, and the assumed loop currents directions of $I_1(s)$ and $I_2(s)$ are indicated.
Part 2
   The two loop equations of the circuit diagram of Figure 4.78 are

$$\text{For loop \#1: } \quad \frac{50}{s} + 2 - \frac{3}{s} = \left[s + 5 + \frac{2}{s} \right] I_1(s) - \left[\frac{2}{s} + 3 \right] I_2(s)$$

$$\text{For loop \#2: } \quad \frac{3}{s} - 9 - \frac{4}{s} - \frac{15}{s^2 + 9} = -\left[\frac{2}{s} + 3 \right] I_1(s) + \left[3s + 7 + \frac{5}{s} \right] I_2(s)$$

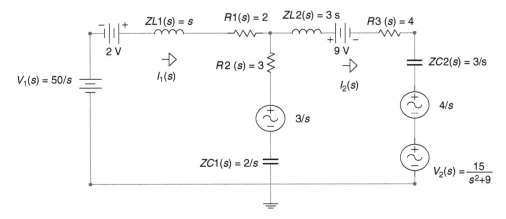

FIGURE 4.78
Equivalent s-domain circuit of Figure 4.77.

Part 3

The Matrix loop equation is given by

$$
\begin{bmatrix} \dfrac{50}{s} + 2 - \dfrac{3}{s} \\[2mm] \dfrac{-1}{s} - 9 - \dfrac{15}{s^2+9} \end{bmatrix} = \begin{bmatrix} s + 5 + \dfrac{2}{s} & \dfrac{-2}{s} - 3 \\[2mm] \dfrac{-2}{s} - 3 & 3s + 4 + \dfrac{5}{s} \end{bmatrix} \begin{bmatrix} I_1(s)\dfrac{2}{s} \\[2mm] I_2(s) \end{bmatrix}
$$

MATLAB Solution

```
% Script file: IC_loop_diff_eqs
syms s Zs Is Vs it y
Zs = [s+5+2/s -2/s-3;-2/s-3 3*s+7+5/s];
Vs = [50/s+2-3/s;-1/s-9-15/(s^2+9)];
Is = inv(Zs)*Vs;
it = ilaplace(Is);
disp('********************************************************')
disp('************ Frequency domain Results ***************')
disp('********************************************************')
disp('The loop currents I1(s) and I2(s) are: ')
disp('I1(s)='),pretty(Is(1))
disp('I2(s)='),pretty(Is(2))
disp('****************************************************')
disp('The values of the currents (in amps) i1(0) and i2(0), and')
disp('i1(t=inf) and i2(t=inf) using the initial and final value
theorems')
disp('are verified below:')
i1_0 = limit(s*Is(1),s,inf)
i2_0 = limit(s*Is(2),s,inf)
i1_inf = limit(s*Is(1),s,0)
i2_inf = limit(s*Is(2),s,0)
disp('****************************************************')
den = [3 22 37 27 6];
disp('The system poles are:')
rr = roots(den);
abs(rr)
disp('****************************************************')
```

```
figure(1)
ezplot(Is(1))
title('I1(s) vs s')
xlabel(' complex frequency s ')
ylabel('I1(s) ')
ylabel(Amplitude')
figure(2)
ezplot(Is(2))
title('I2(s) vs s');
xlabel('complex frequency s');
ylabel('Amplitude')
disp('****************************************************')
disp('****************** RESULTS ********************')
disp('****************************************************')
disp('The loop currents i1(t) and i2(t) (in amps) are: ')
disp('i1(t)='),pretty(it(1))
disp('i2(t)='),pretty(it(2))
disp('****************************************************')
```

The script file *IC_loop_diff_eqs* is executed and the results are as follows (Figures 4.79 and 4.80):

```
>> IC _ loop _ diff _ eqs


****************************************************
*************** RESULTS ********************
****************************************************

The loop currents I1(s) and I2(s) are:
```

$$
I1(s) = \frac{(3s^2 + 7s + 5)s\left(\dfrac{47}{s} + 2\right)}{3s^4 + 22s^3 + 37s^2 + 27s + 6} + \frac{(2 + 3s)s\left(-1/s - 9 - \dfrac{15}{s^2 + 9}\right)}{3s^4 + 22s^3 + 37s^2 + 27s + 6}
$$

$$
I2(s) = \frac{(2 + 3s)s\left(\dfrac{47}{s} + 2\right)}{3s^4 + 22s^3 + 37s^2 + 27s + 6} + \frac{(s^2 + 5s + 2)s\left(-1/s - 9 - \dfrac{15}{s^2 + 9}\right)}{3s^4 + 22s^3 + 37s^2 + 27s + 6}
$$

```
****************************************************
The values of the currents (in amps) i1(0) and i2(0), and
i1(t=inf) and i2(t=inf) using the initial and final value theorems
 are verified below:
i1 _ 0 = 2
i2 _ 0 = -3
i1 _ inf = 0
i2 _ inf = 0
```

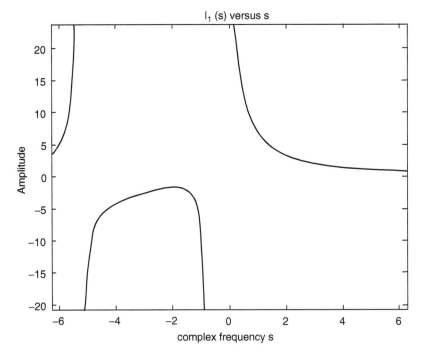

FIGURE 4.79
Plot of $I_1(s)$ of Example 4.19.

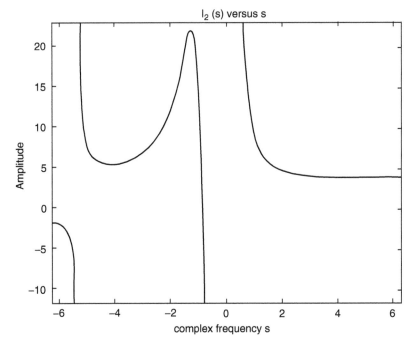

FIGURE 4.80
Plot of $I_2(s)$ of Example 4.19.

```
****************************************************
The system poles are:
ans =
        5.3196
        1.0051
        1.0051
        0.3721
```

Note that all the poles of $I_1(s)$ and $I_2(s)$ are in the left half of the s-plane, then the initial and final value theorems can be used. Note also that the initial currents are verified, and the final currents are zero since the two capacitors for $t = \infty$ act as an open circuit.

Example 4.20

The switch in the circuit diagram of Figure 4.81 closes at $t = 0$, where it remains for $t > 0$. All the storing elements are initially charged as indicated in Figure 4.81, at $t = 0$.

1. Redraw the equivalent circuit in the s-domain
2. Write the s-domain node equations
3. Write the matrix node equation
4. Use MATLAB and obtain expressions for $V_1(s)$ and $V_2(s)$
5. Use MATLAB and obtain plots of $V_1(s)$ versus s and $V_2(s)$ versus s
6. Use MATLAB to evaluate and verify the initial and final voltage for each node by applying the initial and final value theorems

The MATLAB solution is given as follows by the script file *nodal _eqs*:

ANALYTICAL Solution

Part 1

The time domain circuit diagram of Figure 4.80 is redrawn in the s-domain shown in Figure 4.82, indicating the two nodal voltages $V_1(s)$ and $V_2(s)$ and the circuit initial conditions transformed into sources.

Part 2

The two nodal equations are

$$\text{For node } \#V_1: \quad \frac{5}{s} + 2 - \frac{3}{s} = \left[\frac{1}{3} + \frac{1}{4} + \frac{s}{2} + \frac{1}{3s}\right]V_1(s) - \left[\frac{1}{3s} + \frac{1}{4}\right]V_2(s)$$

$$\text{For node } \#V_2: \quad \frac{3}{s} - \frac{2}{s} + \frac{8s}{s^2 + 9} = -\left[\frac{1}{3s} + \frac{1}{4}\right]I_1(s) + \left[\frac{1}{4} + \frac{1}{9} + \frac{1}{3s} + \frac{1}{5s}\right]V_2(s)$$

Part 3

The node matrix equation in the s domain is given by

$$\begin{bmatrix} \dfrac{2}{s} + 2 \\[2ex] \dfrac{3}{s} + \dfrac{8s}{s^2 + 9} \end{bmatrix} = \begin{bmatrix} \dfrac{1}{3s} + \dfrac{7}{12} + \dfrac{s}{2} & \dfrac{-1}{3s} - \dfrac{1}{4} \\[2ex] \dfrac{-1}{3s} - \dfrac{1}{4} & \dfrac{13}{36} + \dfrac{8}{15s} \end{bmatrix} \begin{bmatrix} V_1(s) \\[2ex] V_2(s) \end{bmatrix}$$

MATLAB Solution
```
% Script file: nodal _ eqs
syms s Ys Is Vs it y
Ys = [7/12+s/2+1/(3*s)  -(1/4+1/(3*s));-(1/4+1/(3*s))  13/36+8/(15*s)];
```

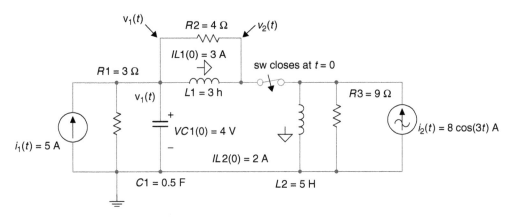

FIGURE 4.81
Circuit diagram of Example 12.20.

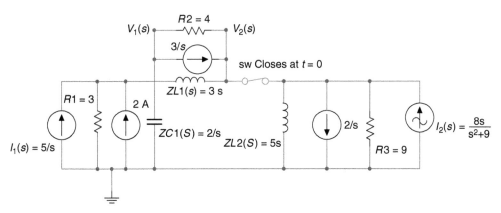

FIGURE 4.82
Equivalent s-domain circuit of Figure 4.81.

```
Is = [2/s+2;8*s/(s^2-9)+1/s];
Vs =inv(Ys)*Is;
disp('*********************************')
disp('********* RESULTS ************')
disp('*********************************')
disp('The node voltages V1(s) and V2(s) are: ')
disp('V1(s)='),pretty(Vs(1))
disp('V2(s)='),pretty(Vs(2))
disp('***********************************')
disp('The initial value voltages V1(t=0) and V2(t=0) (in volts)')
disp('using the initial value theorem are verified returning:')
V1 _ 0 = limit(s*Vs(1),s,inf)
V2 _ 0 = limit(s*Vs(2),s,inf)
disp('***********************************')
disp('The final value voltages V1(t=inf) and V2(t=inf) (in volts)')
```

```
disp('using the final value theorem are verified returning: ')
V1 _ inf =limit(s*Vs(1),s,0)
V2 _ inf =limit(s*Vs(2),s,0)
disp('**********************************')
figure(1)
ezplot(Vs(1))
title('V1(s) vs s')
xlabel(' complex frequency s ');ylabel('V1(s) ')
figure(2)
ezplot(Vs(2))
title('V2(s) vs s');xlabel('complex frequency s');
ylabel('V2(s) ')
```

The script file *nodal _eqs* is executed and the results are as follows (Figures 4.83 and 4.84):

```
>> nodal _ eqs
*******************************
********* RESULTS ************
*******************************
The node voltages V1(s) and V2(s) are:


                                                      /    s         \
                                      (3 s + 4) s |8 ------ + 1/s|
                                                      |     2        |
                 (65 s + 96) s (2/s + 2)            \ s + 9      /
 V1(s) = 6 --------------------------- + 90 ---------------------------
                  2          3                  2          3
           448 s + 286 s + 195 s + 72    448 s + 286 s + 195 s + 72


                                                      2       /   s         \
                                      (7 s + 6 s + 4) s |8 ------ + 1/s|
                                                                |     2        |
                 (3 s + 4) s (2/s + 2)                  \ s + 9      /
 V2(s) = 90 --------------------------- + 90 ---------------------------------
                  2          3                  2          3
           448 s + 286 s + 195 s + 72    448 s + 286 s + 195 s + 72



*******************************************************
The initial value voltages V1(t=0) and V2(t=0) (in volts)
using the initial value theorem are verified returning:

V1 _ 0 = 4
V2 _ 0 = 360/13


*******************************************************
The final value voltages V1(t=inf) and V2(t=inf) (in volts)
using the final value theorem are verified returning:

V1 _ inf = 0
V2 _ inf = 0

*******************************************************
```

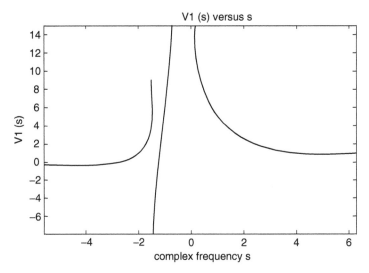

FIGURE 4.83
Plot of $V1(s)$ of Example 4.20.

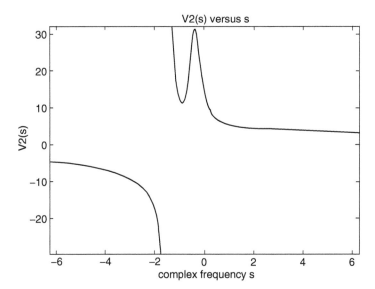

FIGURE 4.84
Plot of $V2(s)$ of Example 4.20.

Example 4.21

The switch in the circuit diagram shown in Figure 4.85 closes at $t = 0$, and the network is unenergized before the switch is closed, with $i_1(t = 0) = i_2(t = 0) = 0.\text{A}$.

1. Write the set of loop equations in the time domain (the assumed directions of $i_1(t)$ and $i_2(t)$ are indicated in Figure 4.85)
2. Transform the equations of part 1 into the s-domain
3. Obtain the matrix loop equations in the s-domain and indicate the impedance matrix $z(s)$ and voltage vector $v(s)$

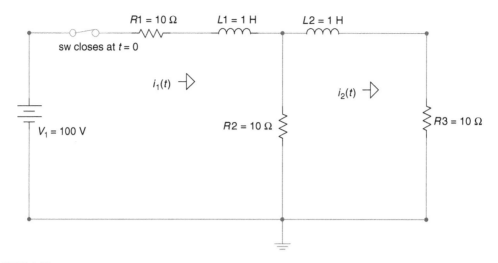

FIGURE 4.85
Circuit diagram of Example 4.21.

4. Use MATLAB and obtain expressions for $I_1(s)$ and $I_2(s)$
5. Use MATLAB to evaluate the initial- and final value of each loop current by applying the initial and final value theorems
6. Use MATLAB to obtain solutions for $i_1(t)$ and $i_2(t)$, for $t \geq 0$
7. Use MATLAB to obtain plots of $i_1(t)$ versus t and $i_2(t)$ versus t, for $t \geq 0$
8. Discuss the results obtained

The MATLAB solution is given as follows by the script file *loop_diffeqs*:

ANALYTICAL Solution

Part 1

The loop equations in the time domain for $t \geq 0$ are

$$100u(t) = \frac{di_1(t)}{dt} + 20i_1(t) - 10i_2(t)$$

$$0 = -10i_1(t) + 20i_2(t) + \frac{di_2(t)}{dt}$$

Part 2

The loop equations in the frequency domain are

$$\frac{100}{s} = (s + 20)I_1(s) - 10I_2(s)$$

$$0 = -10I_1(s) + (s + 20)I_2(s)$$

Part 3

The matrix loop equation of the network in the frequency domain is given by

$$\begin{bmatrix} 100/s \\ 0 \end{bmatrix} = \begin{bmatrix} s + 20 & -10 \\ -10 & s + 20 \end{bmatrix} \begin{bmatrix} I_1(s) \\ I_2(s) \end{bmatrix}$$

where $Z(s) = \begin{bmatrix} s+20 & -10 \\ -10 & s+20 \end{bmatrix}$

and

$$V(s) = \begin{bmatrix} 100/s \\ 0 \end{bmatrix}$$

MATLAB Solution
```
% Script file: loop _ diffeqs
syms s Zs Is Vs
Zs = [s+20 -10;-10 s+20];
Vs = [100/s;0];
Is = inv(Zs)*Vs;
disp('*********************************************')
disp('********* Matlab Symbolic Results *********')
disp('*********************************************')
disp('The loop currents I1(s) and I2(s) are: ')
disp('I1(s)='),pretty(Is(1))
disp('I2(s)='),pretty(Is(2))
disp('*********************************************')
i1 _ t = ilaplace(Is(1));
i2 _ t = ilaplace(Is(2));
disp('The loop currents i1(t) and i2(t)(in amp) are: ')
disp('i1(t)='),
simplify(i1 _ t)
disp('i2(t)='),
simplify(i2 _ t)
disp('*********************************************')
disp('The initial currents:i1(t=0) and i2(t=0)(in amps)')
disp('using the initial value theorem are evaluated below')
I1 _ 0 = limit(s*Is(1),s,inf)
I2 _ 0 = limit(s*Is(2),s,inf)
disp('*********************************************')
disp('The final currents:i1(t=inf) and i2(t=inf)(in amps)')
disp('using the initial value theorem are:')
I1 _ inf = limit(s*Is(1),s,0)
I2 _ inf = limit(s*Is(2),s,0)
disp('*********************************************')

figure(1)
subplot(1,2,1)
ezplot(i1 _ t);axis([0 .5 0 7])
title('i1(t) vs t'),xlabel('time (sec)')
ylabel('i1(t) in amp')
subplot(1,2,2)
ezplot(i2 _ t);axis([0 .5 0 3.5])
title('i2(t) vs t')
xlabel(' time (sec) ')
ylabel('i2(t) in amp ')
```

The script file *loop_diffeqs* is executed and the results are as follows:

```
>> loop _ diffeqs
```

```
**********************************************
********* Matlab Symbolic Results **********
**********************************************
The loop currents I1(s) and I2(s) are:

                     s + 20
  I1(s) = 100  -------------------
                        2
                (s + 40 s + 300) s

                  1000
  I2(s) = -------------------
                  2
            (s + 40 s + 300) s
**********************************************
The loop currents i1(t) and i2(t) (in amp) are:
  i1(t) =
          -5/3*exp(-30*t)-5*exp(-10*t)+20/3

  i2(t) =
          10/3-5*exp(-10*t)+5/3*exp(-30*t)
**********************************************
The initial currents: i1(t=0) and i2(t=0) (in amps)
using the initial value theorem are evaluated below:
I1 _ 0 = 0
I2 _ 0 = 0
**********************************************
The final currents: i1(t=inf) and i2(t=inf) (in amps)
using the final value theorem are:
I1 _ inf = 20/3
I2 _ inf = 10/3
**********************************************
```

Note that at $t = \infty$, the two inductors become shorts, then $i_1(\infty) = 100\ \text{V}/15\ \Omega = 20/3\ \text{A}$, and $i_2(\infty) = i_1(\infty)/2 = 10/3\ \text{A}$.

Observe that the analytical solutions agree completely with the MATLAB results (see Figure 4.86).

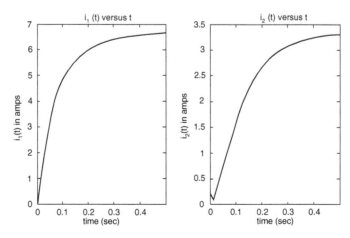

FIGURE 4.86
Current plots of $i_1(t)$ and $i_2(t)$ of Example 4.21.

Example 4.22

Modulated signals are often encountered in telecommunications and consist of mixing (multiplying) two analog signals referred to as the message signal $m(t)$, consisting of low frequencies; and the carrier signal $c(t)$, consisting of a high-frequency sinusoidal.

Let the modulating and carrier signals be given by $m(t) = 3\cos(5t)$ and $c(t) = \cos(50t)$, respectively.

Create the MATLAB script file *telecom_signals* that returns

1. Expressions of
   a. $C(w) = \Im[c(t)]$
   b. $M(w) = \Im[m(t)]$
   c. $M(w) + C(w) = \Im[m(t)] + \Im[c(t)]$
   d. $AM\_DSB\_SC(w) = \Im[m(t) \cdot c(t)] = 1/2\pi(\Im[M(w) \otimes C(w)]^*$
   e. $AM\_DSB\_WC(w) = \Im[(m(t) + 5) \cdot c(t)]$
   f. $FM(w) = \Im[fm(t)] = \Im[\cos(50\,t + m(t))]$
2. The following time and frequency plots:
   a. $m(t)$ versus $t \leftrightarrow M(w)$ versus w
   b. $c(t)$ versus $t \leftrightarrow C(w)$ versus w
   c. $[m(t) + c(t)]$ versus $t \leftrightarrow [M(w) + C(w)]$ versus w
   d. $m(t) \cdot c(t)$ versus $t \leftrightarrow \Im\,[m(t) \cdot c(t)]$
   e. $[m(t)+5\} \cdot c(t)$ versus $t \leftrightarrow \Im\,[(m(t) + 5) \cdot c(t)]$ versus w
   f. $\cos[50\,t + m(t)]$ versus $t \leftrightarrow \Im\{\cos[(50t) + m(t)]\}$ versus w

ANALYTICAL Solution

An AM signal is a signal in which $(m(t)\cos(w_0 t))$, the amplitude of the carrier, is modulated by the message wave $m(t)$; hence the information content of $m(t)$ is carried by the amplitude variations of the carrier.

If the angle of the carrier varies in some manner with respect to the modulation signal $m(t)$, then the modulation technique is referred to as angle modulation given by the following equation:

$$\cos(w_0 t) + f(m(t))$$

The following two methods are used in the generation of angle modulation:

1. PM
2. FM

PM and FM are defined by the following equations:

1. $pm(t) = A\cos[w_0 t + k_a m(t)]$
2. $f_m(t) = A\cos[w_0 t + k_b \int m(t)\,dt(m(t)]$

where A, k_a, and k_b are constants.

It is easy to appreciate that although PM and FM are different forms of angle modulation, they are essentially very similar, specially if the message signal is a sinusoidal. Therefore, it is unnecessary to discuss the two modulation methods, and one is sufficient to illustrate the process. For simplicity, *fm(t)* will be referred as an angular-modulated signal.

AM spectrum expressions can directly be obtained by using the MATLAB function *fourier*.

---

* AM stands for amplitude modulation, DSB for double side band, SC for suppress carrier, WC for with carrier, and FM for frequency modulation.

The FM spectrums can be obtained by the expansion of the time function in terms of the Bessel functions, illustrated as follows:

$$A[\cos(w_0 t + km(t))] = A\{j_0(k)\cos w_0 t + j_1(k)[\cos(w_0 + w_m)t - \cos(w_0 - w_m)t]$$

$$+ j_2(k)[\cos(w_0 + 2w_m)t + \cos(w_0 - 2w_m)t]$$

$$+ j_3(k)[\cos(w_0 + 3w_m)t - \cos(w_0 - 3w_m)t]$$

$$+ j_4(k)[\cos(w_0 + 4w_m)t + \cos(w_0 - 4w_m)t]$$

$$\ldots$$

$$+ j_n(k)[\cos(w_0 + nw_m)t - \cos(w_0 - nw_m)t]\}$$

where w_m is the highest frequency of $m(t)$, and $j_n(k)$ denotes the Bessel function of the first kind and nth order (referred to the command *besselj* in MATLAB).

In general, the $j_n(k)$s are negligible for $n > k$, if $k > 1$.

MATLAB Solution
```
% Script file : telecom _ signals
syms t w m _ t c _ t AM FM fm
m _ t = 3*cos(5*t);
c _ t = cos(50*t);
disp ('****************************************************************')
disp ('************************* R E S U L T S
*****************************')
disp ('The spectrum of the communication signals are:')
disp('****************************************************************')
Cw _ spectrum _ of _ carrier=fourier(c _ t)
disp ('****************************************************************')
Mw _ spectrum _ of _ message = fourier(m _ t)
disp('****************************************************************')
MCw _ spectrum _ carry _ plus _ message = fourier(m _ t+c _ t)
disp('****************************************************************')
AM _ DSB _ SC _ spectrum _ =fourier(c _ t*m _ t)
disp('****************************************************************')
AM _ DSB _ WC _ spectrum _ =fourier(c _ t*(m _ t+5))
disp('****************************************************************')
fm=besselj(3,0)*cos(50*t)+besselj(3,1)*(cos((50+5)*t)-cos((50-5)*t));
fm=fm+besselj(3,2)*(cos((50+5*2)*t)+cos((50-5*2)*t))+besselj(3,3)*(cos
                                ((50+5*3)*t)- cos((50-5*3)*t));
fm= fm+besselj(3,4)*(cos((50+5*4)*t)+cos((50-5*4)*t));
fm=fm+besselj(3,5)*(cos((50+5*5)*t)-cos((50-5*5)*t));
spectrum _ fm=fourier(fm);
FM _ spectrum=vpa(spectrum _ fm,3)
disp('****************************************************************')

figure(1)
subplot(3,2,1);
fplot('3*cos(5*t)',[0 3 -6 6]);
title('TIME DOMAIN');
ylabel('message(t)');
subplot(3,2,2)
w _ mes=[-5 5];
mag _ mes=[3*pi 3*pi];
stem(w _ mes,mag _ mes)
title('FREQUENCY DOMAIN')
```

```
ylabel('mag[M(w)]'),axis([-7 7 0 11]),
subplot(3,2,3)
fplot('cos(50*t)',[0 0.25 -1.5 1.5]);
ylabel('carrier(t)');
subplot(3,2,4)
w _ car=[-50 50];
mag _ car=[pi pi];
stem(w _ car,mag _ car)
ylabel('mag[C(w)]'),axis([-60 60 0 4.5])
subplot(3,2,5)
fplot('cos(50*t)+3*cos(5*t)',[0 2 -4.5 4.5]);
ylabel('carrier(t)+message(t)');xlabel('time (sec)')
subplot(3,2,6)
w _ mes _ car=[-50 -5 5 50];
mag _ mes _ car=[pi 3*pi 3*pi pi];
stem(w _ mes _ car,mag _ mes _ car)
ylabel('mag[M(w)+C(w)]'),axis([-60 60 0 11])
xlabel('frequency w (rad/sec)')
figure(2)
subplot(3,2,1)
fplot('3*cos(50*t)*cos(5*t)',[0 3 -6 6]);
title('TME DOMAIN');
ylabel('am-dsb-sc(t)');
subplot(3,2,2)
w _ AM _ DSB _ SC=[-55 -45 45 55];
mag _ AM _ DSB _ SC=[3/2*pi 3/2*pi 3/2*pi 3/2*pi];
stem(w _ AM _ DSB _ SC,mag _ AM _ DSB _ SC)
title('FREQUENCY DOMAIN')
ylabel('mag[AM-DSB-SC(w)]');
subplot(3,2,3)
fplot('cos(50*t)*(5+cos(5*t))',[0 3 -10 10]);
ylabel('am-dsb-wc(t)');
subplot(3,2,4)
w _ AM _ DSB _ WC=[-55 -50 -45 45 50 55];
mag _ AM _ DSB _ WC=[3/2*pi 5*pi 3/2*pi 3/2*pi 5*pi 3/2*pi];
stem(w _ AM _ DSB _ WC,mag _ AM _ DSB _ WC)
ylabel('mag[AM-DSB-SC(w)]'),axis([-60 60 0 5.5*pi]),
subplot(3,2,5)
fplot('cos((50*t)+3*cos(5*t))',[0 2 -2.5 2.5]);
xlabel('time (sec)');
ylabel('fm(t)');
subplot(3,2,6)
w _ FM=[-75:5:-25 25:5:75];
mag _ FM=[1.15 1.35 0.970 0.405 0.0615 0 0.0615 0.405 0.970 1.35 1.15 1.15
          1.35 0.970 0.405 0.0615 0 0.0615 0.405 0.970 1.35 1.15];
stem(w _ FM,mag _ FM)
ylabel('mag[FM(w)]'),axis([-80 80 0 2]),
xlabel('frequency w (rad/sec)')
```

The script file *telecom_signals* is executed and the results are shown in Figures 4.87 and 4.88.

```
>> telecom _ signals
*****************************************************************
************************R E S U L T S ***************************
The spectrum of the communication signals are:
*****************************************************************
```

```
Cw _ spectrum _ of _ carrier =
 pi*Dirac(w-50)+pi*Dirac(w+50)
**********************************************************************

Mw _ spectrum _ of _ message =
 3*pi*Dirac(w-5)+3*pi*Dirac(w+5)
**********************************************************************
MCw _ spectrum _ carry _ plus _ message =
 3*pi*Dirac(w-5)+3*pi*Dirac(w+5)+pi*Dirac(w-50)+pi*Dirac(w+50)
**********************************************************************
AM _ DSB _ SC _ spectrum _ =
 3/2*pi*Dirac(w-55)+3/2*pi*Dirac(w- 45)+3/2*pi*Dirac(w+45)+3/
 2*pi*Dirac(w+55)
**********************************************************************
AM _ DSB _ WC _ spectrum _ =
3/2*pi*Dirac(w-55)+3/2*pi*Dirac(w-45)+5*pi*Dirac(w-0)+3/2*pi*Dirac(w+45)+
3/2*pi*Dirac(w+55)+5*pi*Dirac(w+50)

**********************************************************************
FM _ spectrum =
.615e-1*Dirac(w-55.)+.615e-1*Dirac(w+55.)-
.615e-1*Dirac(w-45.)-.615e-1*Dirac(w+45.)+
.405*Dirac(w-60.)+.405*Dirac(w+60.)+.405*Dirac(w-40.)+
.405*Dirac(w+40.)+.970*Dirac(w-65.)+
.970*Dirac(w+65.)-.970*Dirac(w-35.)-.970*Dirac(w+35.)+1.35*Dirac(w-70.)+
 1.35*Dirac(w+70.)+
 1.35*Dirac(w-30.)+1.35*Dirac(w+30.)+1.15*Dirac(w-75.)+1.15*Dirac(w+75.)-
 1.15*Dirac(w-25.)-1.15*Dirac(w+25.)
**********************************************************************
```

FIGURE 4.87
Time and frequency domain of parts a, b, and c of Example 4.22.

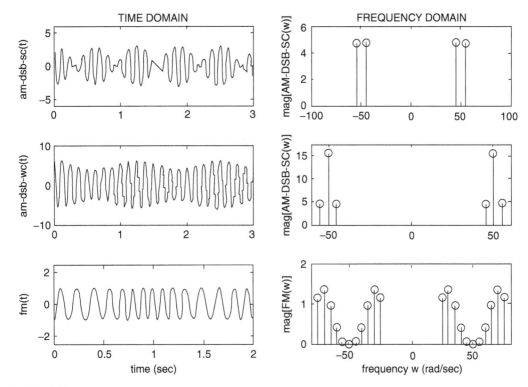

FIGURE 4.88
Time and frequency domain of parts d, e, and f of Example 4.22.

4.5 Application Problems

P.4.1 Given the functions shown in Figure 4.89, draw the even and odd parts for each function by hand, and determine the trigonometric FS of each part using MATLAB.

P.4.2 Write a function file that returns the energy of a given arbitrary function $f(t)$. Test your file for each of the functions shown in Figure 4.89.

P.4.3 Let the FS for the function $f_1(t)$ be given by

$$f_1(t) = \frac{1}{2} - \frac{1}{\pi} \sum_{n=1}^{\infty} \left(\frac{1}{n}\right) \sin(2\pi n t)$$

Evaluate the energy of $f_1(t)$ in the frequency and the time domain.

P.4.4 Using the trigonometric FS given in P.4.3, write a MATLAB program that returns the following plots:

a. $f_1(t)$ vs. t

b. The FS approximation of $f_1(t)$ using the first five terms versus t

c. Repeat part b for the first 15 terms versus t

d. $F_1(w)$ versus w (magnitude and phase)

e. Power of $f_1(t)$ versus t

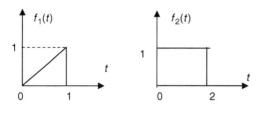

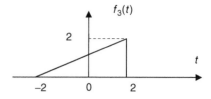

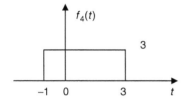

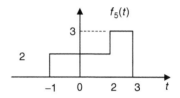

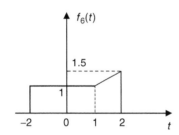

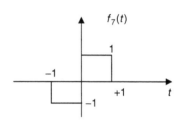

FIGURE 4.89
Plots of P.4.1.

P.4.5 Let

$$f_2(t) = \frac{1}{2} + \frac{1}{\pi} \sum_{n=1}^{\infty} \left(\frac{1}{n}\right) \sin(2\pi nt)$$

Sketch by hand

a. $f_2(t)$ versus t

b. $[f_3(t) = 1 - f_2(t)]$ versus t

c. $mag[F_3(w)]$ versus w

d. $phase[F_3(w)]$ versus w

P.4.6 Repeat P.4.5 using MATLAB.

P.4.7 The script file *drill* returns the FS expansion of the function *f(t)*, as shown in the following, as well as its first four partial FS approximations shown in Figure 4.90.

```
% Script file: drill
w1=1; w3 =3; w5 =5; w7=7;
k1= 4/pi;k3 = 4/(pi*3);
k5 = 4/(pi*5);k7= 4/(pi*7);
k = 0:pi/30:2*pi;
x1 =k1*sin(w1*k);x3 = k3*sin(w3*k);
```

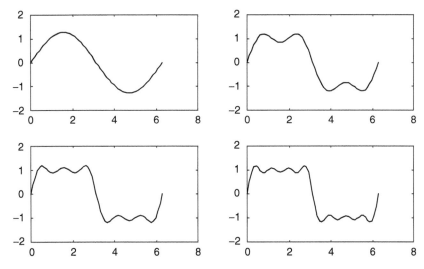

FIGURE 4.90
Plots of P.4.7.

```
x5 = k5*sin(w5*k);x7 = k7*sin(w7*k);
subplot(2,2,1);plot(k,x1)
subplot(2,2,2);plot(k,x1+x3)
subplot(2,2,3);plot(k,x1+x3+x5)
subplot(2,2,4)
plot (k,x1+x3+x5+x7)  11
```

Obtain the following:

a. An expression of $f(t)$

b. The fundamental frequency w_0 and the period T

c. Properly label the axis

d. Estimate the values of the Fourier coefficients F_0 and F_1

e. Estimate the 9th and 10th harmonic

f. A neat flow chart of the program

g. Estimate the discontinuity and the magnitude of $f(t)$ at the discontinuities (Gibbs)

P.4.8   Let the FS of the functions $f_1(t)$ and $f_2(t)$ be given by

$$f_1(t) = \sum_{n=1}^{\infty} \frac{(-1)^{n-1}}{n} \sin(nt)$$

and

$$f_2(t) = \frac{1}{3}\pi^3 + 4\sum_{n=1}^{\infty} \frac{(-1)^n}{n^2} \cos(nt)$$

Determine or obtain

a. The fundamental frequency, w_0, for each of the given functions

b. The plots of $f_1(t)$ versus t and $f_2(t)$ versus t

c. The plots of $F_1(w)$ versus w and $F_2(w)$ versus w

d. The bandwidth of $f_1(t)$ and $f_2(t)$

e. The bandwidth of $[f_1(t) + f_2(t)]$

P.4.9   Evaluate the following FTs by hand, using Table 4.1:

a. $u(t)$

b. $3 + 2u(t)$

c. $\delta(t)$

d. $\delta(t - 2)$

e. $3\delta(t - 4) + 6\delta(t + 5)$

f. $\cos(15t)$

g. $\sin(2t)$

h. $\cos(15t)\sin(2t)$

P.4.10 Repeat P.4.9 by using the MATLAB symbolic toolbox.

P.4.11 Using Table 4.1, determine by hand the following IFTs:

a. $(\text{Sin}(w/2))/(w/2)$

b. $\delta(w)$

c. $\cos(3w)$

d. $2/(jw + 2)$

e. $3/(jw + 2)^2$

f. $-1/w^2$

P.4.12 Repeat P.4.11 by using the MATLAB symbolic toolbox.

P.4.13 Obtain using MATLAB expressions and plots (magnitude and phase) of

a. $F_1(w) = \dfrac{jw}{2 + w^2}$

b. $F_2(w) = \dfrac{1}{(1 + jw)(2 - jw)}$

P.4.14 Determine $f_1(t)$ and $f_2(t)$ for the following frequency functions:

a. $F_1(w) = j/(2 - w + jw)$

b. $F_2(w) = \dfrac{1}{(1 + w^2)(2 - jw)}$

P.4.15 Let

$$f(t) = te^{-2t}u(t) \leftrightarrow F(w) = \frac{1}{(jw + 2)^2}$$

Obtain plots of

a. $f(t)$ versus t

b. $mag[F(w)]$ versus w

c. $phase[F(w)]$ versus w

d. $[power\ of\ f(t)]$ versus w

P.4.16 Given the system shown in Figure 4.91, use MATLAB to obtain the following plots:

a. $x(t)$ versus t

b. $X(w)$ versus w

c. $Y(w)$ versus w

d. $y(t)$ versus t

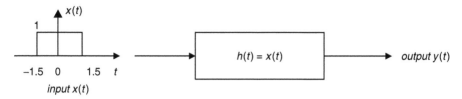

FIGURE 4.91
Block box system diagram of P.4.16.

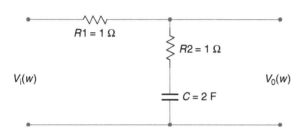

FIGURE 4.92
Circuit diagram of P.4.17.

P.4.17 Verify that the transfer function of the circuit diagram of Figure 4.92 is given by

$$H(w) = \frac{V_0(w)}{V_i(w)} = \left(\frac{1}{2}\right)\frac{(jw + 1/2)}{(jw + 1/4)}$$

a. Obtain plots of *mag[H(w)]* versus w and *angle[H(w)]* versus w
b. Estimate the circuit bandwidth
c. Estimate the regions of nondistortion

P.4.18 Repeat problem P.4.17 for the RC network shown in Figure 4.93, where

$$H(w) = \frac{V_0(w)}{V_i(w)} = \frac{(jw + 0.5)(jw + 0.5)}{(jw)^2 + 1.5jw + 0.25}$$

P.4.19 Verify that the transfer function of the RC network shown in Figure 4.94 is given by

$$H(w) = \frac{V_0(w)}{V_i(w)} = \frac{2jw + 1}{2jw + 2}$$

Estimate its bandwidth and the regions of nondistortion.

P.4.20 Assuming that the input $x(t) = \cos(t)$ for the system diagram shown in Figure 4.91, create a MATLAB script file that returns the following plots:

a. $x(t)$ versus t
b. $X(w)$ versus w
c. $h(t)$ versus t
d. *mag[H(w)]* versus w

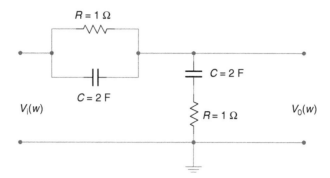

FIGURE 4.93
Circuit diagram of P.4.18.

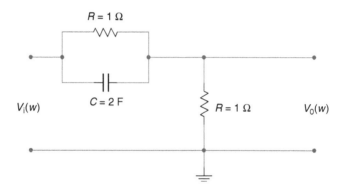

FIGURE 4.94
Circuit diagram of P.4.19.

     e. *angle[H(w)]* versus *w*
     f. *mag[Y(w)]* versus *w*
     g. *angle[Y(w)]* versus *w*
     h. *y(t)* versus *t*

P.4.21 Repeat P.4.20 for the cases where $x(t) = u(t) + u(t-1) - 2u(t-2)$ and $h(t) = 5t$.

P.4.22 Evaluate the following LTs by hand, using Table 4.2:

     a. $u(t)$ i. $e^{-2t}\cos(10t)\,u(t)$
     b. $3 + 2u(t)$ j. $\sinh(10t)\,u(t)$
     c. $\delta(t)$ k. $\cosh(10t - \pi/4)\,u(t)$
     d. $\delta(t - 2)$ l. $2e^{-3t} + 5\cos(3t)\,u(t)$
     e. $3\delta(t - 4) + 6\delta(t + 5)$ m. $e^{-3t}\sin(10t + \pi/3)\,u(t)$
     f. $\cos(15t)\,u(t)$ n. $3e^{-2t}u(-t)$
     g. $\sin(2t)\,u(t)$ o. $\cos(10t)\cos(150t)$
     h. $5e^{-2t}u(t)$ p. $\cos(10t)\sin(150t)$

P.4.23 Repeat P.4.22 using the MATLAB symbolic toolbox.

P.4.24 Using Table 4.2, determine by hand the following ILTs:

     a. $\sin(w/2)/(w/2)$
     b. $\delta(w)$

c. $\cos(3w)$

d. $2/(jw + 2)$

e. $3/(jw + 2)^2$

f. $-1/w^2$

P.4.25 Using the MATLAB symbolic toolbox, repeat P.4.24.

P.4.26 Obtain the following ILTs using Table 4.2, and indicate the Laplace properties used in the process:

a. $(s + 3)/(s^2 + 4s + 5)$

b. $\delta(s)$

c. $\cos(3s)$

d. $(2s^2 + 4s + 4)/(s^2 + 3s + 2)$

e. $s + 3/(s + 2)^2$

f. $-1/s^2$

P.4.27 Verify that $f(t)$ and $F(s)$ given as follows constitute a LT pair

$$f(t) = 3e^{-2t} + 4\sin(4t - 3) + t^2 e^{-3t} \leftrightarrow F(s) = \frac{3}{s+2} + \frac{e^{-3s}}{s^2 + 16} + \frac{2}{(s+3)^3}$$

P.4.28 Find the partial fraction expansion of each of the following functions:

a. $2s/(s^2 + 5s + 4)$

b. $(2s^2 + 1)/(s^2 + 5s + 4)$

c. $2s/(s^2 + 6s + 8)$

d. $2s/(s^2 + 1s + 1)$

e. $s^2/(s^2 + 5s + 4)$

f. $s/(s^2 + 3s + 4)$

P.4.29 Verify the following LT pairs:

a. $f(t) = 5\cos(3t - 2) + 4\sin(3t - 2) \leftrightarrow F(s) = (5s + 12)\dfrac{e^{-2s}}{s^2 + 9}$

b. $f(t) = \sin^5(t) \leftrightarrow F(s) = \dfrac{120}{(s^2 + 1)(s^2 + 3^2)(s^2 + 5^2)}$

c. $f(t) = \sin^6(t) \leftrightarrow F(s) = \dfrac{720}{s(s^2 + 2^2)(s^2 + 4^2)(s^2 + 6^2)}$

d. $f(t) = \dfrac{\sin^2(t)}{t} \leftrightarrow F(s) = \dfrac{1}{4}\ln\left(\dfrac{s^2 + 2^2}{s^2}\right)$

P.4.30 Verify that the current $i(t)$ shown in the circuit diagram of Figure 4.95 is given by

$$i(t) = \mathcal{L}^{-1}\left\{I(s) = \frac{50}{s^2 + 2s + 5}\right\}$$

assuming that all the initial conditions are zero.

Write a MATLAB program that returns the following plots:

a. $i(t)$ versus t

b. $v_R(t)$ versus t

c. $v_L(t)$ versus t

d. $v_C(t)$ versus t

P.4.31 Verify that the current $i(t)$ in the circuit diagram shown in Figure 4.96 is given by

$$i(t) = \pounds^{-1}\left\{I(s) = \frac{s}{s^2 + 3s + 2}\right\}$$

assuming that $v_c(0) = 0$.

Write a MATLAB program that returns the following plots:

a. $i(t)$ versus t

b. $v_c(t)$ versus t

c. $v_R(t)$ versus t

P.4.32 Write a MATLAB program that returns the expressions and plots for the circuit diagram shown in Figure 4.97 for

a. $i(t)$ versus t

b. $v_R(t)$ versus t

c. $v_L(t)$ versus t

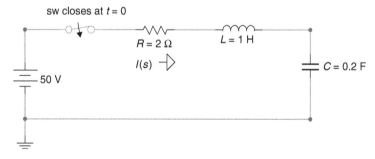

FIGURE 4.95
Circuit diagram of P.4.30.

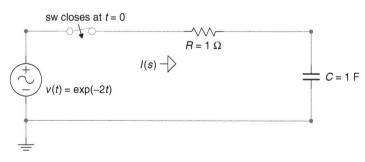

FIGURE 4.96
Circuit diagram of P.4.31.

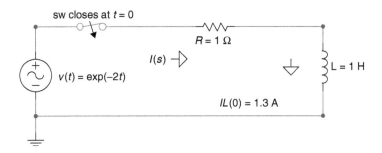

FIGURE 4.97
Circuit diagram of P.4.32.

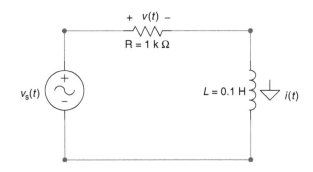

FIGURE 4.98
Circuit diagram of P.4.33.

P.4.33 Use the LT to find the current $i(t)$ and voltage drop $v(t)$ of the circuit diagram shown in Figure 4.98 for the following voltage sources:

a. $v_s(t) = 5\delta(t)$

b. $v_s(t) = 5u(t)$

c. $v_s(t) = 5t\,u(t)$

P.4.34 Use the LT to find the current $i(t)$ and voltage $v(t)$ of the circuit shown in Figure 4.99 for the following voltage sources:

a. $v_s(t) = 5\delta(t)$

b. $v_s(t) = 5u(t)$

c. $v_s(t) = 5t\,u(t)$

P.4.35 For the circuit shown in Figure 4.100, determine expressions and the corresponding plots of $i_1(t)$ versus t and $i_2(t)$ versus t. Evaluate $i_1(\infty)$ and $i_2(\infty)$ by hand computations and by using the final value theorem.

P.4.36 Construct a function of s as the ratio of two polynomials such that the following conditions are satisfied:

$$f(0^+) = \lim_{s \to \infty}[s \cdot F(s)] = 5$$

and

$$f(\infty) = \lim_{s \to 0}[s \cdot F(s)] = 15$$

Obtain an expression of the ILT of $F(s)$ and verify, using MATLAB, that $f(0^+) = \lim_{t \to 0}[f(t)] = 5$ and $f(\infty) = \lim_{t \to \infty}[f(t)] = 15$.

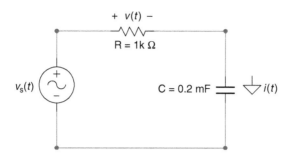

FIGURE 4.99
Circuit diagram of P.4.34.

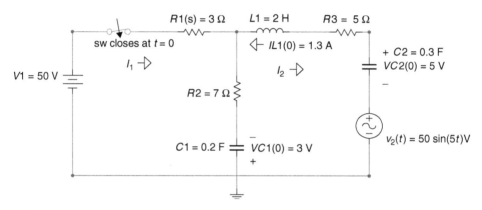

FIGURE 4.100
Circuit diagram of P.4.35.

P.4.37   Let the modulating and carrier signals of a communication signal be given by $m(t) = pul(t/2)$ and $c(t) = \cos(50t)$, respectively.

Create the MATLAB script file *telecom_analysis* that returns

a. Expressions of

     i. $C(w) = \Im[c(t)]$

     ii. $M(w) = \Im[m(t)]$

     iii. $M(w) + C(w) = \Im[m(t)] + \Im[c(t)]$

     iv. $AM\_DSB\_SC(w) = \Im[m(t) \cdot c(t)] = 1/2\pi(\Im[M(w) \otimes C(w)])$

     v. $AM\_DSB\_WC(w) = \Im[\{m(t) + 5\} \cdot c(t)]$

     vi. $FM(w) = \Im[fm(t)] = \Im[\cos(50t + m(t))]$

b. The following time–frequency plots:

     i. $m(t)$ versus $t \leftrightarrow M(w)$ versus w

     ii. $c(t)$ versus $t \leftrightarrow C(w)$ versus w

     iii. $[m(t) + c(t)]$ versus $t \leftrightarrow [M(w) + C(w)]$ versus w

     iv. $[m(t) \cdot c(t)]$ versus $t \leftrightarrow \Im[m(t) \cdot c(t)]$ versus w

     v. $[m(t) + 5 \cdot c(t)]$ versus $t \leftrightarrow \Im[(m(t) + 5) \cdot c(t)]$ versus w

     vi. $\cos[50t + m(t)]$ versus $t \leftrightarrow \Im\{\cos[(50t) + m(t)]\}$ versus w

5

DTFT, DFT, ZT, and FFT

Those who give up liberty for the sake of security deserve neither liberty nor security.

Benjamin Franklin

5.1 Introduction

Recall that a discrete-time system can be defined as a process in which an input sequence(s) $f(n)$ referred as the excitation is transformed into an output sequence(s) $g(n)$ referred to as the response, using a set of predefined rules, algorithms, or in most practical cases a set of difference equations. The discussion in this chapter will be restricted to single input–single output (SISO) systems since they constitute the vast majority of the practical digital systems and the best model to introduce the tools of analysis and synthesis due to its simplicity. The transition from a SISO to a multiple input or output system is a simple process since almost all the systems considered in this chapter are LTI systems and the general principles of superposition hold.

The block diagram of Figure 5.1 is a simple representation of a single-input $\{f(n)\}$ sequence- and single-output $\{g(n)\}$ sequence system, defined by the system transfer function H, where $g(n) = H[f(n)]$.

Some aspects of discrete-time sequences were first introduced, explored, and discussed in Chapter 7 of the book titled *Practical MATLAB® Basics for Engineers* and in Chapter 1 of this book. In this chapter, arbitrary time signals once sampled and converted into discrete-time sequences are studied by using different frequency domain transformations such as discrete-time Fourier transform (DTFT), discrete Fourier transform (DFT), Z-transform (ZT), and fast Fourier transform (FFT).

Recall that the input sequence $f(n)$ can be viewed as a collection of samples of the continuous signal $f(t)$ in which only at the sampling instances the values of $f(t)$ are known, whereas the signal $f(t)$ is not defined for the rest of the time. Consequently, discrete-time signals can be viewed as a sequence of numbers, representing samples where the independent time variable is denoted by the integer n.

The first studies of converting a discrete-time sequence into the frequency domain dates back to the late 1940s and early 1950s. The pioneering research was done at the Bell Telephone Laboratories (New Jersey) and Columbia University (New York) by defining the DTFT, from the analog FT, and then by limiting the discrete-time sequence. The sequence truncation led to the DFT and subsequently to the basis of digital-filter theory design. Pioneering work in the area of digital filters was done at the Bell Telephone Laboratories in the early 1960s, in particular, by J. K. Kaiser.

The DFT became a practical discrete tool with the publication of the research paper by J. W. Cooley and J. K. Tukey (April 1965) titled *Algorithm for Machine Calculations of Complex Fourier Series*.

457

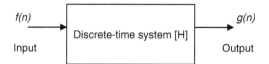

FIGURE 5.1
System block diagram.

Some historians traced the basis of the algorithm used in this research paper to the great nineteenth century German mathematician, Carl Friedrich Gauss. In any case, this paper and the subsequent research and publications led to a set of techniques known collectively as FFT, or radix 2 dissemination, in which the computation efficiency was highly improved from $4N^2 + 2N(2N - 1)$ to $2Nlog_2(N) + 3Nlog_2(N)$ operations in the evaluations of the DFT using the *fft*, where N denotes the length of the discrete sequence.

The other discrete transform, such as the ZT, was developed by the discrete control research group at Columbia University, integrated by professors and graduate students, in the early 1950s and gained acceptance in the 1960s. The ZT is equivalent to the LT for the analog system model, in the sense that a difference equation can be transformed into an algebraic equation.

The discrete signals and systems are analogous to the continuous case in its analysis, synthesis, and representation in the traditional two domains—time and frequency.

Recall that in the continuous case, differential equations were used to describe a network behavior in the time domain, and an algebraic equivalent relation (equation) was used to describe the system in the frequency domain (Fourier and Laplace in Chapter 4 of this book). Discrete systems analysis synthesis follows the same general approach.

The concepts of poles, zeros, transfer function, convergence, impulse response, step response; the convolution process in time and frequency, stability, superposition, power, and energy considerations; and other system concepts and techniques developed for the case of continuous systems can be extended with minor modifications into the analysis and synthesis approach used in discrete-time systems. Principles, concepts, properties, and general analytical tools and techniques of the different transforms and their applications will be stated, tested, and explored in this chapter by using the power of MATLAB®.

5.2 Objectives

After completing this chapter, the reader should be able to

- Express a discrete system in terms of an input and output sequences by means of a sampling rule
- Know the defining characteristics of discrete systems such as linear, time invariance, causal, stable, passive, SISO, and bounded input–bounded output (BIBO)
- Know the discrete-system elements such as unit delay, unit shift, and multiplier
- Transform an input sequence *f(n)* into an output sequence *g(n)* by means of a difference equation
- Derive the discrete-system transfer function from a given difference equation
- Determine the conditions for system stability in both domains—time and frequency
- Use the ZT as the tool to describe the behavior of a discrete-time system or signal

- Know the basic standard discrete signals and their transforms
- Understand the concept of normalized frequencies for discrete-time systems
- Understand the relation between the signal, normalized frequency, and effect of the sampling frequency
- Estimate the spectrum of discrete-time signals and systems
- State the differences between FT and DTFT
- State the differences between DTFT and DFT
- Use a table for the evaluation of DTFT
- Know the most important properties of the different discrete-system transforms
- Know the standard DFT
- Know the meaning and power of FFT
- Understand the computational efficiency of FFT
- Understand when and how FFT is used
- Use the MATLAB command *fft* to evaluate DFT
- Use the power of MATLAB to analyze a variety of discrete-time signals and systems, in both the time and frequency domains

5.3 Background

R.5.1 A discrete-time system transforms an input sequence $f(n)$ into an output sequence $g(n)$ by means of a given algorithm, predefined rules, conversion tables, and most often by a mathematical relation or equation.

Analytically, the preceding statement can be expressed as $g(n) = Lf(n)$, and it is illustrated by means of a block diagram shown in Figure 5.2.

R.5.2 The most common classification and characterization (also referred to as properties) of discrete-time systems are stated as follows:

a. Linearity

Let

$$f_1(n) \to g_1(n) \text{ (where } \to \text{ means transforms to)}$$

and

$$f_2(n) \to g_2(n)$$

Then

$$a_1 f_1(n) + a_2 f_2(n) \to a_1 g_1(n) + a_2 g_2(n)$$

FIGURE 5.2
Discrete-system block diagram.

b. Time- or shift invariance

Let

$$f(n) \rightarrow g(n)$$

Then

$$f(n-k) \rightarrow g(n-k) \text{ for any arbitrary integer } k \ (1, 2, 3, \ldots)$$

c. Causality

If

$$f(n_i) = 0, \text{ over the range } n < n_i$$

Then

$$g(n) = 0, \text{ over the range } n < n_i$$

This property states, in general terms, that the ith output sample $g(ni)$ depends only on the input samples over the range $n \leq ni$, output $g(n)$ of past and present samples, and never on future samples.

d. Stability

There are many accepted definitions of stability for discrete-time systems. In this chapter, the concept of stability is defined simply in the following way: A discrete system is unstable if and only if there exists a finite-bounded input sequence $f(n)$ that causes the output sequence $g(n)$ to blow out or become infinite.

If a given system is not unstable then it is stable. A stable system is a system where a bounded (finite) input $f(n)$ produces a bounded (finite) output $g(n)$. This definition of stability is often referred to as BIBO.

The energy associated with a discrete-time sequence, either an input or output, must be satisfied by the following relations:

$$\text{Energy of } f(n) = \sum_{n=-\infty}^{+\infty} |f(n)|^2 < \infty$$

$$\text{Energy of } g(n) = \sum_{n=-\infty}^{+\infty} |g(n)|^2 < \infty$$

e. Passive

A discrete-time system is passive if the output energy of $[g(n)]$ does not exit the input energy of $[f(n)]$.

R.5.3   The discrete operators or system elements used to transform an input sequence $f(n)$ into an output sequence $g(n)$ are illustrated graphically and defined in Figure 5.3.

R.5.4   This chapter deals mainly with SISO systems. The techniques developed for SISO systems can easily be extended to multiple input–output systems by making use of the superposition principle (assuming that the system considered is linear).

R.5.5   The discussions in this chapter are restricted mainly to discrete systems that are linear, time invariant, and in most cases causal.

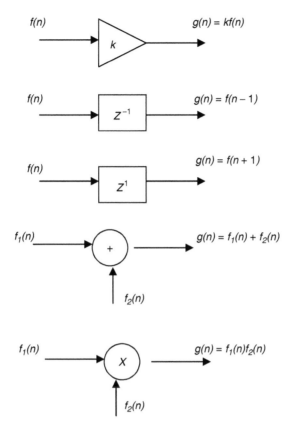

FIGURE 5.3
Discrete operators and system elements.

R.5.6   Time invariant linear systems are completely characterized by its impulse response. If the impulse response is known, then the output sequence $g(n)$ can easily be evaluated for any arbitrary input sequence $f(n)$ (similar to the continuous case).

R.5.7   The notation $h(n)$ is reserved to denote the impulse $\{f(n) = \delta(n)\}$ system response $\{h(n) = g(n) = L[\delta(n)]\}$ (analogous to the continuous case $h(t)$).

R.5.8   When the discrete system is both linear and time invariant (LTI), a number of techniques can be used to transform the linear difference equations that relate the system input to the system output, using the time and frequency domains. A particular useful transform from the time into the frequency domain and vice versa from the frequency into the time domain is the Z transform (ZT).

R.5.9   Before introducing more concepts, let us gain some insight and experience by analyzing some discrete systems. For example, analyze if the following systems, excited by a real input sequence $f(n)$ and producing an output $g_i(n)$, for $i = 1, 2, 3$ are LTI.

The systems are

a. $g_1(n) = f(n)$

b. $g_2(n) = nf(n)$

c. $g_3(n) = f^2(n)$

ANALYTICAL Solution

a. $g_1(n)$ is nonlinear time invariant

b. $g_2(n)$ is linear time-varying

c. $g_3(n)$ is nonlinear

R.5.10 Let us analyze a more complex discrete system, given by the difference equation $g(n) = 4f(n) + 5f(n − 1)$.

a. Draw the block circuit diagram.

b. Determine its impulse response $h(n)$.

ANALYTICAL Solution

Part a. The circuit block diagram is shown in Figure 5.4.

Part b. The impulse response is evaluated by replacing $f(n)$ with $\delta(n)$, yielding the following:

$$g(n) = 4\delta(n) + 5\delta(n − 1)$$

then

$$h(n) = 4\delta(n) + 5\delta(n − 1)$$

Observe that

i. The system is causal since $h(n) = 0$, for $n < 0$.

ii. $g(n)$ depends on $f(n)$ and preceding value of $f(n)$.

iii. The system is stable, BIBO.

iv. The energies associated with the input sequence $f(n)$ and output sequence $g(n)$ are finite.

R.5.11 In general, a causal discrete-time invariance linear systems (LTI) can be described by a difference equation of the form

$$g(n) + a_1 g(n − 1) + a_2 g(n − 2) + \cdots + a_N g(n − N) = b_0 f(n) + b_1 f(n − 1) + b_2 f(n − 2)$$
$$+ \cdots + b_M f(n − M)$$

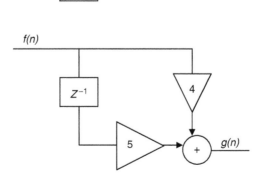

FIGURE 5.4
Discrete-system block diagram of R.5.10.

R.5.12 The concept of causality is an important concept in many systems. Let us consider a simple example where the output of a system depends only on the inputs $f(n), f(n-1), f(n-2), \ldots, f(n-k)$; and let the system equation be given by

$$g(n) = f(n) + \frac{1}{2}f(n-1) + \cdots + \left(\frac{1}{2}\right)^k f(n-k)$$

The system block diagram is shown in Figure 5.5.
Its impulse response would then be given by

$$h(n) = \delta(n) + \left(\frac{1}{2}\right)\delta(n-1) + \left(\frac{1}{2}\right)^2 \delta(n-2) + \cdots + \left(\frac{1}{2}\right)^k \delta(n-k)$$

and for $k \to \infty$

$$h(n) = \sum_{k=0}^{+\infty}\left(\frac{1}{2}\right)^k \delta(n-k) = \left(\frac{1}{2}\right)^n u(n)$$

R.5.13 The output of a discrete-time system $g(n)$ may depend on itself (the output sequence), which may be feedback into the system, as well as the input sequence $f(n)$. The following equation illustrates such a system:

$$g(n) - \left(\frac{1}{2}\right)g(n-1) = f(n)$$

The system block diagram is shown in Figure 5.6.

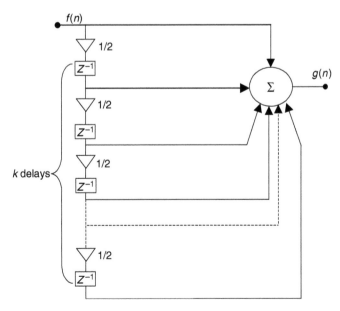

FIGURE 5.5
System block diagram of R.5.12.

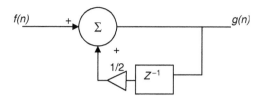

FIGURE 5.6
System block diagram of R.5.13.

The impulse response would then be given by

$$h(n) - \left(\frac{1}{2}\right)h(n-1) = \delta(h)$$

If the system is causal, then $h(n) = 0$, for $n < 0$. Then $h(-1) = 0$, and the recursive difference equation can be evaluated numerically as follows:

$N = 0$	$h(0) = 1$	
$N = 1$	$h(1) - 1/2(1) = 0$	$h(1) = 1/2 = (1/2)^1$
$N = 2$	$h(2) - 1/2(1/2) = 0$	$h(2) = 1/4 = (1/2)^2$
$N = 3$	$h(3) - 1/2(1/4) = 0$	$h(3) = 1/2 = (1/2)^3$
$N = 4$	$h(4) - 1/2(1/8) = 0$	$h(4) = 1/16 = (1/2)^4$

Observe that the coefficients of $h(n)$ are $(1/2)^n$, for $n > 0$, or

$$h(n) = \left(\frac{1}{2}\right)^n u(n) = \sum_{k=0}^{+\infty} \left(\frac{1}{2}\right)^k \delta(n-k)$$

R.5.14 Note that the examples presented in R.5.12 and R.5.13 given by the following difference equations present identical impulse responses. If the impulse responses are identical, then the two given systems are equivalent.

a. $g(n) = f(n) + (1/2) f(n-1) + (1/2)^2 f(n-2) + \cdots + (1/2)^k f(n-k)$

b. $g(n) - (1/2) g(n-1) = f(n)$

Observe that in the first equation, the output depends only on the input sequence, whereas in the second equation, the output depends on its previous output scaled $(g(n))$, delayed by one time unit, as well as by the input sequence.

R.5.15 A memoryless system is a system where the output depends only on the input sequence, and a memory system is a system where the output depends on the previous outputs as well as input.

R.5.16 Recall that any arbitrary discrete sequence can be represented as a superposition of delayed weighted impulses (see Chapter 1 of this book), which is given by

$$f(n) = \sum_{n=-\infty}^{+\infty} f(k)\delta(n-k) = \dots f(-2)\delta(n+2) + f(-1)\delta(n+1) + f(0) + f(1)\delta(n-1)$$

$$+ f(2)\delta(n-2) + f(3)\delta(n-3) + \cdots + f(k)\delta(n-k)\dots$$

R.5.17 Let $f(n) \to g(n)$. Recall that "$\to$" denotes *transforms to*. Then

$$\delta(n) \to h(n)$$

and

$$g(n) \rightarrow H[f(n)]$$

Then

$$g(n) = H\left[\sum_{K=-\infty}^{+\infty} f(k)\delta(n-k)\right], \text{ from R.5.16}$$

$$g(n) = \left[\sum_{K=-\infty}^{+\infty} f(k)H[\delta(n-k)]\right]$$

and

$$g(n) = \left[\sum_{K=-\infty}^{+\infty} f(k)h(n-k)\right]$$

R.5.18 Note that the output of a linear system is given by

$$g(n) = \sum_{n=-\infty}^{+\infty} f(k)h(n-k)$$

or in short

$$g(n) = h(n) \otimes f(n)$$

Recall that $\otimes$ denotes convolution.

R.5.19 For example, let $f(n) = 2^n$ and $h(n) = (1/3)^n u(n)$. Then $g(n) = h(n) \otimes f(n)$ or

$$g(n) = \sum_{k=-\infty}^{+\infty} f(k)h(n-k) = \sum_{k=-\infty}^{+\infty} h(k)f(n-k)$$

$$g(n) = \sum_{k=-\infty}^{+\infty} \left(\frac{1}{3}\right)^k u(k)2^{(n-k)}$$

$$g(n) = \sum_{k=0}^{+\infty} \left(\frac{1}{3}\right)^k 2^n 2^{-k} = 2^n \sum_{k=0}^{+\infty} \left(\frac{1}{3}\right)^k 2^{-k} = 2^n \sum_{k=0}^{+\infty} \left(\frac{1}{3}\right)^k \left(\frac{1}{2}\right)^k$$

$$g(n) = \sum_{n=0}^{+\infty} 2^n \sum_{k=0}^{+\infty} \left(\frac{1}{6}\right)^k = \sum_{n=0}^{+\infty} 2^n \left(\frac{1}{1-(1/6)}\right)$$

Since

$$\sum_{n=0}^{+\infty} a^k = \left(\frac{1}{1-a}\right) \quad \text{for } |a| < 1$$

$$g(n) = 2^n \frac{1}{(5/6)} = \sum_{n=0}^{+\infty}\left(\frac{6}{5}\right)2^n = 2^n\left(\frac{6}{5}\right)u(n)$$

R.5.20 The steps involved in the discrete convolution of $f(n)$ with $h(n)$ are

a. Since

$$g(n) = f(n) \otimes h(n) = \sum_{k=-\infty}^{+\infty} f(k)h(n-k)$$

then the first step in the implementation of the discrete convolution is to change the variable n by k in each of the functions $f(n)$ and $h(n)$.

b. Since $h(k)$ is known, then $h(-k)$ can be represented by reflecting or reversing $h(k)$ with respect to the vertical axis (approximately $k = 0$).

c. To obtain $h(n-k)$, it is necessary to shift $h(-k)n$ places to the left or right when n is negative or positive, respectively, for all possible values of n.

d. For each possible value of $n = n_0$, it is necessary to evaluate the product $h(n_0 - k)$ $f(k)$ and then perform the sum over all values of k ($-\infty \le k \le +\infty$).

e. The steps c and d return just one point in the convolution process, that is, $h(n) \otimes f(n)$, for $n = n_0$.

f. This procedure is to be repeated for all possible values of n, returning for each n a single point of the convolution.

g. The sets of all the connected points obtained in this matter is referred to as the (graphic) convolution of $f(n)$ with $h(n)$.

The example in R.5.21 illustrates graphically the process just described.

R.5.21 For example, evaluate the graphical convolution of $g(n) = h(n) \otimes f(n)$ for the following discrete sequences:

$$h(n) = \left(\tfrac{1}{2}\right)^n u(n) \quad \text{and} \quad f(n) - u(n) - (n-4)$$

ANALYTICAL Solution

The two sequences $h(n)$ and $f(n)$ are illustrated in Figure 5.7, by replacing n with k. Figure 5.8 shows $f(-k)$, that is, $f(k)$ reflected with respect to the vertical axis ($k = 0$). Then $f(0 - k)h(k) = g(0) = 1$, and for all $n < 0$, $g(n) \equiv 0$. Then shifting $f(-k)$ by 1, as indicated in Figure 5.9, we get

$$g(1) = 1 + \frac{1}{2} = \frac{3}{2}$$

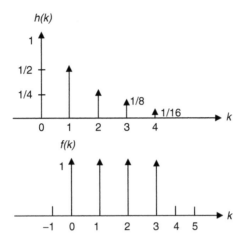

FIGURE 5.7
Discrete sequences *h(n)* and *f(n)* of R.5.21.

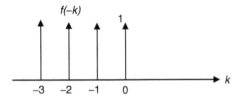

FIGURE 5.8
Discrete sequences *f(−k)* of R.5.21.

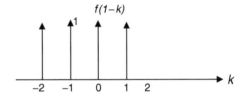

FIGURE 5.9
Discrete sequences *f(1 − k)* of R.5.21.

Observe that the convolution coefficients can be evaluated by following the pattern illustrated as follows:

$$g(0) = h(0) \cdot f(0)$$

$$g(1) = h(1) \cdot f(0) + f(1)h(0)$$

$$g(2) = h(2) \cdot f(0) + h(1)f(1) + h(0)f(2)$$

$$g(3) = h(3) \cdot f(0) + h(2)f(1) + h(1)f(2) + h(0)f(3)$$

...

$$g(k) = h(k) \cdot f(0) + h(k-1)f(1) + \cdots + h(1)f(k-1) + h(0)f(k),$$ for any values of k

R.5.22 Recall that if the impulse response of a given LTI discrete system is known ($h(n)$), then the output sequence $g(n)$ can then be evaluated as the convolution of any arbitrary input $f(n)$ with the system impulse response $h(n)$.

The discrete convolution can be viewed as the system transformation in the time domain of the input sequence $f(n)$ into an output sequence $g(n)$.

R.5.23 Some discrete convolution properties are as follows. Observe that they follow closely the analog case. Let $g(n) = f(n) \otimes h(n)$, then

a. $[f1(n) + f2(n)] \otimes h(n) = f1(n) \otimes h(n) + f2(n) \otimes h(n)$ (linear)

b. $f(n) \otimes h(n) = h(n) \otimes f(n)$ (commutative)

c. $[f1(n) \otimes f2(n)] \otimes h(n) = f1(n) \otimes [f2(n) \otimes h(n)]$ (associative)

R.5.24 The convolution of a sequence $f(n)$ with an impulse $\delta(n - k)$, where k is an integer, results in the sequence $f(n)$ shifted by the integer k. For example

a. $\delta(n) \otimes f(n) = f(n)$

b. $\delta(n - 1) \otimes f(n) = f(n - 1)$

c. $\delta(n - k) \otimes f(n) = f(n - k)$

R.5.25 A direct consequence of the convolution properties of LTI systems with an arbitrary input sequence $f(n)$ are summarized graphically in Figure 5.10.

R.5.26 Often, frequency domain analysis is important to explore the system behavior in the continuous time case, and it is equally important in the analysis and synthesis of discrete-time systems as well. Recall that in the continuous case, the FS and FT were used to represent a time function into representing the same function by the information contained or concentrated at the different frequencies (Chapter 4 of this book).

DTFT is introduced next to play a similar role for the case of a discrete-time sequence or system.

R.5.27 DTFT of $f(n)$ denoted by $F(e^{jW})$ is the Fourier representation of the discrete-time sequence given by the following equation:

$$F(e^{jW}) = \sum_{n=-\infty}^{+\infty} f(n)e^{-jWn}$$

The discrete sequence $f(n)$ can be reconstructed from the coefficients of $F(e^{jW})$ by using IDFT, given by

$$f(n) = \frac{1}{2\pi} \int_{-\pi}^{+\pi} F(e^{jW})e^{jWn}\, dW$$

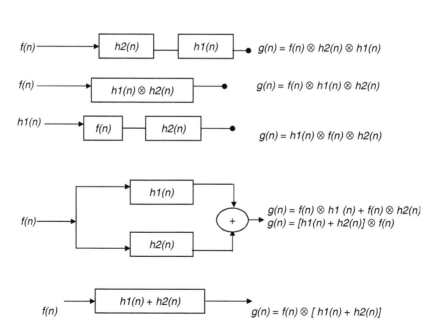

FIGURE 5.10
System examples of R.5.25.

R.5.28 Observe that DTFT closely resembles the analog FT presented in Chapter 4 of this book.

The syntaxes (notation) used to indicate the direct and inverse FT pairs are as follows:

$F(e^{jW}) = \mathcal{F}[f(n)]$ (denotes the discrete time Fourier transform [DTFT])

$f(n) = \mathcal{F}^{-1}[e^{jW}]$ (denotes the inverse discrete Fourier transform [IDFT])

$f(n) \leftrightarrow F(e^{jW})$ (denotes the transform equivalent time–frequency pair)

R.5.29 The DTFT $F(e^{jW})$ is a periodic function of the real variable W, with a period equal to 2π. This is the reason why the integral of the inverse transform is evaluated over the range $-\pi \le W \le +\pi$. Also observe that

$$F(e^{jW}) = \sum_{n=-\infty}^{+\infty} f(n)e^{-jnW}$$

and its expansion is given by

$$F(e^{jW}) = f(0)e^0 + f(1)e^{-jW} + f(2)e^{-j2W} + \cdots + f(n)e^{-jnW}$$

$$+ f(-1)e^{jW} + f(-2)e^{j2W} + \cdots + f(-n)e^{jnW}$$

consisting of a sum of exponentials of the form e^{-jnW}, for $n = 0, \pm1, \pm2, \ldots, \pm\infty$. Observe also that the function $e^{-j2\pi n}$ repeats in frequency of every 2π rad.

R.5.30 Recall that the FT of an analog signal is a function of w, with radian/second as unit, whereas W is used to denote discrete frequencies, with radian as unit.

Some authors differentiate the analog case from the discrete case by using the variable Ω; in this book, W (uppercase) is used to denote discrete frequencies.

R.5.31 Since $F[e^{jW}]$ is a complex function of the real variable W, then $F[e^{jW}]$ can be represented by two parts, real and imaginary in rectangular form, or with a magnitude and phase in polar form (similar to the continuous case) as follows:

$F[e^{jW}] = real\{F[e^{jW}]\} + jimag\{F[e^{jW}]\}$ (rectangular)

$F[e^{jW}] = |F[e^{jW}]|e^{-j\theta(n)}$ (polar)

R.5.32 Let $|F[e^{jW}]|$ represent the magnitude, whereas $\theta(W)$ the phase of DTFT of $f(n)$, then the plot $|F[e^{jW}]|$ versus W is referred to as the magnitude spectrum plot of $f(n)$, whereas $\theta(W)$ versus W is referred to as the phase spectrum plot of $f(n)$.

R.5.33 Recall that the relation between the polar and rectangular forms is

$$real\{F[e^{jW}]\} = F[e^{jW}]cos[\theta(W)]$$

$$imag\{F[e^{jW}]\} = F[e^{jW}]sin[\theta(W)]$$

where

$$|F(e^{jW})| = \sqrt{real^2[F(e^{jW})] + imag^2[F(e^{jW})]}$$

and

$$\tan[\theta(W)] = \frac{imag\{F[e^{jW}]\}}{real\{F[e^{jW}]\}}$$

R.5.34 It can be shown that if $f(n)$ is a real sequence, then $F(e^{jW})$ and $|F(e^{jW})|$ are real functions of W.

R.5.35 Since $\theta(W)$ is a periodic function with a period of 2π rad, then its principal value is defined over the domain $-\pi \le W \le +\pi$.

R.5.36 A sufficient condition for the existence and convergence of DTFT $F(e^{jW})$ is that the following relation must be satisfied:

$$\left|F(e^{jW})\right| = \sum_{n=-\infty}^{+\infty} \left|f(n)\right| < \infty$$

R.5.37 It can be shown that if $f(n)$ is a real sequence, then $F(e^{jW}) = real\{F(e^{jW})\}$ is an even function of W, whereas $imag\{F(e^{jW})\}$ is an odd function of W.

Recall that the spectrum equations indicate that the magnitude spectrum is even, whereas the phase spectrum is odd, indicated by

$$|F(e^{jW})| = |F(e^{-jW})|$$

and

$$\angle F(e^{jW}) = -\angle F(e^{-jW})$$

R.5.38 Let $f_1(n) \leftrightarrow F_1(e^{jW})$ and $f_2(n) \leftrightarrow F_2(e^{jW})$ be transform pairs. Then, some important properties that relate the time and frequency domains are summarized as follows:

a. Linearity (where a_1 and a_2 are arbitrary constants)

$$a_1 f_1(n) + a_2 f_2(n) \leftrightarrow a_1 F_1(e^{jW}) + a_2 F_2(e^{jW})$$

b. Time shifting

$$f(n - n_0) \leftrightarrow e^{-jWn_0} F(ejW)$$

c. Time reverse

$$f(-n) \leftrightarrow F(e^{-jW})$$

d. Differentiation in frequency

$$nf(n) \leftrightarrow -j\frac{d}{dw}\{F(e^{jw})\}$$

e. Frequency shifting (modulation theorem)

$$f(n)[e^{-jW_0 n}] \leftrightarrow F(e^{j(W-W_0)})$$

f. Convolution in time

$$f_1(n) \otimes f_2(n) \leftrightarrow F_1(e^{jW}) \cdot F_2(e^{jW})$$

g. Convolution in frequency

$$f_1(n) f_2(n) \leftrightarrow \frac{1}{2\pi} \cdot \{F_1(e^{jW}) \otimes F_2(e^{jW})\} = \frac{1}{2\pi} \int_{-\pi}^{+\pi} F_1(e^{j\theta}) F_2(e^{j(W-\theta)}) d\theta$$

h. Parseval's theorem is given by

$$\sum_{n=-\infty}^{+\infty} |f(n)|^2 = \frac{1}{2\pi} \int_{-\pi}^{+\pi} |F(e^{jW})^2| dW$$

i. The initial time value is given by

$$f(0) = \frac{1}{2\pi} \int_{-\pi}^{+\pi} F(e^{jW}) dW$$

j. The initial frequency value is given by

$$F(e^{j0}) = \sum_{n=-\infty}^{+\infty} f(n)$$

k. Subtraction of shifted sequences

$$f(n) - f(n-1) \leftrightarrow [1 - e^{jW}] \cdot F(e^{jW})$$

R.5.39 A partial list of the most frequently used transform pairs are summarized in Table 5.1.

TABLE 5.1

DTFT Pairs

Time	Frequency
$f(n)$	$F(e^{jW})$
$\delta(n)$	1
$\delta(n - n_0)$	$e^{-jn_0 W}$
$e^{-jw_0 n}$	$2\pi\delta(W - w_0)$
1	$2\pi\delta(W)$
$\cos(w_0 n)$	$\pi[\delta(W + w_0) + \delta(W - w_0)]$
$\sin(w_0 n)$	$j\pi[\delta(W + w_0) - \delta(W - w_0)]$
$u(n)$	$[1/1 - e^{-jw}] + \pi\delta(W)$
$a^n u(n)$	$1/(1 - ae^{-jW})$, $(|a| < 1)$
$(a + 1)a^n u(n)$	$1/(1 - e^{-jw})^2$, $(|a| < 1)$
$\dfrac{(n + k - 1)!}{n!(k - 1)!} a^n u(n)$	$\dfrac{1}{(1 - ae^{-jW})^k}$, $(|a| < 1)$

R.5.40 Observe that the FT (analog case) and DTFT (discrete case) present similarities, as well as differences. The contrasting characteristics are summarized as follows:

    a. The DTFT is periodic with period equal to 2π, whereas FT is nonperiodic.

    b. The frequency W of DTFT is given in radians, whereas the frequency w of the FT is given in radians/seconds.

    c. The magnitude spectrums in either case are symmetric.

    d. The phase spectrums in either case are asymmetric.

    e. The plots of DTFT are continuous and periodic functions in the frequency domain, whereas discrete in the time domain.

    FT (and FS) of a periodic continuous time function results in a discrete-frequency spectrum.

R.5.41 The process of computing DTFT by hand (or the Inverse Discrete Time Fourier Transform [IDTFT]) can be accomplished by using Table 5.1, a process that may involve some algebraic manipulations as well as applications of some properties, discussed in R.5.38. Let us illustrate the transform process by presenting the following example.

    Evaluate by hand the DTFT of the following discrete function:

$$f(n) = (0.7)^n u(n - 3)$$

ANALYTICAL Solution

    a. The transform of (from the Table 5.1)

$$(0.7)^n u(n) \leftrightarrow \frac{1}{1 - 0.7e^{-jW}}$$

    b. The shift in time of the step function by three time units returns the following in frequency (from the time-shift property of R.5.38):

$$\frac{1}{1 - 0.7e^{-jW}} e^{-jW3}$$

    c. The coefficient of $f(n)$ must also be shifted by $(0.7)^3$, resulting in the following transform:

$$(0.7)^n u(n - 3) \leftrightarrow \frac{(0.7)^3 e^{-jW3}}{1 - 0.7e^{-jW}}$$

R.5.42 Let the discrete-system transfer function be $H(e^{jW})$, given by

$$H(e^{jW}) = \frac{G(e^{jW})}{F(e^{jW})} = \frac{g_0 + g_1 e^{-jW} + g_2 e^{-j2W} + \cdots + g_n e^{-jnW}}{f_0 + f_1 e^{-jW} + f_2 e^{-j2W} + \cdots + f_m e^{-jmW}}$$

    Recall that the MATLAB signal processing toolbox includes the function *freqz*, which can be used to evaluate DTFT of $H(e^{jW})$. Some of the most common forms of the *freqz* function are listed as follows (they were introduced in Chapter 7 of *Practical MATLAB® Basics for Engineers*).

    a. *[H, W] = freqz(G, F)*, where G and F are row vectors defined by their coefficients (g and f) arranged in descending power of e^{jnW} and e^{jmW}

    b. *H = freqz(G, F)*

c. $[H, W] = freqz(G, F, W)$, over the range $-\pi \leq W \leq \pi$

d. $[H, W] = freqz(G, F, W, Fs/2)$, where $0 \leq W \leq Fs/2$ and Fs denotes the sampling frequency

e. $[H, W] = freqz(G, F, K)$, where K is the number of equally spaced points over the domain $0 \leq W \leq \pi$

Observe that if $H(e^{jW})$ represents the transfer function of a given system, then $G(e^{jW})$ and $F(e^{jW})$ represents the DTFT of its output and input, respectively.

R.5.43 Recall that DTFT was defined by

$$F(e^{jW}) = \sum_{n=-\infty}^{+\infty} f(n)e^{-jWn}$$

And replacing e^{-jw} by the (new) variable z, DTFT is transformed into what is referred to as the two-sided or bilateral ZT. The ZT of a discrete sequence $f(n)$ is then defined by

$$F(z) = \sum_{n=-\infty}^{+\infty} f(n)z^{-n}$$

R.5.44 The ZT is the preferred tool in the analysis and synthesis of discrete-time systems, defined by a linear constant coefficient difference equation. The ZT is for a discrete system what LT is for a continuous system.

The drawback of DTFT is that the convergence and existence conditions may not exist for many sequences, and therefore no frequency representation is possible for these systems. The ZT, however, may exist for many sequences for which DTFT does not exist.

R.5.45 Note that the equation that defines the ZT is in the form of an infinite power series of the complex variable z, and its coefficients are the discrete samples of $f(n)$.

R.5.46 Observe that the resulting series expansion of ZT is a Laurent series*.

The notation used to denote the ZT of $f(n)$ is as follows:

$$Z[f(n)] = F(z) = \sum_{n=-\infty}^{+\infty} f(n)z^{-n}$$

where $Z[f(n)]$ denotes the ZT of $f(n)$.

R.5.47 Since z is a complex variable, z can be represented as a point on the complex plane and expressed in rectangular and polar forms as

$$z = real(z) + jimag(z) \quad \text{(rectangular form)}$$

$$z = |z|e^{jW} = re^{jW} \quad \text{(exponential form)}$$

where $|z| = r = 1$.

Observe that the complex variable $z = e^{jW}$ maps into a unit radius circle centered at the origin of the complex plane.

---

* Since most transforms are named after a scientist such as Laplace, Fourier, or Hilbert, it would be logical to call the ZT the Laurent transform.

R.5.48 The power of ZT is that any difference equation can be easily transformed into an algebraic expression, which can then be easily solved by algebraic manipulations. This transformation can be accomplished based on two important properties, which can be easily learned and applied. They are

a. Linearity

b. Time shift

It is obvious that ZT is linear, since DTFT is also linear. Then, based on the linearity principles

$$Z[a_1f_1(n) + a_2f_2(n)] = Z[a_1f_1(n)] + Z[a_2f_2(n)] = a_1F_1(z) + a_2F_2(z)$$

and the time-shift property yields

$$Z[f(n - n_0)] = z^{-n_0}F(z) \quad \text{for any integer } n_0$$

R.5. 49 Let us consider the case where the input to a discrete system is $f(n) = r(n) = r^n$ (a ramp), and its system impulse response is $h(n)$. Then the output $g(n)$ can be evaluated by $g(n) = f(n) \otimes h(n)$ (convolution of $f(n)$ with $h(n)$ in time), analyzed as follows:

$$g(n) = \sum_{k=-\infty}^{+\infty} r^k h(n - k)$$

or

$$g(n) = \sum_{k=-\infty}^{+\infty} r^{n-k} h(k)$$

and

$$g(n) = r^n \sum_{k=-\infty}^{+\infty} h(k) r^{-k}$$

where clearly

$$\sum_{k=\infty}^{+\infty} r^k h(k) = H(z)$$

Then $g(n) = r^n H(z)$. Its block diagram is illustrated in Figure 5.11.

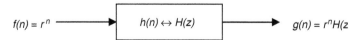

$$f(n) = r^n \longrightarrow \boxed{h(n) \leftrightarrow H(z)} \longrightarrow g(n) = r^n H(z$$

FIGURE 5.11
System block diagram of R.5.49.

R.5.50 $H(z)$ is called the discrete-system function or discrete-system transfer function. Recall that in the continuous case, the notation used for the transfer function was $H(w)$ or $H(s)$. Note that the discrete case is similar and given by

$$H(z) = \frac{G(z)}{F(z)} = \frac{\text{ZT of the output sequence}}{\text{ZT of its input sequence}}$$

$$H(z) = \frac{\sum\limits_{n=-\infty}^{+\infty} b_k z^{-n}}{\sum\limits_{n=-\infty}^{+\infty} a_k z^{-m}}$$

where a and b are the polynomial coefficients of the input and output sequences, respectively, arranged in descending powers of z, assuming that the initial conditions are set to zero.

For example, let the difference equation of a causal system be given by

$$g(n) - 7g(n-1) + 3g(n-2) = f(n) + 3f(n-2) - 5f(n-3)$$

Evaluate by hand, the expression of the system transfer function $H(z)$.

ANALYTICAL Solution

$$G(z) - 7z^{-1}G(z) + 3z^{-2}G(z) = F(z) + 3z^{-2}F(z) - 5z^{-3}F(z)$$

$$G(z)[1 - 7z^{-1} + 3z^{-2}] = F(z)[1 + 3z^{-2} - 5z^{-3}]$$

$$H(z) = \frac{G(z)}{F(z)} = \frac{1 + 3z^{-2} - 5z^{-3}}{1 - 7z^{-1} + 3z^{-2}}$$

R.5.51 Since the ZT was obtained by a simple variable substitution in the DTFT, it is logical to think that the properties of ZT and DTFT are similar. For completeness, the ZT properties are summarized as follows:

a. Linearity

$$a_1 f_1(n) + a_2 f_2(n) \leftrightarrow a_1 F_1(z) + a_2 F_2(z)$$

b. Time shifting

$$f(n - n_0) \leftrightarrow z^{-n_0} F(z)$$

c. Time reversal

$$f(-n) \leftrightarrow F(z^{-1})$$

d. Time multiplication by a power sequence

$$a^n f(n) \leftrightarrow F(z/a)$$

e. Conjugate of a sequence

$$f*(n) \leftrightarrow F*(z) \quad \text{(where } * \text{ denotes conjugate)}$$

f. Differentiation in frequency

$$nf(n) \leftrightarrow -z \frac{d}{dz}\{F(z)\}$$

g. Frequency shifting (modulation theorem)

$$f(n)[e^{-jwo\,n}] \leftrightarrow F(e^{jwo}z)$$

h. Convolution in time

$$f_1(n) \otimes f_2(n) \leftrightarrow F_1(z)\,F_2(z)$$

i. Convolution in frequency

$$f(n)\,f_2(n) \leftrightarrow \frac{1}{2\pi j} \cdot [F_1(z) \otimes F_2(z)]$$

R.5.52 Some useful observations about the ZT are summarized as follows:
   a. Let $f(n) \leftrightarrow F(z)$, then

$$f(n-1) \leftrightarrow z^{-1}F(z)$$
$$f(n-2) \leftrightarrow z^{-2}F(z)$$
$$f(n-b) \leftrightarrow z^{-b}F(z) \quad \text{for any integer } b$$
$$f(n+a) \leftrightarrow z^{a}F(z) \quad \text{for any integer } a$$

Note that delaying a sequence $f(n)$ by b time units in the time domain is equivalent to multiplying its transform $F(z)$ by z^{-b}.
   b. Since $Z[\delta(n)] = 1$, then by definition of the ZT

$$Z[\delta(n)] = \sum_{n=-\infty}^{+\infty} \delta(n)z^{-n}$$

and recall that

$$\delta(n) = \begin{cases} 1 & n = 0 \\ 0 & n \neq 0 \end{cases}$$

then $Z[\delta(n)] = 1$, since $z^0 = 1$.
   c. Recall (from Chapter 1 of this book) that any arbitrary sequence $f(n)$ can be expressed as

$$f(n) = \sum_{k=\infty}^{+\infty} f(k)\delta(n-k)$$

and since the transform of an impulse is known, as well as a shifted impulse, then ZT of any arbitrary sequence $f(n)$ can be easily evaluated.

R.5.53 For example, let an arbitrary discrete-time sequence be given by

$$f(n) = 3\delta(n) + 5\delta(n - 2) - 6\delta(n - 5) + 2\delta(n + 3)$$

Then its ZT[Z{f(n)}] can easily be evaluated by applying the relations presented in R.5.52 yielding the following analytical solution:

ANALYTICAL Solution

$$Z[f(n)] = Z[3\delta(n)] + Z[5\delta(n - 2)] - Z[6\delta(n - 5)] + Z[2\delta(n + 3)]$$

$$Z[f(n)] = 3Z[\delta((n)] + 5Z[\delta(n - 2)] - 6Z[\delta(n - 5)] + 2Z[\delta(n + 3)]$$

$$Z[f(n)] = 3[1] + 5z^{-2} - 6z^{-5} + 2z^{3}$$

$$Z[f(n)] = 2z^{3} + 3 + 5z^{-2} - 6z^{-5}$$

R.5.54 Recall that it is possible in many cases to express the ZT of an infinitely long sequence in a close compact form by using the following relation:

a. $\sum_{n=0}^{+\infty} a^{n} = 1/1 - a \quad$ for $|a| < 1$

b. $\sum_{n=0}^{N-1} a^{n} = 1 - a^{N}/1 - a \quad$ for any a

c. $\sum_{n=0}^{+\infty} na^{n} = a/(1 - a)^{2} \quad$ for $|a| < 1$

R.5.55 By using the properties given in R.5.51 and R.5.52, any LTI system difference equation that relates the input sequence to its output sequence can easily be transformed from the time domain to the z (frequency) domain.

For example, let us analyze the following system difference equation:

$$g(n) - \left(\frac{1}{2}\right)g(n - 1) = f(n)$$

where $f(n)$ and $g(n)$ represent the input and output sequences of a given system. A compact representation of the system transfer function $H(z) = G(z)/F(z)$ can be easily obtained and the condition for convergence can then be stated.

ANALYTICAL Solution

$$G(z) - \frac{1}{2}G(z)z^{-1} = F(z)$$

$$G(z)\left[1 - \frac{1}{2}z^{-1}\right] = F(z)$$

and

$$H(z) = \frac{G(z)}{F(z)} = \frac{1}{1 - (0.5)z^{-1}}$$

convergence occurs for $|0.5z^{-1}| < 1$ or $|z| > |0.5|$.

Recall that if

$$H(z) = \frac{G(z)}{F(z)} = \frac{A}{1 - (a)z^{-1}} = \frac{Az}{z - a}$$

then the ROC is given by $|z| > |a|$, where a may be real or complex. The ROC is revisited later in this chapter.

R.5.56 Simple-system structures consisting of two cascaded blocks are analyzed in the time domain and z-domain (frequency), in Figures 5.12 and 5.13, respectively.

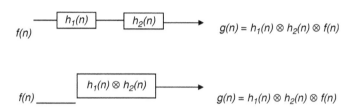

FIGURE 5.12
System-time block diagram of R.5.56.

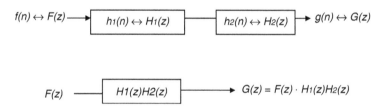

FIGURE 5.13
System block diagram in the z-domain of R.5.56.

ANALYTICAL Solution

R.5.57 Let us revisit the simple system shown in Figure 5.14 and review the relations between the time and z-domains.

$$Z[f(n)] = F(z) = \sum_{n=-\infty}^{+\infty} f(n)z^{-n}$$

$$H(z) = \sum_{n=-\infty}^{+\infty} h(n)z^{-n}$$

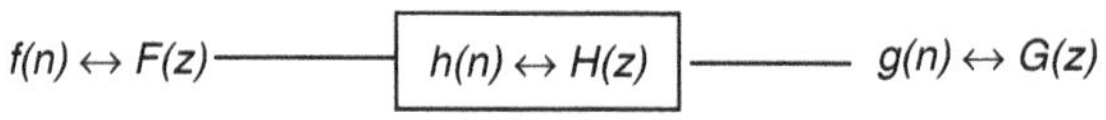

$$f(n) \leftrightarrow F(z) \boxed{\quad h(n) \leftrightarrow H(z) \quad} g(n) \leftrightarrow G(z)$$

FIGURE 5.14
System diagram of R.5.57.

$$G(z) = \sum_{n=-\infty}^{+\infty} g(n)z^{-n}$$

$$G(z) = F(z) \cdot H(z)$$

$$H(z) = \frac{G(z)}{F(z)}$$

and recall that

$$g(n) = \sum_{k=0}^{+\infty} f(k)h(n-k)$$

R.5.58 Since the ZT of $f(n)$ is given by

$$F(z) = \sum_{n=-\infty}^{+\infty} f(n)z^{-n}$$

then ROC defines the set of values or regions of z (in the complex plane) that makes the summation $\left(\sum_{n=-\infty}^{+\infty} f(n)z^{-n}\right)$ converge (finite).

The ROC is denoted by R and is given by a ring defined by $z1 < |z| < z2$, whose inner and outer radii are $r1$ and $r2$, respectively, which define the behavior of $f(n)$ as n approaches plus or minus infinity.

R.5.59 For example, let us revisit the causal exponential function $f_1(n)$ given by

$$f_1(n) = \sum_{n=-\infty}^{+\infty} u(n)r_1^n$$

Then ZT is given by

$$Z[f_1(n)] = F_1(z) = \sum_{n=-\infty}^{+\infty} r_1^n u(n)z^{-n} = \sum_{n=0}^{+\infty} r_1^n z^{-n}$$

$$Z[f_1(n)] = F_1(z) = \sum_{n=0}^{+\infty} (r_1 z^{-1})^n$$

Recall that the preceding series converges if it is of the form

$$\sum_{n=0}^{+\infty} a^n = \frac{1}{1-a} \quad \text{for } |a| < 1$$

Then

$$F_1(z) = \frac{1}{1 - r_1 z^{-1}}$$

and

$$\left| r_1 z^{-1} \right| < 1$$

or

$$|z| > r_1$$

Then the convergence conditions exist for the values of z that are located outside the circle of radius r_1, centered at the origin of the complex plane.

R.5.60 The region given in R.5.59 is the ROC and must be specified as part of ZT, defining in this way a one-to-one-relation between $f(n)$ and $F(z)$.

R.5.61 For example, let us explore the ROC of the noncausal exponential function $f_2(n)$, given by

$$f_2(n) = \sum_{n=-\infty}^{+\infty} r_2^n(-n-1)$$

ANALYTICAL Solution

$$Z[f_2(n)] = F_2(z) = \sum_{n=-\infty}^{+\infty} r_2^n u(-n-1)z^{-n}$$

$$F_2(z) = \sum_{n=-\infty}^{-1} r_2^n z^{-n} = \sum_{n=-\infty}^{-1} (r_2 z^{-1})^n$$

$$F_2(z) = \sum_{n=1}^{+\infty} (r_2^{-1} z)^n = \sum_{n=1}^{+\infty} \left(\frac{z}{r_2} \right)^n = -1 + \sum_{n-0}^{+\infty} \left(\frac{z}{r_2} \right)^n$$

$$F_2(z) = -1 + \frac{1}{1 - (z/r_2)} = -1 + \frac{r_2}{r_2 - z} = \frac{z}{z - r_2}$$

or

$$r_2^n u(-n-1) \Leftrightarrow \frac{z}{z - r_2}$$

with the ROC given by $|z| < r_2$.

R.5.62 Let us evaluate the ZT and the ROC of the discrete-step function $f_3(n) = u(n)$. Recall that

$$f_3(n) = u(n) = \begin{cases} 1 & n \geq 0 \\ 0 & n < 0 \end{cases}$$

ANALYTICAL Solution

$$Z[f_3(n)] = F_3(z) = \sum_{n=-\infty}^{+\infty} u(n)z^{-n} = \sum_{n=0}^{+\infty} z^{-n} = \frac{1}{1-z^{-1}}$$

then

$$f_3(n) = u(n) \Leftrightarrow \frac{z}{z-1} \quad \text{for } |z^{-1}| < 1 \text{ or } |z| > 1$$

R.5.63 To gain some insight and experience in the evaluation of the ZT, additional transformation examples of various functions and their ROC are presented as follows. For example, evaluate ZT and its ROC of the discrete function $f_4(n)$ given by

$$f_4(n) = -u(-n-1) = \begin{cases} -1 & n < 0 \\ 0 & n \geq 0 \end{cases}$$

ANALYTICAL Solution

$$F_4(z) = \sum_{n=-\infty}^{+\infty} -u(-n-1)z^{-n}$$

$$F_4(z) = \sum_{n=-\infty}^{-1} (-1)z^{-n} = -\sum_{n=1}^{+\infty} z^n$$

$$F_4(z) = -\frac{1}{1-z} = \frac{z}{z-1} \quad \text{for } |z| < 1$$

R.5.64 Let us evaluate the ZT and ROC of the discrete function $f_5(n)$ given by

$$f_5(n) = 2^n u(n) - 3^n u(-n-1)$$

ANALYTICAL Solution

$$Z[f_5(n)] = Z[2^n u(n)] - Z[3^n u(-n-1)]$$

$$Z[2^n u(n)] = \frac{z}{z-2} \quad \text{for } |z| > |2|$$

and

$$Z[3^n u(-n-1)] = \frac{z}{z-3} \quad \text{for } |z| < |3|$$

Then ROC is given by $|2| < |z| < |3|$.

R.5.65 The ROC for a finite length sequence is the entire z-plane. Only two points in the z-plane may be excluded from the ROC. They are $z = 0$ and $z = \infty$.

R.5.66 Let us turn our attention now and explore the effects on the z-domain of the process of differentiation in the time domain of the causal sequence given by $f_6(n) = a^n u(n)$.

ANALYTICAL Solution

Since

$$f_6(n) = a^n u(n) \Leftrightarrow \frac{z}{z-a} \quad \text{for } |z| > a$$

$$\frac{d}{da}[f_6(n)] \Leftrightarrow \frac{d}{da}\left[\frac{z}{z-a}\right]$$

Then

$$n a^{n-1} u(n) \Leftrightarrow \frac{z}{(z-a)^2} \quad \text{for } |z| > a$$

In general

$$\frac{d^n}{da^n}[f(n)] \Leftrightarrow \frac{d^n}{da^n}[F(z)]$$

or

$$\binom{n}{m} a^{n-m} u(n) \Leftrightarrow \frac{z}{(z-a)^{m+1}} \quad \text{for } |z| > a$$

R.5.67 The result obtained for the causal sequence analyzed in R.5.66 can be used to evaluate ZT of the noncausal sequence $f_7(n) = a^n u(-n-1)$.
   The resulting time–frequency relation is given by

$$\binom{n}{m} a^{n-m} u(-n-1) \Leftrightarrow \frac{-z}{(z-a)^{m+1}} \quad \text{for } |z| < a$$

R.5.68 Recall that the zero and pole plot of the system transfer function $H(z) = G(z)/F(z)$ is referred to as the pole/zero plot. Recall that the values of z that make $H(z)$ go to zero or infinity are the system zeros and poles, respectively.
   It is customary to indicate the zeros by the character "o," whereas the poles are indicated by the character "x," when represented by a plot on the complex plane.
   The location of the poles define the ROC of $H(z)$ in the following way:

a. The ROC never includes a pole since a pole makes the denominator (or $F(z)$) of $H(z)$ go to zero, therefore $H(z)$ becomes infinity.

b. The ROC may be either the exterior of a circle, a bounded ring defined by two circles, or the interior of a circle, where poles are allowed only on its boundaries.

TABLE 5.2

Table of ZT Pairs

Discrete Time Signals	Bidirectional Relation	ZT
$f[n]$	⟺	$F(z)$
$\delta[n]$	⟺	1
$\delta[n - n_0]$	⟺	z^{-n_0} for $z \neq 0$ and $z \neq \infty$
$u[n]$	⟺	$1/(1 - z^{-1})$ for $|z| > 1$
$\alpha^n u(n)$	⟺	$1/(1 - \alpha z^{-1})$ for $|z| > |\alpha|$
$-\alpha^n u[-n - 1]$	⟺	$1/1 - \alpha z^{-1}$ for $|z| < |\alpha|$
$n\alpha^n u[n]$	⟺	$\alpha z^{-1}/(1 - \alpha z^{-1})^2$ for $|z| > |\alpha|$
$-n\alpha^n u[-n - 1]$	⟺	$\alpha z^{-1}/(1 - \alpha z^{-1})^2$ for $|z| < |\alpha|$
$\alpha^n \cos(\omega_0 n) u[n]$	⟺	$\dfrac{1 - \alpha^n \cos(\omega_0 n) z^{-1}}{\{1 - [2\alpha \cos(\omega_0)] z^{-1} + \alpha^2 z^{-2}\}}$ $|z| > |\alpha|$
$\alpha^n \sin(\omega_0 n) u[n]$	⟺	$\dfrac{\alpha^n \sin(\omega_0) z^{-1}}{\{1 - [2\alpha^n \cos \omega_0] z^{-1} + \alpha^2 z^{-2}\}}$ $|z| > |\alpha|$

c. For the causal signal $f(n)$, if $f(n) = 0$ for $n < 0$, then the exterior of the circle defined by the pole with the largest magnitude constitutes the ROC.

d. For the noncausal signal $f(n)$, if $f(n) \neq 0$ for $n < 0$, the ROC is inside the circle defined by the pole with the smallest magnitude.

R.5.69 The ROC of a stable, LTI system includes the unit circle since that fact would imply that the impulse response is absolutely summable.

R.5.70 A summary of the most-often used ZT pairs is given in Table 5.2.

R.5.71 The evaluation of the inverse ZT is given by the following equation:

$$f(n) = \oint_c F(z) z^{n-1} dz$$

where the symbol $\oint_c$ denotes an integration over the close contour c that includes all the poles of $F(z)$.

The contour integral can be evaluated by using the Cauchy's residue theorem, resulting in

$$f(n) = \sum \text{ [residues of } F(z) z^{n-1} \text{ at the poles inside } c]$$

R.5.72 Observe that the inverse ZT as defined in R.5.71 involves a difficult computational process, and thereby it is not practical and therefore seldom used. Simpler methods based on conversion transform tables, including partial-fraction expansion, and long division are the preferred methods.

Of course, it is far simpler to employ the MATLAB symbolic toolbox to evaluate the ZT and the inverse ZT of arbitrary functions, as illustrated in the next sections.

Recall that the system impulse response $h(n)$ of an LTI system is critical to evaluate the response of any arbitrary sequence. Note that the system impulse response can be evaluated by taking the inverse ZT of the system transfer function $H(z)$.

R.5.73 For example, evaluate $f_1(n)$ and $f_2(n)$ by taking the inverse ZT of $F_1(z)$ and $F_2(z)$ given by

a. $F_1(z) = 4z + 3z^{-3} - 12z^{-8}$, using Table 5.2

b. $F_2(z) = z^3/z - 1$, first using long division, then Table 5.2

ANALYTICAL Solutions

a. Using Table 5.2

$$f_1(n) = Z^{-1}[F_1(z)] = Z^{-1}[4z + 3z^{-3} - 12z^{-8}] = 4\delta(n+1) + 3\delta(n-3) - 12\delta(n-8)$$

b. Using long division

$$f_2(n) = Z^{-1}[F_2(z)] = Z^{-1}\left[\frac{z^3}{z-1}\right] = Z^{-1}\left[z^2 + z + \frac{z}{z-1}\right]$$

Then from Table 5.2 $f_2(n) = Z^{-1}[F_2(z)] = \delta(n+2) + \delta(n+1) - u(n)$

R.5.74 As an additional example, let $F(z)$ be given. Determine then $f(n)$ and its ROC for

$$F(z) = \frac{z+2}{2z^2 - 7z + 3}$$

ANALYTICAL Solutions

By partial fraction expansion

$$F(z) = \frac{2}{3} - \frac{z}{z-(1/2)} + \left(\frac{1}{3}\right)\frac{z}{z-3}, \text{ the resulting poles are } z_1 = 1/2 \text{ and } z_2 = 3$$

Then the following three ROC are defined:

a. Region number 1, $|z| > 3$, outside the two poles then

$$f(n) = (2/3)\delta(n) - (1/2)^n u(n) + (1/3)3^n u(n)$$

b. Region number 2, $(1/2) < |z| < 3$, bounded by the two poles, then

$$f(n) = (2/3)\delta(n) - (1/2)^n u(n) - (1/3)3^n u(-n-1)$$

c. Region number 3, $|z| < (1/2)$, region inside the smallest pole, then

$$f(n) = (2/3)\delta(n) + (1/2)^n u(-n-1) - (1/3)3^n u(-n-1)$$

R.5.75 The MATLAB symbolic toolbox command *ztrans(f)* returns the ZT of the scalar symbolic object f with the default independent variable n.

R.5.76 For example, let

$$f(n) = \sum_{n=0}^{+\infty} (1/3)^n = 1 + 1/3 + 1/9 + 1/27 + \cdots + 1/3^n = (1/3)^n u(n)$$

Evaluate the *Z[f(n)]* in the following ways:

a. Analytically

b. Symbolically by using MATLAB

ANALYTICAL Solution

$$F(z) = Z[f(n)] = \sum_{n=0}^{+\infty} f(n)z^{-n} = 1 + \left(\frac{1}{3}\right)z^{-1} + \left(\frac{1}{9}\right)z^{-2} + \left(\frac{1}{27}\right)z^{-3} + \cdots + \frac{1}{3}^{n} z^{-n}$$

$$F(z) = \frac{z}{1 - (1/3)z^{-1}} = \frac{z}{z - (1/3)} \quad \text{for} \quad |z| > \frac{1}{3}$$

MATLAB Solution
```
>> syms z n
>> Fz = ztrans(1/3^n)

    Fz =
         3*z/(3*z-1)

>> pretty(Fz)
              z
      3  -------
          3 z - 1
```

R.5.77 Let us gain some experience by evaluating the ZT of the following functions by using MATLAB:

a. $f_1(n) = u(n)$

b. $f_2(n) = a^n u(n)$

c. $f_3(n) = cos(an)u(n)$

d. $f_4(n) = sin(an)u(n)$

e. $f_5(n) = b^n sin(an)u(n)$

MATLAB Solution
```
>> syms a b n
>> f1 _ n = 1^n;                         % part(a)
>> F1 _ z = ztrans(f1 _ n)

    F1 _ z =
             z/(z-1)

>> pretty(F1 _ z)

              z
             -----
             z - 1

>> f2 _ n = a^n;                         % part(b)
>> F2 _ z =ztrans(f2 _ n)
```

```
        F2 _ z =
                z/a/(z/a-1)


>> pretty(F2 _ z)


                        z
                     ----------
                     a (z/a - 1)


>> f3 _ n =1^n*cos(a*n);                              % part (c)
>> F3 _ z = ztrans(f3 _ n)


        F3 _ z =
                (z-cos(a))*z/(z^2-2*z*cos(a)+1)


>> pretty(F3 _ z)


                      (z - cos(a)) z
                     -------------------
                        2
                     z - 2 z cos(a) + 1


>> f4 _ n =1^n*sin(a*n);                              % part(d)
>> F4 _ z = ztrans(f4 _ n)


        F4 _ z =
                z*sin(a)/(z^2-2*z*cos(a)+1)


>> pretty(F4 _ z)


                        z sin(a)
                     -------------------
                        2
                     z - 2 z cos(a) + 1


>> f5 _ n = b^n*sin(a*n);                             % part(e)
>> F5 _ z =ztrans(f5 _ n)


        F5 _ z =
                z/b*sin(a)/(z^2/b^2-2*z/b*cos(a)+1)


>> pretty(F5 _ z)


                           z sin(a)
                     ----------------------------
                       /   2                 \
                       |  z        z cos(a)  |
                     b |---- - 2 -------- + 1 |
                       |   2         b        |
                       \  b                  /
```

R.5.78 Write the script file *sym_z_plots* that returns the plots of *F1_z, F2_z, F3_z,* and *F4_z* versus *z,* defined in R.5.77 for *a* = 0.5

MATLAB Solution

```
%script file: sym_z_plots
syms a b n
f1_n = 1^n;
F1_z = ztrans(f1_n)
f2_n = a^n;
F2_z = ztrans(f2_n)
f3_n = 1^n*cos(a*n);
F3_z = ztrans(f3_n)
f4_n = 1^n*sin(a*n);
F4_z = ztrans(f4_n)

figure(1)
subplot(2,2,1)
ezplot(F1_z)
subplot(2,2,2);
F2_z=subs(F2_z,a,0.5);
ezplot(F2_z);
subplot(2,2,3);
F3_z=subs(F3_z,a,0.5);
ezplot(F3_z);
subplot(2,2,4);
F4_z = subs(F4_z,a,0.5);
ezplot(F4_z);
>> sym_z_plot
```

The script file *sym_z_plots* and the results are shown in Figure 5.15.

See Figure 5.15.

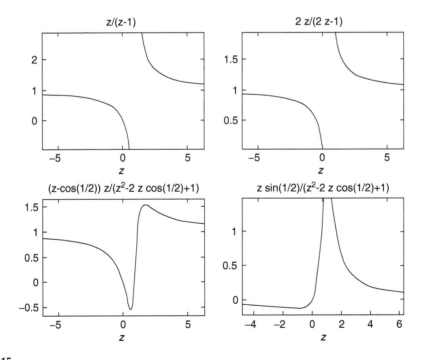

FIGURE 5.15
Plots of the ZTs of R.5.78.

R.5.79 The MATLAB symbolic toolbox command *iztrans(Fz)* returns the inverse ZT of the symbolic object *Fz*, as a function of *n*. For example, let

$$Fz = \frac{z}{z - a} \quad \text{for } |z| > a$$

Use MATLAB to obtain an expression for *f(n)*.

ANALYTICAL Solution

$$f(n) = Z^{-1}[F(z)] = a^n u(n)$$

MATLAB Solution
```
>> syms a z
>> Fz = z/(z-a);
>> fn = iztrans(Fz)

        fn =
              a^n
```

R.5.80 Let us gain some additional experience by evaluating the inverse ZT of the following functions by using MATLAB:

a. $F_1(z) = z/z - 2$

b. $F_2(z) = 2z/(z - 2)^2$

MATLAB Solution
```
>> syms z n
>> F1z = z/(z-2);
>> f1n = iztrans(F1z)

        f1n =
              2^n

>> F2z = 2*z/(z-2)^2;
>> f2n = iztrans(F2z)

        f2n =
              2^n*n
```

R.5.81 The implicit assumption in the preceding discussion is that the system transfer function *H(z)* is a proper rational function of the independent complex variable *z*.

R.5.82 Recall that a partial fraction expansion of *H(z)* can be performed only if *H(z)* is a proper rational function. If *H(z)* is not a proper rational function, then *H(z)* consists of a finite length sequence that can be extracted from *H(z)* by synthetic division (dividing the polynomials of the numerator by the polynomial of the denominator of *H(z)*), plus a rational function $H_1(z)$ as illustrated as follows:

Let $H(z) = z^k + H_1(z)$, where *k* is an integer. Then the remainder $H_1(z)$ becomes a rational function and can be decomposed into a partial-fraction expansion.

R.5.83 Recall that a number of useful MATLAB discrete functions that relate the transfer function *H(z)* to its poles and zeros were presented in Chapter 7 of *Practical MATLAB® Basics for Engineers*. For completeness, and as a quick review, some commands are revisited as follows:

Let $H(z) = G(z)/F(z)$ then

a. The MATLAB command [*zeros, poles, k*] = *tf2zp(G, F)* returns the zeros, poles, and gain *k* of a given transfer function given by *G* and *F*, two row vectors consisting of coefficients of the polynomials of *G(z)* and *F(z)* arranged in descending powers of *z*, where *H(z)* = *G(z)/F(z)*.

b. The MATLAB command [*G, F*] = *zp2tf(zeros, poles, k)* returns the coefficients of the polynomials representing the numerator *(G)* and denominator *(F)* of *H(z)* arranged in descending powers of *z*, given the zeros, poles, and gain *k* of *H(z)*.

R.5.84 Recall that the plot of the *poles* and *zeros* are shown on the *z*-plane (complex plane), including the unit circle drawn automatically as reference, by using the following MATLAB commands:

a. *zplane(zeros, poles)*, where the zeros and poles are given as column vectors

b. *zplane(G, F)*, where *G* and *F* are row vectors consisting of the coefficients of the polynomials of *G(z)* and *F(z)* arranged in descending powers of *z* representing the numerator (output) and denominator (input) of *H(z)*, respectively

R.5.85 Recall that the command [*res, poles, k*] = *residuez(G, F)* returns the partial fraction expansion coefficients of *H(z)*, where *G* and *F* are row vectors consisting of the coefficients of the polynomials of *G(z)* and *F(z)* arranged in descending powers of *z*, representing the numerator and denominator of *H(z)*, respectively.

R.5.86 Recall that the command [*G, F*] = *residuez(res, poles, k)* returns *G* and *F*, in the form of row vectors consisting of the coefficients of the polynomials of *G(z)* and *F(z)* arranged in descending powers of *z*, representing the numerator and denominator, respectively, of *H(z)*; given as column vectors *res, poles,* and *k*, referred in R.5.85.

R.5.87 Recall that the output of a discrete system can be evaluated by using the following two MATLAB commands:

a. [*h, n*] = *impz(G, F, leng)* returns the discrete system impulse response *h*, given the vectors *G* and *F*, consisting of the coefficients of the polynomials of *G(z)* and *F(z)* arranged in descending powers of *z* (representing the numerator and denominator of *H(z)*, respectively), and *leng* is an optional parameter that defines the desired length of *h*.

Recall that the impulse response is given by $Z^{-1}[H(z)]$.

b. *gn* = *filter(G, F, fn)* returns the discrete system output *gn*, where *G* and *F* are the coefficients of the polynomials *G(z)* and *F(z)* of *H(z)*, respectively, arranged in descending powers of *z*.

R.5.88 DTFT as well as ZT deals with arbitrary length sequences, including infinite length sequences. DFT is a computational transform when spectral representation is desirable, dealing with a finite length sequence. There is a simpler computational relation between the time samples and its corresponding frequencies in the DFT.

For a finite sequence of length *N*, the *DTFT{F(e^{jW})}* of a discrete-time sequence *f(n)* returns a set of *N* frequencies, which may be sufficient to approximate the spectrum of *f(n)*.

R.5.89 Let *f(n)* be an *N*-length finite complex value sequence with nonzero values over the range $0 \le n \le N - 1$, then the DFT, denoted by *F(k)*, is the *N*-point sequence defined by

$$F(k) = \sum_{n=0}^{N-1} f(n)e^{-j(2\pi/N)nk} \quad \text{for } k = 0, 1, 2, 3, \dots, N - 1$$

Let

$$w_N = e^{-j(2\pi/N)}$$

then

$$F(k) = \sum_{n=0}^{N-1} f(n)\, W_N^{nk} \quad \text{for } k = 0, 1, 2, 3, \ldots, N-1$$

R.5.90 The IDFT is given by

$$f(n) = \left(\frac{1}{N}\right) \sum_{n=0}^{N-1} F(k)\, W_N^{-nk} \quad \text{for } k = 0, 1, 2, 3, \ldots, N-1$$

Observe that because of the exponential term $e^{-j2\pi nk/N}$, $f(n)$ becomes periodic outside the interval $0 \le n \le N-1$.

R.5.91 Observe that DFT is related to DTFT and ZT by making the following substitutions:

$F(k) = F(e^{-jW})$, where $W = 2\pi k/N$ (DTFT)

$F(k) = F(z)$, where $z = e^{-j(2\pi k/N)}$ (ZT)

R.5.92 DFT can be viewed as a finite sum representation of discrete sinusoidal of the periodic sequence $f(n)$, with a period or length N.

R.5.93 Observe that DFT can be viewed as a simple and practical way to approximate DTFT by a finite sequence N.

R.5.94 The MATLAB command *fft(fn, M)** returns the DFT of *fn* consisting of *M* elements, where *fn* is given by a sequence of *N* elements, where $M > N$. In its simplest form, *fft(fn)* returns the *N* points of DFT of *fn*. The magnitude and phase plots of *fn* can be obtained by using the MATLAB command [*fft_mag, fft_phase*] = *fftplot(fn, Ts)*, where *Ts* denotes the sampling rate. The default value is *Ts* = 1.

R.5.95 The MATLAB function *fn* = *ifft(F, N)* returns the *N* points of the IDFT of *F*, where the length [*fn*] = length [*F*].

R.5.96 DFT provides an efficient approach to the numerical complexity of the evaluation of the DTFT of a finite length sequence. The computational efficiency is based on the argument that the sequence $f(n,)$ with length N must be an integer power of 2. If N is not an integer power of 2, then the sequence $f(n)$ is augmented to a length of M by concatenating $M - N$ zero value samples, where M is an integer power of 2. The DFT of the new sequence *f_aug(n)* with length M can now be computed efficiently by using the *fft* algorithm.

R.5.97 The DFT of an *N*-point real or complex discrete-time sequence is an *N*-complex sequence. If the length required for the DFT is M (integer power of 2), where $M > N$,

---

* The *fft* algorithm was first presented by J.W. Cooley and J.W. Tukey in their paper *An Algorithm for the Machine Calculations of the Complex Fourier Series,* Math. Comp, April 1965.

then by increasing the length N no distortion is introduced. The new augmented sequence $f\_aug(n)$ is then given by

$$f\_aug(n) = \begin{cases} f(n) & \text{for } n < N \\ 0 & \text{for } N+1 < n < M \end{cases}$$

The operation of expanding $f(n)$ by adding zeros is called zero-padding. Note that for computational efficiency, the relation $M = 2^k$ must be satisfied, for the smallest integer k. Observe also that zero-padding is employed when a high-resolution spectrum is required given by a short time sequence.

Recall that in its simplest version, the MATLAB command *fft(fn)* returns the DFT of the sequence *fn* evaluated point by point. The k-length sequence of the DFT corresponding to the time sequence *fn* can be evaluated by using the command *fft(f, k)*.

R.5.98 Let us evaluate the eight DFT points of the discrete sequence $f(t) = cos(2\pi t)$ over one period.

MATLAB Solution
```
>> t = [0:1:7];                          % N = 8
>> fn = cos(2*pi*t/8);
>> fft _ of _ fn = (fft(fn))'

    fft _ of _ fn =
                    -0.0000
                     4.0000 + 0.0000i
                     0.0000
                    -0.0000 + 0.0000i
                     0.0000
                    -0.0000 - 0.0000i
                     0.0000
                     4.0000 - 0.0000i
```

Observe that eight time points return eight transform points, in which the frequencies are ordered $[0\ 1\ 2\ 3\ 4\ -3\ -2\ -1]$, called *reverse-wrap-around*. Note that the first-half of the frequencies is positive and the second-half negative.

R.5.99 Observe also that the real and imaginary parts of the transforms correspond to the cosine (even) components and sine (odd) components, respectively. Note also that the set of frequencies obtained are given as complex conjugate pairs.

R.5.100 Recall that the energy of the discrete-time sequence $f(n)$ is given by

$$\text{Energy} = \sum_{n=-\infty}^{+\infty} [f(n) \cdot f(n)^*] = \sum_{n=-\infty}^{+\infty} |f(n)|^2$$

R.5.101 Recall also that the average power P_{ave} of the sequence $f(n)$ with length N is given by

$$P_{ave} = \frac{1}{N} \sum_{n=0}^{N-1} |f(n)|^2$$

R.5.102 Note that the power content of a signal is the square of the absolute value of its DFT.

For the signal *f(t) = cos(2πt)* just considered (R.5.98), the power spectrum for the positive frequency range is illustrated as follows:

MATLAB Solution
```
>> p = abs(fft(fn))/4;
>> power = (p(1:4).^2)'
```

        power =
                    0.0000
                    1.0000
                    0.0000
                    0.0000

Note that the power content is 1 at the frequency of 1.

R.5.103 Let us use the script file *sample_cos* to illustrate the process of obtaining the power-spectrum plot (Figure 5.16) of a practical case consisting of 3000 samples taken from a sinusoidal signal with a frequency of 25 Hz, over a time interval of 2.3 s.

MATLAB Solution
```
% Script file: sample_cos
N = 3000; T = 2.3;
t = [0:N-1]/N;
t = T*t;
fn = cos(2*pi*25*t);
p = abs(fft(fn))/(N/2);
power = p(1:N/2).^2;
freq = [0:N/2-1]/T;
plot(freq(1:100),power(1:100));
xlabel('frequency in Hertz'), ylabel('power')
title('Power spectrum')
>> sample_cos
```

The script file *sample_cos* is executed and the result is shown in Figure 5.16.

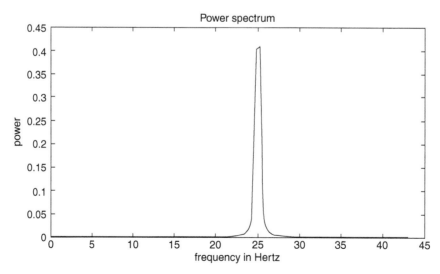

FIGURE 5.16
Power-spectrum plot of R. 5.103.

TABLE 5.3

Properties of the DFT

N-Point Time Sequence	N-Point DFT
$F_1(n)$	$F_1(k)$
$f_2(n)$	$F_2(k)$
$af_1(n) + bf_2(n)$	$aF_1(k) + bF_2(k)$
$F(n - n_o)$	$W_N^{Kn_o} F(k)$
$W_N^{-K_o n} f(n)$	$F(k - k_0)$
$\sum\limits_{m=0}^{N-1} [f_1(m)f_2(n-m)]_N$	$F_1(k) \cdot F_2(k)$
$f_1(n) \cdot f_2(n)$	$(1/N) \sum\limits_{m=0}^{N-1} [F_1(m)F_2(K-m)]$
$(1/N)F(n)$	$f(-k)$
$f^*(n)$	$F^*(-k)_N$
$f^*(-n)_N$	$F^*(k)$

R.5.104 The properties of DFT are similar to DTFT and ZT. For completeness, the most important properties are summarized in Table 5.3.

Note that the notation $f(n)_N = f(n)$ denotes the interval $(0, N - 1)$, and for any n outside the interval $f(n)_N = f(n + rN)$, for $r = 1, 2, 3, \ldots$. Recall that the symbol $*$ is used to denote the complex conjugate.

Note also that the convolution associated with DFT is the circular convolution, even when in most cases, the linear convolution is the desired end result.

Recall that if $f_1(n)$ is of length N_1 and $f_2(n)$ of length N_2, then the length of the linear convolution $[f_1(n) \otimes f_2(n)]$ is of length $N_1 + N_2 - 1$. This would suggest that the sequences $f_1(n)$ and $f_2(n)$ should be appended with $N_2 - 1$ and $N_1 - 1$ zeros, respectively, to make the two sequences of equal length. Then the circular convolution of $f_1(n)$ with $f_2(n)$ of the augmented vectors returns the linear convolution given by $f_1(n) \otimes f_2(n)$.

R.5.105 The DFT and DTFT share many similarities. One of the differences is that DFT assumes that the operations are circular, which is limited over the range 0 to $N - 1$. Recall that DFT is an N-point to N-point transformation.

If the input time sequence is over the range $0 \le n \le N - 1$, then its transform is defined over the same range $0 \le k \le N - 1$.

The time shift cannot be implemented in the conventional way, since the finite sequence N is defined over a given domain and everything else outside that domain is assumed nonexistent. If $f(n)$ is shifted by n_0 creating $f(n - n_0)$, then it would no longer be in the range $0 \le n \le N - 1$. A new type of shift must then be defined where the shifted resulting sequence is over the range $0 \le n \le N - 1$. This new shifting operation is called the circular shift.

R.5.106 For example, let the sequence $f(n)$ be given by the sequence $f(n) = [1\ 2\ 3\ 4\ 5\ 6\ 7]$. Then the circular shift is illustrated by the following two examples, which are self-explanatory:

a. $f(n - 1) = [7\ 1\ 2\ 3\ 4\ 5\ 6]$
b. $f(n - 3) = [5\ 6\ 7\ 1\ 2\ 3\ 4]$

R.5.107 Let the notation $f(n - 1)_7$ denote that the finite sequence $f(n)$ consisting of the elements $[1\ 2\ 3\ 4\ 5\ 6\ 7]$ is shifted by one element to the right, returning the sequence $f(n - 1)_7 = [7\ 1\ 2\ 3\ 4\ 5\ 6]$; and the expression $f(n + 1)_7$ indicate that the elements of $f(n)$ are rotated (or shifted) to the left by one element, returning the sequence $f(n + 1)_7 = [2\ 3\ 4\ 5\ 6\ 7\ 1]$.

R.5.108 Let us next compare the linear convolution with the circular convolution. Recall that the linear discrete convolution of the sequences $f_1(n)$ with $f_2(n)$ is indicated by $\otimes$ and defined by the following equation:

$$f_L(n) = f_1(n) \otimes f_2(n) = \sum_{m=-\infty}^{+\infty} f_1(m)f_2(n - m)$$

which consists of a sequence of length $2N - 1$.

   The assumption is that each of the sequences $f_1(n)$ and $f_2(n)$ are of equal length N. However, the circular convolution of the two sequences given by $f_1(n)$ and $f_2(n)$ is indicated by the character $\circ$ and it is defined by the following equation:

$$f_C(n) = f_1(n) \circ f_2(n) = \sum_{m=0}^{N-1} f_1(m)f_2(n - m)$$

returning the sequence $f_C(n)$ which consists of (length) N elements.

R.5.109 Let us gain some insight of the circular process by evaluating the circular convolution given by the sequence

$$f_1(n) = [1\ 0\ 2\ 4\ 2] \quad \text{with} \quad f_2(n) = [2\ 1\ 2\ 3\ 5]$$

ANALYTICAL Solution

The elements of the convolution $f_C(n) = f_1(n) \circ f_2(n)$ can be computed as follows:

$$f_C(0) = f_1(0)f_2(0) + f_1(1)f_2(4) + f_1(2)f_2(3) + f_1(3)f(2) + f_1(4)f_2(1)$$

$$f_C(0) = 1(2) + (0)(5) + 2(3) + 4(2) + 2(1)$$

$$f_C(0) = 2 + 0 + 6 + 8 + 2 = 18$$

$$f_C(1) = f_1(0)f_2(1) + f_1(1)f_2(0) + f_1(2)f_2(4) + f_1(3)f_2(3) + f_1(4)f_2(2)$$

$$f_C(1) = 1(1) + 0(2) + 2(5) + 4(3) + 2(2)$$

$$f_C(1) = 1 + 0 + 10 + 12 + 4 = 27$$

$$f_C(2) = f_1(0)f_2(2) + f_1(1)f_2(1) + f_1(2)f_2(0) + f_1(3)f_2(4) + f_1(4)f_2(3)$$

$$f_C(2) = 1(2) + 0(1) + 2(2) + 4(5) + 2(3)$$

$$f_C(2) = 2 + 0 + 4 + 20 + 6 = 32$$

$$f_C(3) = f_1(0)f_2(3) + f_1(1)f_2(2) + f_1(2)f_2(1) + f_1(3)f_2(0) + f_1(4)f_2(4)$$

$$f_C(3) = 1(3) + 0(2) + 2(1) + 4(2) + 2(5)$$

$$f_C(3) = 4 + 0 + 2 + 8 + 10 = 24$$

$$f_C(4) = f_1(0)f_2(4) + f_1(1)f_2(3) + f_1(2)f_2(2) + f_1(3)f_2(1) + f_1(4)f_2(0)$$

$$f_C(4) = 1(5) + 0(3) + 2(2) + 4(1) + 2(2)$$

$$f_C(4) = 5 + 0 + 4 + 4 + 4 = 17$$

Then the circular convolution of the sequence $f_1(n) = [1\ 0\ 2\ 4\ 2]$ with $f_2(n) = [2\ 1\ 2\ 3\ 5]$ is given by $f_C(n) = f_1(n) \circ f_2(n)$, which results in $f_C(n) = [1\ 0\ 2\ 4\ 2] \circ [2\ 1\ 2\ 3\ 5] = [18\ 27\ 32\ 24\ 17]$.

R.5.110 Let us now illustrate the process of evaluating DFT by hand for the sequence $f(n) = \{1\ 0\ 2\ 4\ 2\}$.

ANALYTICAL Solution

Recall that

$$F(k) = \sum_{n=0}^{N-1} f(n)\, e^{-j2\pi nk/N}$$

then

$$F(k) = f(0) + f(1)e^{-j2\pi k/5} + f(2)e^{-j2\pi 2k/5} + f(3)e^{-j2\pi 3k/5} + f(4)e^{-j2\pi 4k/5}$$

$$F(k) = 1 + 2e^{-j4\pi k/5} + 4e^{-j2\pi 3k/5} + 2e^{-j2\pi 4k/5}$$

$$F(0) = 1 + 2 + 4 + 2 = 9$$

$$F(1) = 1 + 2e^{-j4\pi 2/5} + 4e^{-j6\pi/5} + 2e^{-j8\pi/5} = -3.2361 + 3.0777i$$

$$F(2) = 1 + 2e^{-j4\pi(2)/5} + 4e^{-j6\pi(2)/5} + 2e^{-j\pi 8(2)/5} = 1.2361 + 0.7265i$$

$$F(3) = 1 + 2e^{-j4\pi(3)/5} + 4e^{-j6\pi(3)/5} + 2e^{-j\pi 8(3)/5} = 1.2361 - 0.7265i$$

$$F(4) = 1 + 2e^{-j4\pi(4)/5} + 4e^{-j6\pi(4)/5} + 2e^{-j\pi 8(4)/5} = -3.2361 + 3.0777i$$

R.5.111 The coefficients of the DFT of the sequence $f(n) = [1\ 0\ 2\ 4\ 2]$ just evaluated in R.5.110 are verified by reevaluating the DFT equations by using MATLAB as well as *fft* command, indicated as follows:

MATLAB Solution
```
>> fn = [1 0 2 4 2];          % time sequence
>> DFT_Fk = (fft(fn))'        % DFT using fft

    DFT_Fk =
            9.0000
           -3.2361 - 3.0777i
            1.2361 + 0.7265i
            1.2361 - 0.7265i
           -3.2361 + 3.0777i
```

```
>>                                    % evaluation of DFT using equations

>> F0 =1+2*exp(-j*4*pi*0/5)+4*exp(-j*6*pi*0/5)+2*exp(-j*8*pi*0/5)

   F0 =
        9

>> F1=1+2*exp(-j*4*pi/5)+4*exp(-j*6*pi/5)+2*exp(-j*8*pi/5)

   F1 =
        -3.2361 + 3.0777i

>> F2 =1+2*exp(-j*4*pi*2/5)+4*exp(-j*6*pi*2/5)+2*exp(-j*8*pi*2/5)

   F2 =
        1.2361 - 0.7265i

>> F3 =1+2*exp(-j*4*pi*3/5)+4*exp(-j*6*pi*3/5)+2*exp(-j*8*pi*3/5)

   F3 =
        1.2361 + 0.7265i

>> F4 =1+2*exp(-j*4*pi*4/5)+4*exp(-j*6*pi*4/5)+2*exp(-j*8*pi*4/5)

   F4 =
        -3.2361 - 3.0777i
```

Note that the DFT coefficients obtained by using the *fft* command are identical to the analytical results using the equations of R.5.110.

R.5.112 Recall that the linear convolution of a finite sequence $h(n)$ of length N with an arbitrary sequence $f(n)$ can be evaluated by using two algorithms known as

a. Overlap-add method

b. Overlap-save method

R.5.113 The overlap-add method is based on segmenting the sequence $f(n)$ into a finite number of continuous segments each of length M.

The output sequence $g(n)$ is obtained by convoluting the impulse response $h(n)$ of length N with each of the segments of $f(n)$, and the partial results are added, as follows:

$$g(n) = f(n) \otimes h(n)$$

and since

$$f(n) = f_1(m) + f_2(m) + \cdots = \sum_{i=1}^{+\infty} fi(m)$$

$$g(n) = \left[\sum_{i=1}^{+\infty} f_i(m) \right] \otimes h(n)$$

$$g(n) = \sum_{i=1}^{+\infty} [f_i(m) \otimes h(n)]$$

Observe that if each segment $f_i(m)$, for $i = 1, 2, 3, \ldots$, is of length M, then each convolution $[h(n) \otimes f_i(m)]$ returns a sequence of length $N + M - 1$. Therefore, the first convolution $[f_1(n) \otimes h(n)]$ returns the first N points of $g(n)$ and the remaining $M - 1$ points are added to the next convolution $[h(n) \otimes f_2(m)]$. This procedure is repeated until all the short convolutions are added to obtain the final convolution resulting in $g(n)$.

The building up of the sequence $g(n)$ for the first two segments of $f(n)[f_1(m) + f_2(m)]$ are analyzed as follows:

$$g(n) = \begin{cases} f_1(m) \otimes h(n) & \text{for } 1 \leq n \leq N \\ f_1(m) \otimes h(n) + f_2(m) \otimes h(n) & \text{for } N + 1 \leq n \leq 2N + M - 1 \end{cases}$$

Note that the convolution points $2N + 1 \leq n \leq 2N + M - 1$ are added to the convolution of $[f_3(m) \otimes h(n)]$ and so on.

R.5.114 Note that the overlap-add method is based on the summations of linear convolutions. The overlap-save method is based on circular convolutions and consists of segmenting $f(n)$ into overlapping blocks of $f(n)$ and performing the circular convolution of the segments and keeping only the terms that correspond to the linear portion of the convolution and discarding all the other terms.

R.5.115 The MATLAB function $gn = fftfilt(hn, fn)$ or $gn = fftfilt(hn, fn, seg)$ returns gn that represents the output sequence $g(n)$, as a result of performing the $conv(hn, fn)$, where hn is the system impulse response $h(n)$ and fn is the system input sequence $f(n)$ segmented into successive sections of length of 512 samples, when not specified, or by defining the segmentation length by seg.

Note that the term $filt$ stands for a rectangular window or filter that is used to limit the range of the convolutions involved.

R.5.116 A comparison between the linear and circular convolutions is useful to provide an insight of the overlap-save method.

Consider the following example: Assume that $f(n)$ is of length 5 and $h(n)$ of length 4. Then the evaluation of the coefficients of the linear and circular convolutions is illustrated as follows:

The linear convolution analysis for $g_L(n) = f(n) \otimes h(n)$ and its corresponding eight points are as follows:

$$g_L(0) = f(0)h(0)$$

$$g_L(1) = f(1)h(0) + h(1)f(0)$$

$$g_L(2) = f(0)h(2) + f(1)h(1) + f(2)h(0)$$

$$g_L(3) = f(0)h(3) + f(1)h(2) + f(2)h(1) + f(3)h(0)$$

$$g_L(4) = f(1)h(3) + f(2)h(2) + f(3)h(1) + f(4)h(0)$$

$$g_L(5) = f(2)h(3) + f(3)h(2) + f(4)h(1)$$

$$g_L(6) = f(3)h(3) + f(2)h(4)$$

$$g_L(7) = f(4)h(3)$$

The analysis of the circular convolution for $f_1(n)$ with $h(n)$, denoted by $g_C(n)$, is presented as follows:

First $h(n)$ is converted from a length 4 sequence into a length 5 sequence by setting the element $h(4) = 0$, then the five terms of the circular convolution are evaluated as follows:

$$g_C(0) = f(0)h(0) + f(1)h(4) + f(2)h(3) + f(3)h(2) + f(4)h(1)$$

$$g_C(1) = f(0)h(1) + f(1)h(0) + f(2)h(4) + f(3)h(3) + f(4)h(2)$$

$$g_C(2) = f(0)h(2) + f(1)h(1) + f(2)h(0) + f(3)h(4) + f(4)h(3)$$

$$g_C(3) = f(0)h(3) + f(1)h(2) + f(2)h(1) + f(3)h(0) + f(4)h(4)$$

$$g_C(4) = f(0)h(4) + f(1)h(3) + f(2)h(2) + f(3)h(1) + f(4)h(0)$$

R.5.117 Let us compare the elements of the linear convolution sequence $g_L(n)$ with the elements of the circular convolution $g_C(n)$. The observation yields

$$g_L(0) \neq g_C(0)$$

$$g_L(1) \neq g_C(1)$$

$$g_L(2) \neq g_C(2)$$

$$g_L(3) = g_C(3)$$

$$g_L(4) = g_C(4)$$

R.5.118 From R.5.117, it can be stated that if $h(n)$ is a sequence of length M and $f(n)$ is a sequence of length N where $N > M$, then the first $M - 1$ values of the circular convolution do not correspond with the linear convolution, whereas the remaining $N - M + 1$ elements of the circular convolution correspond to the values of the linear convolutions.

R.5.119 For example, if $h(n)$ is a sequence of length 4 and $f(m)$ of length 5, the first $4 - 1$ coefficients of the circular convolution should be discarded although coefficients 3 and 4 are identical to the linear convolution values.

R.5.120 Recall that DFT is often used to evaluate the linear convolutions. Efficient algorithms for the evaluation of DFT have been topics of intense research during the 1960s and 1970s. The results of this research led to a collection of efficient computational algorithms known collectively as the *fft*s.

The earliest algorithm goes back to the paper published by Good in the late 1950s. These algorithms are also referred as the prime factor algorithms. The first Cooley–Tukey algorithms were published in the early 1960s and one of the main assumptions in its implementation was that the DFT with length N is a power of two, leading in this way into an efficient computational implementation.

To evaluate the DFT of an N-point sequence, the number of complex multiplications is approximately of the order of N^2. Since a complex multiplication involves four real multiplications and two additions, then the number of real multiplications is of the order of $4N^2$ and the number of real additions is approximately $4N^2 - 2N$.

The efficiency of the DFT algorithms consists of taking advantage of the nature of the complex number (W_N^{nk}), in particular its periodicity and symmetry.

R.5.121 The *fft* algorithms are based in decomposing the N-point DFT computations into smaller size DFT. For example, the DFT of an N-point sequence can be decomposed into two $(N/2)$ point sequences reducing the computational process from $4N^2 - 2N$ to $2N + N^2$ real additions.

In addition, since $W_N^0 = 1$ and $W_n^{(N/2)+k} = -W_n^k$, the total number of computations is further reduced by half.

R.5.122 The linear convolutions can be evaluated by using discrete transforms that employ the FFT. This technique is referred to as the fast convolution. The evaluation of the convolution in the time domain for a sequence of N samples requires $4N^2$ real multiplications.

For the case of an N-point DFT, both the time as well as frequency functions are periodic. The points $n = 1$ to $N/2$ are symmetric with respect to the points between $(N/2) + 1$ through $N - 1$.

If the time sequence is real, then the DFT has real and imaginary components, which are symmetric and asymmetric, respectively.

R.5.123 The term *fast* refers in general to a number of techniques that use the discrete transforms in an efficient computational environment. For example, the convolution sum requires N^2 multiplications as compared with $4Nlog_2(N) + 4N$ multiplications when using the *fft*. If the number of discrete elements $N = 1000$, then the convolution requires 1,000,000 multiplications, whereas the transform method to evaluate the convolution requires about 44,000 multiplications.

R.5.124 For example, to sample music at a rate that is pleasant to the ear, the sampling rate accordance to Nyquist would be about 40,000 Hz. A few seconds of music quickly builds up into a large amount of data.

R.5.125 If a continuous time signal $f(t)$ is band-limited, the spectra characteristics of $f(n)$ constitute a good approximation of $f(t)$ if the sampling rate was performed at least at the Nyquist rate (or higher). If $f(t)$ is not band-limited (defined over $-\infty < t < +\infty$), then $f(n)$ becomes an infinite sequence over $-\infty < n < +\infty$. Practical limits must be imposed on the length of the sequence $f(n)$ if a digital computer is used, by windowing or truncating the sequence $f(n)$ to a finite number of samples (N). The DTFT of the signal $f(n)$ is given by $F(e^{jW}) = \sum_{n=-\infty}^{+\infty} f(n)e^{-jWn}$ and is truncated by the use of the window $w(n)$, over the range $-N/2 \le n \le N/2$, yielding the following relation:

$$F_w(e^{jW}) = \sum_{n=-N/2}^{N/2} f(n)w(n)e^{-jWn}$$

R.5.126 To provide a good resolution, the DFT of length L is usually chosen to be greater than the window length N, by adding $L - N$ zero value samples. In addition, the relation $L = 2^k > N$ for the smallest integer value of k must be maintained to gain computational efficiency.

R.5.127 The MATLAB command $k = nextpow2(N)$ returns the smallest integer k so that $2^k \ge N$. It is a useful command for finding the nearest power of 2, given an arbitrary sequence length, when the *fft* command is used.

R.5.128 There are a number of other efficient algorithms to evaluate the DFT, especially when a few time samples are available, such as the Goertzel algorithm. For practical reasons, only the *fft* command is used in this book.

R.5.129 The discrete sequence $f(n)$ is in most cases a 1-D array, where the elements of $f(n)$ are the order samples of the continuous time function $f(t)$. In general, f can be a matrix denoted by $f(n, m)$ with n rows and m columns. In this case, the MATLAB function $fft(f)$ returns the DFT of each of the m columns of f.

R.5.130 The MATLAB command *fft2(f)* where *f* is an *n* by *m* matrix returns the 2-D *fft* of the elements of *f*, where *f* may represent, for example, a 2-D matrix representing a picture (video) where each element may be a dot (pixel) with varying degrees of intensities (white–gray–black), for a white-and-black image. The *fft2* returns first the *fft* of each column of *f*, then the *fft* of each row of the resulting matrix, resulting in a complex *n* by *m* DFT matrix. The command *ifft2(F)* returns the 2-D IDFT.

R.5.131 The MATLAB command $F = fftn(f)$ returns the *n*-dimension DFT of the *n*-dimension array *f*.

R.5.132 The MATLAB command $f = iffn(F)$ returns the *n*-dimensional IDFT of the *n*-dimension array *F*.

R.5.133 Since DFT is a linear transform, the *fft2* can also be viewed as taking first the transform of each of the rows and then the transforms of the columns of the resulting matrix.

R.5.134 Let us summarize some facts about the *fft*.

The *fft* is not a new transform but rather an efficient computational algorithm that returns the DFT of an arbitrary finite discrete-time sequence. The computational efficiency of the *fft* is based on the following facts:

a. The DFT is periodic with the period extending over 2π, from $-\pi$ to $+\pi$.

b. If the number of samples is N, and they are all real, then the required range of DFT is over $N/2$.

c. When the discrete-time sequence $f(n)$ is a real sequence, the transform coefficients through F_m are complex conjugate of F_{N-m}, for any $0 \le m \le N/2$.

d. It can be shown that all the values outside the set of transforms $F_0, F_1, F_2, ..., F_{N/2}$ are redundant.

R.5.135 The MATLAB command *fftshift(F)* returns the frequency-spectrum components centered in the middle in which half of the frequencies is positive and the other half negative.

R.5.136 The MATLAB command *ifftshift(F)* undoes the effects of *fftshift(F)*.

R.5.137 For example, let *fn_8* = [1 2 3 4 5 6 7 8] be an eight-sample sequence.

a. Evaluate the DFT of *fn_8* by using MATLAB.

b. Observe that $F_3 = F_5^*$ (recall that the character * denotes the complex conjugate).

c. Let *fn_4* = [1 2 3 4], and evaluate its DFT and observe its symmetry.

d. Inverse the sequence *fn_8*, and call the new sequence as *fn_8_i* = [8 7 6 5 4 3 2 1]. Verify that reversing the sequence in time is equivalent of evaluating the complex conjugate of the DFT of the original time sequence *fn_8*.

e. Create the sequence *fn_4_0* = [1 2 3 4 0 0 0 0], evaluate its DFT, and compare the results with DFT of *fn_4*.

f. Let *f_3by3* = *magic(3)* matrix and evaluate its DFT using the *fft* command.

g. Let *f_trans_3by3* = (*magic(3)*)' matrix and evaluate its DFT by using the *fft* function, and compare the results obtained with part f.

h. Repeat parts g and f by using the *fft2* function, verifying the linearity property.

```
MATLAB Solution
>> fn _ 8 = [1 2 3 4 5 6 7 8];
>> DFT _ fn _ 8 = (fft(fn _ 8))'                          % part(a)

   DFT _ fn _ 8 =
                36.0000
                -4.0000 - 9.6569i
                -4.0000 - 4.0000i
                -4.0000 - 1.6569i
                -4.0000
                -4.0000 + 1.6569i
                -4.0000 + 4.0000i
                -4.0000 + 9.6569i

>> % note that F(3)= -4.0000 - 1.6569i
>> % F(5)= -4.0000 + 1.6569i, observe   F(3)=F(5)*,        part(b)

>> fn _ 4 = [1 2 3 4];                                    % part(c)
>> DFT _ fn _ 4 = (fft(fn _ 4))'

   DFT _ fn _ 4 =
                10.0000
                -2.0000 - 2.0000i
                -2.0000
                -2.0000 + 2.0000i

>> fn _ 8 _ I=fliplr(fn _ 8);            %  fn _ 8 _ i =[8 7 6 5 4 3 2 1];
>> DFT _ fn _ 8 _ i = (fft(fn _ 8 _ i))'               % part (d)

   DFT _ fn _ 8 _ i =
                36.0000
                 4.0000 + 9.6569i
                 4.0000 + 4.0000i
                 4.0000 + 1.6569i
                 4.0000
                 4.0000 - 1.6569i
                 4.0000 - 4.0000i
                 4.0000 - 9.6569i

>> fn _ 8 _ 0 = [1 2 3 4 0 0 0 0];                       % part(e)
>> DFT _ fn _ 8 _ 0 = (fft(fn _ 8 _ 0))'

   DFT _ fn _ 8 _ 0 =
                10.0000
                -0.4142 + 7.2426i
                -2.0000 - 2.0000i
                 2.4142 + 1.2426i
                -2.0000
                 2.4142 - 1.2426i
                -2.0000 + 2.0000i
                -0.4142 - 7.2426i

>> % observe that DFT _ fn _ 8 _ 0 = DFT _ fn _ 4 at four points
```

```
>> f _ 3by3 = magic(3)                                    % part(f)

   f _ 3by3 =
             8   1   6
             3   5   7
             4   9   2


>> DFT _ f _ 3by3 = (fft(f _ 3by3))'

   DFT _ f _ 3by3 =
                    15.0000      4.5000 - 0.8660i    4.5000 + 0.8660i
                    15.0000     -6.0000 - 3.4641i   -6.0000 + 3.4641i
                    15.0000      1.5000 + 4.3301i    1.5000 - 4.3301i
>> %Note that the results obtained in part(g) are not equal to part(f)
>> f _ trans _ 3by3 = (magic(3))'                         % part(g)

   f _ trans _ 3by3 =
                    8   3   4
                    1   5   9
                    6   7   2


>> DFT _ f _ trans _ 3by3 = (fft(f _ trans _ 3by3))'      % part(h)

   DFT _ f _ trans _ 3by3 =
                    15.0000      4.5000 - 4.3301i    4.5000 + 4.3301i
                    15.0000     -3.0000 - 1.7321i   -3.0000 + 1.7321i
                    15.0000     -1.5000 + 6.0622i   -1.5000 - 6.0622i


>> DFT _ f _ trans _ 3by3 = (fft2(f _ trans _ 3by3))'

   DFT _ f _ trans _ 3by3 =
                    45.0000             0                    0
                    0           13.5000 - 7.7942i      0 + 5.1962i
                    0                0 - 5.1962i   13.5000 + 7.7942i


>> DFT _ f _ 3by3 = (fft2(f _ 3by3))'

   DFT _ f _ 3by3 =
                    45.0000          0 - 0.0000i         0 + 0.0000i
                    0           13.5000 - 7.7942i    0.0000 - 5.1962i
                    0            0.0000 + 5.1962i   13.5000 + 7.7942i
```

R.5.138 The command *fft* is used to evaluate the correlation function, which is used in signal detection and estimation, in applications such as radar and sonar.

The autocorrelation of $f_1(n)$ is defined as:

$$R_{11}(s) = \sum_{n=-\infty}^{+\infty} f_1(n)f_1(n+s) \quad \text{over the range} -\infty \le s \le +\infty$$

where s is referred to as the scanning or searching parameter. Observe that the autocorrelation is very similar to the convolution. Recall that the convolution is given by

$$f_1(s) \otimes f_2(s) = \sum_{n=-\infty}^{+\infty} f_1(n) f_2(s-n)$$

R.5.139 The cross-correlation of two sequences $f_1(n)$ with $f_2(n)$ is given by

$$xcorr(f_1(s), f_2(s)) = R_{12}(s) = \sum_{n=-\infty}^{+\infty} f_1(n)f_2(n + s) \quad \text{over the range } -\infty \le s \le +\infty$$

whereas the convolution is given by

$$conv(f_1(s), f_2(s)) = \sum_{n=-\infty}^{+\infty} f_1(n) \, f_2(s - n) \quad \text{over the range } -\infty \le s \le +\infty$$

R.5.140 The relation between the convolution and cross-correlation is given by

$$f_2(-n) \otimes f_1(n) = R_{12}(n)$$

R.5.141 The MATLAB command *xcorr(f)* returns the autocorrelation of *f*, and the command *xcorr(f₁, f₂)* returns the cross-correlation of the sequences f_1 with f_2. Observe that $xcorr(f_1, f_1) = xcorr(f_1)$.

R.5.142 Since the convolution can be evaluated using DFT, it is logical to think that DFT can be used to evaluate the cross-correlation. The relation is given by

$$DFT[R_{12}(n)] = DFT * [f_1] \cdot DFT[f_2]$$

or

$$R_{12}(n \Leftrightarrow)F_1 * (k)F_2(k)$$

R.5.143 Some properties of the correlation function are summarized as follows:
   a. $R_{11}(n) = R_{11}(-n)$
   b. The $max[R_{11}(n)]$ occurs at $n = 0$
   c. $R_{12}(n) = R_{21}(-n)$
   d. $xcorr(f_1, f_2) = f_1(n) \otimes f_2(-n)$

R.5.144 Let *fn1* = [0 1 2 −1 −2 0 1 0] and *fn2* = [0 1 2 3 4 5 6 7].
   a. Evaluate by hand the first three coefficients $R_{11}(0)$, $R_{11}(1)$, and $R_{11}(2)$ of the autocorrelation for *fn1* (Figure 5.17).
   b. Evaluate by hand the first three coefficients $R_{12}(0)$, $R_{12}(1)$, and $R_{12}(2)$ of the cross-correlation for *fn1* with *fn2*.
   c. Verify the preceding results by using the MATLAB commands *xcorr(fn1)* and *xcorr(fn1, fn2)*.
   d. Compare the results obtained in parts a and b, with part c.
   e. Obtain the plots of the autocorrelation function of *fn1* and it is an even function and its maximum value occurs at $n = 0$.
   f. Obtain plots of the cross-correlation *xcorr(fn1, fn2)* and *xcorr(fn2, fn1)*, and verify that *xcorr(fn1, fn2)* ≠. *xcorr(fn1, fn2)*

ANALYTICAL Solution

$$R_{11}(0) = [0\,1\,2\,1\,2\,0\,1\,0].\ast[0\,1\,2\,1\,2\,0\,1\,0] = [0\,1\,4\,1\,4\,0\,1\,0] = 11$$

$$R_{11}(1) = [0\,1\,2\,1\,2\,0\,1\,0].\ast[1\,2\,1\,2\,0\,1\,0\,0] = [0\,2\,2\,2\,0\,0\,0\,0] = 6$$

$$R_{11}(2) = [0\,1\,2\,1\,2\,0\,1\,0].\ast[2\,1\,2\,0\,1\,0\,0\,0] = [0\,1\,4\,0\,2\,0\,0\,0] = 7$$

$$R_{12}(0) = [0\,1\,2\,1\,2\,0\,1\,0].\ast[0\,1\,2\,3\,4\,5\,6\,7] = [0\,1\,4\,3\,8\,0\,6\,0] = 22$$

$$R_{12}(1) = [0\,1\,2\,1\,2\,0\,1\,0].\ast[1\,2\,3\,4\,5\,6\,7\,0] = [0\,2\,6\,4\,10\,0\,7\,0] = 29$$

$$R_{12}(2) = [0\,1\,2\,1\,2\,0\,1\,0].\ast[2\,3\,4\,5\,6\,7\,0\,0] = [0\,3\,8\,5\,12\,0\,0\,0] = 28$$

MATLAB Solution

```
% Script file: auto _ cross
fn1=[0 1 2 1 2 0 1 0]; fn2 = [0 1 2 3 4 5 6 7];
auto _ corr=xcorr(fn1);cross _ corr = xcorr(fn1,fn2);cross _ corrx =
  xcorr(fn2,fn1);
disp('****************** R E S U L T S ***************************')
disp('The auto-correlation coefficients of f1(n)are:')
auto _ corr
disp('The cross-correlation coefficients of f1(n) with f2(n) are:')
cross _ corr
disp('**************************************************************')
n=-7:7;
subplot(3,1,1)
bar(n,auto _ corr)
title('Auto-correlation of f1(n)')
xlabel('time index n')
ylabel('Amplitude')
subplot(3,1,2)
bar(n,cross _ corr)
title('Cross-correlation of f1(n) with f2(n)')
xlabel('time index n'); ylabel('Amplitude')
Subplot(3,1,3)
bar(n,cross _ corrx)
title('Cross-correlation of f2(n) with f1(n)')
xlabel('time index n'); ylabel('Amplitude')
********************** R E S U L T S **************************
The auto-correlation coefficients of f1(n)are:

auto _ corr =
  Columns 1 through 8
  -0.0000  -0.0000   1.0000   2.0000   3.0000   7.0000   6.0000   11.0000
  Columns 9 through 15
   6.0000   7.0000   3.0000   2.0000   1.0000  -0.0000  -0.0000
The cross-correlation coefficients of f1(n) with f2(n) are:
cross _ corr =
  Columns 1 through 8
  -0.0000   7.0000  20.0000  24.0000  34.0000  28.0000  29.0000   22.0000
  Columns 9 through 15
   15.0000   9.0000   5.0000   2.0000   1.0000  -0.0000   0
**************************************************************
```

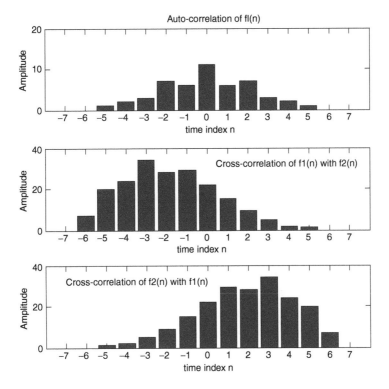

FIGURE 5.17
Plots of R.5.144.

5.4 Examples

Example 5.1

Let $f(n) = 0.6^n u(n)$.

a. Evaluate the DTFT of $f(n)$ by hand.
b. Create the script file DTFT that returns the magnitude and phase plots of the DTFT of $f(n) = 0.6^n u(n)$ of part a, over the range $-\pi < W < \pi$.
c. Repeat part b by using the MATLAB function freqz.
d. Evaluate the magnitude and phase errors of the two techniques employed (part b and part c).

ANALYTICAL Solution

Part a

$$F(e^{jW}) = \sum_{n=-\infty}^{+\infty} f(n)\, e^{-jWn}$$

$$F(e^{jW}) = \sum_{n=-\infty}^{+\infty} 0.6^n u(n)\, e^{-jWn}$$

$$F(e^{jW}) = \sum_{n=0}^{+\infty} 0.6^n\; e^{-jWn} = \sum_{n=0}^{\infty} (0.6\, e^{-jW})^n = \frac{1}{1 - 0.6 e^{-jW}}$$

MATLAB Solution

```
% Script file : DTFT
% part(b)
W = linspace(-pi,pi,21);
DTFT = 1./(1-0.6*exp(-j.*W));

figure(1
subplot (2,1,1)
plot (W,abs(DTFT));title(' DTFT [0.6^n u(n)] = 1./(1-0.6*exp(-j.*W))');
xlabel ('frequency W in rad.'), ylabel('Magnitude')
subplot (2,1,2)
plot (W,angle(DTFT));
xlabel ('frequency W in rad.'), ylabel('Phase in rad.')

figure(2)
% part(c)
P = [1 0]; Q = [1 -0.6];
h = freqz (P,Q,W);
subplot (2,1,1)
plot (W,abs(h));title('DTFT[0.6^n u(n)]  using  freqz');
xlabel ('frequency W in rad.'), ylabel ('Magnitude')
subplot (2,1,2)
plot (W,angle(h));
xlabel ('frequency W in rad.'), ylabel ('Phase in rad.')

figure(3)
error _ mag=abs(DTFT)-abs(h);
error _ phase=angle(DTFT)-angle(h);
subplot(2,1,1);
bar(W,error _ mag);
title('Magnitude error');ylabel('Amplitude')
subplot(2,1,2);
bar(W,error _ phase);xlabel('frequency W in rad. ');
title('Phase error'), ylabel('Phase in rad.');
```

The script file *DTFT* is executed and the results are shown in Figures 5.18 through 5.20.

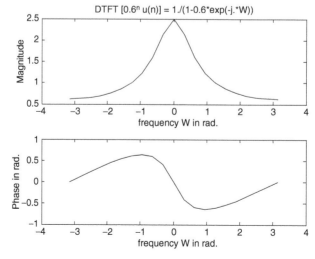

FIGURE 5.18
DTFT of *f(n)* of Example 5.1.

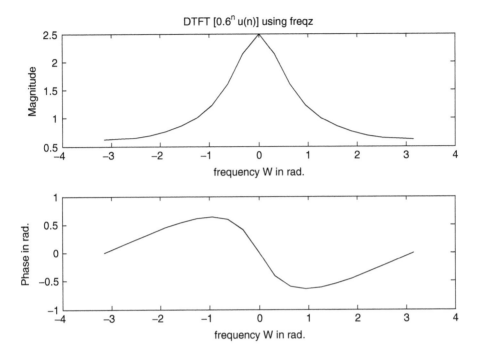

FIGURE 5.19
DTFT of *f(n)* using *freqz* of Example 5.1.

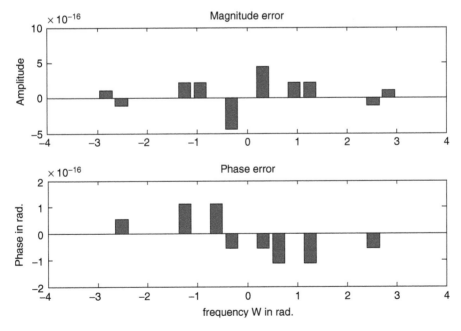

FIGURE 5.20
Error evaluation (parts b and c) of Example 5.1.

Example 5.2

Given the sequences $f_1(n) = [0.3^n u(n)]$ and $f_2(n) = [0.9^n u(n)]$

   a. Evaluate by hand the DTFT of $f_1(n)$ and $f_2(n)$.

Create the script file *DTFT_properties* that returns the following:

   b. Plots of $f_1(n)$ versus n and $f_2(n)$ versus n, over the range $0 \le n \le 9$.

   c. The magnitude and phase plots of DTFT of $f1(n)$ and $f_2(n)$, over the range $-\pi < W < +\pi$.

   d. Verify the even symmetry of the relation $real[F_1(e^{jW})] = real[F_1(e^{-jW})]$ and odd symmetry of $imagl[F_1(e^{jW})] = imagl[-F_1(e^{-jW})]$.

   e. Use the linear property of DTFT and evaluate and plot the time–frequency relation $2f1(n) + 3f_2(n) \Leftrightarrow 2F_1(e^{jW}) + 3F_2(e^{jW})$.

   f. Verify that $F_1(e^{jW})$ and $F_2(e^{jW})$ are periodic functions with periods of 2π, by plotting each function over the range $-2\pi < W < 2\pi$.

   g. Compare the DTFT of Example 5.1 with the DTFT of part c and observe that if $f(n) = a^n u(n)$, then as a increases the BW decreases, as long as $|a| < 1$.

ANALYTICAL Solution

Part a

$$F_1(e^{jW}) = \sum_{n=-\infty}^{+\infty} f_1(n)e^{-jWn}$$

$$F_1(e^{jW}) = \sum_{n=-\infty}^{+\infty} 0.3^n e^{-jWn}$$

$$F_1(e^{jW}) = \sum_{n=0}^{+\infty} 0.3^n e^{-jWn} = \sum_{n=0}^{\infty} (0.3e^{-jW})^n = \frac{1}{1 - 0.3e^{-jW}} = \frac{e^{jW}}{e^{jW} - 0.3}$$

$$F_2(e^{jW}) = \sum_{n=-\infty}^{+\infty} f_2(n)e^{-jWn}$$

$$F_2(e^{jW}) = \sum_{n=0}^{+\infty} 0.9^n e^{-jWn} = \sum_{n=0}^{\infty} (0.9e^{-jW})^n = \frac{1}{1 - 0.9e^{-jW}} = \frac{e^{jW}}{e^{jW} - 0.9}$$

MATLAB Solution

```
% Script file: DTFT _ properties
n = 0:9;
f1 = (0.3.*ones(1,10)).^n;
f2 = (0.9.*ones(1,10)).^n;

figure(1)
nn= [(-4:-1) n];f1nn =[zeros(1,4) f1];
subplot (2,1,1)
```

```
stem (nn,f1nn)
ylabel ('Amplitude of f1(n)');xlabel ('time index n');
title('f1(n) vs. n')
axis ([-4 10 0 1.1])
subplot (2,1,2)
f2nn =[zeros(1,4) f2];
stem(nn,f2nn)
title('f2(n) vs. n')
ylabel ('Amplitude of f2(n)');
xlabel ('time index n'); axis([-4 10 0 1.1])
W=linspace(-pi,pi,100);
DTFT _ f1=1./(1-0.3*exp(-j.*W));
DTFT _ f2=1./(1-0.9*exp(-j.*W));

figure(2)
subplot (2,1,1)
plot (W,abs(DTFT _ f1));title(' DTFT[f1(n)]=1./(1-0.3*exp(-j.*W))');
xlabel ('frequency W '), ylabel('Magnitude')
subplot (2,1,2)
plot (W,angle(DTFT _ f1));
xlabel ('frequency W'), ylabel('Phase angle in rad.')

figure(3)
subplot (2,1,1)
plot (W,abs(DTFT _ f2));title(' DTFT[f2(n)]=1./(1-0.9*exp(-j.*W))');
xlabel('frequency W '), ylabel('Magnitude')
subplot(2,1,2)
plot (W,angle(DTFT _ f2));
xlabel ('frequency W '), ylabel('Phase angle in rad.')

figure(4)
subplot (2,1,1)
plot (W,real(DTFT _ f1));title(' real [DTFT[f1(n)] vs. W');
ylabel ('Amplitude')
subplot (2,1,2)
plot (W,imag(DTFT _ f1));title(' imag [DTFT[f1(n)] vs. W');
xlabel ('frequency W '), ylabel('Amplitude')

figure(5)
f12 = 2*f1nn+3*f2nn;
DTFT _ 12 = 2*DTFT _ f1+3*DTFT _ f2;
subplot (3,1,1)
stem(nn,f12);
ylabel ('Amplitude')
title ('[2f1(n)+3f2(n)] vs. time index n')
subplot (3,1,2)
plot (W,abs(DTFT _ 12)), title('abs[DTFT(2f1(n)+3f2(n))] vs. W');
ylabel ('Magnitude')
subplot (3,1,3)
plot (W,angle(DTFT _ 12));title('angle[DTFT(2f1(n)+3f2(n))] vs. W');
ylabel (' Phase in rad.');
xlabel (' frequency W')
```

```
figure(6)
W=linspace(-2*pi,2*pi,100);
DTFT _ f1=1./(1-0.3*exp(-j.*W));
DTFT _ f2=1./(1-0.9*exp(-j.*W));
subplot (2,1,1)
plot (W,abs(DTFT _ f1));title(' DTFT[f1(n)]=1./(1-0.3*exp(-j.*W))');
xlabel ('frequency W '), ylabel('Magnitude')
subplot (2,1,2)
plot (W,angle(DTFT _ f1));
xlabel ('frequency W'), ylabel('Phase angle in rad.')

figure(7)
subplot(2,1,1)
plot(W,abs(DTFT _ f2));title(' DTFT[f2(n)]=1./(1-0.9*exp(-j.*W))');
xlabel('frequency W '), ylabel('Magnitude')
subplot(2,1,2)
plot(W,angle(DTFT _ f2));
xlabel('frequency W '), ylabel('Phase angle in rad.')
```

The script file *DTFT_properties* is executed and the results are shown in Figures 5.21 through 5.27.

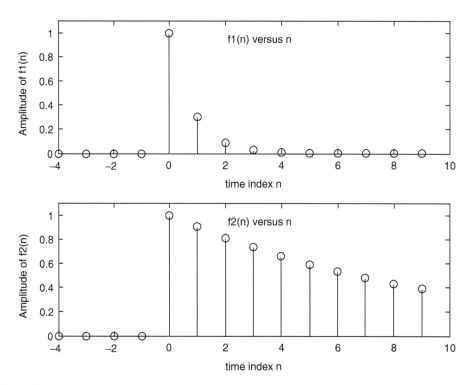

FIGURE 5.21
Discrete plots of $f_1(n)$ and $f_2(n)$ of Example 5.2.

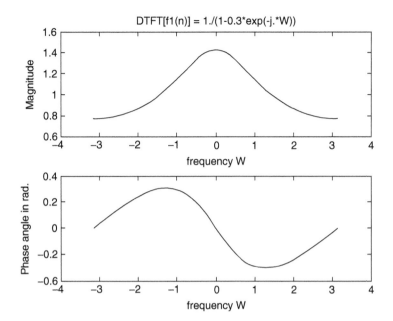

FIGURE 5.22
Plot of DTFT[$f_1(n)$] of Example 5.2.

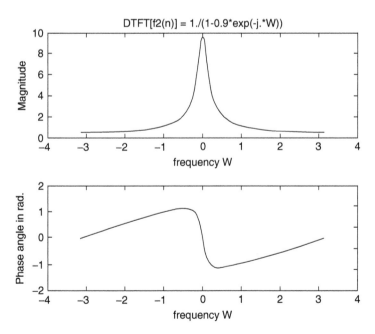

FIGURE 5.23
Plot of DTFT[$f_2(n)$] of Example 5.2.

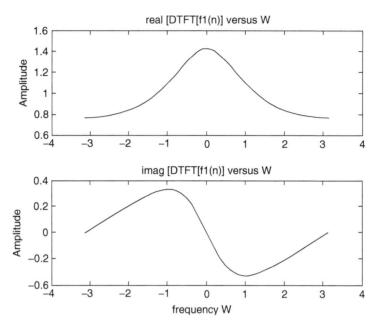

FIGURE 5.24
Plot of the real and imaginary parts of DTFT[$f_1(n)$] of Example 5.2.

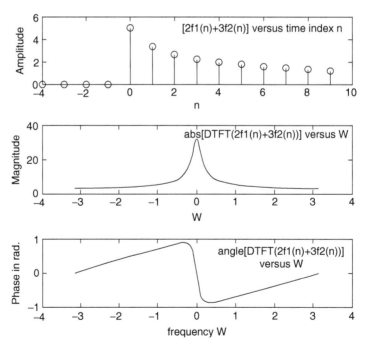

FIGURE 5.25
Plot of *2f1(n) + 3f₂(n)*, *abs(DTFT[2f1(n) + 3f₂(n)])*, and *angleDTFT[2f1(n) + 3 f₂(n)]* of Example 5.2.

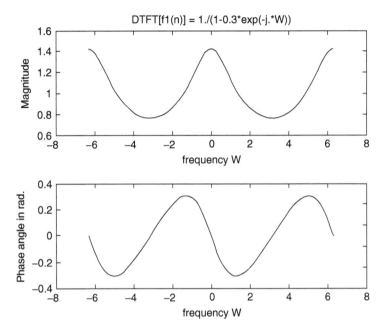

FIGURE 5.26
Plot of $F_1(e^{jW})$ over the range $-2\pi < W < 2\pi$ of Example 5.2.

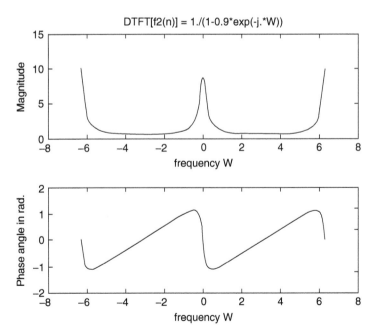

FIGURE 5.27
Plot of $F_2(e^{jW})$ over the range $-2\pi < W < 2\pi$ of Example 5.2.

Example 5.3

Let the real sequence $h(n)$ be given by its DTFT as follows:

$$H(e^{jW}) = \frac{0.007 + 0.034e^{-jW} + 0.04e^{-2jW} + 0.04e^{-3jW} + 0.03e^{-4jW} + 0.009e^{-j5W}}{1 + 2.63e^{-jW} + 2.81e^{-j2W} + 1.5e^{-j3W} + 0.4e^{-j4W}}$$

Create the script file *DTFT_H* that returns the magnitude and phase plots of the DTFT of $H(e^{jW})$ versus W, using 128 discrete points over the range $0 \leq W \leq 4\pi$.

MATLAB Solution
```
% Script file: DTFT _ H
P = [0.007 0.034 0.04 0.04 0.03 0.009];
Q = [1 2.63 2.81 1.5 0.4];
W = linspace(0,4*pi,128);
H = freqz(P,Q,W);
subplot (2,1,1)
plot(W,abs(H))
title ('Magnitude of H[exp(jW)] vs. W')
ylabel ('Magnitude');xlabel('frequency W');
subplot (2,1,2)
plot(W,angle(H))
title('Phase of H[exp(jW)] vs. W')
xlabel('frequency W');ylabel('Phase in rad.')
```

The script file *DTFT_H* is executed and the results are shown in Figure 5.28.

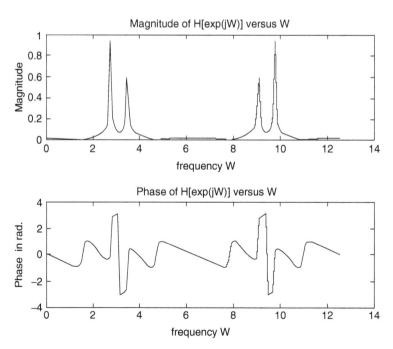

FIGURE 5.28
Magnitude and phase plots of the DTFT of $H(e^{jW})$ of Example 5.3.

Example 5.4

Given the discrete sequence $f(n) = [1\ 2\ 3]$,

    a. Evaluate by hand the DTF coefficients of $f(n)$

    b. Create the script file *DFT_IDFT* that returns the DTF of $f(n)$

    c. Recover the sequence $f(n)$ by evaluating the IDTF of part b

ANALYTICAL Solution

$$F(k) = \sum_{n=0}^{N-1} f(n)e^{-j\frac{2\pi nk}{N}}$$

$$F(0) = \sum_{n=0}^{2} f(n) = 1 + 2 + 3 = 6$$

$$F(1) = \sum_{n=0}^{2} f(n)e^{-j\frac{2\pi n}{3}} = 1 + 2e^{-j2\pi/3} + 3e^{-j4\pi/3} = -1.5000 + 0.8660i$$

$$F(2) = \sum_{n=0}^{2} f(n)e^{-j\frac{4\pi n}{3}} = 1 + 2e^{-j4\pi/3} + 3e^{-j8\pi/3} = -1.5000 - 0.8660i$$

MATLAB Solution
```
% Script file: DFT _ IDFT
fn = [1 2 3]
DTF _ fn = fft(fn);
disp('*************** R E S U L T S **************** ')
disp('The given time sequence is f(n) = [1 2 3]')
disp('The DTF of f(n)=[1 2 3] using fft is given by :')
disp(DTF _ fn)
fnn = ifft(DTF _ fn);
disp('The IDTF of the DFT of f(n)=[1 2 3] using ifft is given by:')
disp(fnn)
disp('************************************************** ')
```

The script file *DFT_IDFT* is executed and the results are as follows:

```
****************** R E S U L T S ***********************
The given time sequence is f(n) = [1 2 3]
The DTF of f(n) = [1 2 3] using fft is given by:
  6.0000          -1.5000 + 0.8660i     -1.5000 - 0.8660i
The IDTF of the DFT of f(n) = [1 2 3] using ifft  is given by:
  1    2    3
***********************************************************
```

Example 5.5

Create the script file *DTFT_DFT* that returns the magnitude and phase plots of DTFT as well as DFT of the following discrete-time sequence:

$$f(n) = \sum_{n=0}^{k-1} \cos\left(\frac{3\pi n}{N}\right) \quad \text{for } N = 16$$

over the ranges $0 \leq k \leq 32$ and $0 \leq k \leq 64$.

MATLAB Solution

```
% Script file: DTFT _ DFT
k = linspace(0,16,32);
fn32 = cos(3*pi*k/16);
kk = linspace(0,16,64);
fn64 = cos(3*pi*kk/16);
DFT _ fn32 = fft(fn32);
DFT _ fn64 = fft(fn64);
DFT _ fn = fft(fn32,256);
W = 0:1:255;

figure(1)
subplot (2,1,1)
plot (W,abs(DFT _ fn));
title (' Magnitudes of the DTFT and DFT')
ylabel ('Amplitude')
axis ([0 259 0 20])
hold on
WW= linspace(0,255,32);
plot(WW,abs(DFT _ fn32),'*',WW,abs(DFT _ fn32))
axis ([0 259 0 20])
subplot (2,1,2)
plot (W,abs(DFT _ fn));
ylabel ('Amplitude')
hold on
WWW= linspace(0,255,64);
plot (WWW,abs(DFT _ fn64),'o',WWW,abs(DFT _ fn64));
xlabel (' Index k');axis([0 259 0 29]);

figure(2)
subplot(2,1,1)
plot (W,angle(DFT _ fn));
title (' Phase of the DTFT and DFT')
ylabel ('Phase angle')
axis ([0 259 -4 4])
hold on
plot (WW,angle(DFT _ fn32),'*',WW,angle(DFT _ fn32))
subplot (2,1,2)
plot (W,angle(DFT _ fn))
axis ([0 259 -4 4])
hold on
plot (WWW,angle(DFT _ fn64),'o',WWW,angle(DFT _ fn64));
axis([0 259 -4 4])
ylabel('Phase angle')
xlabel('Index k')
```

The script file *DTFT_DFT* is executed and the results are indicated in Figures 5.29 and 5.30.

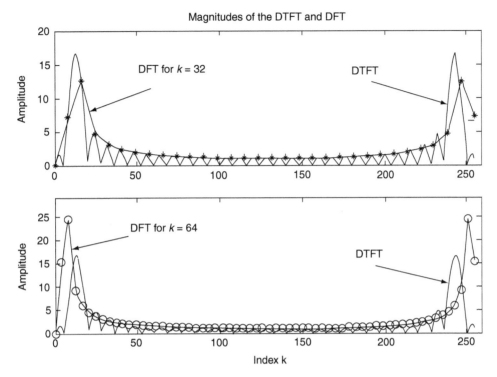

FIGURE 5.29
Magnitude plots of DTFT and DFT of the discrete sequence *f(n)* of Example 5.5.

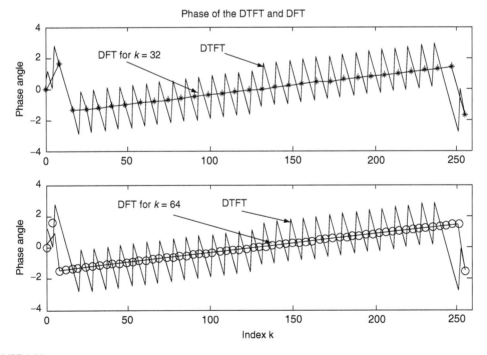

FIGURE 5.30
Phase plots of DTFT and DFT of the discrete sequence *f(n)* of Example 5.5.

Example 5.6

Let the input sequence to a system be given by $f(n) = [1\ 0\ 3\ 2]$, and its impulse system response be $h(n) = [-1\ 0\ 3\ 4\ 6\ 3]$.

Create the script file *convolutions* that returns

   a. The system output $g(n) = f(n) \otimes h(n)$, using the direct time convolution, as well as the DFT of $g(n)$

   b. The system output $g(n)$ by evaluating the IDFT of the product of the DFT of $f(n)$ with the DFT of $h(n)$

   c. The estimate of the error of the methods used in part a with respect to part b by means of a plot

ANALYTICAL Solution

Note that DFT involves circular convolution, therefore the sequences $f(n)$ and $h(n)$ must be augmented by appending zeros to conform to a length of nine $(6 + 4 - 1 = 9)$ for each sequence.

MATLAB Solution

```
% Script file: convolutions
fn = [1 0 3 2];
hn = [-1 0 3 4 6 3];
gn = conv(fn,hn);                  % direct convolution
n = 0:8;
fn _ aug = [fn zeros(1,5)];
hn _ aug = [hn zeros(1,3)];
DTF _ fn _ aug = fft(fn _ aug);    % DFTs by zero padding
DTF _ hn _ aug = fft(hn _ aug);
DTF _ aug = DTF _ fn _ aug.*DTF _ hn _ aug;
ggn = ifft(DTF _ aug);             % linear convolution

figure(1)
subplot (3,1,1)
stem (n,gn); hold on; plot(n,gn); ylabel ('Amplitude');
xlabel('time index n')
title(' Linear convolution [g(n)=conv(f(n),h(n))] vs. n')
subplot(3,1,2)
DFT _ lin = fft(gn,9);
stem (n,abs(DFT _ lin));hold on; plot(n,abs(DFT _ lin));
xlabel ('frequency W');
ylabel('Magnitude');
title ('Magnitude of DFT[g(n)] vs.W')
subplot(3,1,3)
stem (n, angle(DFT _ lin)); hold on; plot(n, angle(DFT _ lin));
title ('Phase of DFT[g(n)] vs. W'); xlabel('frequency W')
ylabel('Angle in rad.')

figure(2)
subplot(3,1,1)
```

```
stem (n,ggn);hold on; plot(n,ggn); ylabel('Amplitude');
xlabel ('time index n')
title(' Circular convolution g(n)=ifft[DTF(f(n)*DTF(h(n)] vs. n')
subplot (3,1,2)
DFT _ cir = fft(ggn,9);
stem(n,abs(DFT _ cir));hold on; plot(n,abs(DFT _ cir));
xlabel('frequency W');
ylabel('Magnitude');
title('Magnitude of DFT[g(n)] vs. W')
subplot (3,1,3)
stem(n, angle(DFT _ cir));hold on; plot(n, angle(DFT _ cir));
xlabel('frequency W');
title ('Phase of DFT[g(n)] vs. W');ylabel('Angle in rad.')

figure(3)
error = gn-ggn;
bar(n,error); title('[Errors] vs. time index n')
ylabel ('Amplitude')
xlabel (' time index n')
```

The script file *convolutions* is executed and the results are indicated in Figures 5.31 through 5.33.

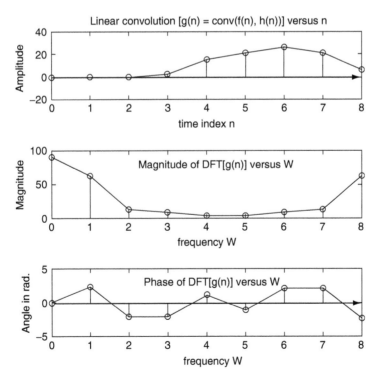

FIGURE 5.31
Plots of part a of $g(n)$ and DFT[$g(n)$] by evaluating the linear convolution of Example 5.6.

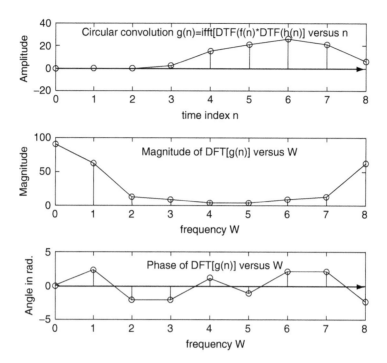

FIGURE 5.32
Plots of part b of $g(n)$ and DFT$[g(n)]$ evaluated by circular convolution of Example 5.6.

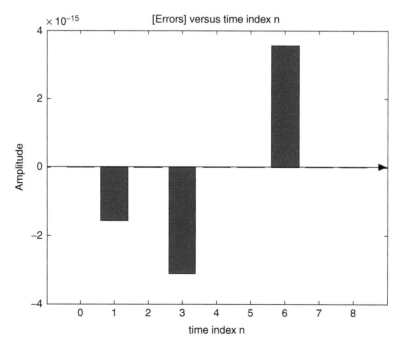

FIGURE 5.33
Error plot of part a with respect to part b of Example 5.6.

Note that the error in the evaluation of $g(n)$ performed by comparing the linear convolution with respect to the circular convolution, is extremely small, in the order of 10^{-15}.

Example 5.7

Let

$$f(n) = \begin{cases} 0 & n < 16 \\ 1 & 16 \le n \le 32 \\ 0 & 33 \le n < 64 \end{cases}$$

Create the script file *rectang* that returns the following plots:

a. $f(n)$ versus n
b. The magnitude and phase plots of *DFfT[f(n)]* versus k, and compare it with the magnitude and phase plots *of DFT[f(n)]* versus k, over the ranges $0 \le k \le 32$ and $0 \le k \le 64$

MATLAB Solution

```
% Script file: rectang
k = linspace(0,63,64);
fn = [zeros(1,16) ones(1,16) zeros(1,32)];
DFT _ fn = fft(fn,256);
DFT _ fn64 = fft(fn,64);
DFT _ fn32 = fft(fn,32);
W=0:1:255;

figure(1)
plot (k,fn,'o',k,fn),axis([0 63 0 1.1]); title(' f(n) vs. t')
xlabel ('time index n'); ylabel('Amplitude')

figure(2)
subplot (2,1,1)
plot (W,abs(DFT _ fn));
title(' Magnitudes of the DTFT and DFT')
ylabel('Amplitude')
%axis([0 259 0 20])
hold on
WW=linspace(0,255,32);
plot(WW,abs(DFT _ fn32),'*',WW,abs(DFT _ fn32))
axis([0 259 0 20])
subplot(2,1,2)
plot(W,abs(DFT _ fn));
ylabel('Amplitude')
hold on
WWW= linspace(0,255,64);
plot(WWW,abs(DFT _ fn64),'o',WWW,abs(DFT _ fn64));
xlabel(' Index k');axis([0 259 0 20]);

figure(3)
subplot(2,1,1)
plot(W,angle(DFT _ fn));
title(' Phase of the DTFT and DFT')
ylabel('Phase angle')
%axis([0 259 -4 4])
hold on
plot(WW,angle(DFT _ fn32),'*',WW,angle(DFT _ fn32))
```

```
subplot(2,1,2)
plot(W,angle(DFT _ fn))
hold on
plot(WWW,angle(DFT _ fn64),'o',WWW,angle(DFT _ fn64));
ylabel('Phase angle')
xlabel('Index k')
```

The script file *rectang* is executed and the results are indicated in Figures 5.34 through 5.36.

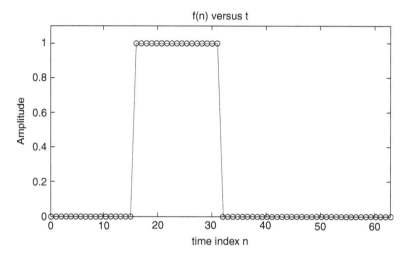

FIGURE 5.34
Plot of *f(n)* of Example 5.7.

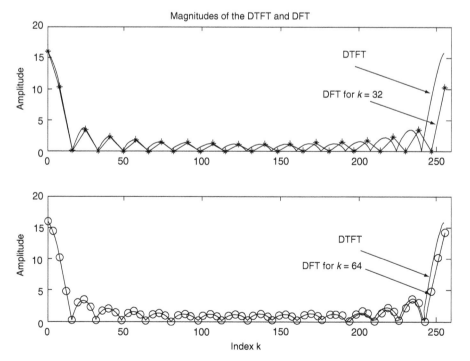

FIGURE 5.35
Magnitude plots of DTFT and DFT of *f(n)* of Example 5.7.

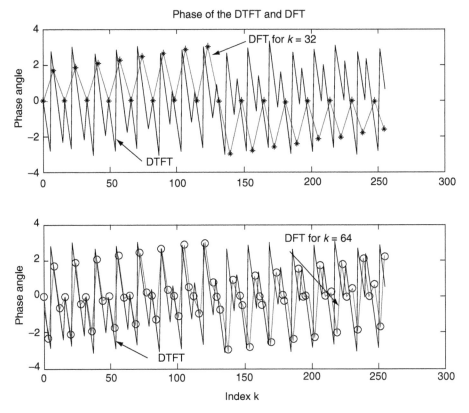

FIGURE 5.36
Phase plots of DTFT and DFT of *f(n)* of Example 5.7.

Example 5.8

Let a communication signal consist of a random noise signal plus an information signal *f(t)* consisting of three frequencies—1000, 2000, and 3000 Hz.

The noise signal and information signal *f(t)* are defined as follows:

$$noise = [0 \ 0.5*randn(1, 255)]$$

$$f(t) = 0.7sin(2\pi1000 \ t) + 0.93cos(2\pi \ 2000 \ t) + 0.69sin(2\pi3000 \ t)$$

Let the communication signal be given by *signal(t)* = *noise* + *f(t)*, over the range $0 < t < (1/8000)255$, where the sampling period is *Ts* = (1/8000) s and its sequence length of *N* = 256.

Create the script file *signal_n* that returns the contaminated signal *signal(t)* in the time and frequency domains. Observe that in the time domain, the frequencies 1000, 2000, and 3000 Hz are not clearly evident, while in the frequency domain its present is clearly visible.

MATLAB Solution
```
% Script file: signal _ n
fs = 8000;Ts =1/fs;
N=256;t = linspace(0,Ts*255,256);noise = [0 0.5*randn(1,255)];
```

```
signal _ n = noise+0.7*sin(2*pi*1000*t)+0.93*cos(2*pi*2000*t)+0.69*sin
                                                          (2*pi*3000*t);


figure(1)
plot (t(1:33),signal _ n(1:33))
title('[noise+0.3*sin(2*pi*1000*t)+0.93*cos(2*pi*2000*t)+0.69*sin(2*pi*
3000*t)] vs t')
xlabel('time'); ylabel('Amplitude');

figure(2)
DFT _ mag = abs(fft(signal _ n,256));
DFT _ phase = angle(fft(signal _ n,256));
subplot (2,1,1)
DFTX = DFT _ mag(1:1:128);
W = [1:1:N/2].*fs/N;
plot (W,DFTX); title('DFT Magnitude and Phase')
ylabel ('Magnitude');xlabel('frequency in Hertz')
subplot (2,1,2)
DFTY= DFT _ phase(1:1:128);
W = [1:1:N/2].*fs/N;
plot (W,DFTY);xlabel('frequency in Hertz'); ylabel ('Angle')
```

The script file *signal_n* is executed and the results are indicated in the Figures 5.37 and 5.38.

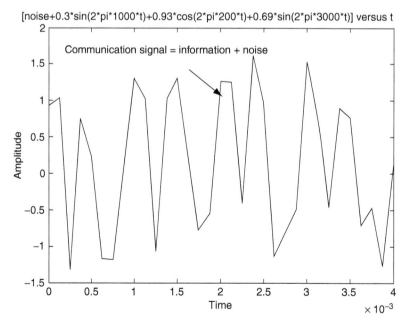

FIGURE 5.37
Plot of *signal(t) = noise + f(t)* of Example 5.8.

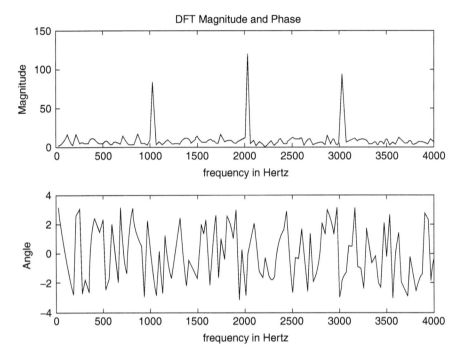

FIGURE 5.38
Plots of DFT of *signal(t)* = *noise* + *f(t)* of Example 5.8.

Example 5.9

Verify the linearity property of DFT that states, in short, the following:

The sum of $DFT[f_1(n)] + DFT[f_2(n)] + DFT[f_3(n)]$ ... is equal to the DFT of the sums of $(f_1(n) + f_2(n) + f_3(n) ...)$, for the following specific case:

$$DFT[u(n)] + DFT\left[\left(\frac{1}{3}\right)^n u(n)\right] + DFT\left[7\cos\left(\frac{n\pi}{3}\right)u(n)\right] =$$

$$DFT\left[u(n) + \left(\frac{1}{3}\right)^n u(n) + 7\cos\left(\frac{n\pi}{3}\right)u(n)\right] \quad \text{for } 1 \le n \le 16$$

where

$$f_1(n) = u(n)$$

$$f_2(n) = \left(\frac{1}{3}\right)^n u(n)$$

$$f_3(n) = 7\cos\left(\frac{n\pi}{3}\right)u(n)$$

and

$$f(n) = f_1(n) + f_2(n) + f_3(n)$$

Create the script file *linearity* that returns the following plots:

   a. figure(1): *[f(n) = f1(n) + f2(n) + f3(n)]* versus *n*, *abs{DFT[f(n)]}* versus *W*, and *angle{DFT[f(n)]}* versus *W* (Figure 5.39)
   b. figure(2): *abs{DFT[f1(n)}* versus *W*, *abs{DFT[f2(n)}* versus *W*, and *abs{DFT[f3(n)}* versus *W* (Figure 5.40)
   c. figure(3): *angle{DFT[f1(n)}* versus *W*, *angle{DFT[f2(n)}* versus *W*, and *angle{DFT[f3(n)}* versus *W* (Figure 5.41)
   d. figure(4): Let *DFT[f(n)] = DFT[f1(n)] + DFT[f2(n)] + DFT[f3(n)]*; obtain plots of *abs {DFT[f(n)] = abs{[fft(fn, 16)]}* and *angle{DFT[f(n)] = angle{fft(fn, 16)}* (Figure 5.42)
   e. figure(5): Let us estimate the error in time and frequency by using the following relations (Figure 5.43):
      i. The error in time = *[f1(n) + f2(n) + f3(n)] − ifft{DFT[f(n)]}*
      ii. *error_magnitude_DFT = abs{DFT[f1(n) + f2(n) + f3(n)]} − abs{DFT[f1(n)]} − abs{DFT[f2(n)]} − abs{DFT [f3(n)]}*
      iii. *error_phase_DFT = angle{DFT[f1(n) + f2(n) + f3(n)]} − {angle[DFT[f1(n)] + angle{DFT[f2(n)]} + angle{DFT[f3(n)]}*

MATLAB Solution
```
% Script file: linearity
n = 0:1:15;
f1n = [ones(1,16)];       % discrete time sequences
f2n = (1/3).^n;
f3n = 7*cos(n*pi/3);
fn = f1n + f2n + f3n;
DFT _ fn = fft(fn,16);

figure(1)
subplot (3,1,1)
stem (n,fn);hold on; stairs(n,fn); ylabel('Amplitude');
title(' [f(n) = f1(n) + f2(n) + f3(n)] ')
subplot(3,1,2)
stem(n,abs(DFT _ fn));hold on; stairs(n,abs(DFT _ fn));
ylabel('abs [ DFT[f(n)]]');
subplot (3,1,3)
stem (n,angle(DFT _ fn));hold on;stairs(n,angle(DFT _ fn));
ylabel ('angle[  DFT [f(n)]]'); xlabel('frequency W in rad.')

figure(2)
DFT _ f1= fft(f1n,16);    % magnitude of the DFT's of the discrete
                                              sequences
DFT _ f2 = fft(f2n,16);
DFT _ f3 = fft(f3n,16);
subplot (3,1,1)
stem(n,abs (DFT _ f1));hold on; stairs(n,abs(DFT _ f1));
ylabel ('abs [ DFT[f1(n)]]');
title ('[Magnitudes of the DFTs of f1(n), f2(n) & f3(n)] vs. W ')
```

```
subplot (3,1,2)
stem(n,abs(DFT _ f2));hold on; stairs(n,abs(DFT _ f2));
ylabel('abs [ DFT[f2(n)]]');
subplot (3,1,3)
stem (n, abs(DFT _ f3));hold on;stairs(n,abs(DFT _ f3));
xlabel ('frequency W in rad.');ylabel('abs [ DFT[f3(n)]]')


figure(3)                    % phases of the DFT's of the discrete
sequences
subplot(3,1,1)
stem (n,angle(DFT _ f1));hold on; stairs (n,angle(DFT _ f1));
ylabel ('angle [ DFT[f1(n)]]');
title ('[Phases of the DFTs of f1(n), f2(n) & f3(n)] vs. W')
subplot (3,1,2)
stem(n,angle(DFT _ f2)); hold on; stairs (n,angle(DFT _ f2));
ylabel('angle [ DFT[f2(n)]]');
subplot (3,1,3)
stem (n, angle(DFT _ f3));hold on;stairs(n,angle(DFT _ f3));
xlabel ('frequency W in rad.'); ylabel('angle[ DFT[f3(n)]]');


figure(4)                    % magnitude and phase of the sum of DFTs
DFT _ fn = DFT _ f1+ DFT _ f2 + DFT _ f3;
DFTfn = fft(fn,16);
Subplot (2,1,1)
stem (n,abs(DFT _ fn)); hold on; stairs(n,abs(DFT _ fn));
ylabel ('abs [ DFT[f(n)] ]');xlabel('frequency W in rad.');
title(' DFT[f1(n)] + DFT[f2(n)] + DFT[f3(n)] vs. W')
subplot (2,1,2)
stem (n,angle(DFT _ fn));hold on; stairs (n,angle(DFT _ fn));
xlabel ('frequency W in rad.'); ylabel ('angle [ DFT[fn(n)] ]');


figure(5)                    % error analysis in time and frequency
IDFT = ifft(DFT _ fn);
error _ time = fn-IDFT;
error _ mag _ DFT = abs(DFTfn)-abs(DFT _ fn);
error _ phase _ DFT = angle(DFTfn)-angle(DFT _ fn);
subplot(3,1,1)
bar (n, error _ time);
title (' Error analysis in time and frequency ')
ylabel('Amplitude in time')
subplot (3,1,2)
bar (n,error _ mag _ DFT);
ylabel ('[abs(DFTs)]');
subplot(3,1,3)
bar (n,error _ phase _ DFT)
ylabel (' [angle(DFTs)]')
xlabel ('frequency W in rad.');
```

The script file *linearity* is executed, and the results are shown in Figures 5.39 through 5.43.

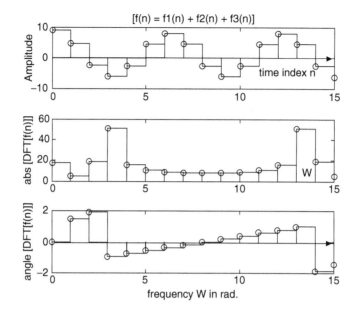

FIGURE 5.39
Plots of $[f(n) = f1(n) + f2(n) + f3(n)]$ and $abs\{DFT[f(n)]\}$ of Example 5.9.

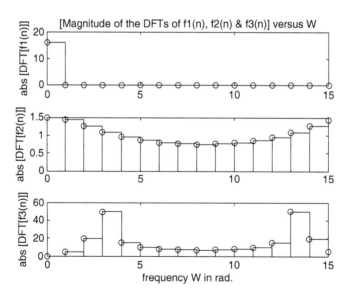

FIGURE 5.40
Plots of $absDFT$ of $f_1(n)$, $f_2(n)$, and $f_3(n)$ of Example 5.9.

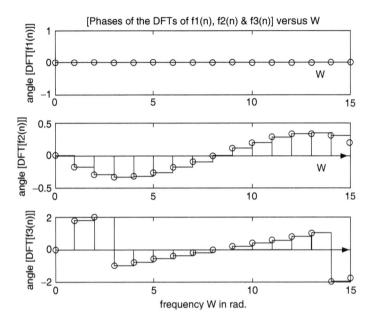

FIGURE 5.41
Plots of *angleDFT* of $f_1(n)$, $f_2(n)$, and $f_3(n)$ of Example 5.9.

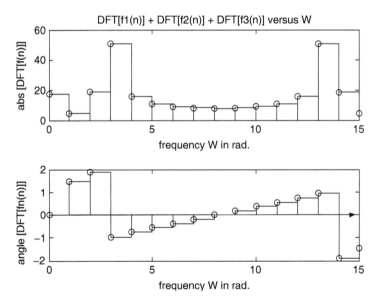

FIGURE 5.42
Plots of *magnitude* and *phase angle* of the sums of DFT of $f_1(n)$, $f_2(n)$, and $f_3(n)$ of Example 5.9.

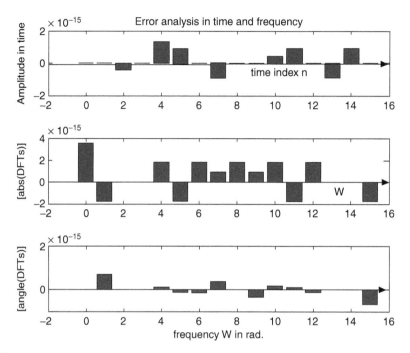

FIGURE 5.43
Time and frequency error plots of Example 5.9.

Example 5.10

Let the audio analog signal $f(t)$ be given by

$$f(t) = \begin{cases} 0 & 0 \le t \le 64Ts \\ 7\cos(2\pi 1000\, t) & 64Ts \le t < 128Ts \\ 0 & 128Ts \le t < 192Ts \\ 7\cos(2\pi 1000\, t) & 192Ts \le t \le 256Ts \end{cases}$$

and sampling rate be $Ts = 1/8000$ samples/s.

Convert the continuous time function $f(t)$ into a discrete sequence $f(n)$.

Create the script file *shift* that uses the function $f(n)$ to verify the property that states that a shift in time only affects the phase spectrum and not the magnitude spectrum by shifting $f(n)$ by 32 time samples and return the following plots:

a. figure(1): $f(n)$ versus n, $abs\{fft[f(n)]\}$ versus f, and $angle\{fft[f(n)]\}$ versus f (Figure 5.44)

b. figure(2): Repeat part a for $f(n - 32)$ (Figure 5.45)

c. figure(3): [*error in time* = $f(n) - f(n - 32)$] versus t, [*error_mag* = $abs\{fft[f(n)]\} - abs\{fft[f(n - 32)]\}$] versus f, and [*error_phase* = $angle\{fft[f(n)]\} - angle\{fft[f(n - 32)]\}$] versus f (Figure 5.46)

d. Estimate the cumulative magnitude and phase errors in the frequency domain

MATLAB Solution

```
% Script file: shift
N=256; fs = 8000;Ts =1/fs; shift =32;
t = linspace(0,Ts*255,256);
window = [zeros(1,64) ones(1,64) zeros(1,64) ones(1,64)];
f = 7*cos(2*pi*1000*t);fn=f.*window;
DFT _ fn = fft(fn,256);
fn _ shift = 7*cos(2*pi*1000*(t-shift).*window);
DFT _ fn _ shift = fft(fn _ shift,256);

figure(1)
subplot (3,1,1);              % analysis of f(n) = 7cos(2.pi.1000.t)*window
plot (t,fn); ylabel('Magnitude');
title ('f(n) = 7cos(2.pi.1000.n)*window, analyzed in the time and
                                    frequency domains')
subplot (3,1,2)
W= [1:1:N/2].*fs/N;
DFT _ mag = abs(DFT _ fn);
DFT _ phase = angle(DFT _ fn);
DFT _ mage = DFT _ mag(1:1:128);
DFT _ phasee = DFT _ phase(1:1:128);
plot(W,DFT _ mage); ylabel('abs [ DFT[f(n)]]');
subplot (3,1,3)
plot (W,DFT _ phasee);
ylabel ('angle[ DFT [f(n)]]'); xlabel (' frequency in Hz')

figure(2);             % Analysis of  7cos(2.pi.(n-shift))
DFT _ mags = abs(DFT _ fn _ shift);
DFT _ phases = angle(DFT _ fn _ shift);
DFT _ magess = DFT _ mag(1:1:128);
DFT _ phasess =DFT _ phase(1:1:128);
subplot (3,1,1)
plot (t,fn _ shift);
title(' f(n-32) analyzed in the time and frequency domains')
ylabel('Amplitude in time');
subplot(3,1,2)
plot(W,DFT _ magess);
ylabel('Magnitude DFT[f(n-32)]');
subplot(3,1,3)
plot(W,DFT _ phasess);
xlabel('frequency in Hz '); ylabel('Phase of DFT[f(n-32)]');

figure(3)               % error analysis
subplot(3,1,1)
error _ time = fn-fn _ shift;
plot(t, error _ time);
title(' Error analysis in the time and frequency domains ')
ylabel('Magnitude in time')
subplot(3,1,2)
error _ mag=abs(DFT _ fn)-abs(DFT _ fn _ shift);
bar(W,abs(error _ mag(1:128)));ylabel('Magnitude ');
subplot(3,1,3)
error _ phase=angle(DFT _ fn)-angle(DFT _ fn _ shift);
bar(W,abs(error _ phase(1:128)));ylabel('Phase ');
```

```
xlabel('frequency in Hz')
disp('******************************************************************')
disp('********************* R E S U L T S ***************************')
fprintf('For a shift of:%f\n',shift);
error _ mag _ DFT= sum(abs(DFT _ fn)-abs(DFT _ fn _ shift));
error _ phase _ DFT=sum(angle(DFT _ fn)-angle(DFT _ fn _ shift));
fprintf('Cumulative magnitude of the error in the frequency
                                    domain:%f\n',error _ mag _ DFT);
fprintf('Cumulative phase error in the frequency
                                    domain:%f\n',error _ phase _ DFT);
disp('******************************************************************')
```

The script file *shift* is executed, and the results are as follows:

```
>> shift

****************************************************************************
************************* R E S U L T S ***************************
For a shift of: 32.000000
Cumulative magnitude of the error in the frequency domain:-2245.168919
Cumulative phase error in the frequency domain:3.141593
****************************************************************************
```

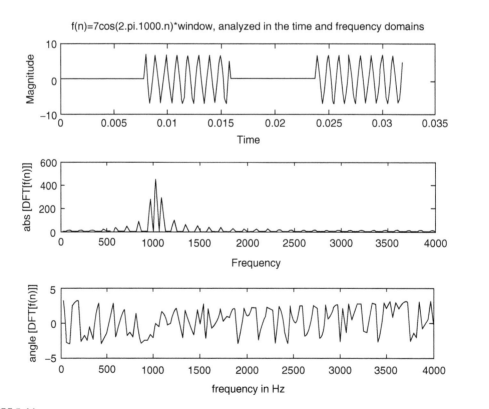

FIGURE 5.44
Plots of *f(n)* in time and frequency of Example 5.10.

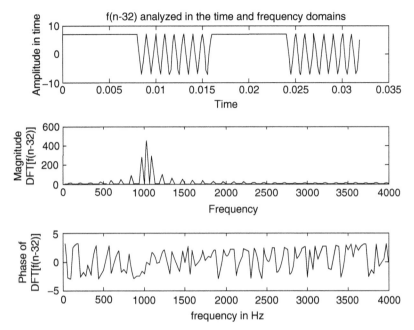

FIGURE 5.45
Plots of *f(n − 32)* in time and frequency of Example 5.10.

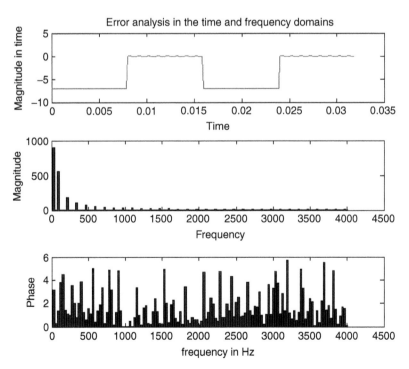

FIGURE 5.46
Error plots in time and frequency of Example 5.10.

Observe that the magnitude spectrum for *f(n)* and *f(n − 32)* are similar. The phase-error plot shows variations that are proportional to the shift (32). Note that the error in time is large; but in frequency, approximately 1000 Hz, the magnitude and phase errors are small.

Example 5.11

This example explores and verifies some properties that relate the correlation with the convolution process.

Let $f_1(n) = sin(2\pi n/16)$ and $f_2(n) = cos(2\pi n/16)$, over the range $n = 0, 1, 2, 3, 4, ..., 16$.
Create the script file *conv_corr* that returns the following:

a. figure(1): $f_1(n)$ versus *n* and $f_2(n)$ versus *n*, over the range $0 \leq n \leq 31$, using the *stairs* command

b. figure(2), explore the *conv* and *xcorr* functions in time by obtaining and comparing the following plots:

    i. *[corr = xcorr(f1n, f2n)]* versus *n*

    ii. *[conv1 = conv(f1n, f2(−n))]* versus *n*

    iii. *[conv2 = conv(f1(−n), f2n)]* versus *n*

c. figure(3) verifies that the maximum point occurs at the midrange, and this point corresponds to $R_{11}(0)$ and $R_{22}(0)$, by obtaining the following plots:

    i. *[R_{11} = aut_corr1 = xcorr(f1n, f1n)]* versus *n*

    ii. *[R_1 = aut_xcorr(f1n)]* versus *n*

    iii. *[R_{22} = aut_corr2 = xcorr(f2n, f2n)]* versus *n*

    Observe that $R_1 \neq R_{22}$, but $R_{11} = R_{22}$.

d. figure(4), normalized frequency analysis (verify that the two following plots are identical)

    i. *abs[conj[fft(f1n)] * fft(f2n)]* versus *W*

    ii. *abs[fft(xcorr(f1(n), f2(n)))]* versus *W*

e. figure(5), error analysis in time

    i. *[error1(n) = xcorr(f1(n), f2(n)) − conv(f1(n), f2(−n))]* versus *n* and

    ii. *[error2(n) = abs(abs(xcorr(f1(n), f2(n))) − abs(conv(f1(−n), f2(n))))]* versus *n*

MATLAB Solution

```
% Script file: conv _ corr
n=0:1:31;
f1n=sin(2*pi*n./16);
f2n=cos(2*pi*n./16);
f1n _ rev=fliplr(f1n);
f2n _ rev=fliplr(f2n);

figure(1)                    % discrete time plots of f1(n) and f2(n)
subplot(2,1,1)
stairs(n,f1n), axis([0 33 -1.1 1.1]),
title(' Stair plot of f1(n) vs. n');
ylabel('Amplitude ');
subplot(2,1,2)
stairs(n,f2n); axis([0 33 -1.1 1.1]),
title(' Stair plot of  f2(n) vs. n');
xlabel('time index n');ylabel('Amplitude');
```

```
figure(2)
corr = xcorr(f1n,f2n);
N= length (corr);
m = 0:1:N-1;
subplot (3,1,1)
stem(m,corr);title('Convolution and correlation plots');
ylabel('Amplitude');
subplot(3,1,2)
conv1=conv(f1n,f2n _ rev);
stem(m,conv1)
xlabel('time index n');ylabel('Amplitude');
subplot(3,1,3)
conv2=conv(f1n _ rev,f2n);
stem(m,conv2);
xlabel('time index n');ylabel('Amplitude');

figure(3)                % auto-correlations
mm=-31:31;
aut _ corr1=xcorr(f1n,f1n);aut=xcorr(f1n);
aut _ corr2=xcorr(f2n,f2n);
plot(mm,aut _ corr1,mm,aut _ corr2,mm,aut);
title('Auto-correlation plots');
ylabel('Amplitude');
legend('R11','R22','R1')

figure(4)                % frequency analysis
DFT _ corr = fft(corr,256);
DFT _ f1n = fft(f1n,256);
DFT _ f2n = fft(f2n,256);
DFT _ 12 = conj(DFT _ f1n).*DFT _ f2n;
k=1:1:256;
subplot (2,1,1)
plot (k/256,abs(DFT _ 12));title('abs[conj[fft(f1n)]*fft(f2n)] vs. W ')
ylabel('Magnitude')
subplot(2,1,2)
plot(k/256,abs(DFT _ corr));title('abs[fft(xcorr(f1(n),f2(n)))] vs. W')
xlabel('W'); ylabel('Magnitude')

figure(5)
subplot (2,1,1)
error1 = corr-conv1;
bar(m,error1);title('error1=corr[f1(n),f2(n)]-conv[f1(n),f2(-n)]')
ylabel('Amplitude');
subplot(2,1,2)
error2 = abs(abs(corr)-abs(conv2));
bar(m,error2)
title ('error2=abs[abs[corr[f1(n),f2(n)]]-abs[conv[f1(-n),f2(n)]]]')
xlabel ('time index n'); ylabel('Amplitude')
```

The script file *conv_corr* is executed, and the results are as follows:

```
>> conv _ corr
```

See Figures 5.47 through 5.51.

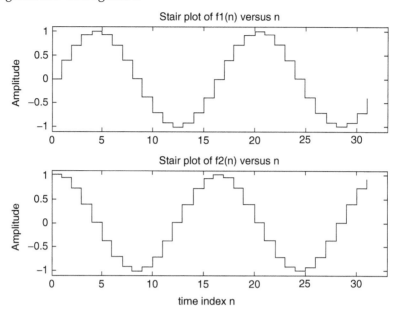

FIGURE 5.47
Plots of $f_1(n)$ and $f_2(n)$ of Example 5.11.

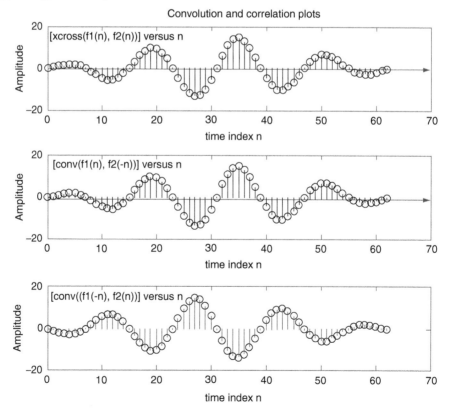

FIGURE 5.48
Convolution and correlation plots of part b of Example 5.11.

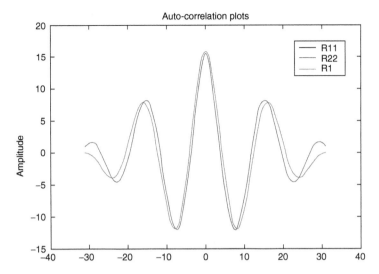

FIGURE 5.49
(See color insert following page 374.) Autocorrelation plots of part c of Example 5.11.

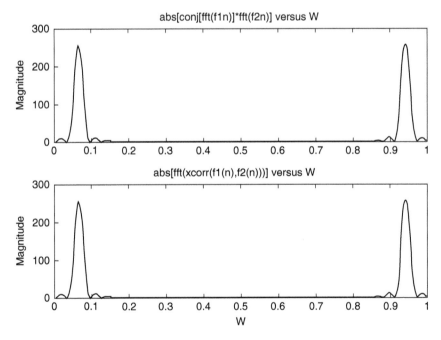

FIGURE 5.50
Plots of part d—*abs[conj[fft(f1n)] * fft(f2n)]* and *abs[fft(xcorr(f1(n), f 2(n)))]*—of Example 5.11.

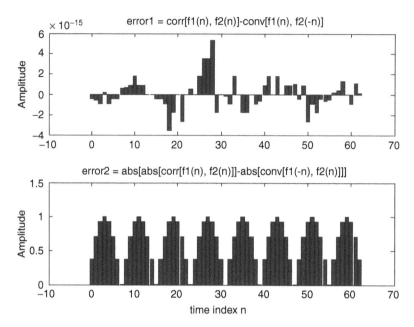

FIGURE 5.51
Error plots of part e of Example 5.11.

Example 5.12

The objective of this example is to explore the relation between the FT of a continuous time function and its discrete version.

Let $f(t) = 3e^{-5t}u(t)$.

Create the script file *FT_DFT* that returns the following plots:

a. $f(t)$ versus t, over the range $0 \le t \le 1$ (Figure 5.52)

b. Magnitude and phase of $[DFT[f(n)]]$ versus w (Figure 5.53)

c. Magnitude and phase of $[FT[f(t)]]$ versus w (Figure 5.54)

d. *error_mag* = magnitude of $[FT[f(t)]]$ − magnitude of $[DFT[f(n)]]$ versus w and *error_phase* = phase of $[FT[f(t)]]$ − phase of $[DFT[f(n)]]$ versus w (Figure 5.55)

MATLAB Solution

```
% Script file: FT _ DFT
N = 526;t = linspace(0,5,N);
fn =3*exp(-5*t); fs = 8000;Ts = 1/fs;
ws = 2*pi/Ts;
F = fft(fn);
Fdc = fftshift(F)*Ts;
W = ws*(-N/2:(N/2)-1)/N;

figure(1)
plot (t(1:100),fn(1:100));title ('[f(t)=3*exp(-5*t)] vs t')
ylabel ('Amplitude'); xlabel ('time');axis([0 1 0 3.5]);

figure(2)                              % discrete frequency analysis
subplot (2,1,1)
plot (W,abs(Fdc));title('abs[DFT[f(n)]] vs w');ylabel('Magnitude');
```

```
axis([-1e4 1e4 0 0.01])
subplot (2,1,2)
plot (W,angle(Fdc));title('angle[DFT[f(n)]] vs w');ylabel('Phase');
xlabel ('frequency w ');

figure(3)                                % analog frequency analysis
F _ table = 3./(5+j*W);
subplot (2,1,1)
absF _ table = abs(F _ table);
plot(W(200:400),absF _ table(200:400));
title('abs[FT[f(t)]] vs w');
ylabel('Magnitude');axis([-.1e4 .1e4 0 0.8]);
subplot (2,1,2)
plot (W,angle(F _ table));title('angle[FT[f(t)]] vs w');ylabel('Phase');
xlabel ('frequency w ');axis([-1e4 1e4 -2 2]);

figure(4)                                % error analysis in frequency
error _ mag = abs(F _ table)-abs(Fdc);
error _ phase = angle(F _ table)-angle(Fdc);
subplot (2,1,1)
plot (W(200:400) ,error _ mag(200:400));axis([-.2e4 .2e4 0 0.8]);
title ('error mag.=abs[FT[f(t)]]-abs[DFT[f(n)]] vs w n');
ylabel ('Magnitude'); xlabel ('index n')
subplot(2,1,2)
bar(W,error _ phase);title('error phase=angle[FT[f(t)]-angle[DFT[f(n)]]
                                                             vs w');
ylabel('Phase');xlabel('frequency w');
```

The script file *FT_DFT* is executed, and the results are as follows:

```
>> FT _ DFT
```

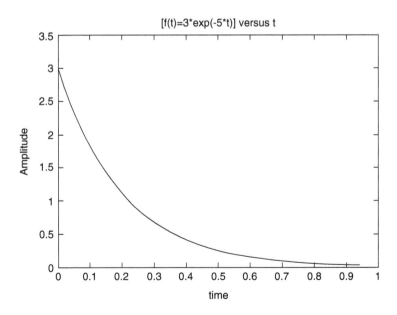

FIGURE 5.52
Plot of *f(t)* over the range $0 \leq t \leq 1$ of Example 5.12.

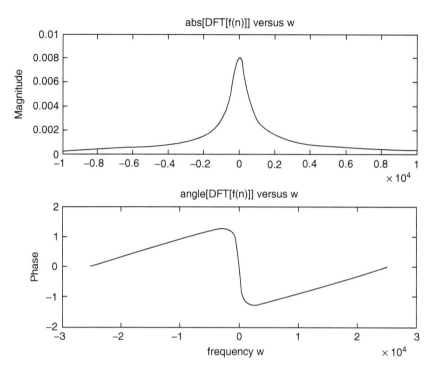

FIGURE 5.53
Plot of DFT[*f(n)*] of Example 5.12.

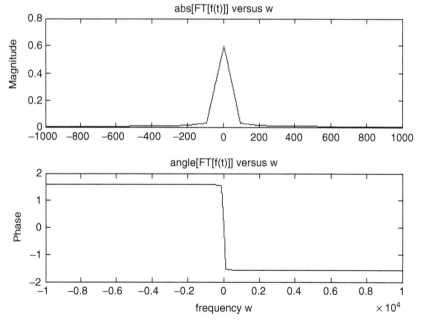

FIGURE 5.54
Plot of FT[*f(t)*] of Example 5.12.

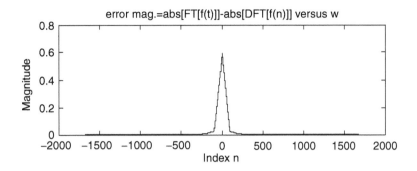

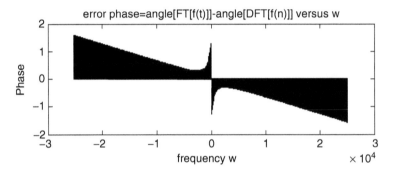

FIGURE 5.55
Error plot of FT[*f(n)*] with respect to DFT[*f(n)*] of Example 5.12.

Example 5.13

The system difference equation of an LTI system is given by

$$g(n) = 0.7g(n-1) - 0.3g(n-2) + f(n) + 0.9f(n-1)$$

a. Determine by hand the discrete-transfer function $H(z) = G(z)/F(z)$.
b. Determine by hand $H(e^{jW})$.
c. Obtain the plots of the input $f(n)$ versus n and output $g_f(n)$ versus n at steady state (Figure 5.56), over the range $1 \le n \le 50$, assuming that the input is $f(n) = (0.5)^n sin(0.033\pi n)u(n)$, using the MATLAB command *filter*.
d. Obtain the plot of the impulse response $h(n)$ by using the MATLAB command *impz* and $g(n)$ by using convolution (Figure 5.57), over the range $1 \le n \le 50$. Let $g_c(n)$ denote the convolution of $h(n)$ with $f(n)$ (time domain) and $g_f(n)$ be the system output obtained by the *filter* command (frequency domain).
e. Evaluate and plot the magnitude and phase of $fft[g_f(n)]$ and $fft[g_c(n)]$, over the range $1 \le k \le 50$ (Figure 5.58).
f. Estimate and plot (Figure 5.59) the error of the above two methods ($g_f(n)$ and $g_c(n)$) in the time domain defined by the following relation:

$$error(n) = g_f(n) - g_c(n), \quad over\ 1 \le n \le 50$$

g. Estimate and plot (Figure 5.59) the magnitude and phase errors of the two methods in the frequency domain defined by the following relations:

$$mag\_error = abs\{fft[g_f(n)]\} - abs\{fft[g_c(n)]\}$$

$$phase\_error = angle\{fft[g_f(n)]\} - angle\{fft[g_c(n)]\}$$

over the range $1 \le k \le 50$.

The MATLAB solution for Example 5.13 is given by the script file *discrete_syst* as follows:

ANALYTICAL Solution

Part a

$$g(n) = 0.7g(n-1) - 0.3g(n-2) + f(n) + 0.9f(n-1)$$

$$g(n) - 0.7g(n-1) + 0.3g(n-2) = f(n) + 0.9f(n-1)$$

Taking the ZT yields

$$G(z) - 0.7z^{-1}G(z) + 0.3z^{-2}G(z) = F(z) + 0.9z^{-1}F(z)$$

$$G(z)[1 - 0.7z^{-1} + 0.3z^{-2}] = F(z)[1 + 0.9z^{-1}]$$

$$H(z) = \frac{1 + 0.9z^{-1}}{1 - 0.7z^{-1} + 0.3z^{-2}}$$

Part b

Replacing $z = e^{jW}$

$$H(e^{jW}) = \frac{G(e^{jW})}{F(e^{jW})} = \frac{1 + 0.9e^{-jW}}{1 - 0.7e^{-jW} + 0.3e^{-j2W}}$$

MATLAB Solution
```
% Script file: discrete _ syst
num = [1 0.9];
den = [1 -0.7 0.3];
n = 1:1:50;
fn = 0.9.^n.*sin(0.033*pi.*n);
gnf = filter(num,den,fn);

figure(1)                                    % input and output plots
subplot(2,1,1)
plot(n,fn,'o',n,fn);
title('[System input sequence f(n)] vs n')
xlabel('time index n');ylabel('Magnitude')
subplot(2,1,2)
plot(n,gnf,'s',n,gnf);
```

```
title('[System output sequence g(n)] vs n, using filter')
xlabel('time index n');ylabel('Magnitude')

figure(2)                               % impulse analysis
hn = impz(num,den,n);
subplot (2,1,1)
stem(n,hn);
title(' [System impulse response h(n)] vs n, using impz')
ylabel('Amplitude'); xlabel('time index n')
subplot(2,1,2)
nn=0:1:50;
gnc=conv(hn,fn);
stem(nn(1:1:50),gnc(1:1:50))
title(' [System output g(n)] vs n, using convolution')
ylabel('Amplitude');xlabel('time index n')

figure(3)                               % frequency analysis
DFT _ gnf = fft(gnf); abs _ DFT _ gnf = abs(DFT _ gnf);
ang _ DFT _ gnf=angle(DFT _ gnf);
DFT _ gnc = fft(gnc);abs _ DFT _ gnc=abs(DFT _ gnc);
ang _ DFT _ gnc=angle(DFT _ gnc);
subplot(2,2,1); stem(nn(1:1:50),abs _ DFT _ gnf(1:1:50));
ylabel('abs[fft(gf(n))]');
title('abs[fft(gf(n))] vs k')
axis([0 50 0 20]);
subplot(2,2,2)
stem(nn(1:1:50),ang _ DFT _ gnf(1:1:50));title('angle[fft(gf(n))] vs k');
ylabel('angle[fft(gf(n))]');
axis([0 50 -4 4]);
subplot(2,2,3);
stem(nn(1:1:50),abs _ DFT _ gnc(1:1:50));title('abs[fft(gc(n))] vs k');
xlabel('index k');ylabel('abs[fft(gc(n))]');axis([0 50 0 20])
subplot(2,2,4)
stem(nn(1:1:50),ang _ DFT _ gnc(1:1:50));title('angle[fft(gc(n))] vs k');
xlabel('index k');ylabel('angle[fft(gc(n))]');
axis([0 50 -4 0])

figure(4)                               % error analysis
subplot(3,1,1)
error _ time = gnf(1:1:50)-gnc(1:1:50);
bar(nn(1:1:50),error _ time);axis([0 50 -0.5 0.5]);
title(' error plots in time and frequency')
ylabel('Amplitude')
subplot(3,1,2)
mag _ error = abs _ DFT _ gnf(1:1:50)- abs _ DFT _ gnc(1:1:50);
phase _ error = ang _ DFT _ gnf(1:1:50)- ang _ DFT _ gnc(1:1:50);
bar(nn(1:1:50),mag _ error); ylabel('Magnitude error')
axis([0 50 -10 10])
subplot(3,1,3)
bar(nn(1:1:50),phase _ error);ylabel('phase error');
axis([0 50 -10 10]); xlabel('index k');
```

The script file *discrete_syst* is executed, and the results are as follows:

```
>> discrete _ syst
```

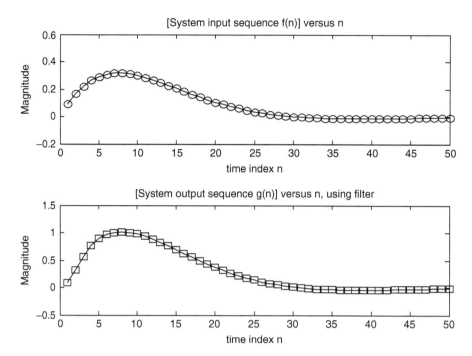

FIGURE 5.56
Plots of the system input *f(n)* and system output $g_f(n)$ of Example 5.13.

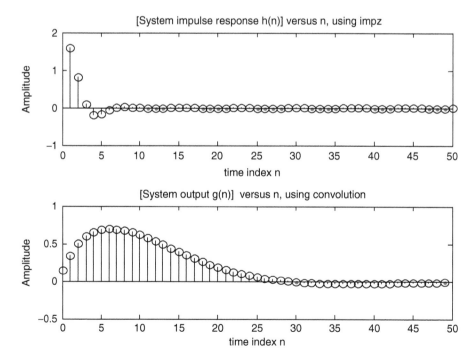

FIGURE 5.57
Plots of system impulse *h(n)* and system output $g_c(n)$ of Example 5.13.

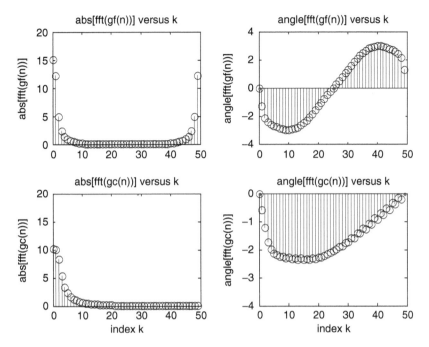

FIGURE 5.58
Magnitude and phase plots of *fft[gf(n)]* and *fft[gc(n)]*, over the range $1 \leq k \leq 50$ of Example 5.13.

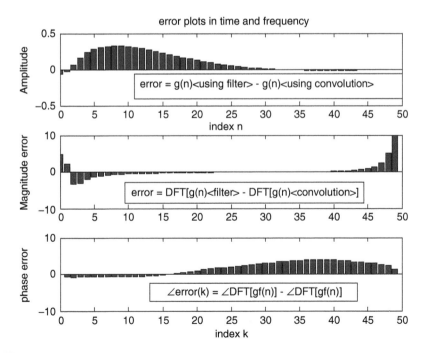

FIGURE 5.59
Error plots in the time and frequency domains of Example 5.13.

Example 5.14

Let the impulse response of an LTI system be given by $h(n) = \sum_{n=0}^{15} 0.5^n$ and the system input sequence be $f(n) = \sum_{n=0}^{15} 15\cos(5\pi n/3)$.

Create the script file *syst_anal* that returns the following plots:

a. *f(n)* versus *n* and *h(n)* versus *n*, using the *stem* command (Figure 5.60)

b. *g(n)* = {IDFT of the products of DFT [*f(n)*] times DFT [*h(n)*]} versus *n* (Figure 5.61)

c. *gnn(n)* = convolution of *h(n)* with *g(n)* versus *n* (Figure 5.61)

d. Estimate the error plot (in the time domain) of the two approaches (parts b and c), given by (Figure 5.62)

$$\text{error}(n) = [g(n) - gnn(n)] \text{ versus } n$$

MATLAB Solution
```
% Script file: syst _ anal
n = 0:15;
fn = cos(5*pi*n/3);
hn = 0.5.^n;                          % the length of convolution is 16*2-1
DFT _ F = fft(fn,31);
DFT _ H = fft(hn,31);
gn = ifft(DFT _ F.*DFT _ H);

figure(1)
subplot(2,1,1)
stem(n,fn)
ylabel('Amplitude');axis([0 15 -1.2 1.2]);
title(' [Input sequence f(n)] vs n ')
subplot(2,1,2)
stem(n,hn)
ylabel('Amplitude ');axis([0 15 -0.5 1.2]);
xlabel('time index n'); title('[Impulse response h(n)] vs n')

figure(2)                            % output plots
subplot(2,1,1)
nn=0:1:30;
plot(nn,gn);title('[IDFT of the products of the DFTs ] vs n');
xlabel('time index n '); ylabel('Amplitude');axis([0 25 -1.5 2]);
subplot(2,1,2)
gnn=conv(fn,hn);
plot(nn,gnn);axis([0 25 -1.5 2]);
xlabel('time index n'), ylabel('Amplitude')
title('[Direct convolution in the time domain] vs. n');

figure(3)
error = gn-gnn;
bar(nn,error);title(' Error =[part(b)-part(c)] vs. n' );
xlabel('time index n'), ylabel('Magnitude');
```

The script file *syst_anal* is executed, and the results are as follows:

```
>> syst _ anal
```

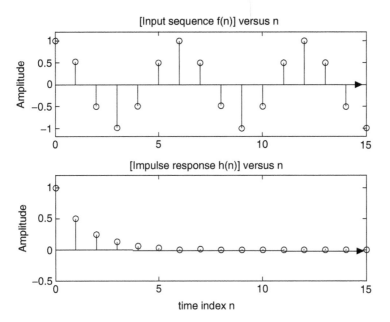

FIGURE 5.60
Plots of *f(n)* and *h(n)* of Example 5.14.

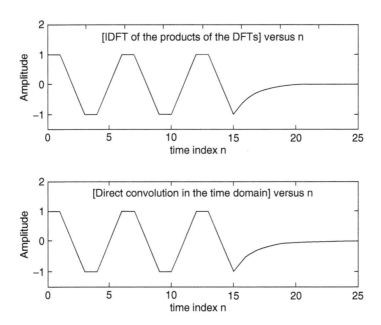

FIGURE 5.61
Plots of *g(n)* = IDFT of the products of DFT[*h(n)*] with DFT[*f(n)*], and time convolution of *h(n)* with *f(n)* of Example 5.14.

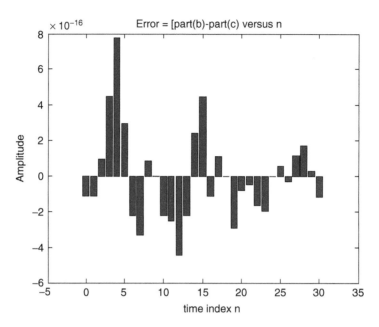

FIGURE 5.62
Time-domain error given by error = part b − part c of Example 5.14.

Example 5.15

Create the script file *analys* that analyzes the discrete system transfer function given by

$$H(e^{jW}) = \frac{G(e^{jW})}{F(e^{jW})} = \frac{5 + 0.9e^{-jW} + e^{-2jW}}{1 - 0.7e^{-jW} + 0.3e^{-j2W}}$$

and returns the following:

a. The pole/zero plot of $H(e^{jW})$ (evaluates the system stability) (Figure 5.63)
b. The plot of *abs{$H(e^{jW})$}* versus W (Figure 5.64)
c. The plot of angle *{$H(e^{jW})$}* versus W
d. List all the poles, zeros, and stand alone term of the transfer function $H(z=e^{jW})$ and compare the results obtained with the results of part a
e. Determine the ROC of $H(z = e^{jW})$
f. The magnitude of the system poles (length from origin)
g. The partial fraction expansion of $H(z)$
h. $h(n)$ by using the symbolic MATLAB inverse ZT of $H(z)$

MATLAB Solution
```
%Script file = analys
num = [5 0.9 1];
den = [1 -0.7 0.3];

figure(1)
zplane(num,den); title('Pole and zero plot of H(z)')

figure(2)
W = -pi:pi/99:pi;
[H,W] = freqz(num,den,W);
```

```
magH = abs(H);phaseH = angle(H);
subplot (2,1,1);
plot (W,magH);title ('abs[H(exp(-jW))] vs W');ylabel('Magnitude');
subplot (2,1,2);
plot (W,phaseH);title('angle[H(exp(-jW))] vs W');
ylabel ('Phase in rad.');xlabel('frequency W in rad');
[z,p,k] = tf2zp(num,den);
m = abs(p);
disp('****************************************');
disp('*********** R E S U L T S ************');
disp('****************************************')
disp('The zeros of H(z) are:');disp(z);
disp('The poles of H(z) are:');disp(p);
disp('The gain  of H(z) is:');disp(k);
disp('The magnitude of the poles are:');
disp(m);disp('The ROC is outside the circle with radius:');
disp(max(m))
% partial fraction expansion
[r,pp,kk] = residuez (num,den);
disp('Partial fraction coefficients of H(z)')
disp('the residues are:')
disp(r);disp('the stand alone term is :');disp(kk);
syms z Fz;
disp('The partial fraction expansion is:')
Fz = r(1)/(z-pp(1))+r(2)/(z-pp(2))+ kk
hn = iztrans(Fz);disp('The impulse response h(n) is:');hn
disp('^^^^^^^^^^^^^^^^^^^^^^^^^^^^^^^^^^^^^^^^^^')
```

The script file *analys* is executed, and the results are as follows:

```
>> analys

****************************************
*********** R E S U L T S ************
****************************************

The zeros of H(z) are:
                         -0.0900 + 0.4381i
                         -0.0900 - 0.4381i
The poles of H(z) are:
                         0.3500 + 0.4213i
                         0.3500 - 0.4213i

The gain of H(z) is:
                          5
The magnitude of the poles are:
                                  0.5477
                                  0.5477
The ROC is outside the circle with radius:
                                      0.5477

Partial fraction coefficients of H(z)
the residues are:
                    0.8333 - 4.5295i
                    0.8333 + 4.5295i

the stand alone term is:
  kk =
       3.3333
```

```
The partial fraction expansion is:
Fz =
(5/6-1/426*i*3723311^(1/2))/(z-7/20-1/20*i*71^(1/2))+
(5/6+1/426*i*3723311^(1/2))/(z-7/20+1/20*i*71^(1/2))+10/3

The impulse response h(n) is:
hn =
     127/9*charfcn[0](n)-97/18*(-6/(i*71^(1/2)-7))^n-979/1278*i*(-6/(i*71^
(1/2)-7))^n*71^(1/2)97/18*6^n*(1/(7+i*71^(1/2)))^n+979/1278*i*6^n*(1/(7+i*71^
(1/2)))^n*71^(1/2)
^^^^^^^^^^^^^^^^^^^^^^^^^^^^^^^^^^^^^^^^^^^^^^^^^^^^^^^^^^^^^^^^^^^^^^^^^^^^
```

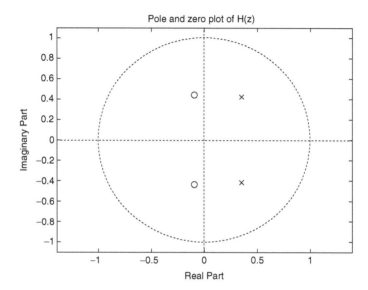

FIGURE 5.63
Pole/zero plot of $H(e^{jW})$ of Example 5.15.

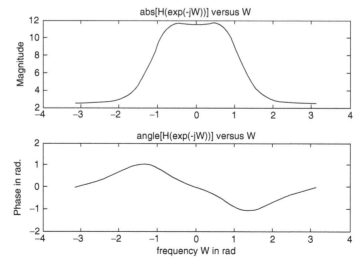

FIGURE 5.64
Plot of $abs\{H(e^{jW})\}$ and angle $\{H(e^{jW})\}$ of Example 5.15.

Example 5.16

Create the script file *pole_zeroes* that returns the rational transfer function $H(z)$, the numerator and denominator coefficients of $H(z)$, and the first 11 coefficients of their system impulse response $h(n)$ given the following system's zeros, poles, and gain.

$z_1 = 0.21$ $p_1 = -0.45$
$z_2 = 3.14$ $p_2 = 0.67$
$z_3 = -0.3 + j5$ $p_3 = 0.81 + j0.72$
$z_4 = -0.3 - j5$ $p_4 = 0.81 - j0.72$
$k = 2.3$ $p_5 = -0.33$

MATLAB Solution
```
% Script file: pole _ zeroes
z1 = 0.21 ; p1 =-0.45;
z2 = 3.14 ; p2 = 0.67;
z3 =-0.3 + 5j; p3 = 0.81+0.72j;
z4 = -0.3-5j; p4 = 0.81-0.72j;
k =2.3; p5 = 0.33;
zr = [z1 z2 z3 z4];
pl = [p1 p2 p3 p4 p5];
z = zr'; p=pl';
[num,den] = zp2tf(z,p,k);
n =0:10;
disp ('*************************************************')
disp ('*************** R E S U L T S ***************')
disp ('*************************************************')
tf(num,den,1)
disp('********************')
disp ('Coefficients of the numerator polynomial of H(z)'); disp(num');
disp('********************')
disp ('Coefficients of the denominator polynomial of H(z)');
disp(den');L=11; [y,t] = impz(num,den,L);
disp('********************')
disp ('The first 11 coefficients of the impulse response h(n) are:');
disp('********************')
disp(' coef. num.      h(n)')
disp('********************')
[n' y]
disp ('*************************************************')
```

The script file *pole_zeroes* is executed, and the results are as follows:

```
>> pole _ zeroes

************************************************************
****************R E S U L T S ****************************
************************************************************
Transfer function:
2.3 z^4 - 6.325 z^3 + 54.6 z^2 - 192.4 z + 38.05
--------------------------------------------------------
z^5 - 2.17 z^4 + 1.837 z^3 - 0.1757 z^2 - 0.43 z + 0.1169
Sampling time: 1
**************************
Coefficients of the numerator polynomial of H(z)
```

```
        0
    2.3000
   -6.3250
   54.6006
 -192.4085
   38.0520
***************************
Coefficients of the denominator polynomial of H(z)
    1.0000
   -2.1700
    1.8366
   -0.1757
   -0.4300
    0.1169
***************************
The first 11 coefficients of the impulse response h(n) are:
***************************
   coef. num.        h(n)
***************************
ans =
      0               0
    1.0000          2.3000
    2.0000         -1.3340
    3.0000         47.4817
    4.0000        -86.5192
    5.0000       -236.1448
    6.0000       -346.0348
    7.0000       -311.8158
    8.0000       -125.3485
    9.0000        148.4517
   10.0000        376.3726
***************************
```

Note that the system transfer function is given by

$$H(z) = \frac{2.3z^{-1} - 6.325z^{-2} + 54.6006z^{-3} - 192.4085z^{-4} + 38.0520z^{-5}}{1 - 2.17z^{-1} + 1.8366z^{-2} - 0.1757z^{-3} - 0.43z^{-4} + 0.1169z^{-5}}$$

Note also that the corresponding system impulse response $h(n) = Z^{-1}[H(z)]$ is given by

$$h(n) = 0 + 2.3\delta(n-1) - 1.3403\delta(n-2) + 47.4817\delta(n-3) - 86.5192\delta(n-4) - 236.1448\delta(n-5) - 346.0348\delta(n-6) - 311.8158\delta(n-7) - 155.3485\delta(n-8) + 148.4517\delta(n-9) + 376.3726\delta(n-10)$$

Example 5.17

Let the system difference equation be given by

$$g(n) = 0.5g(n-1) + 2f(n)$$

Create the script file *diff_eq_res*, that simulates the given difference equation and returns the response plots for $n \leq 10$, for each of the following inputs:

   a. $f(n) = u(n)$ (step response)
   b. $f(n) = \delta(n)$ (impulse response)

MATLAB Solution

```
% Script file: diff_eq_res
gnpast_step =0; m=1;
for n =0:15;
    gnstep(m) = 0.5*gnpast_step+2*1;
    gnpast_step = gnstep(m);
    m = m+1;
end
gnstep = [zeros(1,5) gnstep];

figure(1)
subplot(2,1,1)
mm =-5:1:15;
stem(mm,gnstep);title('Simulation of step response')
xlabel('time index n');ylabel('Amplitude'),axis([-5 10 0 4.3])
subplot(2,1,2)
dirack=1;h_past=0;k=1;
for n =0:15;
    hn(k) = 0.5*h_past + 2*dirack;
    dirack = 0;
    h_past = hn(k);
    k = k+1;
end;
hn = [zeros(1,5) hn];
stem(mm,hn)
xlabel('time index n');ylabel('Amplitude');axis([-5 10 0 2.1])
title('Simulation of impulse response')
```

The script file *diff_eq_res* is executed, and the simulated results are shown in Figure 5.65.

```
>> diff_eq_res
```

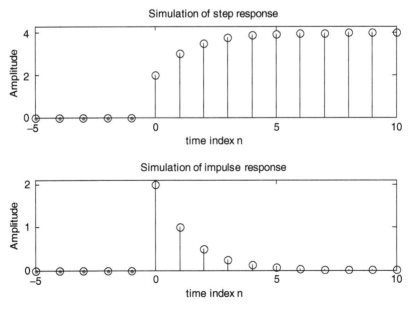

FIGURE 5.65
Simulation plots of system step and impulse responses of Example 5.17.

Example 5.18

Evaluate by hand the transfer function of the difference equation of Example 5.17, given by $g(n) = 0.5g(n - 1) + 2f(n)$, and then create the script file *step_imp* that returns the coefficients of discrete-step and impulse responses using system simulates and the MATLAB commands *dstep* and *impz*, respectively. Also verify that the solutions obtained (Example 5.17) fully agree with the solutions obtained using the MATLAB commands by evaluating the cumulative errors by comparing the two methods.

ANALYTICAL Solution

$$g(n) - 0.5g(n - 1) = 2f(n)$$

$$G(z)[1 - 0.5z^{-1}] = 2F(z)$$

$$H(z) = \frac{G(z)}{F(z)} = \frac{2}{1 - 0.5z^{-1}}$$

MATLAB Solution
```
% Script file: step _ imp
num = [0 2];
den = [1 -0.5];
nn = 16;

figure(2)
subplot(2,1,1)
dstep(num,den)
subplot(2,1,2)
impz(num,den)
nnn = 0:10;
[Y,X] = dstep(num,den); [YY,XX] = impz(num,den);
a = gnstep(6:16)';b =(Y(2:12))';error _ step=a-b;
c =hn(6:16)';d=(YY(2:12))';error _ imp=c-d;
disp('************************************************************')
disp('*********************R E S U L T S**********************')
disp('************************************************************')
disp(' The first 11 coefficients of the step response')
disp('using the simulation of the difference equation are :')
a
disp('using the command dstep are :')
b
disp(' The first 11 coefficients of the impulse response')
disp('using the simulation of the difference equation are :')
c
disp('using the command impz are :')
d
disp(' The cumulative step error [a-b] is :')
sum(error _ step)
disp(' The cumulative impulse error [c-d] is :')
sum(error _ imp)
disp('************************************************************')
```

The script file *step_imp* is executed, and the results are as follows:

```
>> step _ imp
****************************************************************
********************** R E S U L T S *************************
****************************************************************
 The first 11 coefficients of the step response
using the simulation of the difference equation are :
a =
    2.0000    3.0000    3.5000    3.7500    3.8750    3.9375    3.9688
    3.9844    3.9922    3.9961    3.9980
using the command dstep are :
b =
    2.0000    3.0000    3.5000    3.7500    3.8750    3.9375    3.9688
    3.9844    3.9922    3.9961    3.9980
 The first 11 coefficients of the impulse response
using the simulation of the difference equation are :
c =
    2.0000    1.0000    0.5000    0.2500    0.1250    0.0625    0.0313
    0.0156    0.0078    0.0039    0.0020
using the command impz are:
 d =
    2.0000    1.0000    0.5000    0.2500    0.1250    0.0625    0.0313
    0.0156    0.0078    0.0039    0.0020
 The cumulative step error [a-b] is :
 ans =
        0
 The cumulative impulse error [c-d] is :
 ans =
        0
****************************************************************
```

Example 5.19

The objective of this example is to illustrate the computational efficiency of evaluating the *fft* when the sequence length is a power of 2.

Create the script files *time_analysis_512* and *time_analysis_513*, that returns the time statistics by evaluating the *fft* of the signal $f(t) = cos(2\pi850t)$, twice using 512 and 513 samples, respectively, with a sampling rate of $Ts = 1/8000$ in both.

MATLAB Solution

```
% Script file: time _ analysis _ 512
% statistic analysis using 512 points; a power of 2
clear;
tstart = clock; tic;
ta = cputime;
fs = 8000;Ts=1/fs;N=512;
t = linspace(0,Ts*511,512);signal _ 512 = cos(2*pi*850.*t);
DFT _ 512 = fft(signal _ 512,512);
estimea = etime(clock,tstart);
tictoca = toc;
t1= cputime-ta;
disp('******************** R E S U L T S ********************')
disp('**********   T I M E    S T A T I S T I C S **********')
disp('   f(n)  consist of 512 samples   ')
fprintf('est.time =   %6.6f\n',estimea);
```

```
fprintf('cpu.time =    %6.6f\n',t1);
fprintf('time elapse =%6.6f\n',tictoca);
disp('************************************************************')

% Script file: time _ analysis _ 513
% statistic analysis using 513 points
clear;
tstarta = clock; tic;
taa = cputime;
fs = 8000;Ts=1/fs; N= 513;
t = linspace(0,Ts*512,513); signal _ 513 = cos(2*pi*850.*t);
DFT _ 613 = fft(signal _ 513,513);
estimeaa = etime(clock,tstarta);
tictocaa = toc;
t1a = cputime-taa;
disp ('************** T I M E    S T A T I S T I C S ********')
disp ('   f(n)  consist of 513 samples   ')
fprintf ('est.time =   %6.6f\n',estimeaa);
fprintf ('cpu.time =   %6.6f\n',t1a);
fprintf ('time elapse =%6.6f\n',tictocaa);
disp ('************************************************************')

>> time _ analysis _ 512

******************R E S U L T S ************************
********** T I M E    S T A T I S T I C S *******
f(n)  consist of 512 samples
est.time =   0.010000
cpu.time =   0.020000
time elapse =0.020000
*********************************************************

>> time _ analysis _ 513

********** T I M E    S T A T I S T I C S ******
f(n)  consist of 513 samples
est.time =   0.020000
cpu.time =   0.030000
time elapse =0.030000
*********************************************************
```

Observe from the preceding results that the time efficiency increases when using 512 samples (power of 2) instead of 513 samples, when evaluating (executing) the *fft* of the (arbitrary) function $f(t) = cos(2\pi 850t)$.

5.5   Application Problems

P.5.1   Determine if the following systems are LTI:

  a. $g(n) = f(n) \cdot f(n)$

  b. $g(n) = log[f(n)]$

  c. $g(n) = cos[f(n)]$

d. $g(n) = f(n) + f(n - 1) - f(n + 1)$

e. $g(n) = \sqrt{f(n)}$

f. $g(n) = |f(n)|$

P.5.2 Let the impulse response of an LTI system be $h(n)$. Evaluate in the time and frequency domains if the following impulse responses correspond to stable systems:

a. $h(n) = 0.5^n u(n)$

b. $h(n) = 0.5^n[u(n - 3) - u(n - 10)]$

c. $h(n) = 1.05^n u(n)$

d. $h(n) = 0.5^n \cos(n)u(n)$

e. $h(n) = 1.05^n u(n)$

f. $h(n) = 1/5^{-n} u(n)$

g. $h(n) = 1.05^n u(-n)$

P.5.3 Use MATLAB to create a script file that returns each of the signals of P.5.2 over the range $0 \le n \le 31$ and their respective energies.

P.5.4 Let $f(n)=[1\ 2\ 3\ -1\ -2\ -3\ 4\ 5\ 6]$ and $g(n) = [4\ 3\ 2\ 1\ 1\ 2\ 3\ 4\ 5]$. Evaluate

a. The linear convolution

b. The circular convolution

P.5.5 Verify that the circular convolution is commutative using the sequences of P.5.4.

P.5.6 Let $h(n) = 0.5^n u(n)$ and $f(n) = \delta(n) + 2\delta(n - 1) - 3\delta(n - 3)$.

a. Sketch both the sequences by hand.

b. Obtain the plot of the convolution given by $f(n) \otimes h(n)$.

c. Obtain the autocorrelation plot of the function $f(n)$ with $h(n)$.

d. Obtain the cross-correlation plot of $f(n)$ with $h(n)$.

P.5.7 Let $f_1(n) = \delta(n) + 2\delta(n - 1) - 3\delta(n - 3)$, $f_2(n) = u(n) - u(n - 10) - 3\delta(n - 3)$, and $f_3(n) = u(n + 10) - u(n - 10) + \delta(n - 3)$.

a. Determine the sequence length of the following convolutions by hand:

i. $g_1(n) = f_1(n) \otimes f_2(n)$

ii. $g_2(n) = f_1(n) \otimes f_3(n)$

iii. $g_3(n) = f_2(n) \otimes f_3(n)$

b. Determine by hand the first and last nonzero element of part a

c. Determine by hand $g_1(0)$, $g_2(-1)$, and $g_3(3)$

d. Verify the results of parts a, b, and c by using MATLAB

P.5.8 Let $f_1(n) = 0.5^n u(n)$ and $f_2(n) = 1.5^n u(-n)$. Obtain the infinite length convolution sequence $g_3(n) = f_1(n) \otimes f_2(n)$.

P.5.9 Let the causal LTI system equation be given by $g(n) = g(n - 1) + f(n) + 3f(n - 3)$, with $g(n) = 0$, for $n \le 0$ and $f(n) = 0$, for $n \le 0$. Use MATLAB and simulate the output $g(n)$ over the range $0 \le n \le 10$.

P.5.10 Let $f(n) = \sin(\pi n/5)u(n) + \cos(\pi n/10)u(n)$ and $h(n) = 0.3^n u(n - 5)$. Verify that the cross-correlation of $f(n)$ with $h(n)$ is identical to the convolution given by $f(n) \otimes h(-n)$.

P.5.11 Create a script file that returns a random sequence $f(n)$ with length 10 and its convolution with itself 10 times ($f(n) \otimes f(n) \otimes f(n)$ 10 times). Discuss the shape of the resulting wave and compare it with the standard Gaussian function.

P.5.12 Determine the discrete FS coefficients of the following periodic series:

a. $f_1(n) = cos(\pi n/2)$

b. $f_2(n) = sin(\pi n/2) + cos(\pi n/3)$

P.5.13 Let the causal LTI system equation be $g(n) = (1/3)g(n - 1) + 3f(n)$, with $g(n) = 0$ for $n \leq 0$.

a. Determine and plot its impulse response $h(n)$ over the range $0 \leq n \leq 31$ by simulating the difference equation.

b. Repeat part a for the case of a step response

c. Determine its transfer system function given by $H(z) = G(z)/F(z)$

d. Repeat parts a and b using the commands *impz* and *dstep*

e. Evaluate and plot the system poles and zeros of $H(z)$

f. Discuss and state if the system is stable by observing the locations of its system poles

g. Discuss if the system is stable by verifying the BIBO criteria

P.5.14 Use MATLAB and verify the following relations:

a. $\sum_{n=-\infty}^{+\infty} [f(n)]^2 = \dfrac{1}{2\pi} \int_{-\pi}^{+\pi} |F(e^{jW})| dW$ Parseval's theorem

b. $f(0) = \dfrac{1}{2\pi} \int_{-\pi}^{+\pi} F(e^{jW}) dW$ Initial value theorem

c. $F(e^{j0}) = \sum_{n=-\infty}^{+\infty} f(n)$

using the following sequence $f(n) = [1\ 2\ 3\ 4\ 5\ 6\ 7\ 8\ 9\ 10]$.

P.5.15 Let $f(n) = 0.3^n$, for $n = 0, 1, 2, 3, …, 8, 9, 10$. Use MATLAB to verify the following relations:

a. $f(n - 2) \Leftrightarrow e^{jW2}F(e^{-jW})$

b. $f^*(n) \Leftrightarrow F^*(e^{-jW})$ (where * denotes complex conjugate)

c. $f(-n) \Leftrightarrow F(e^{-jW})$

P.5.16 Evaluate the DTFT, DFT, and ZT of the following sequences:

a. $f_1(n) = \delta(n) + 2\delta(n - 1) - 3u(n)$

b. $f_2(n) = u(n) - u(n - 10) - 3r(n)$

c. $f_3(n) = 3\cos(3n/5)u(n) + 2\sin(2n/5) + 5\delta(n - 2)$

P.5.17 Given the difference equation $g(n) = -3g(n - 1) + 2g(n - 2) + 5$, determine its system transfer function and indicate if the system is causal and stable.

P.5.18 Given the system transfer function

$$H(e^{jW}) = \frac{0.5 - e^{-jW}}{1 + 0.5e^{-jW} - 0.5e^{-j2W}}$$

determine its system difference equation, system realization, and indicate if the system is causal and stable.

P.5.19 The ZT of the right-sided sequence $f(n)$ is given by

$$F(z) = \frac{1 + 0.2z^{-1}}{(1 + 031z^{-1})(1 - 0.43z^{-1})}$$

Use MATLAB to determine the sequence $f(n)$ by a partial fraction expansion and verify the result obtained.

P.5.20 Let the input sequence of an LTI system be given by $f(n) = 2(1/6)^n u(n) - u(n-1)$ and its output sequence by $g(n) = (1/6)^n u(n)$. Determine
  a. The impulse response
  b. The step response
  c. The system difference equation
  d. The system transfer function
  e. The system realization (in terms of delays, adders, and multipliers)

P.5.21 A system consists of two cascaded subsystems with the following transfer functions:

$$H_1(e^{jW}) = \frac{1 - e^{-jW}}{1 + e^{-jW}}$$

and

$$H_2(e^{jW}) = \frac{0.5 - e^{-jW}}{1 + 0.5e^{-jW} - 0.5e^{-j2W}}$$

Determine
  a. The difference equation of each subsystem
  b. The poles and zeros for each subsystem
  c. The difference equation of the overall system
  d. The poles and zeros of the overall system
  e. The overall system transfer function
  f. A system realization in terms of delays, adders, and multipliers
  g. The ROC of each subsystem and overall system

P.5.22 Let the system transfer function of an LTI system be given by

$$H(z) = \frac{1 + 0.2z^{-1}}{1 + 031z^{-1} + 0.43z^{-2}}$$

Use MATLAB to obtain
  a. the system impulse response $h(n)$
  b. its partial fraction expansion of $H(z)$
  c. a pole/zero plot of the transfer function $H(z)$

    d. the ROC

    e. the magnitude and phase plots of $H(z)$

    f. the magnitude of its poles

    g. its output if its input is $f(n) = 0.3^n u(n)$

    h. the state space equations using the command *tf2ss*

P.5.23 An input sequence consisting of the sum of two sinusoidal waves with frequencies of 0.3 and 0.7 Hz, respectively, and unity amplitude, and a length of 30 samples, sampled at its Nyquist rate use MATLAB and obtain plots of the input and output sequences in the time and frequency domains for the following cases:

    a. Up sample by a factor of 2

    b. Down sample by a factor of 3

6

Analog and Digital Filters

The world we have created today has problems which cannot be solved by thinking the way we thought when we created them.

Albert Einstein

6.1 Introduction

Since the discovery of the electric waves by Campbell and Wagner (1915), filters have gained importance in almost every phase of electrical engineering.

The term *filter* describes a variety of different systems. They all have as the main objective the removal of unwanted parts of a given signal such as noise, or extract useful parts of the signal such as information.

Filter theory was first studied in the 1920s (based on the image parameters) primarily by Zobel. In the 1930s, filter theory evolved into what is known today as *modern filter theory*, based on well-defined mathematical equations, models, and specs.

Over time the concept of *system transfer* or *gain function* became the foundations of filter theory, and a systematic approach was developed through the efforts of a number of mathematicians, scientists, and engineers, to name just a few: Cauer, Foster, Bode, Darlington, and many others.

Filters can be classified based on different criteria such as

- Circuit configuration (lattice/ladder/etc.)
- Elements used (active/passive/digital/analog/etc.)
- Physical characteristics of the element used (electrical/electronics/optical/mechanical/electromechanical/etc.)
- Frequency response (pass band/low pass [LP]/high pass [HP]/etc.)
- Impulse response (infinite/finite)

In this chapter, the criteria used is based on the system frequency response, meaning that these devices (filters) allow certain bands of frequencies to pass while attenuating others, by using a variety of elements and configurations in their implementation.

In most cases, filters are specified by the frequency response characteristics in the pass and stop bands. In its simplest version, the filter specs must include a nonzero amplitude associated with the pass band, a region with a zero amplitude associated with the stop band, and a transition region between the pass and stop bands.

An ideal filter presents only two states in its output, either it allows to *pass* or to *stop* a range of frequencies present in its input.

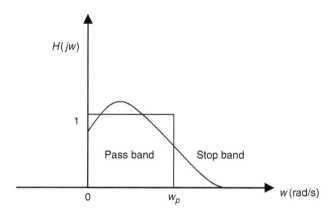

FIGURE 6.1
Frequency characteristics of an LPF.

In practice, the magnitude response characteristics in the pass and stop band(s) are not constant and are specified within some acceptable tolerances. Moreover, the ideal transition bands between the pass and stop bands is supposed to be as abrupt as possible.

Practical filters are approximations of ideal filters within acceptable tolerances. The more closely a real filter resembles its ideal model, the better its design and performance.

The quality of a filter implementation can be evaluated in terms of the deviation from its ideal characteristics.

Figure 6.1 illustrates the characteristics of a low-pass filter (LPF). The ideal case is shown by constant values with an abrupt discontinuity, whereas the real case is shown by a continuous line.

6.2   Objectives

After completing this chapter the reader should be able to

- Define filter characteristics and specs
- Understand the basic filter prototypes
- Define and describe the characteristics of the basic filter prototypes
- Normalize and denormalized a prototype in terms of the desire magnitude and frequency specs
- Define the basic filters in terms of equations, plots, or sketches (LP, HP, band pass [BP], band reject [BR])
- Identify and discuss filter terminology and specs such as cutoff frequencies, pass band, stop band, transition band, bandwidth (BW), and ripples
- Understand the characteristics of the most common filter prototypes used in industry such as Butterworth, Chebyshev, Cauer or Elliptic, and Bessel
- Understand the steps involved in the analysis and synthesis processes involving filters
- Know the meaning of the Q of a filter and how it affects its complexity

- Transform a prototype to any filter specs (denormalized process)
- Understand the concept of distortion (magnitude, phase, or delay)
- Understand the concepts of linear phase
- Understand the difference between active and passive filters
- Understand the difference between analog and digital filters
- Understand the differences between finite impulse response (FIR) and infinite impulse response (IIR) digital filter
- Understand the (practical) importance of using windows in the design of filters
- Understand that there is a trade-off when windows are used, such as ripple amplitude and the transition time
- Develop simple filter prototype sections such as T and π (used in electrical communications)
- Understand the effects of using polynomials in the process of modeling and system implementation
- Use MATLAB® in the analysis and synthesis process involving filters

6.3 Background

R.6.1 Filters are devices that are specified in the frequency domain. The synthesis and analysis process is generally done in the frequency domain.

R.6.2 An ideal filter displays two main bands in the frequency domain

- A pass band
- A stop band

R.6.3 Any sinusoidal input with any one of the frequencies of the pass-band region emerges or is present in the output with its magnitude almost (little attenuation) unchanged if the filter is implemented with passive elements (RLC).

R.6.4 Any applied sinusoidal input with frequencies in the stop-band region is not present in its output or if it is its magnitude is drastically attenuated.

R.6.5 In the pass-band region, the magnitude frequency response of an ideal passive filter is unity, whereas in the stop band it is zero.

R.6.6 An ideal filter allows any sinusoidal input to either pass or be rejected.

R.6.7 Assuming the inputs to a filter are sinusoidals and its frequencies are in the pass-band region of the filter, then they emerge as outputs. If the output is amplified by a constant $k > 1$ over the pass band, then the filter presents a gain. The type of filters with a gain $k > 1$ is called active.

R.6.8 Assuming the inputs to a filter are sinusoidals and its frequencies are in the pass-band region of the filter, and if they emerge as outputs with different gains, then the output suffers from amplitude distortion and the output is a distorted version of its input.

R.6.9 Strictly, ideal filters that present two regions (pass band and stop band) cannot be implemented in practice, since they are noncasual devices (the output is returned before the input is applied).

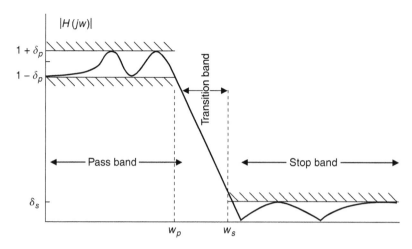

FIGURE 6.2
Filter's bands.

R.6.10 A real filter presents three bands (for an LPF), which is illustrated in Figure 6.2
- Pass band, for $w < w_p$
- Stop band, for $w > w_s$
- Transition band, for $w_p < w < w_s$

    The plot or sketch shown in Figure 6.2 may be used to spec a real (LP) filter.

R.6.11 A real filter creates a phase shift between the input and its output (where phase indicates the propagation delay).

R.6.12 If a phase shift (delays) of all the sinusoidals that make up the input with respect to their output is constant in the range of frequencies defined by the pass band, then all the sinusoidal inputs are delayed by the same amount of time and the filter is said to present no *delay distortion*, and the output waveform is a replica of the input as far as phase is concerned.

R.6.13 If the phase shift of a filter is linear with respect to their frequencies in the pass-band region, then the filter is said to present no phase or delay distortion.

R.6.14 Filters are best described in terms of the transfer function.

    Recall that the *transfer function* is the ratio of the output $V_o(w)$ voltage to the input voltage $V_i(w)$ (in the frequency domain).

    Recall that the transfer function is also known as the *gain* or *system function*.

R.6.15 Recall that the transfer function is denoted by $H(w)$, and it is a function of the complex frequency ($s = jw$). Therefore, $H(w)$ or $H(jw)$ is a complex function consisting of two parts:

- A real part
- An imaginary part

    The description of the transfer function $H(w)$ translates graphically into two plots:

- Plot $H(w)$ versus w called the magnitude or system gain plot
- Plot $\angle[H(w)]$ versus w called the phase gain plot

R.6.16 The magnitude plot is usually given in decibels. Recall that a positive value in decibels indicates amplification and a negative value indicates attenuation.
Recall also that the dB is defined as

$$dB = 20log_{10}[H(jw)]$$

R.6.17 Real filters are specified by means of tolerances (or fluctuations) in the pass and stop bands, referred to as ripples, which is also expressed in decibels, in terms of its peak or the maximum value in the pass-band region denoted by α_p, and with the minimum attenuation in the stop band denoted by α_s defined in R.6.19.

R.6.18 For the case of the LPF illustrated in Figure 6.2, the equations below are the type used to define the tolerances in the pass, stop, and transition bands.

a. $1 - \delta_p \le |H(jw)| \le 1 + \delta_p$ for $w \le w_p$ (pass-band region)

b. $|H(jw)| < \delta_s$ for $w_s < w < \infty$ (stop-band region)

c. $\delta_s|H(jw)|<1 - \delta_p$ for $w_p < w < w_s$ (transition-band region)

where δ_p is called the pass-band ripple, δ_s the stop-band ripple, w_p the filter's cutoff frequency, and w_s the filter's stop-band frequency.

R.6.19 The ripple tolerances in the pass and stop bands denoted by δ_p and δ_s are specified in decibels as

a. $\alpha_p = -20log_{10}(1 - \delta p)$ dB

b. $\alpha_s = -20log_{10}(\delta s)$ dB

R.6.20 The filter characteristic may be specified in terms of the loss function $L(jw)$ instead of the gain function $H(jw)$.

R.6.21 The loss function is defined as the inverse of the gain or transfer function. Then $L(jw) = \frac{1}{H(jw)}$, or $\frac{V_i(jw)}{V_o(jw)}$, given in decibels as

$$-20log_{10}|H(jw)|$$

Note that a 0 dB (gain or loss) implies no gain (or loss), meaning that the output is equal to its input, or the filter (system) acts like a short circuit.

R.6.22 Filters are classified according to the pass-band regions as

a. LPF

b. High-pass (HPF)

c. Band-reject (BRF) or band-stop (BSF)

d. Band-pass (BPF)

The magnitude and phase plots of these filters are shown in Figures 6.3 and 6.4.

R.6.23 Note that the concept of BW applies only to the LPF and BPF and is defined as follows:

$$BW \ (LPF) = w_p$$

$$BW \ (BPF) = w_u - w_l.$$

R.6.24 More complicated filters with multiple pass- or stop-band regions can be created using the basic four prototype structures defined in R.6.22 (LPF, HPF, BPF, BRF).

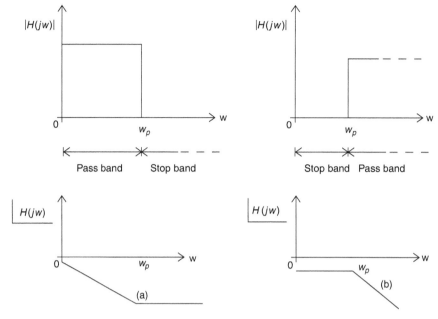

FIGURE 6.3
LP and HP filter's (magnitude and phase) characteristics.

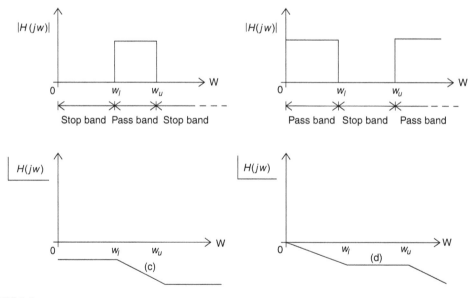

FIGURE 6.4
BP and BR filter's (magnitude and phase) characteristics.

R.6.25 The design of any of the filters defined in R.6.22 is accomplished by first designing a normalized LPF prototype. A normalized LPF is a filter with a cutoff frequency of $w_p = 1$ rad/s connected to a load of $R = 1\ \Omega$.

R.6.26 The normalized LPF will be implemented employing electrical elements using the notations R_n, L_n, and C_n (referred to as normalized resistors, inductors, and capacitors).

R.6.27 Once a normalized LPF prototype is obtained (w_{np} = 1 rad/s and load R_0 = 1 Ω), the denormalized LPF process starts by adjusting the $w_p = k_0$ rad/s, where k_0 is the desired BW by scaling the values of the C's, L's, and R's. This adjustment is referred to as an LP-to-LP transformation, and it is accomplished by what is known as a frequency and impedance scaling.

R.6.28 The frequency scaling transformation, LP_n (low-pass normalized) to LP (denormalized) is accomplished by using the following relation:

$$H(jw) = H_n(jw/k_0)$$

R.6.29 In terms of circuit elements, frequency scaling is accomplished by dividing each normalized capacitor and inductor labeled C_n and L_n by k_0, and by leaving the resistances unchanged, as follows:

$$C_{f_s} = \frac{C_n}{k_0}$$

$$L_{f_s} = \frac{L_n}{k_0}$$

$$R_{f_s} = R_n \text{ (unchanged)}$$

where the index f_s denotes a frequency scale transformation.

R.6.30 The relation indicated in R.6.28 translates into the following:
- if $k_0 > 1$ the BW of $H(jw)$ spreads
- if $k_0 < 1$ the BW of $H(jw)$ contracts

R.6.31 By using frequency scaling, the transformation of a normalized LPF to an arbitrary LPF with an arbitrary BW is easily accomplished by simple divisions. Filter handbooks, design tables, and graphs were used in the past to accomplish these transformations. Tables are still used, but they are gradually being replaced by sofisticated computer software packages such as MATLAB.

R.6.32 Once frequency scaling was attained, the next objective in the synthesis process is the impedance scaling, that is, the load connected to the filter output can be an arbitrary load R (not necessarily R_n = 1 Ω).

R.6.33 Impedance scaling is attained by multiplying all inductors and resistors by R and dividing each capacitor by R as follows:

$$R \leftarrow R_n R$$

$$L \leftarrow R L_{f_s}$$

$$C \leftarrow \frac{C_{f_s}}{R}$$

R.6.34 The transformation of a normalized LPF prototype into another denormalized filter type can be accomplished by the variable substitution in $H(s)$, shown in Table 6.1 and the corresponding circuit structures shown in Table 6.2. The variable substitution translates into a mathematical mapping or a segment (elements) or block substitution.

TABLE 6.1

Frequency Transformations

• LP-to-HP	$s \Rightarrow w_o/s$
• LP-to-BP	$s \Rightarrow \dfrac{(s^2 + w_o^2)}{(s * \mathrm{BW})}$
• LP-to-BR	$s \Rightarrow \dfrac{(s * \mathrm{BW})}{(s^2 + w_o^2)}$

Note: BW = $w_u - w_l$ (bandwidth) and $w_o = \sqrt{w_u - w_e}$, where w_o denotes the filter's center frequency and represents the geometric mean of w_u (upper cutoff) and w_e (lower cutoff).

TABLE 6.2

Element Transformation

Normalized Elements	HPF Elements	BPF Elements	BRF Elements

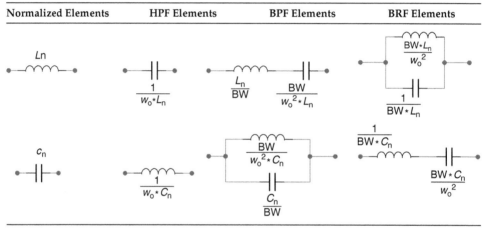

Note: L_n and C_n are normalized LPF prototypes.

For example, the LP-to-HP transformation means that the s's in the transfer function $H(s)$ of the LP normalized filter are replaced by w_c/s and in terms of circuit elements that translates that the inductor L_n be replaced by a capacitor

$$C = \frac{1}{(w_c * L_n)}$$

R.6.35 The designing specs of a filter can best be visualized by a sketch in which the bands of frequencies that are highly attenuated or blocked within certain limits as well as the band of frequencies that pass with little or no attenuation are indicated within limits.

Analytically, this translates into a set of parameters that specify a filter given by

a. The deviation parameters δ_p and δ_s

b. The frequencies w_p and w_s and transition width $w_s - w_p$

c. The filter's order given by n

R.6.36 The number of poles of the filter's transfer function $H(s)$, given by its denominator, defines the filter's order n.

R.6.37 Clearly, the higher the order (n), the better the approximation. Higher-order filters imply higher cost translated in terms of number of elements and filter complexity.

R.6.38 First- and second-order filters are often the starting prototypes in most designs, the reason for which they are presented in the following.

First-order system transfer functions for LP and HP filters are as follows:

a. $H_{LP}(s) = \dfrac{w_o}{s + w_o}$

b. $H_{HP}(s) = \dfrac{s}{s + w_o}$

Note that there are no first-order BP or BR filters.

R.6.39 Second-order system transfer functions for LP, HP, BP, and BR filters are as follows:

1. $H_{LP}(s) = \dfrac{A_1}{s^2 + B_1 s + w_o^2}$

2. $H_{HP}(s) = \dfrac{A_2 s^2}{s^2 + B_2 s + w_o^2}$

3. $H_{BP}(s) = \dfrac{A_3 s}{s^2 + B_3 s + w_o^2}$

4. $H_{BR}(s) = \dfrac{A_4 s + A_5}{s^2 + B_4 s + w_o^2}$

where the As, Bs, and w_os are constants that control the transfer function range and domain.

R.6.40 For purely practical reasons, the filters presented in this book are divided into

a. Analog

where the analog filters are subdivided into

i. Passive

ii. Active

b. Digital

whereas digital filters can be

i. FIR

ii. IIR

R.6.41 Passive analog electrical filters exclusively employ R, L, and C elements. Therefore, its poles and zeros for simple RC or RL filters are always real, whereas complex poles are produced when both inductors and capacitors are employed.

R.6.42 In addition to the passive elements (R_s, L_s, and C_s), active filters use active components such as operational amplifiers (OA). OA are electronic devices that use an external power source to amplify and filter at the same time the input signal.

By using feedback amplification, the effects of inductors can be emulated by employing capacitors. It is appropriate to mention that capacitors are much cheaper than inductors and follow more closely the theoretical conditions, creating in this way efficient and inexpensive filter implementations.

These filters are used in diverse applications such as equalizers, noise-reducing devices, compression devices, and expansion and enhancement devices.

In all the devices, the assumption is that the signal being processed (filter) is either an electrical voltage or a current carrying information such as sound, data, or video.

R.6.43 Once an OA is used in a filter, the filter is referred to as an active filter.

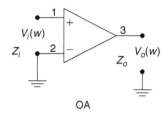

Voltage gain = 200,000 (106 db)
$Z_o = 75\ \Omega$
$Z_i = 2\ M\Omega$

Output voltage swing = +13 V, –13 V
Small signal BW = 1 MHz

FIGURE 6.5
The operational amplifier.

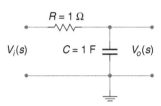

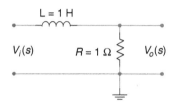

The transfer function of either circuit is $H(s) = \dfrac{V_o(s)}{V_i(s)} = \dfrac{1}{s+1}$

FIGURE 6.6
Normalized passive first-order LPF.

The OA is basically a voltage amplifier with a large open-loop gain (about 10^6), a large input impedance (about 2 MΩ), and a small output impedance (about 75 Ω). The symbol used to represent an OA as well as its typical electrical specs are indicated in Figure 6.5.

R.6.44 Digital filters, in contrast, use digital processors to perform numerical operations on a sequence of discrete (sampled) values representing an analog signal.

The processor can be a powerful computer, a personal computer (PC), or a digital signal processing (DSP) special chip.

Samplers, analog-to-digital converters and encoders, are used to convert and process the analog signal into a discrete signal.

This signal is then represented by a sequence of numbers that are stored and then processed.

R.6.45 Some of the key features that distinguish digital from analog filters are summarized as follows:

- Digital filters are programmable.
- Digital filters are implemented using software packages.
- Digital filters can easily be changed or updated without affecting their circuitry (hardware).
- Digital filters can be designed, tested, and implemented on a general-purpose computer.
- Digital filters are independent of physical variables such as temperature, tolerances of elements, noise, fluctuations, or interference.
- Analog filters use elements that are environment dependent, and any filter change is usually hard to implement and a complete redesign is often required.

R.6.46 Let us get back to the normalized passive first-order filters. Some simple implementations of LPF are shown in Figure 6.6.

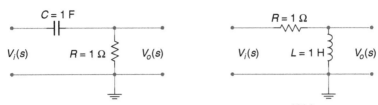

The transfer function of either network is given by $H(s) = \dfrac{V_o(s)}{V_i(s)} = \dfrac{s}{s+1}$

FIGURE 6.7
Normalized passive first-order HPF.

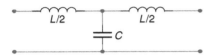

FIGURE 6.8
T and π LPF prototype segments.

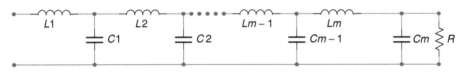

FIGURE 6.9
All-pole prototype LPF.

Observe that the effect of the inductor in an LPF can be replaced by a capacitor.

R.6.47 Examples of normalized passive first-order HPFs are illustrated in Figure 6.7. Observe again that the effect of the inductor in the HPF can be replaced by employing a capacitor.

R.6.48 The preceding simple prototype filter types (LP and HP) are typically implemented as segments (eliminating the resistors), and are referred to as T and π sections, discussed in R.6.49.

R.6.49 Typical T and π LPF prototype segments, also referred to as sections or blocks, are illustrated in Figure 6.8.

An all-pole prototype LPF can be realized by connecting T sections as illustrated in Figure 6.9. The order of the filter is then given by the total number of independent capacitors and inductors. For the network shown in Figure 6.9, the order n of the LPF is given by $n = 2m$.

R.6.50 Typical HP T and π sections are illustrated in Figure 6.10, and are obtained by replacing the inductors by capacitors and capacitors by inductors of the LP sections shown in Figure 6.8.

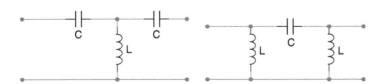

FIGURE 6.10
T and π HPF prototype segments.

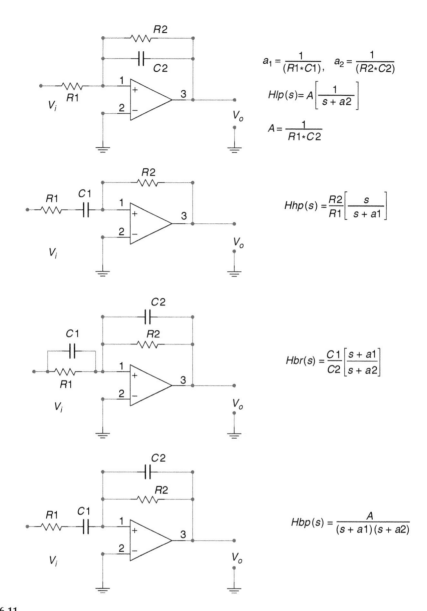

FIGURE 6.11
Typical active filter structures and its respective transfer functions.

R.6.51 Let us turn our attention to the most common active filter structures. Configurations as well as the respective transfer functions are indicated in Figure 6.11.

R.6.52 The loading effects associated with passive filters are overcome by using OA. Many complex filter realizations can then be implemented by connecting prototypes resulting into products or sum structures of simple (first- or second-order) filter types.

R.6.53 Figure 6.12 illustrates a second-order BRF employing simple first-order filters (LP and HP, with $w_2 > w_1$). Observe that by changing the summation by a product (by cascading an LP with an HP, with $w_1 > w_2$), a BPF may be implemented.

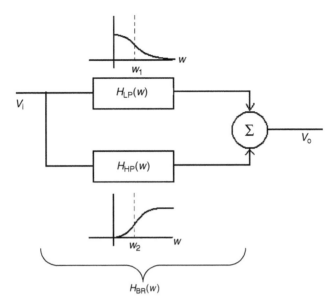

FIGURE 6.12
Second-order BRF employing simple first-order LP and HP filters.

R.6.54 Practical filters often used in applications (appliances) are named after the research-ers, often mathematicians who studied the polynomials that make up the transfer function that best matches the ideal filter characteristics.

The most common practical filters are referred to as

a. Butterworth

b. Chebyshev

c. Cauer (elliptic)

d. Bessel

These filters return characteristics based on the facts that the system transfer function presents regions with smooth or ripple (oscillations) behavior.

R.6.55 The Butterworth filter is probably the most popular practical filter. The filter char-acteristics consist of maximum flat response in the pass-band region, with no rip-ples either in the pass band or stop band.

The basic design approach consists of choosing an appropriate model of the transfer function $H(jw)$. From this transfer function, the complex conjugate $H(-jw)$ is then obtained, and from the product $H(jw) * H(-jw) = |H(jw)^2|$ the poles of the transfer function can be evaluated.

The Butterworth magnitude-squared response of an n-order LPF is given by the following equation:

$$H(jw) * H(-jw) = |H_a(jw)^2| = \cfrac{1}{\left[1 + \left(\cfrac{w}{w_p}\right)^{2n}\right]}$$

(The subindex a in H stands for analog.)
The normalized Butterworth magnitude is given by

$$|H_a(jw)| = \frac{1}{\left[1 + \left(\dfrac{w}{w_p}\right)^{2n}\right]^{1/2}}$$

Table 6.3 provides the transfer function $H(jw)$ for the case of LP normalized filters of the order $n = 1, 2, 3, 4,$ and 5.

The denominators of $H(jw)$ are commonly referred to as the Butterworth polynomial of order n. Figure 6.13 illustrates the magnitude response of the normalized Butterworth LPFs for the following orders $n = 1, 2, 3, 4$ and 10.

Observe that as the order increases (n), the pass-band region becomes flatter and the transition region narrows.

Observe also that the maximum flatness occurs at low frequencies (the limiting case is $w = 0$).

TABLE 6.3

Butterworth Polynomials

n	$H(jw)$
1	$[jw + 1]^{-1}$
2	$[(jw)^2 + 1.41(jw) + 1]^{-1}$
3	$[(jw)^3 + 2(jw)^2 + 2(jw) + 1]^{-1}$
4	$[(jw)^4 + 2.613(jw)^3 + 3.414(jw)^2 + 2.613(jw) + 1]^{-1}$
5	$[(jw)^5 + 3.236(jw)^4 + 5.236(jw)^3 + 5.2361(jw)^2 + 3.2361(jw) + 1]^{-1}$

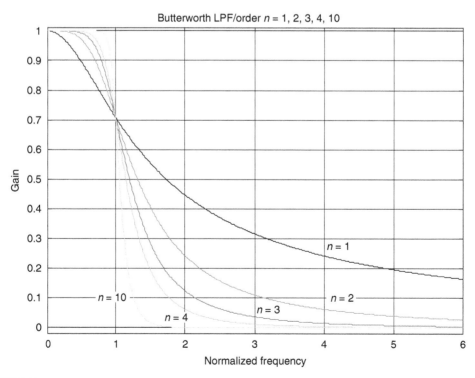

FIGURE 6.13
Magnitude plots of normalized Butterworth LPFs of orders $n = 1, 2, 3, 4,$ and 10.

R.6.56 Observe also that the two parameters that completely specify the Butterworth filter are

- The 3 dB cutoff frequency w_p
- The order n

The Butterworth filter uses the Taylor series approximation to model the ideal LPF prototypes for the two key frequencies: $w = 0$ and $w = \infty$ (for any order n).

R.6.57 Analog-Chebyshev filters trade off flatness in the pass band for a shorter transition band. The main objective of this filter is to minimize the design error, where the error is defined as the difference between the ideal LPF (brick wall) filter and the implementation of the actual response of the filter over a prescribed band of frequencies.

The magnitude is either given by an equiripple gain in the pass band and a monotic behavior in the stop band or a monotonic behavior in the pass band and equiripple in the stop band, depending on the filter type.

Chebyshev filters are classified according to their ripple and monotonicity as

- Type 1
- Type 2

R.6.58 The type-1 Chebyshev transfer function minimizes the difference between the ideal (brick wall) and the actual frequency response over the entire pass band by returning equal magnitude ripple in the pass band and a smooth decrease with maximum flatness gain in the stop band (see Figure 6.14).

R.6.59 The type-2 Chebyshev filter transfer function, also referred to as the inverse Chebyshev transfer function filter minimizes the difference between the ideal (brick wall) and the actual frequency response over the entire stop band by returning an equal

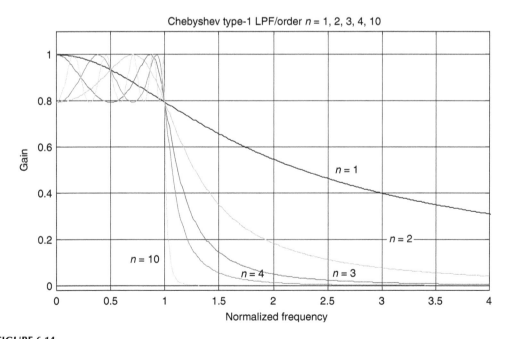

FIGURE 6.14
(See color insert following page 374.) Magnitude plots of normalized analog Chebyshev type-1 LPFs of orders $n = 1, 2, 3, 4,$ and 10.

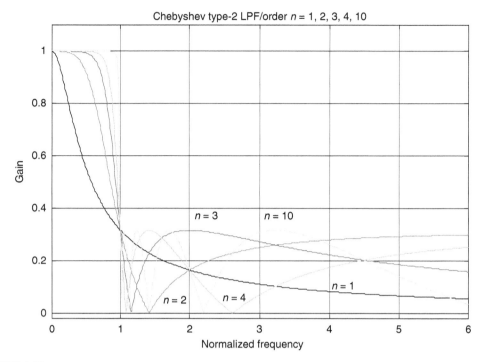

FIGURE 6.15
(See color insert following page 374.) Magnitude plots of normalized analog Chebyshev type-2 LPFs of orders $n = 1, 2, 3, 4,$ and 10.

ripple in the stop band and a smooth gain with maximum flatness in the pass band (see Figure 6.15).

The stop band does not approach zero as fast as type 1.

The ripple effect becomes more evident as the order of the filter increases.

R.6.60 The magnitude square of the transfer function $H(jw)$ of a Chebyshev type-1 filter is given by

$$|H_a(w)|^2 = \frac{1}{1 + \varepsilon^2 C_n^2(w/w_p)}$$

where $C_n(x)$ is referred to as the Chebyshev polynomial of order n, in the region defined by

$$1 < |H_a(w)|^2 = \frac{1}{(1 + \varepsilon^2)^{0.5}} \quad \text{for } w < w_p$$

where ε denotes the maximum stop band deviation.
The Chebyshev polynomial coefficients $C_n(x)$ of order n are defined by

$$C_n(x) = \begin{cases} \cos(n\cos^{-1}(x)) & \text{for } 0 \le x \le 1 \\ \cos(n\cosh^{-1}(x)) & \text{for } x > 1 \end{cases}$$

TABLE 6.4

Chebyshev Polynomials

n	$C_n(w)$
1	w
2	$2w^2 - 1$
3	$4w^3 - 3w$
4	$8w^4 - 8w^2 + 1$

The Chebyshev polynomial coefficients $C_n(x)$ oscillate between -1 and 1, and can be generated by using the recursion equation given as follows:

$$C_{n+1}(x) = 2C_n(x) - C_{n-1}(x)$$

with $C_0(x) = 1$ and $C_1(x) = x$.

Table 6.4 shows the Chebyshev polynomials for $n = 1, 2, 3,$ and 4.

Figure 6.14 displays the magnitude responses of an analog Chebyshev type-1 LPF for $n = 1, 2, 3, 4,$ and 10.

R.6.61 The Chebyshev type-2 filter is the complement of the Chebyshev type-1 filter, in the sense that the magnitude response is smooth in the pass band and presents equal ripple in the stop band.

Unlike the type-1 filter, which consists only of poles, the type-2 filters consist of both poles and zeros.

The magnitude square of the response is given by

$$|H_a(jw)|^2 = \frac{\varepsilon^2 C_n(w_s/w)}{1 + \varepsilon^2 \left[\dfrac{C_n(w_s/w_p)}{C_n(w_s/w)} \right]^2}$$

where ε defines the maximum stop-band deviation, w_p denotes the pass-band cutoff frequency, w_s denotes the stop-band cutoff frequency, and $C_n(x)$ are the Chebyshev polynomials (defined in Table 6.4).

Figure 6.15 shows the magnitude responses of an analog Chebyshev type-2 filters for $n = 1, 2, 3, 4,$ and 10.

R.6.62 Elliptic filters, also known as Cauer filters, present ripples in the pass as well as in the stop-band regions. The main characteristic of the elliptic filter is that the transition band is the narrowest with respect to the other two filter types, the Butterworth and Chebyshev filters.

The square of the magnitude of the transfer function of the Cauer filters is given by

$$|H_a(jw)|^2 = \frac{1}{1 + \varepsilon^2 J_n^2(w,h)}$$

where ε is the pass-band deviation, $J_n(w, h)$ is the n-order Jacobian, and h indicates the ripple's height (in the pass- and stop-band regions).

Figure 6.16 shows the magnitude response of an elliptic filter for $n = 1, 2, 3, 4,$ and 10.

R.6.63 Butterworth, Chebyshev (types 1 and 2), and elliptic filters deal basically with the magnitude response, with little concern about their phase response.

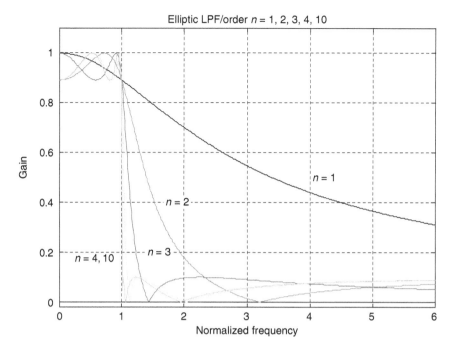

FIGURE 6.16
(See color insert following page 374.) Magnitude plots of normalized analog elliptic LPFs of orders $n = 1, 2,$ 3, 4, and 10.

R.6.64 Bessel filters, also known as linear-phase filters, present a linear-phase response in the pass band, in addition they try to approximate the magnitude response in the pass-band region satisfying the desired specs.

The system transfer function of an nth-order Bessel filter is given by the transfer function shown as follows:

$$H_a(jw) = \frac{(2n)!}{2^n n! B_n(jw)}$$

where $B_n(jw)$ is the n-order Bessel polynomial defined by

$$B_n(jw) = \sum_{l=0}^{n} \frac{(2n-1)!(jw)^l}{2^{n-1} l!(n-1)!}$$

Figure 6.17 shows the magnitude response and Figure 6.18 shows the phase response of the Bessel filters for orders $n = 1, 2, 3, 4,$ and 10.

R.6.65 It is obvious that the mathematics involved in either the analysis or synthesis of any type of filter is generally a complicated and complex proposition, which is difficult and labor intensive. This book attempts to define the filter types often used in real application and provides an overview of the filter theory and a simple and user friendly approach to the analysis and synthesis process. The mathematical complexities of the power of filter theory, in either the synthesis or analysis processes, can be avoided by using MATLAB.

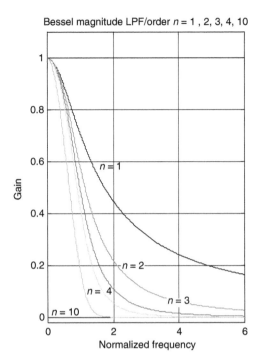

FIGURE 6.17
(See color insert following page 374.) Magnitude plots of normalized analog Bessel LPFs of orders $n = 1, 2, 3,$ 4, and 10.

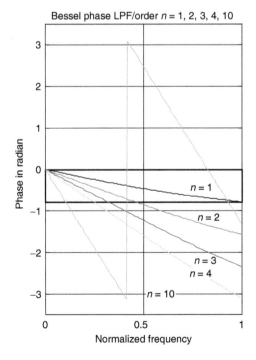

FIGURE 6.18
(See color insert following page 374.) Phase plots of normalized analog Bessel LPFs of orders $n = 1, 2, 3, 4,$ and 10.

MATLAB offers a number of functions that can be used in the analysis and synthesis of almost any filter type. The corresponding MATLAB function files are presented and discussed in R.6.66 through R.6.72.

R.6.66 The following function files can be used in the analysis and synthesis of analog *Butterworth* filters:

a. *[z, p, k] = buttap(n)* returns the *zeros (z)*, *poles (p)*, and *gain factor (k)* of the normalized analog *Butterworth* LPF transfer function of order n (with a cutoff frequency given by $w_p = 1$ rad/s).

b. *[num, den] = butter(n, w_n, 's')* returns the numerator and denominator polynomial coefficients as row vectors in descending powers of s, of an n-order transfer function of an LPF with a cutoff frequency at w_n rad/s. If w_n consists of a two-element vector $[w_l, w_u]$, where $w_l < w_u$, the function returns a second-order band-pass transfer function with drops at the w_l and w_u edges, and a BW given by $w_u - w_l$.

c. *[num, den] = butter(n, w_n, 'filtertype', 's')* returns an n-order HP or band-stop (BS) filter of order $2n$, where *'filtertype'* is specified as *'high'* for an HPF, or *'stop'* for a BSF with the (3 dB) stop-band edges given by $w_n = [w_l, w_u]$, where $w_l < w_u$.

d. *[n, w_n] = buttord(w_p, w_s, R_p, R_s, 's')* returns the lowest-order Butterworth filter (n) given by the filter parameters defined by

w_p = Pass-band edge (rad/s)

w_s = Stop-band edge (rad/s)

R_p = Maximum pass-band attenuation (dB)

R_s = Minimum stop-band attenuation (dB)

n = Order of the filter

w_n = Cutoff frequency (rad/s)

R.6.67 The MATLAB function *[num, den] = maxflat (a, b, w_p)* returns a Butterworth-type filter with the transfer function having two orders given by a and b, where a defines the degree of the *num* (numerator) and b the *den* (denominator) of the filter's transfer function instead of just using one order, and where w_p is the pass-band edge that takes a value over the range 0–1.

When the orders of the *num* and *den* are the same $(a = b)$, the functions *maxflat* and *butter* return the same filter results.

R.6.68 For example, verify that the *maxflat* and *butter* MATLAB functions return the same filter specs for the case when $a = b = 4$ and $w_p = 0.3$.

```
MATLAB Solution
>> [num,den] = maxflat (4,4,0.3)

    num =
            0.0186   0.0743   0.1114   0.0743   0.0186
    den =
            1.0000 -1.5704   1.2756  -0.4844   0.0762

>> [num,den] = butter(4,0.3)

    num =
            0.0186   0.0743   0.1114   0.0743   0.0186
    den =
            1.0000 -1.5704   1.2756  -0.4844   0.0762
```

R.6.69 The following example illustrates the case where the desired transfer function of an LP Butterworth filter is given by a denominator polynomial of order 6, whereas the numerator polynomial is of order 4, and $w_p = 0.3$.

MATLAB Solution
```
>> [num,den] = maxflat (4,6,0.3)

    num =
          0.0086   0.0344   0.0515   0.0344   0.0086
    den =
          1.0000  -2.4633   3.0480  -2.1861   0.9427  -0.2278   0.0239
```

R.6.70 Once the *poles, zeros,* and *gain* of the desired filter transfer function are known (indicated by the variables p, z, and k), the numerator and denominator polynomial coefficients of the transfer function can be obtained by using the MATLAB function

$$[num, den] = zp2tf(z, p, k)$$

R.6.71 The following function files can be used in the analysis and synthesis of analog Chebyshev type-1 filters:

a. [z, p, k] = *cheblap(n, R_p)* returns the *zeros (z), poles (p),* and *gain factor (k)* of the normalized analog type-1 Chebyshev LPF transfer function of order n with a pass-band ripple R_p expressed in decibels.

b. [num, den] = *cheby1(n, R_p, w_n, 's')* returns the *num* and *den* of the LPF transfer function of order n (numerator and denominator) as vectors consisting of the polynomial coefficients of the transfer function in descending powers of s, given R_p (pass-band ripple in decibel) and w_n (pass-band edge in radians/second).
   If w_n is specified as a two-element vector such as $w_n = [w_l, w_u]$, where $w_l < w_u$, the function returns the transfer function of a second-order BPF with pass-band edges given by w_l and w_u.

c. [num, den] = *cheb1(n, R_p, w_n, 'filtertype', 's')* returns the *num* and *den* of the transfer function of order n (numerator and denominator) as vectors consisting of the polynomial coefficients arranged in descending powers of s. When *'filtertype'* is specified as *'high'* for an HPF, or *'stop'* for a BSF, then $w_n = [w_u, w_l]$ is given by two values, for $w_u > w_l$, given the band-pass ripple R_p (in decibels).

d. [n, w_n] = *cheb1ord(w_p, w_s, R_p, R_s, 's')* returns the lowest LPF order *(n)* of the Chebyshev type-1 analog filter, where w_n indicates the cutoff frequency (radians per second), given the following input variables:

w_p = Pass-band edge (rad/s)
w_s = Stop-band edge (rad/s)
R_p = Maximum pass-band attenuation (dB)
R_s = Minimum stop-band attenuation (dB)

R.6.72 The function files used for the analysis and synthesis of analog type-2 Chebyshev filters are summarized as follows:

a. [z, p, k] = *cheb2ap (n, R_s)*

b. [num, den] = *cheby2 (n, R_s, w_n, 's')*

c. [num, den] = *cheby2 (n, R_s, 'filtertype', 's')*

d. [n, w_n] = *cheb2ord (w_p, w_s, R_p, R_s, 's')*

The variables used in the preceding function files are defined as follows:

z = Zeros of the transfer function

p = Poles of the transfer function

k = Gain of the transfer function

num = Numerator coefficients of the transfer function given as vectors in descending powers of s

den = Denominator coefficients of the transfer function given as vectors in descending powers of s

n = Order of filter

R_s = Minimum stop-band ripple attenuation (dB)

w_n = Pass-band edge (rad/s)

R_p = Maximum pass-band ripple attenuation (dB)

w_p = Pass-band edge (rad/s)

'filtertype' = High or stop (for either HP or BP)

The type-2 Chebyshev filter functions are self-explanatory since they are similar in syntax and content to the function defined in R.6.71 for the Chebyshev type-1 filter.

R.6.73 The following function files can be used in the analysis and synthesis of the analog elliptic filters:

a. *[z, p, k] = elliap (n, R_s)*

b. *[num, den] = ellip (n, R_s, w_n, 's')*

c. *[num, den] = ellip (n, R_s, 'filtertype', 's')*

d. *[n, w_n] = ellipord (w_p, w_s, R_p, R_s, 's')*

Note that the elliptic filter functions follow the syntax and content similar to the function files defined for the Butterworth and Chebyshev filters, and no further function description is given. The workout examples provide insight of the specific function and applications.

R.6.74 The function files used in the analysis and synthesis of the Bessel filters are defined as follows:

a. *[z, p, k] = besselap (n)*

b. *[num, den] = besself (n, w_n)*

c. *[num, den] = besself (n, w_n, 'filtertype')*

The functions and variables used in the Bessel functions are self-explanatory and in case of doubt refer to the Chebyshev and elliptic filter function descriptions (R.6.71).

R.6.75 In most filter problems, the MATLAB functions that return the *zeros*, *poles*, and *gain* forms provide a more accurate response than the ones using the transfer function.

For filters of order 15 or less, the function files that return the *num* (numerator) and *den* (denominator) of the transfer function can safely be used, but for filters of orders greater than 15, the *zeros*, *poles*, and *gain* returns a more reliable version.

R.6.76 Some important observations of the different filter's phase prototypes are summarized as follows:

a. Butterworth and Chebyshev filters present a linear-phase response over a good portion of the pass-band region.

b. Bessel filters are generally used when the desired phase response is linear over a larger portion of the pass band.

R.6.77 The MATLAB function $[num_L, dem_L] = lp2lp\ (num_N, den_N, w_c)$ is used to transform a normalized LPF prototype given by the transfer function $H(s) = num_N(s)/den_N(s)$ into a denormalized LPF with a numerator num_L, denominator den_L, and cutoff frequency w_c. This function file returns the coefficients of the polynomials num_L and den_L as a row vector arranged in descending powers of s.

R.6.78 Similarly, the MATLAB function $[num_H, den_H] = lp2hp\ (num_N, den_N, w_c)$ is used to transform a low-pass normalized filter into a denormalized HPF with the pass band given by w_c.

R.6.79 The function file $[numbp, denbp] = lp2bp\ (num_N, den_N, w_o, BW)$ transforms an LP normalized filter into a BPF or BSF ($lp2bs$), where w_o is its center frequency with BW.

R.6.80 Some practical observations of the resulting filter complexity and their associated costs are summarized as follows:

   a. The complexity of a filter is generally given by its order n. The higher the order, the greater the complexity and cost in term of elements.

   b. Generally, for a particular set of filter specs, the elliptic filter implementation is less complex than the equivalent Chebyshev (type 1 or 2).

   c. The Chebyshev (type 1 or 2) cost is considerably less than its Butterworth equivalent counterpart.

   d. An estimate of the filter complexity is given by the following example: an elliptic filter of order 6 requires an equivalent Chebyshev filter (type 1 or 2) of order 10 and a Butterworth equivalent implementation of order 29.

R.6.81 Recall that digital filters as well as analog filters have as their primary objective to filter, meaning to pass or reject, certain band of frequencies. Analog filters are expressed as functions of w (radians per second), whereas digital filters are functions of $W = w * T$ called the scaled or digital frequency, where T is the sampling period (sometimes indicated by T_s).

R.6.82 Recall that by using the digital frequency W (some textbooks use Ω), the frequency response of the digital filter is always periodic with period 2π.

R.6.83 Digital filters can be classified based on their impulse response length as follows:

   a. FIR

   b. IIR

R.6.84 An FIR filter returns an impulse response with an amplitude that decays to zero in a finite number of steps (samples) or duration.

R.6.85 An IIR filter returns an impulse response that has, in theory, an infinite duration or continues forever because of its recursive nature (feedback).

   This definition is not accurate because the impulse response of any system including IIR filters in real life eventually decay to zero in a finite time, which may be long.

R.6.86 IIR filters are used to closely approximate the gain and phase response of analog filters and the specs are usually given in terms of analog variables such as BW, minimum pass-band ripple (R_p), and maximum stop-band ripple (R_s). These types of filters can substitute analog filters.

R.6.87 The approach followed in the design of IIR filters are as follows:

   a. Design an analog prototype with the desired frequency response (transfer function) given by its transfer function $H(s)$.

b. Transform the analog filter into its digital version by obtaining the discrete filter transfer function $H(z)$ or its impulse response $h(n)$.

R.6.88 One fundamental difference exists between $H(s)$ and its discrete mapping $H(z)$. Since the phase function of the digital prototype is given by

$$\angle H(z) = \tan^{-1}[Imag(H(z)/Real(H(z))] + \pi/2(1 + sign(Real(H(z))))$$

where $z = {}^{jW}$.

Then the gain presents a periodic behavior (with period 2π), then the principal value is limited over the digital frequencies over the range $-\pi \le W \le +\pi$.

R.6.89 The steps involved in the design of IIR filters using MATLAB are summarized as follows:

1. Design an analog LP normalized prototype filter

2. Use frequency transformation to transform the LP analog prototype into the desired filter type: LP, HP, BP, or BS filter depending on the requirements of the design (using the MATLAB functions: *lp2lp, lp2hp*, etc.)

3. Transform the analog filter into its digital version by

   a. Bilinear transformation

   b. Impulse invariant method

   c. Yule–Walker algorithm

These techniques are presented and discussed in the following.

R.6.90 The bilinear transformation is a one-to-one mapping of the complex analog frequency $s(s = \sigma + jw)$ of the transfer function $H(s)$ (s-plane) into the complex digital frequency $z(z = e^{sT})$ (z-plane).

R.6.91 Mathematically, the mapping s (analog) $\rightarrow z$ (digital) or $z \rightarrow s$ is accomplished by the following substitution:

$$s \rightarrow A\frac{z-1}{z+2}$$

or

$$z \rightarrow \frac{A+s}{A-s}$$

where $A = 2/T$, in which T is the sampling period.

The objective of the mapping is to move the poles and zeros of the desired analog transfer function $H(s)$ from the s-plane into the z-plane.

R.6.92 The effect of the bilinear transformation results in moving the poles and zeros from the left half of the s-plane into the unit circle in the z-plane, where the imaginary axis in the s-plane is mapped into the outside of the unit circle of the z-plane and the right half of the s-plane is mapped into the outside of the unit circle of the z-plane guaranteeing stability in this way.

Therefore, a stable analog filter $H(s)$ results in a stable digital filter $H(z)$.

Recall that to analyze the stability of a digital filter it is sufficient to compute the magnitude of the poles, and if they are smaller in magnitude than 1 (inside the unit circle), then the system is stable otherwise the system is unstable.

R.6.93 Since the analog frequency w (radians per second) is related to the digital frequency W (radians) by the relation $W = wT$, the infinite range of $w(-\infty$ to $+\infty)$ in the s-plane is compressed into the range $-\pi \le W \le \pi$ (referred to as frequency wrapping).

R.6.94 The resulting relation between the analog frequency w and the discrete frequency W is given by

$$W = (2/T) * tan(W/2) \text{ (a nonlinear frequency transformation)}$$

or

$$W = 2 * arctan(w * T/2)$$

R.6.95 The MATLAB function *unwrap(phase)* unwraps the *phase* angle (in radians) and limits the *phase* over the range of interest given by $-\pi \le W \le +\pi$.

R.6.96 The other transformations used in the design of LP, HP, BP, or BS filters from the analog prototypes into the digital ones are given as follows:

a. LP: $s = (2/T) * [(z - 1)/(z + 1)]$, $w = (2/T) * tan(W/2)$

b. HP: $s = (2/T) * [(z + 1)/(z - 1)]$, $w = -(2/T) * cot(W/2)$

c. BP: $s = (2/T) * [(z^2 - 2 * c * z + 1)/(z^2 - 1)]$, $w = (2/T) * [(c - cos(W))/(sin(W))]$

d. BS: $s = (2/T) * [(z^2 - 1)/(z^2 - 2 * c * z + 1)]$, $w = (2/T) * [(c - cos(W))/(sin(W))]^{-1}$

R.6.97 An example of a bilinear transformation is provided as follows: Let us transform the first-order analog (RC) LPF illustrated by the circuit diagram of Figure 6.6 into a digital filter. Recall that the filter transfer function is given by

$$H(w) = \frac{V_o(jw)}{V_i(jw)} = \frac{\dfrac{1}{jwC}}{\dfrac{1}{jwC} + R} = \frac{\dfrac{1}{R * C}}{j * w + \dfrac{1}{R * C}}$$

without any loss of generality let $R = C = 1$ and $s = jw$, then

$$H(s) = \frac{V_o(s)}{V_i(s)} = \frac{1}{s + 1}$$

Assuming that the sampling period is $T = 2$ and applying the corresponding bilinear transformation defined in R.6.96 yields

$$H\left[s = \frac{z - 1}{z + 1}\right] = \frac{1}{\dfrac{z - 1}{z + 1} + 1}$$

then the discrete transform becomes

$$H(z) = 0.5 + 0.5 * z^{-1}$$

Note that in this simple example the resulting digital filter is an FIR. A more complex analog filter transformation could result in an IIR filter. Hand computations can be avoided by using MATLAB as indicated in R.6.98.

R.6.98 The MATLAB function *[numz, denz] = bilinear (numa, dena, 1/T_s)* returns the discrete polynomial coefficients *numz* and *denz* (numerator and denominator of $H(z)$) arranged in descending powers of z, where *numa* and *dena* are the analog numerator and denominator of $H(s)$, where T_s indicates its sampling rate.

R.6.99   The function *[pz, zz, kz] = bilinear (pa, za, ka, fs, fc)* returns the digital *poles, zeros,* and *gain* given by *pz, zz,* and *kz* of its discrete transfer function *H(z)*, where *pa, za,* and *ka* are the analog poles, zeros, and gain of *H(s)*, where *fs* = $1/T_s$ and *fc* = $1/(2\pi)$.

         The bilinear transformation with the optional frequency prewarping is given by *[pz, zz, kz] = bilinear (pa, za, ka, fs)* returns the *s*-domain analog transfer function specified by *pa, za, ka, fs* into the *z* discrete equivalent transfer function given by the following substitutions:

$$H(z) = H(s)\bigg|_{s=2*fs\frac{(z-1)}{(z+1)}}$$

where the column vectors *za* and *pa* are the analog zeros and poles of *H(s)*, and *fs* represents its sample frequency in hertz.

R.6.100   The *impulse invariant response* method converts the impulse response of the desired analog filter *h(t)* into an equivalent digital FIR *h(n)*, by properly sampling *h(t)*, with the hope that the analog response would closely follow its digital response. Hence, the digital filter preserves the impulse response of the original analog filter.

R.6.101   Observe that the *bilinear transformation* is a process done entirely in the frequency domain, whereas the impulse response method is done in the time domain.

R.6.102   The impulse invariant IIR filter is realized by the following steps:

a. $h(t) = £^{-1}[H(s)]$

b. Substitute *t* by *nT*

c. Obtain the digital transfer function $H(z) = \{Z[h(nT)]\} * T$

         Observe that the analog filter impulse response would then have the same values as the sample digital filter impulse response at the sampling instances *nT*.

         Recall that when the inverse Laplace transform of *H(s)* cannot be obtained directly, then a partial fraction expansion of *H(s)* may be required for obtaining in this way a parallel structure implementation.

R.6.103   The impulse invariant method suffers from aliasing, and this is the main reason that it is not widely used. The bilinear transformation does not suffer from aliasing and is by far more popular than the impulse invariance method.

R.6.104   The *step invariant* technique is an alternate way in the implementation of an IIR filter by working with the step response instead of the impulse response. The step response of the analog filter would then have the same values at the sampling instances as the digital filter step response.

R.6.105   Mathematically, the impulse invariance response transforms *H(s)* into *H(z)* [*H(s)* → *H(z)*] by making the following substitution:

$$\frac{1}{s+a} \Rightarrow \frac{1}{1 - z^{-1}e^{-aT}}$$

(transformation used for simple poles of *H(s)*).

$$\frac{s+a}{s^2 + 2as + a^2 + b^2} \Rightarrow \frac{1 - z^{-1}e^{-aT}\cos(bT)}{1 - 2z^{-1}e^{-aT}\cos(bT) + z^{-2}e^{-2aT}}$$

$$\frac{b}{s^2 + 2as + a^2 + b^2} \Rightarrow \frac{1 - z^{-1}e^{-aT}\sin(bT)}{1 - 2z^{-1}e^{-aT}\cos(bT) + z^{-2}e^{-2aT}}$$

(transformations used for the case of complex poles of $H(s)$).

R.6.106 For example, let the analog transfer function of a second-order Butterworth low pass-band filter be given by

$$H(s) = \frac{100}{s^2 + 14.14s + 100}$$

Use the impulse invariant method to transform the analog filter into a digital one if the sampling rate is given by $T = 0.0314$ s.

ANALYTICAL Solution

Rearranging the terms of $H(s)$ into the format given in R.6.105 results in

$$H(s) = \frac{100}{s^2 + 2(14.14/2)s + (14.14/2)^2 + 100 - (14.14/2)^2}$$

$$H(s) = \frac{100}{s^2 + 2(14.14/2)s + (7.07)^2 + 100 - (7.07)^2}$$

$$H(s) = \frac{2*50}{s^2 + 2(14.14/2)s + (50) + 100 - (50)}$$

$$H(s) = \frac{2*(50)^{1/2}(50)^{1/2}}{s^2 + 2(14.14/2)s + (50) + (50)}$$

$$H(s) = 2*(50)^{1/2}\frac{(50)^{1/2}}{s^2 + 2(14.14/2)s + (7.07)^2 + (7.07)^2}$$

$$H(s) = 2*(50)^{1/2}\frac{7.07}{s^2 + 2(14.14/2)s + (7.07)^2 + (7.07)^2}$$

Then, let $a = 7.07$, $b = 7.070$, and $T = 0.0314$, and using the substitutions of R.6.105 $H(s) \rightarrow H(z)$ the following discrete system is obtained: The numerator of $[H(z)] = [2*(50)^{1/2}]*[z^{-1}e^{-(7.07)*(0.0314)}*\sin(7.07*0.0314)]$ and the denominator of $[H(z)] = [1 - 2z^{-1}e^{-(7.07)*(0.0314)}*\cos(7.07*0.0314) + z^{-2}e^{-2*7.07*0.0314}]$.

Then, after performing the corresponding algebraic manipulations the following simplified expression is obtained:

$$H(z) = \frac{Y(z)}{X(z)} = 2*\sqrt{50}\frac{0.0882z^{-1}}{1 - 1.5626z^{-1} + 0.6414z^{-2}}$$

As a consequence of the sampling process (Nyquist theorem), the filter magnitude is multiplied by $1/T$. Therefore, after the transformation process from the continuous to the discrete domain is done, the filter transfer function has to be compensated by multiplying the response by *T*.

For this example, the final transfer function would then be the preceding expression for *H(z)* multiplied by *T* = 0.0314.

The reader can appreciate that the substitution and algebraic manipulations involved in the transformation process just presented is labor-intensive.

This computational effort can be avoided by using the power of MATLAB as illustrated in the following points.

R.6.107 The MATLAB function *[numz, denz] = impinvar (nums, dens, f_s)* maps the analog transfer function *H(s)* into the discrete transfer function *H(z)* using the substitutions presented in R.6.105 and returns the numerator and denominator of *H(z)* = *numz/denz*, given the numerator and denominator of *H(s)* = *num(s)/den(s)*, with a sampling frequency given by f_s. The function *impinvar* used without specifying f_s takes the default value of f_s = 1 Hz.

R.6.108 The MATLAB function *[numz, denz] = yulewalk (n, f, mag)* returns the numerator *numz* and denominator *denz*, where *H(z)* = *numz(z)/denz(z)* as the polynomial coefficients that best fit the points defined by the vectors *f* and *mag*, where *n* denotes the filter order. *f* is a normalized vector specifying the set of key frequencies starting at 0 and ending at 1 and *mag* is the corresponding vector with the desired magnitude at the frequencies defined by *f*.

R.6.109 Unlike the transform of the analog prototype method, the *yulewalk* algorithm is a direct method to design a digital filter based on specs in the discrete domain with no constrains imposed by the standard filter prototypes. The digital filter coefficients are evaluated by obtaining the inverse *fft* of the desired, ideal power spectrum and then by solving the *modified Yule–Walker equations*.

R.6.110 The FIR equivalent of this function is the MATLAB command *fir2*, which also returns a filter based on a linear piecewise magnitude approximation.

R.6.111 MATLAB's signal processing toolbox includes a number of functions that can directly be used in the analysis and synthesis of IIR filters based on the *bilinear* transformation.

The *bilinear* transformation functions based on the standard prototype filters are listed as follows:

a. *[numz, denz] = butter(n, w_n)*

b. *[z, p, k] = butter(n, w_n)*

c. *[numz, denz] = cheby1(n, R_p, w_n)*

d. *[z, p, k] = cheby1(n, R_p, w_n)*

e. *[numz, denz] = cheby2(n, R_s, w_n)*

f. *[z, p, k] = cheby2(n, R_s, w_n)*

g. *[numz, denz] = ellip(n, R_p, R_s, w_n)*

h. *[z, p, k] = ellip(n, R_p, R_s, w_n)*

where *numz* and *denz* denote the numerator and denominator coefficients of *H(z)*, R_p and R_s denote the pass- and stop-band ripples, *n* is the order of the filter, w_n is the pass-band edge; a number between 0 and 1, *z* denotes discrete zeros, *p* denotes discrete poles, and *k* is the gain factor of the discrete transfer function while the sampling frequency is assumed to be 2 Hz or a sampling rate of *T* = 0.5 s.

R.6.112 Recall that an IIR filter can be defined in the time domain by a difference equation. For example, let the system differential equation be given by

$$y(n) - 0.02y(n-1) + 0.009y(n-2) = 0.1x(n) + 0.2x(n-1) + 0.3x(n-2)$$

where x represents the input sequence and y is its output sequence.

Observe that the output sequence $y(n)$ is a function of the input sequence $x(n)$ as well as the previous (delayed) output sequences.

R.6.113 An FIR filter can be defined by a difference equation with a finite input sequence x. For example, let the system differential equation be given by

$$y(n) = 0.036x(n) - 0.02x(n-1) + x(n-2)$$

Observe that the output sequence $y(n)$ is a function of its input sequence $x(n)$ (scaled and delayed).

R.6.114 Some practical IIR design considerations and observations are summarized as follows:

a. There is a ripple problem to consider.

b. There is no guarantee that the design leads to a stable implementation.

c. There is a stability problem to consider.

d. Differences may exist among the different types of IIR filters.

e. When the magnitude response is the main concern IIR is generally a good choice.

f. IIR filters require 5–10 times less coefficients than the equivalent FIR implementation.

R.6.115 An FIR filter is a filter that has a finite response for $n < N$ and a linear phase, where N is the length of the filter response. Such filters generally require N constant multipliers, $N - 1$ two port input adders and N stages of delays resulting in shift register–type structure.

R.6.116 All FIR filters present linear phase and an impulse response with some sort of symmetry that satisfies the general relation given by

$$h(n) = \pm h(N - n)$$

This sequence ($h(n)$) presents an even symmetry sequence for the positive case and an odd symmetry for the negative case, assuming the sequence $h(n)$ consists of real coefficients.

R.6.117 The symmetry condition for the impulse response is a necessary condition for the linear phase.

R.6.118 The symmetry condition depends on the length of $h(n)$ and whether N is odd or even.

R.6.119 Four types of symmetry conditions for FIR filters with real coefficients and lengths N result. They are summarized as follows:

a. *Type 1.* Symmetric impulse response with $N = odd$.

b. *Type 2.* Symmetric impulse response with $N = even$.

c. *Type 3.* Asymmetric impulse response with $N = odd$.

d. *Type 4.* Asymmetric impulse response with $N = even$.

R.6.120 Type-1 FIR filter is defined by the following relation:

$$h(n) = h(N - n) \quad \text{for } 0 \le n \le N - 1, \text{ with } N = odd.$$

Then the amplitude response is given by

$$H_1(W) = e^{-jNW/2} \sum_{n=0}^{N/2} a_1(n)\cos(W_n)$$

where $a_1(0) = h(N/2)$ is the mid-point and $a_1(n) = 2h((N/2) - n)$ for $1 \le n \le N/2$.
  Type-1 FIR filters can be used to design any filter type. The following sequence illustrates a type-1 FIR filter:

$$h(n) = -0.09\delta(n) + 0.30\delta(n-1) + 0.3593\delta(n-2) + 0.30\delta(n-3) - 0.09\delta(n-4)$$

R.6.121 The type-2 FIR filter is defined by

$$h(n) = h(N - n) \quad \text{for } 0 \le n \le N - 1, \text{ with } N = even$$

then

$$H_2(W) = e^{-jNW/2} \sum_{n=0}^{(N+1)/2} a_2(n)\cos[W(n - 1/2)]$$

where

$$a_2(n) = 2h\left(\frac{N+1}{2} - n\right) \quad \text{for } n = 1, 2, \dots (N+1)/2$$

Since the response is defined for $-\pi \le W \le \pi$, then

$$H_2(W = \pi) = 0$$

Therefore, type-2 FIR filters cannot be used to implement an HP or a BP filter. The following sequence illustrates a type-2 FIR filter:

$$h(n) = -0.09\delta(n) + 0.30\delta(n-1) + 0.3593\delta(n-2) + 0.3593\delta(n-3)$$
$$+ 0.30\delta(n-4) - 0.09\delta(n-5)$$

R.6.122 The type-3 FIR filter is defined by

$$h(n) = -h(N - n) \quad \text{for } 0 \le n \le N, \text{ with } N = odd$$

$$H_3(W) = e^{-jNW/2}e^{j\pi/2} \sum_{n=1}^{N/2} [a_3(n)\sin(Wn)]$$

where $a_3(n) = 2h\left[\dfrac{N}{2} - n\right]$ for $n = 1, 2, \dots N/2)$

Observe that at $W = 0$.

$H_3(W = 0) = 0$. Therefore, this function cannot be used to implement an LPF (H_3 ($W = \pi$) = 0). As a result, this function can be used to implement either a BS or an HP filter.

The following sequence illustrates a type-3 FIR filter:

$$h(n) = -0.019\delta(n) + 0.3593\delta(n-1) + 0.023\delta(n-2)$$
$$- 0.3593\delta(n-3) + 0.019\delta(n-4)$$

R.6.123 The type-4 FIR filter is defined by

$$h(n) = -h(N-n) \quad \text{for } 1 \le n \le N, \text{ with } N = even$$

Then

$$H(W) = e^{-jNW/2}e^{j\pi/2}\sum_{n=1}^{N/2} a_4(n)\sin[W(n-1/2)]$$

where

$$a_4(n) = 2h\left(\frac{N+1}{2} - n\right) \quad \text{for } n = 1,\, 2, \ldots \frac{N+1}{2}$$

Observe that at $W = 0$, $H(W = 0) = 0$; therefore, this transfer function cannot be used to implement an LPF.

The following sequence illustrates a type-4 FIR filter:

$$h(n) = -0.019\delta(n) + 0.3593\delta(n-1) - 0.3593\delta(n-3) + 0.019\delta(n-3)$$

Analytical examples of each FIR filter type are presented in the following points.

R.6.124 For example, analyze by hand a type-1 FIR filter with $N = 5$.

ANALYTICAL Solution

Then

$$H(z) = h(0) + h(1)z^{-1} + h(2)z^{-2} + h(3)z^{-3} + h(4)z^{-4}$$

Note that $h(0) = h(4)$ and $h(1) = h(3)$.

Then

$$H(z) = [h(0) + h(4)z^{-4}] + [h(1)z^{-1} + h(3)z^{-3}] + h(2)z^{-2}$$
$$H(z) = h(0)z^{-2}(z^2 + z^{-2}) + h(1)z^{-2}[z^1 + z^{-1}] + h(2)z^{-2}$$

substituting z by $e^{jW}(z \to e^{jW})$

$$H(e^{jW}) = h(0)e^{-2jW}(e^{jW2} + e^{-jW2}) + h(1)e^{-jW2}(e^{-jW} + e^{jW}) + h(2)e^{-2jW}$$

$$H(e^{jW}) = h(0)e^{-2jW}2cos(2W) + h(1)e^{-jW}2cos(W) + h(2)e^{-2jW}$$

$$H(e^{jW}) = e^{-2jW}[h(0)2cos(2W) + 2h(1)cos(W) + h(2)]$$

$$|H(e^{jW})| = |2h(0)cos(2W) + 2h(1)cos(W) + h(2)| \quad \text{for } 0 \le n \le 4$$

then

$$\angle H(e^{jW}) = -2W$$

Observe that the magnitude of $H(e^{-jW})$ is a real function of W, whereas the phase is a linear function of W over the range of $-\pi \le W \le +\pi$.

R.6.125 Now consider the sequence

$$h(n) = -0.09\delta(n) + 0.29\delta(n-1) + 0.53\delta(n-2) + 0.29\delta(n-3) - 0.09\delta(n-4)$$

which represents the impulse response of a type-1 FIR filter with $N = 5$.

ANALYTICAL Solution

The magnitude and phase response is then given by (using R.6.124)

$$|H(e^{jW})| = 2(-0.09)cos(2W) + 2 * (0.29)cos(W) + (0.53)$$

$$\text{for } n = 0, 1, 2, 3, 4 \quad \text{and} \quad \angle H(e^{jW}) = -2W$$

R.6.126 The example below illustrates the analysis of a transfer function of an FIR type-2 filter with $N = 8$.

ANALYTICAL Solution

$$H(z) = h(0) + h(1)z^{-1} + h(2)z^{-2} + h(3)z^{-3} + h(4)z^{-4} + h(5)z^{-5} + h(6)z^{-6} + h(7)z^{-7}$$

Then applying symmetry conditions, the following relations hold:

$$h(0) = h(7)$$
$$h(1) = h(6)$$
$$h(2) = h(5)$$
$$h(3) = h(4)$$

Then, its spectrum is given by

$$H(e^{jW}) = [2h(0)cos(7W/2) + 2h(1)cos(5W/2) + 2h(2)cos(3W/2)$$
$$+ 2h(3)cos(W/2)]e^{-jW7/2}$$

Clearly, the (phase) $\angle H(e^{jW}) = -(7/2)W$ and the quantity inside the brackets represents its magnitude spectrum.

R.6.127 Let us analyze an FIR type-3 filter with $N = 9$ given by

$$H(z) = h(0) + h(1)z^{-1} + h(2)z^{-2} + h(3)z^{-3} + h(4)z^{-4} + h(5)z^{-5}$$
$$+ h(6)z^{-6} + h(7)z^{-7} + h(8)z^{-8}$$

ANALYTICAL Solution

Then using the symmetry conditions and the fact that $h(4)$ must be zero the following frequency spectrum is obtained:

$$H(e^{jW}) = j[2h(0)sin(4W) + 2h(1)sin(3W) + 2h(2)sin(2W) + 2h(3)sin(W)]e^{-j4W}$$

Clearly, $\angle H(e^{jW}) = -4W$ and the quantity inside the brackets represents its magnitude spectrum.

R.6.128 Let us consider the frequency response of an FIR type-4 filter with $N = 8$, given as follows:

$$H(e^{jW}) = j[2h(0)sin(7W/2) + 2h(1)sin(5W/2) + 2h(2)sin(3W/2)$$
$$+ 2h(3)sin(W/2)]e^{-jW7/2}$$

Then $\angle H(e^{jW}) = -(7/2)W$ and the quantity inside the brackets represents the magnitude spectrum.

R.6.129 The zeros on the z-plane of an FIR filter having real impulse response coefficients occur in (complex conjugate) pairs. Applying symmetry and some mathematical considerations, that are left as an exercise, the following general observations can be stated about the locations of the filter's zeros:

   a. *Type-1 filter,* zeros occur anywhere (inside, on, and outside the unit circle)

   b. *Type-2 filter,* one zero occurs at $z = -1$ and the remaining zeros anywhere

   c. *Type-3 filter,* one zero occurs at $z = 1$ and one at $z = -1$ and the remaining zeros occur anywhere

   d. *Type-4 filter,* one zero occurs at $z = 1$, and the remaining zeros anywhere

R.6.130 FIR filters can be designed by choosing an ideal frequency select filter with an IIR, and then limiting or truncating its impulse response.

R.6.131 Recall that the process of truncating or limiting a sequence is referred to as *windowing*. A number of window models are accepted standards currently used by engineers. Recall also that the standard windows were introduced and discussed in Chapter 1.

R.6.132 FIR filters are characterized by having a finite number of coefficients in its impulse response that affect the output of the filter in the following way:

   a. By increasing the number of coefficients, the transition region decreases and the filter approaches ideal *brickwall* conditions, introducing discontinuities in the transition band (region between the pass and stop bands).

   b. Any discontinuity when approximated by a Fourier series expansion shows the *Gibb's* effect in the form of oscillations. This effect can be reduced by using *windows*, in which case the transition region becomes smoother (see Chapter 1).

   c. The use of windows generally causes a reduction of the ripple magnitude, but also causes an increase of the transition band.

   d. There are other ways to reduce the Gibb's effects, but the simplest, easiest, cost-effective, and the one generally preferred by engineers is the windowing method.

R.6.133 The window effect in the time domain is translated in limiting the number of terms of the impulse response of a filter.

R.6.134 The infinite sequence response $h_{IIR}(n)$ can be truncated in the time domain by limiting the sequence by an appropriate window function labeled *win(n)*.

This fact translates into a finite response sequence $h_{FIR}(n)$.

Therefore, analytically

$$h_{FIR}(n) = h_{IIR}(n) * win(n)$$

R.6.135 Recall that the product of two sequences in the time domain such as $h_{IIR}(n)$ times *win(n)* translates in a convolution of their respective transform of each of the sequences in the frequency domain (Chapter 5).

R.6.136 The following family of MATLAB functions:

$$hn = fir1\ (L,\ Wn)$$

$$hn = fir1\ (L,\ Wn,\ 'filtertype')$$

$$hn = fir1\ (L,\ Wn,\ window)$$

$$hn = fir1\ (L,\ Wn,\ 'filtertype',\ window)$$

are multipurpose functions used in the design of LP, HP, BP, and BS FIR filters employing the Fourier series approach.

The *fir1* command returns the impulse response coefficients *hn* with length $N = L + 1$.

LPF are designed with a normalized cutoff frequency over the range $0 \leq Wn \leq 1$.

The function $hn = fir1\ (L,\ Wn)$, for example, can be used to design a BPF by defining *Wn* as a row vector consisting of two elements, $Wn = [W_l\ W_u]$, where W_l represents the low-frequency edge, whereas W_u is its high-frequency edge.

The MATLAB function $hn = fir1\ (L,\ Wn,\ 'high')$ returns the HPF impulse response of length L (where L is an even integer).

The same function can be used to design an SBF by replacing *'high'* by *'stop'*, where *Wn* is a two-element row vector.

If no window is specified, the *Hamming window* is the default option. If the *filtertype* is not specified an LPF is the default option.

R.6.137 The MATLAB functions:

$$hn = fir2(L, f, mag)$$

$$hn = fir2(L, f, mag, window)$$

are used in the design of the magnitude response of arbitrary-shaped LPF, where L is the filter order with the magnitude specified by the vector *mag* at the frequencies given by the vector *f*.

The output of the vector *hn* is of length $N = L + 1$ and the coefficients are arranged in descending order of *z*. Obviously, the *length (f)* must agree with the *length (mag)* and *f* must have as its first element a 0 and a 1 as the last.

A jump condition at f_x is indicated by two samples, both with the same frequency f_x but with different magnitudes.

Any window model can be used as long as its length is $L + 1$.
The MATLAB default window is the *Hamming window*.

R.6.138 The Parks–McClellan algorithm is a popular alternative to the windowing synthesis algorithms followed for the FIR filter case. The process is based on the
Chebyshev polynomials and its implementation resulting in a highly efficient
iterative procedure called the *remez exchange algorithm* developed during the 1970s.
MATLAB implements the *remez* algorithm by calling one of the *remez* family of
functions indicated as follows:

$$hn = remez\ (L, f, mag)$$

$$hn = remez\ (L, f, mag, wei)$$

$$hn = remez\ (L, f, mag, 'filtertype')$$

$$hn = remez\ (L, f, mag, wet, 'filtertype')$$

R.6.139 The Parks–McClellan algorithm is frequently used in the design of FIR multiband
linear-phase filters. The output is the impulse response coefficients of length
$N = L + 1$ in descending powers of z.

The vector f specifies the frequencies, over the range $0 \le f \le 1$, with a sampling
frequency of 2 Hz, where f must be specified as an increasing sequence (over the
range of frequencies), whereas the corresponding magnitudes are given by *mag*,
and where the frequency spacing of f must be at least 0.1.

Recall that the vector *mag* corresponds to the magnitudes at the frequencies
specified by the vector f (also defined over the range $0 \le mag \le 1$) and can be further controlled by an additional weighted vector *wei*.

R.6.140 Observe that the *remez* functions use no window since recursive optimization
techniques are used in the algorithm and the windowing technique would undo
the optimization process.

R.6.141 For example, a normalized multilevel filter spec is shown by the sketch in Figure 6.19. This filter spec is implemented below by the *remez* function defined by
the row vectors f and *mag* given by

$$f = [0\ .3\ .5\ .7\ .9\ 1]$$

$$mag = [.4\ .7\ .5\ 0\ .8\ .8]$$

Observe that the range $0 \le f \le 0.3$ is specified by the elements $f(1)$ and $f(2)$ resulting in a magnitude of 0.4 specified by the first element of *mag*.

When a band of frequencies is irrelevant, the magnitude in that band constitutes
a *don't care* condition. The simplest and efficient way to implement a *don't care* is by
repeating the previous magnitude entry in the vector *mag*.

For example, let us assume that the range $0.5 < f < 0.7$ constitutes a *don't care* in
the frequency sketch of Figure 6.19. Then this condition can be implemented by
substituting the value of *mag(2)* (0.7 instead of 0.5) for the entry *mag(3)*.

R.6.142 Some practical observations and a summary for the two discrete types of filters
IIR and FIR are given as follows:

a. FIR filters have no feedback loops and therefore are always stable, easy to design,
usually requiring a large number of components; in the order of 5–10 times more

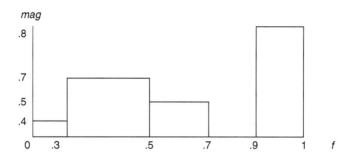

FIGURE 6.19
Normalized multilevel filter characteristics of R.14.141.

than the equivalent IIR such as adders, multipliers, and delay elements; and is the preferred filter choice when the required phase response is linear.

b. IIR filters are recursive in nature, employing feedback loops, which makes them highly efficient in terms of components, but may create unstable conditions that may result in outputs that may grow without bound (independent of its inputs).

c. Butterworth/IIR filters present no in-band ripple and they are the closest to the FIR filters in terms of stability and phase response, with no sharp cutoff characteristic.

d. Chebyshev/FIR filters present ripple either in the pass- or stop band depending on the type (1 or 2), and generally the filter's Chebyshev order is less than the equivalent Butterworth/FIR filters.

e. Elliptic/FIR filters present ripples in the pass band as well as in the stop band and their order is the lowest with respect to the Chebyshev and Butterworth filter implementation.

R.6.143 Once the desirable transfer function is obtained, the next step in the design process is to implement the filter in terms of circuit elements. Recall that this process is referred to as synthesis. Recall that synthesis is the inverse of analysis.

R.6.144 For example, let us synthesize the following analog transfer function:

$$H(s) = \frac{4s^2 + 6s}{6s^2 + 9s} \tag{6.1}$$

The steps followed in the process of synthesis is illustrated below:
Let us assume that the desired structure for $H(s)$ is indicated in Figure 6.20. Then

$$H(s) = \frac{Z_2(s)}{Z_1(s) + Z_2(s)}$$

and

$$H(s) + \frac{1}{\dfrac{Z_1(s)}{Z_2(s)} + 1} \tag{6.2}$$

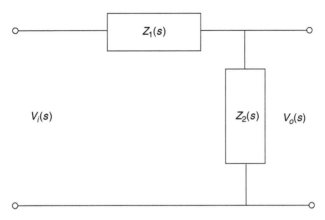

FIGURE 6.20
Structure for $H(s)$ of R.14.144.

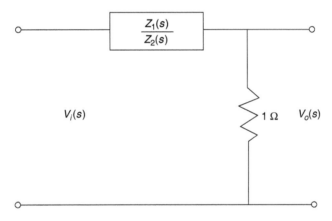

FIGURE 6.21
Reduction of $H(s)$ of Figure 6.20.

and the resulting network structure is illustrated in Figure 6.21.

Let us equate Equation 6.1 with Equation 6.2 resulting in the following:

$$H(s) = \cfrac{1}{\cfrac{6s^2 + 9s}{4s^2 + 6s}} = \cfrac{1}{\cfrac{4s^2 + 6s}{4s^2 + 6s} + \cfrac{2s^2 + 3s}{4s^2 + 6s}}$$

$$H(s) = \cfrac{1}{\cfrac{6s^2 + 9s}{4s^2 + 6s}} = \cfrac{1}{1 + \cfrac{2s^2 + 3s}{4s^2 + 6s}}$$

Let

$$Z = \frac{2s^2 + 3s}{4s^2 + 6s}, \text{note that } Z = \frac{Z_1(s)}{Z_2(s)}$$

where

$$Z = \frac{2s^2}{4s^2 + 6s} + \frac{3s}{4s^2 + 6s}$$

Let

$$Za = \frac{2s^2}{4s^2 + 6s}$$

and

$$Zb = \frac{3s}{4s^2 + 6s}$$

where $Z = Za + Zb$.

Clearly, Za and Zb are impedances in series. Its admittances are

$$Ya = \frac{4s^2 + 6s}{2s^2} = \frac{4s^2}{2s^2} + \frac{6s}{2s^2} = 2 + \frac{3}{s}$$

and

$$Ya = 2 + \frac{1}{\frac{1}{3}s} \quad \text{or} \quad Za = \frac{1}{2} \bigg\| \frac{1}{3}s$$

The impedance Za is implemented in terms of simple electrical elements in Figure 6.22.

The admittance

$$Yb = \frac{4s^2 + 6s}{3s} = \frac{4s^2}{3s} + \frac{6s}{3s} = \frac{4}{3}s + 2$$

or

$$Zb = \left[\frac{1}{(4/3)s}\right] \bigg\| \left[\frac{1}{2}\right] \text{ implemented in terms of electrical elements in Figure 6.23.}$$

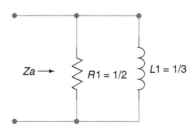

FIGURE 6.22
Synthesis of Za.

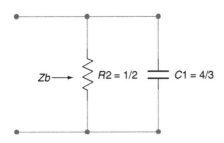

FIGURE 6.23
Synthesis of *Zb*.

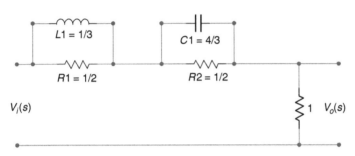

FIGURE 6.24
Synthesis of $H(s) = \dfrac{4s^2 + 6s}{6s^2 + 9s}$ of R.6.144.

Then, $H(s) = \dfrac{4s^2 + 6s}{6s^2 + 9s}$ is synthesized into the network shown in Figure 6.24.

R.6.145 The synthesis process for digital filters is simpler compared to the analog case. Once the transfer function $H(z)$ is known, the difference equation can be obtained and an implementation can be realized (see Chapter 5 for details).

6.4 Examples

Example 6.1

Given the first-order analog LPF shown in the circuit diagram of Figure 6.25, evaluate by hand the following:

  a. The transfer function $H(w) = V_o(w)/V_i(w)$
  b. Obtain expressions for $|H(w)|$ and $\angle H(w)$

Create the script file *LP_filter_analysis* that returns the following plots:

  c. $|H(f)|$ versus f and $\angle H(f)$ versus f using the MATLAB function *freqs*
  d. $|H(f)|_{dB}$ versus f and $\angle H(f)$ versus f using a semilog scale and the MATLAB function *freqs*
  e. Bode plots of $|H(w)|$ and $\angle H(w)$ over the range $-5*(1/(R*C)) \le w \le 5*(1/(R*C))$
  f. Bode plots of $|H(w)|$ and $\angle H(w)$ with no argument (w)

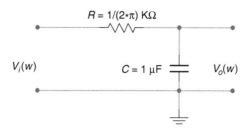

FIGURE 6.25
Circuit diagram of Example 6.1.

g. Repeat $|H(f)|$ versus f
$\quad \angle H(w)$ versus f
$\quad |H(f)|_{dB}$ versus f (semilog scale)
$\quad \angle H(f)$ versus f (semilog scale)

using standard circuit techniques.

ANALYTICAL Solution

Part (a)

$$H(w) = \frac{V_o(w)}{V_i(w)} = \frac{\dfrac{1}{jwC}}{\dfrac{1}{jwC} + R} = \frac{\dfrac{1}{R*C}}{j*w + \dfrac{1}{R*C}}$$

Part (b)

$$|H(w)| = \frac{\dfrac{1}{R*C}}{\sqrt{w^2 + \left(\dfrac{1}{R*C}\right)^2}}$$

$$\angle H(w) = tan^{-1}(w*R*C)$$

MATLAB Solution
```
% Script File: LP_filter_analysis
R=1e3/(2*pi);
C=.1e-6;
a=1/(R*C);
w =-5*a:10000:5*a;
num=a;
den=[1 a];
Hlp = freqs(num,den,w);

figure(1);
subplot(2,1,1);
plot(w/(2*pi),abs(Hlp));
xlabel ('frequency in Hrz');
ylabel ('magnitude');
title('Magnitude of H(f) vs. frequency');
```

```
axis ([-5e4 5e4 0 1.1])
subplot (2,1,2);
plot (w/(2*pi),angle(Hlp)*180/pi);
xlabel ('frequency in Hrz');
ylabel ('phase in degrees');
title('Phase of H(f) vs. frequency');

figure(2)
db = 20.*log(abs(Hlp));
subplot(2,1,1);
semilogx(w/(2*pi),db);
ylabel('mag.in dbs');
title(' magnitude (dbs) vs. freqs.');
axis([1e3 1e5 -40 5]);grid on;
subplot(2,1,2);
semilogx(w/(2*pi),angle(Hlp)*180/pi);
title('phase ang. vs. freqs.');
ylabel('phase in degrees');
xlabel('frequency, in Hrz');
axis([1e3 1e5 -80 0]);grid on;

figure(3);
bode(num,den,w)
title('Bode plots with argument w');

figure(4);bode(num,den)
title('Bode plots with no w');
figure(5)
% using standard circuit approach (complex algebra)
XC = 1./(w.*C);
Z = sqrt(R^2+XC.^2);
magHele = abs(XC./Z); phaseHele =-atan(R./XC);
subplot(2,2,1)
plot(w./(2*pi),magHele);
title('mag.vs. freq.');
ylabel('magnitude')
axis([-5e4 5e4 0 1.1]); grid on;
subplot(2,2,2)
plot(w./(2*pi),phaseHele*180/pi);
title('phase vs. freq.');
ylabel('phase (in degrees)');grid on;
subplot(2,2,3);
dbele = 20.*log10(magHele);
semilogx(w./(2*pi),dbele);title('mag.(in db) vs. freq. ')
ylabel('magnitude(db)');
xlabel('frequency in Hrz');
axis([1e3 1e6 -15 1.0]);grid on;
subplot(2,2,4);
semilogx(w./(2*pi),phaseHele*180/pi);title('phase vs. freq. ')
xlabel('frequency in Hrz');
ylabel('phase(degrees)'); grid on;
```

Back in the command window, the script file *LP_filter_analysis* is executed and the results are as follows (Figures 6.26 through 6.30):

```
>> LP _ filter _ analysis
```

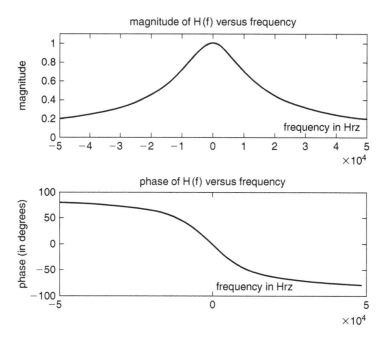

FIGURE 6.26
Plots of *H(f)* using *freqs* of Example 6.1.

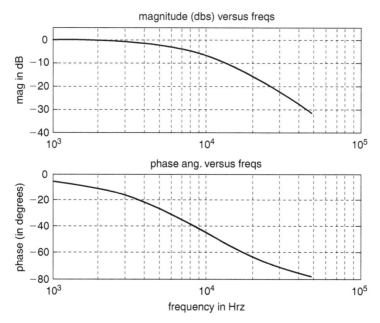

FIGURE 6.27
Plots of $|H(f)|_{dB}$ and $\angle H(f)$ using a semilog scale of Example 6.1.

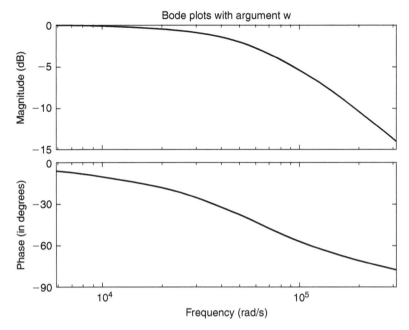

FIGURE 6.28
Bode plots of $|H(w)|$ and $\angle H(w)$ over the range $-5*(1/(R*C)) \le w \le 5*(1/(R*C))$ of Example 6.1.

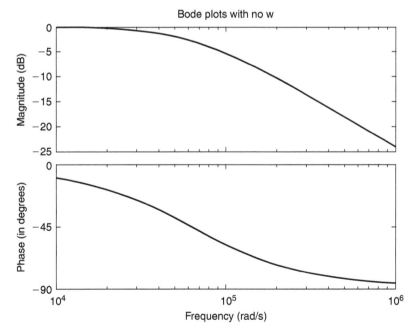

FIGURE 6.29
Bode plots of $|H(w)|$ and $\angle H(w)$ with no argument (w) of Example 6.1.

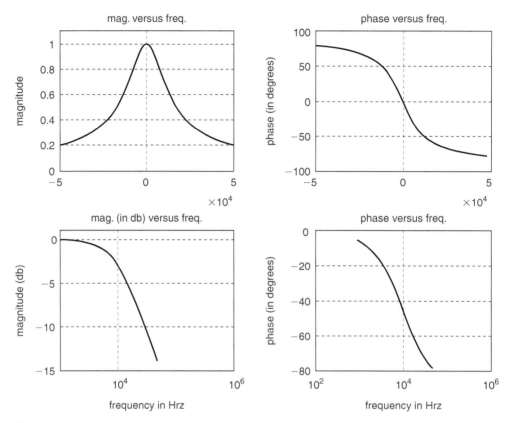

FIGURE 6.30
Plots of $|H(f)|$, $\angle H(f)$, $|H(f)|_{dB}$, and $\angle H(f)$ using standard circuit techniques of Example 6.1.

Example 6.2

Given the first-order analog HPF shown in the circuit diagram of Figure 6.31, evaluate by hand the following:

a. The transfer function $H(w) = V_o(w)/V_i(w)$
b. Obtain expressions for $|H(w)|$ and $\angle H(w)$

Create the script file *HP_filter_analysis* that returns the following plots:

c. $|H(f)|$ versus f and $\angle H(f)$ versus f using the MATLAB function *freqs*
d. $|H(f)|_{dB}$ versus f and $\angle H(f)$ versus f using a semilog scale and the MATLAB function *freqs*
e. Bode of $|H(w)|$ and $\angle H(w)$ over the range $-5*(1/(R*C)) \leq w \leq 5*(1/(R*C))$
f. Bode of $|H(w)|$ and $\angle H(w)$ with no argument (w)
g. $|H(f)|$ versus f
   $\angle H(w)$ versus f
   $|H(f)|_{dB}$ versus f (semilog scale)
   $\angle H(f)$ versus f (semilog scale)
   using standard circuit techniques.

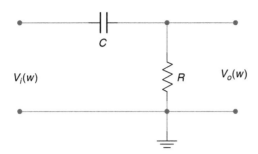

FIGURE 6.31
Circuit diagram of Example 6.2.

ANALYTICAL Solution

Part (a)

$$H(w) = \frac{V_o(w)}{V_i(w)} = \frac{R}{\dfrac{1}{jwC} + R} = \frac{j*w}{j*w + \dfrac{1}{R*C}}$$

Part (b)

$$|H(w)| = \frac{w}{\sqrt{w^2 + \left(\dfrac{1}{R*C}\right)^2}}$$

$$\angle H(w) = (\pi/2) - \tan^{-1}(w*R*C)$$

MATLAB Solution

```
% Script File: HP_filter_analysis
% RC-High Pass Filter Analysis
R=1e3/(2*pi);
C=.1e-6;
a = 1/(R*C);
w = -5*a:10000:5*a;
num = [1 0];
den = [1 a];
Hhp = freqs(num,den,w);

figure(1);
subplot(2,1,1);
plot(w/(2*pi),abs(Hhp));
ylabel('magnitude');
xlabel('frequency in Hrz');
title('Magnitude of H(f) vs. frequency');
subplot(2,1,2);
plot(w/(2*pi),angle(Hhp)*180/pi);
xlabel('frequency in Hrz');
ylabel('phase in degrees');
title('Phase of H(f) vs. frequency');

figure(2)
db=20.*log(abs(Hhp));
```

```
subplot(2,1,1);
semilogx(w/(2*pi),db);
ylabel('magnitude in dbs');
title('magnitude vs. freqs.');
axis([1.9e2 1.4e4 -50 2]);
grid on;
subplot(2,1,2);
semilogx(w/(2*pi),angle(Hhp)*180/pi);
title('phase ang. vs. freqs.');
ylabel('phase in degrees');
xlabel('frequency in Hrz');
axis([1.9e3 1.4e4 0 90]);
grid on;

figure(3);
bode(num,den,w)
title('Bode plots with argument w');
axis([5e3 2e5 0 100])

figure(4);
bode(num,den)
title('Bode plots with no w');

figure(5)
% using standard circuit approach (complex algebra)
XC=1./(w.*C);
Z=sqrt(R^2+XC.^2);
magHele=(R./Z);
phaseHele=(atan(XC./R)).*180./pi;
subplot(2,2,1)
plot(w./(2*pi),magHele);
title('magnitude vs. freq.');
ylabel('magnitude')
grid on;
subplot(2,2,2)
plot(w./(2*pi),phaseHele);
title('phase vs. freq.');
ylabel('phase in degrees');
grid on;
subplot(2,2,3);
dbele=20.*log10(magHele);
semilogx(w./(2*pi),dbele);title(' magnitude vs. freq.')
ylabel('magnitude(db)');
xlabel('frequency in Hrz');
grid on;
subplot(2,2,4);
semilogx(w./(2*pi),phaseHele);title(' phase vs. freq.')
xlabel('frequency in Hrz');
ylabel('phase(degrees)');
grid on;
```

Back in the command window, the script file *HP_filter_analysis* is executed and the results are shown as follows (Figures 6.32 through 6.36):

```
>> HP _ filter _ analysis
```

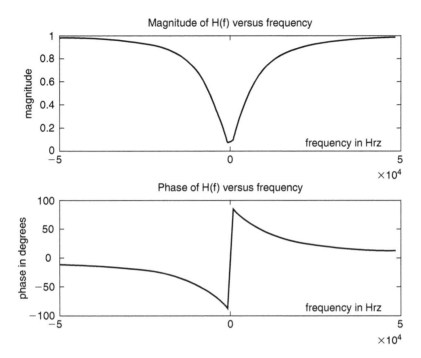

FIGURE 6.32
Plots of *H(f)* of Example 6.2.

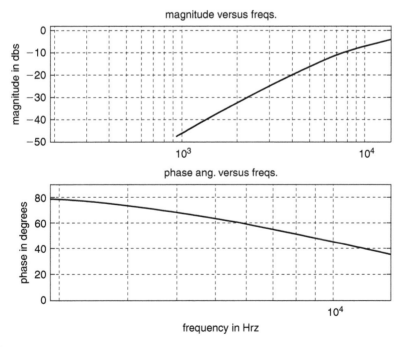

FIGURE 6.33
Plots of $|H(f)|_{dB}$ and $\angle H(f)$ using a semilog scale of Example 6.2.

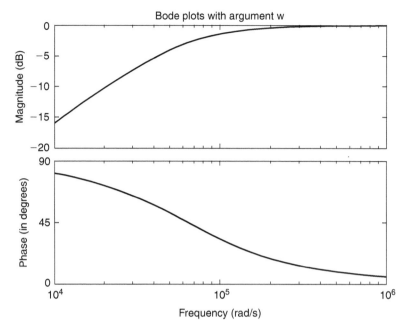

FIGURE 6.34
Bode plots of $|H(w)|$ and $\angle H(w)$ over the range $-5*(1/(R*C)) \leq w \leq 5*(1/(R*C))$ of Example 6.2.

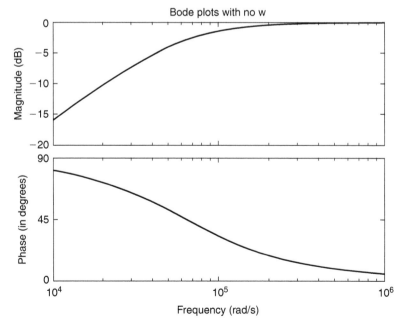

FIGURE 6.35
Bode plots of $|H(w)|$ and $\angle H(w)$ with no argument *(w)* of Example 6.2.

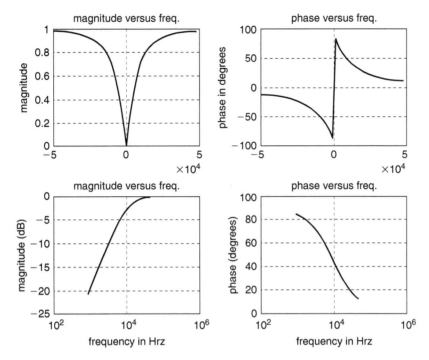

FIGURE 6.36
Plots of $|H(f)|$, $\angle H(f)$, $|H(f)|_{dB}$, and $\angle H(f)$ using standard circuit techniques of Example 6.2.

Example 6.3

Given the second-order filter shown in Figure 6.37.

a. Determine by hand the transfer function $H(s) = V_o(s)/V_i(s)$.

b. Verify using MATLAB that the circuit of Figure 6.37 corresponds to a normalized LPF by obtaining the magnitude and phase plots of $H(w)$.

c. Make the following substitutions corresponding to the frequency shifting property: $R = R_N$, $L = L_N/2$, and $C = C_N/2$ and obtain its transfer function $H(s)$, its magnitude and phase plots, and its numerator and denominator polynomial coefficients.

d. Using the transformations of part (c), obtain magnitude and phase plots of the nonnormalized filter $H(w)$ corresponding to an LPF and verify that the filter's cutoff frequency is $w_c = 2\,\text{rad/s}$.

e. Use the MATLAB function *lp2lp* on the filter transfer function of part (b) to shift the cutoff frequency to $w_c = 2\,\text{rad/s}$. Obtain the system transfer function $H(s)$, its magnitude and phase plots, and its numerator and denominator polynomial coefficients.

f. Implement the filter specs of part (b) using a Butterworth, second-order normalized LPF. Obtain its transfer function $H(s)$, its magnitude and phase plots, and its numerator and denominator polynomial coefficients.

g. Using the MATLAB function *lp2lp* transform the normalized LPF of part (f) into a denormalized LPF with cutoff frequency of $w_c = 2\,\text{rad/s}$. Obtain its transfer function $H(s)$, its magnitude and phase plots, and its numerator and denominator polynomial coefficients.

h. Implement the normalized Butterworth LPF of part (f) using electrical elements.

i. Implement the denormilized Butterworth LPF of part (g) using electrical elements (RLC).

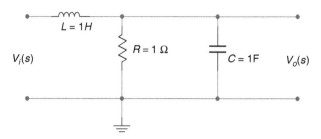

FIGURE 6.37
Circuit diagram of Example 6.3.

The solutions of Example 6.3 are given by the script file *shifting_LPF* as follows:

ANALYTICAL Solution

$$H(s) = \cfrac{\cfrac{R\cfrac{1}{sC}}{R+\cfrac{1}{sC}}}{sL + \cfrac{R\cfrac{1}{sC}}{R+\cfrac{1}{sC}}}$$

$$H(s) = \frac{V_o(s)}{V_i(s)} = \cfrac{\cfrac{R}{sC}}{\left(R+\cfrac{1}{sC}\right)sL + \cfrac{R}{sC}}$$

$$H(s) = \frac{V_o(s)}{V_i(s)} = \cfrac{\cfrac{R}{sC}}{sRL + \cfrac{R}{sC} + \cfrac{L}{C}}$$

$$H(s) = \frac{V_o(s)}{V_i(s)} = \cfrac{R}{sC\left(sRL + \cfrac{R}{sC} + \cfrac{L}{C}\right)}$$

$$H(s) = \frac{V_o(s)}{V_i(s)} = \frac{R}{s^2RLC + Ls + R} \quad \text{for } R = C = L = 1$$

$$H(s) = \frac{V_o(s)}{V_i(s)} = \frac{1}{s^2 + s + 1}$$

MATLAB Solution
```
% Script file: shifting _ LPF
% part(b), LPF with wₙ =1,R=L=C=1
num1=1;
den1= [1 1 1];
w = [0:.2:4];
H1 = freqs(num1,den1,w);
```

```
mag1= abs(H1);
phase1=angle(H1);

figure(1)
subplot(1,2,1)
plot(w,mag1)
xlabel('frequency in rad/sec');
ylabel('Magnitude');
title('Magnitude of H(w) vs. w [for R=L=C=1]');
subplot(1,2,2)
plot(w,phase1*180/pi)
xlabel('frequency in rad/sec.');
ylabel('Angle in degrees');
title('Phase of H(w) vs. w [for R=L=C=1]')

% part(c)
L=1/2;C=1/2; R=1; a=R*L*C;
num2 = R;
den2 = [a L R];
H2 = freqs(num2,den2,w);
mag2 = abs(H2);
phase2 = angle(H2);

figure(2)
subplot(1,2,1)
plot(w,mag2)
xlabel('frequency in rad/sec.');
ylabel('Magnitude');
title('Magnitude of H(w) vs. w [for L=C=1/2,R=1]');
subplot(1,2,2)
plot(w,phase2*180/pi)
xlabel('frequency in rad/sec');
ylabel('Angle in degrees');
title('Phase of H(w) vs. w [for L=C=1/2,R=1]')
disp('**************************************************')
disp('******* M A T L A B ***** R E S U L T ***********')
disp('**************************************************')
disp('Solution to part (c) ')
disp('Coefficients of numerator of trans. function [LPF/ with
L=C=1/2,R=1]')
disp(num2)
disp('Coefficients of denominator of trans. function [LPF/lp2lp/wc=2]')
disp(den2)
tf(num2,den2)
disp('**************************************************')

figure(3)
[P,Q]=lp2lp(num1,den1,2);
Hlp2lp=freqs(P,Q,w);
mag3=abs(Hlp2lp);
phase3=angle(Hlp2lp);
subplot(1,2,1)
plot(w,mag3)
xlabel('frequency in rad/sec.');
```

```
ylabel('Magnitude');
title('Magnitude of H(w) vs. w [using lp2lp]');
subplot(1,2,2)
plot(w,phase3*180/pi)
xlabel('frequency in rad/sec');
ylabel('Angle in degrees');
title('Phase of H(w) vs. w [using lp2lp]')
disp('**************************************************')
disp('Solution to part (e) ')
disp('Coefficients of the numerator of the trans. function [wc=2]')
disp(P)
disp('Coefficients of the denominator of the trans. function [wc=2]')
disp(Q)
tf(P,Q)
disp('**************************************************')

%part(f),solution using Butter/norm.LPF
[z,p,k] = buttap(2);
[Pn,Qn] = zp2tf(z,p,k);
disp('**************************************************')
disp('Solution to part (f) ')
disp('Coefficients of the numerator of the trans. function
[butter/LPF/wc=1]')
disp(Pn)
disp('Coefficients of the denominator of the trans. function
[butter/LPF/wc=1]')
disp(Qn)
tf(Pn,Qn)
disp('**************************************************')

figure(4)
w=0:.1:3;
Hbut=freqs(Pn,Qn,w);
mag4 = abs(Hbut);
phase4 = angle(Hbut);
subplot(1,2,1)
plot(w,mag4)
xlabel('frequency in rad/sec.');
ylabel('Magnitude');
title('Magnitude of H(w) vs. w [Butter/wc=1]');
subplot(1,2,2)
plot(w,phase4*180/pi)
xlabel('frequency in rad/sec');
ylabel('Angle in degrees');
title('Phase of H(w) vs. w [Butter/wc=1]')

% part(g),transforming the butter/LPF using lp2lp
% with cut off frequency,wc=2 rad/sec
[Pl,Ql] = lp2lp(Pn,Qn,2);
disp('**************************************************')
disp('Solution to part (g) ')
disp('Coefficients of the numerator of the transf. function
[butter/wc=2rad/sec]')
disp(Pl)
```

```
disp('Coefficients of the denominator of the transf. function
[butter/wc=2rad/sec]')
disp(Q1)
tf(P1,Q1)
disp('**************************************************')

figure(5)
w =0:.1:5;
H1 =freqs(P1,Q1,w);
mag5=abs(H1);
phase5=angle(H1);
subplot(1,2,1)
plot(w,mag5)
xlabel('frequency in rad/sec');
ylabel('Magnitude');
title('Magnitude of H(w) vs. w [Butter/wc=2]');
subplot(1,2,2)
plot(w,phase5*180/pi)
xlabel('frequency in rad/sec');
ylabel('Angle in degrees');
title('Phase of H(w) vs. w [Butter/wc=2]')
```

Back in the command window, the script file *shifting_LPF* is executed and the results are shown as follows (Figures 6.38 through 6.42):

```
>> shifting _ LPF
   **************************************************
   ******* M A T L A B ***** R E S U L T ************
   **************************************************

   Solution to part (c)
   Coefficients of numerator of trans. function [LPF/ with
   L=C=1/2,R=1]
      1
   Coefficients of denominator of trans. function [LPF/lp2lp/wc=2]
      0.2500   0.5000   1.0000

   Transfer function:
         1
   --------------------
   0.25 s^2 + 0.5 s + 1


   **************************************************
   **************************************************
   Solution to part (e)
   Coefficients of the numerator of the trans. function [wc=1]
      4

   Coefficients of the denominator of the trans. function [wc=1]
      1.0000   2.0000   4.0000
   Transfer function:
        4
   -------------
   s^2 + 2 s + 4
   **************************************************
```

```
****************************************************
Solution to part (f)
Coefficients of the numerator of the trans. function
[butter/LPF/wc=1]
   0   0   1
Coefficients of the denominator of the trans. function
[butter/LPF/wc=1]
   1.0000   1.4142   1.0000
Transfer function:
      1
   -----------------
   s^2 + 1.414 s + 1
****************************************************
****************************************************
Solution to part (g)
Coefficients of the numerator of the transf. function
[butter/wc=2rad/sec]
   4
Coefficients of the denominator of the transf. function
[butter/wc=2rad/sec]
   1.0000   2.8284   4.0000

Transfer function:
      4
   -----------------
   s^2 + 2.828 s + 4
****************************************************
```

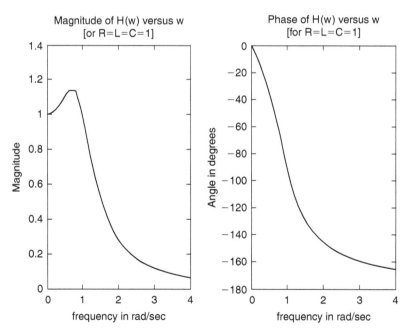

FIGURE 6.38

Plots of $H(w)$ of Example 6.3 with $R = L = C = 1$.

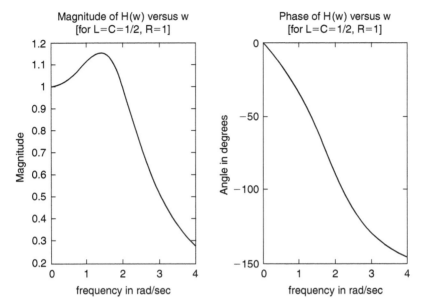

FIGURE 6.39
Plots of *H(w)* of Example 6.3 with *R* = 1 and *L* = *C* = 1/2.

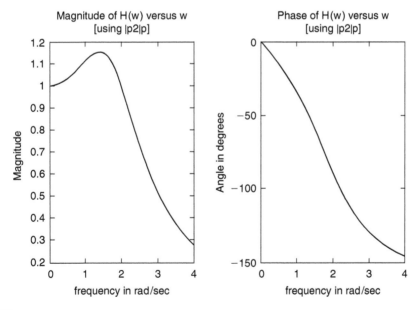

FIGURE 6.40
Plots of *H(w)* using *lp2lp* of Example 6.3.

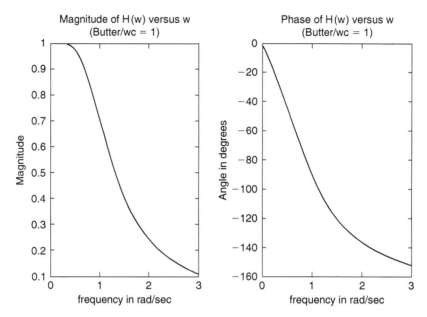

FIGURE 6.41
Normalized LP Butterworth filter plots of *H(w)* of Example 6.3.

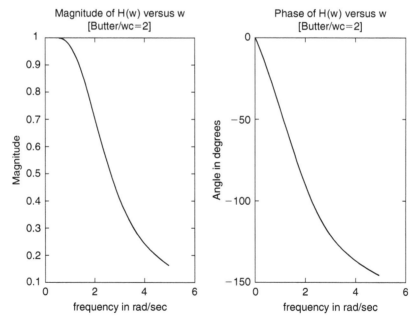

FIGURE 6.42
LP Butterworth filter plots of *H(w)* with w_c = 2 rad/s of Example 6.3.

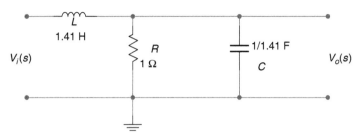

FIGURE 6.43
Synthesis of Butterworth normalized (w_c = 1 rad/s) LPF prototype of Example 6.3.

The implementation of a Butterworth normalized (w_c = 1 rad/s) LPF prototype using electrical components is illustrated below, obtained by equating the coefficients of the obtained transfer function with the second-order Butterworth polynomial as indicated in the following (the solution in term of electrical components is shown in Figure 6.43).

$$R = 1\,\Omega, RLC = 1 \quad \text{and} \quad L = 1.412\,\text{H}$$

then

$$C = 1/(R * C) = 1/1.41\,\text{F}$$

The implementation of a Butterworth denormalized (w_c = 2 rad/s) LPF prototype in terms of electrical components is illustrated in Figure 6.44, where R = 1 Ω, L = 2.8284 H, then C = 1/(R * C) = 1/(4 * 2.8284) F.

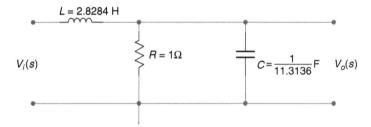

FIGURE 6.44
Synthesis of Butterworth denormalized (w_c = 2 rad/s) LPF prototype of Example 6.3.

Example 6.4

Test the performance of the LPF and HPF of Examples 6.1 and 6.2 by applying the following input: $x(t) = 5 * cos(2 * pi * 1000 * t) + 12.5 * sin(2 * pi * 35000 * t)$ to each filter and by obtaining and observing the resulting plots of their respective outputs. Discuss the results.

The solution of Example 6.4 is given by the script file *test_filter* as follows:

MATLAB Solution
```
% Script File: test _ filter
% testing LPF & HPF
R = 1e3/(2*pi);                    % electrical elements
C =.1e-6;
a = 1/(R*C);
```

```
w =-5*a:10000:5*a;
t =linspace(0,2e-3,1000);
f0 = 1000;                                  % input frequencies
f1 = 35000;
input0 =5.*cos(2*pi*f0.*t);input1=12.5*sin(2*pi*f1.*t);
input = input0+input1;                      % input signal
H0lp = a./(j*2*pi*f0+a);                    % mag of LPF at f = f0
H1lp = a./(j*2*pi*f1+a);                    % mag of LPF at f = f1
resp0lp = abs(H0lp).*(cos((2*pi*f0.*t)+angle(H0lp)));
resp1lp = abs(H1lp).*(sin((2*pi*f1.*t)+angle(H0lp)));
outputlp = resp0lp+resp1lp;                 % output of LPF
H0hp = j*2*pi*f0./(j*2*pi*f0+a);            % mag of HPF at f = f0
H1hp = j*2*pi*f1./(j*2*pi*f1+a);            % mag of HPF at f = f1
resp0hp = abs(H0hp).*(cos((2*pi*f0.*t)+angle(H0hp)));
resp1hp = abs(H1hp).*(sin((2*pi*f1.*t)+angle(H0hp)));
outputhp = resp0hp+resp1hp;                 % output of HPF


figure(1)
subplot(3,1,1)
plot(t,input);
axis([0 2e-3 -18 18]);
ylabel('Input Amplitude');
subplot(3,1,2)
plot(t,outputlp);
ylabel('Output Amplitude');
axis([0 2e-3 -1.5 1.5]);
subplot(3,1,3)
plot(t,input0);
axis([0 2e-3 -6.5 6.5]);
ylabel('Amplitude');
xlabel('time in sec')


figure(2)
subplot(3,1,1)
plot(t,input);
axis([0 2e-3 -18 18]);
ylabel('Input Amplitude');
subplot(3,1,2)
plot(t,outputhp);
ylabel('Output Amplitude');
axis([0 .1e-3 -1.2 1.2]);
subplot(3,1,3)
plot(t,input1);
axis([0 .1e-3 -12.5 12.5]);
ylabel('Amplitude');
xlabel('time in sec')
```

Back in the command window, the script file *test_filter* is executed and the filters' outputs are shown in Figures 6.45 and 6.46.

Note that the LPF output consists of a 1000 Hz sinusoidal wave plus a small high-frequency 35 kHz ripple. The HPF passes the 35 kHz frequency, while completely removing the low-frequency component of 1000 Hz.

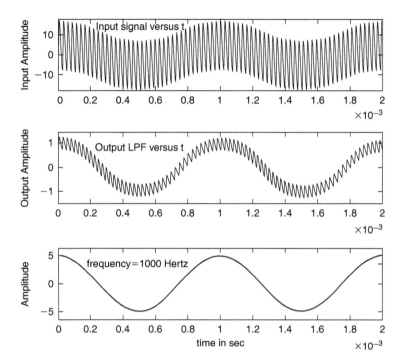

FIGURE 6.45
Input and output plots of the LPF of Example 6.4.

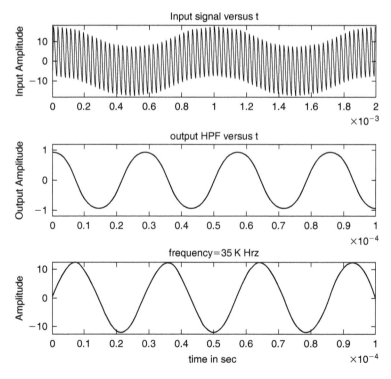

FIGURE 6.46
Input and output plots of the HPF of Example 6.4.

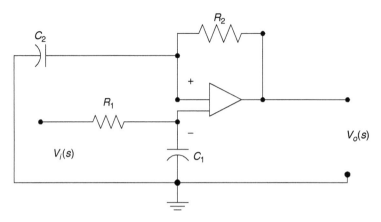

FIGURE 6.47
Circuit diagram of Example 6.5.

Example 6.5

Let $H(s) = \dfrac{V_o(s)}{V_i(s)} = \dfrac{a_2 s + 1}{a_2 s + 1}$ be the transfer function of the active filter shown in Figure 6.47,

where $s = jw$, with the following time constants $a_2 = R_2 * C_2 = 5 * 10^{-2}$ s and $a_1 = R_1 * C_1 = 3 * 10^{-4}$s. Create the script file *active_filter* that returns the following plots:

a. $|H(w)|$ (in decibels) versus w
b. $\angle H(w)$ (in degrees) versus w
c. Bode plots of $H(w)$ (magnitude and phase)
d. Evaluation and plot the zero–pole of the transfer function $H(w)$

MATLAB Solution
```
% Script file: active _ filter
a1 = 3e-4;
a2 = 5e-2;
w = 1e1:1e3:1e8;
num = [a2 1];
den = [a1 1];
H = freqs(num,den,w);
mag = 20*log10(abs(H));

figure(1)
subplot(2,1,1)
plot(w,mag);axis([1e2 1e4 0 50]);
ylabel('Magnitude');xlabel('w in rad/sec');
title(' Magnitude of H(w) vs. w')
subplot(2,1,2);
phase=angle(H)*180/pi;
plot(w,phase);
ylabel('Angle in degrees');
axis([1e2 1e4 0 90]);
xlabel('w in rad/sec')
```

```
title('Phase of H(w) vs. w')
figure(2);bode(num,den);
title('Bode plots:magnitude and phase')

figure(3);zplane(num,den);grid on;
title('zero-pole plot')
disp('********************************');
disp('***** Numerical evaluations ****');
disp('********************************');
disp('The system zero is at :');roots(num)
disp('The system pole is at : ');roots(den)
disp('********************************');
```

Back in the command window, the script file *active_filter* is executed and the results are shown as follows (Figures 6.48 through 6.50):

```
>> active _ filter
  ********************************
  ****** Numerical evaluations ******
  ********************************
  The system zero is at :
   ans =
         -20
  The system pole is at :
  ans =
         -3.3333e+003
  ********************************
```

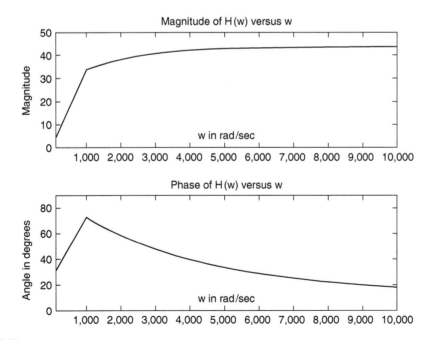

FIGURE 6.48
Magnitude and phase plots of *H(w)* of Example 6.5.

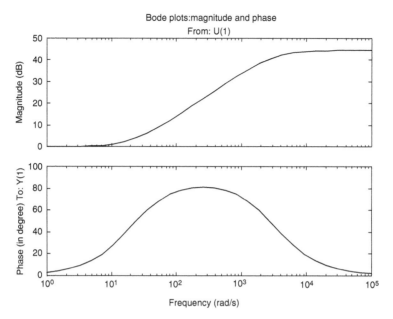

FIGURE 6.49
Bode plots of *H(w)* of Example 6.5.

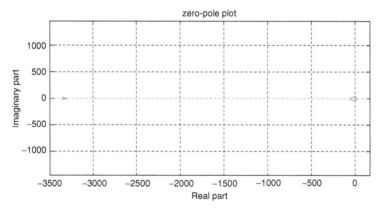

FIGURE 6.50
Zero/pole plots of *H(w)* of Example 6.5.

Example 6.6

Create the script file *butt_chev* that returns the magnitude and phase plots of the following analog LPFs:

  a. Butterworth
  b. Chebyshev/type 1
  c. Chebyshev/type 2

with the following filter's specs:

1. The filter's order of $N = 5$
2. Pass-band edge at $W_p = 1$ rad/s
3. Pass-band ripple of $R_p = 1$ dB
4. Stop-band ripple of $R_s = 1$ dB

MATLAB Solution

```
% Script file: butt_chev
Rp = 1 ;                              % in db (passband ripple)
Rs = 1 ;                              % in db (stopband ripple)
Wp = 1 ;                              % pass band edge
                                      % Butterworth LPF.

n = 5; % order
% transfer function, But. LPF.
[num1, den1] = butter (n, Wp, 's');
% plot the frequency response Butterworth. LPF
w = [0:.1: 5];
h1 = freqs (num1, den1, w);
gain1 = 20 * log10 (abs (h1));

figure(1);
subplot(1,3,1);
plot(w,gain1);axis([0 5 -80 20]);
title ('Butterworth LPF, N = 5')
xlabel (' frequency w (rad/sec)')
ylabel ('gain, db')
                                       % Chebyshev LPF, type 1
[num2, den2] = Cheby1 ( n, Rp, Wp, 's');
% plot the frequency response, Chebyshev 1 LPF
h2 = freqs (num2,den2,w);
gain2 = 20 * log10 (abs(h2));
subplot (1, 3, 2);
plot (w, gain2);
title (' Chebyshev 1 LPF, N=5'); axis([0 5 -100 20]);
xlabel ('frequency w (rad/sec)');
ylabel ('gain, db')
                                       % Chebyshev LPF, type 2.
Wn = 1;% cut off norm. freq.
[num3, den3] = Cheby2 (n, Rs, Wn, 's');
h3 = freqs (num3, den3, w);
gain3 = 20 * log10 (abs(h3));
subplot (1, 3, 3);
plot (w, gain3);
title ('Chebyshev 2 LPF, N=5');axis([0 2 -50 10]);
xlabel (' frequency w (rad/sec)');
ylabel ('gain, db')
figure(2)                             % Butterworth phase
phase1=angle(h1);
subplot(1,3,1);
plot(w,phase1);axis([0 3 -4 4]);
xlabel ('freq w(rad/sec)');
ylabel('phase in rad');
title('Butterworth, phase plot')
                                      % Chebyshev1 phase
phase2=angle(h2);
subplot(1,3,2)
plot(w,phase2);axis([0 2 -4 4]);
xlabel('freq w(rad/sec)');
ylabel('phase in rad');
title('Chebyshev 1, phase plot ')
                                      % Chebyshev2 phase
```

```
phase3=angle(h3);
subplot(1,3,3);
plot(w,phase3);axis([0 3 -2 2]);
xlabel('freq w(rad/sec)');
ylabel('phase in rad');
title('Chebyshev 2, phase plot')
```

Back in the command window, the script file *butt_chev* is executed and the magnitude and phase plots are shown in Figures 6.51 and 6.52.

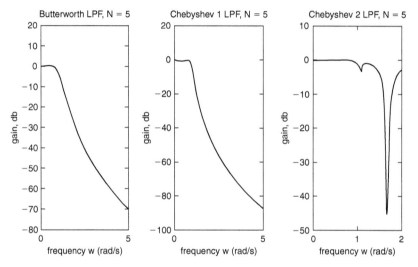

FIGURE 6.51
Magnitude plots of the analog LP Butterworth, Chebyshev/type-1, and Chebyshev/type-2 filters of Example 6.6.

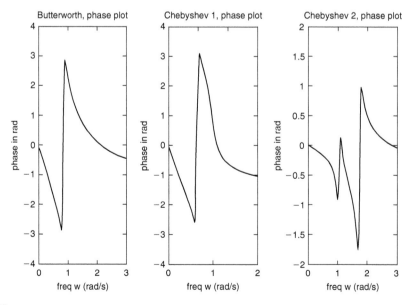

FIGURE 6.52
Phase plots of the analog LP Butterworth, Chebyshev/type-1, and Chebyshev/type-2 filters of Example 6.6.

Example 6.7

Create the script file *analog_digital* that returns the following:

a. The third-order normalized analog low-pass Butterworth filter's magnitude and phase plots of $H(w)$.
b. Use *lp2lp* to transform the normalized filter of part (a) to an analog LPF with $w_c = 3$ rad/s, and obtain the magnitude and phase plots of $H(w)$.
c. With a sampling rate of $T = 0.3$ obtain the numerator and denominator coefficients of the discrete IIR transfer function $H(z)$ filter for $w_c = 3$ rad/s of part (b).
d. Obtain its zero/pole plot of $H(z)$ of part (c) and verify that the system implementation is stable.
e. Obtain its zero/pole plot of $H(s)$ of part (b) and verify that the system is stable and the three analog poles map into the three discrete poles of $H(z)$.
f. Obtain the magnitude and phase plots of the discrete transfer function $H(z)$.

MATLAB Solution
```
% Script file: analog _ digital
[z,p,k]  = buttap(3);              % create an analog normalized filter
[Pn,Qn]= zp2tf (z,p,k);            % analog normalized transfer function
T=.3;Wc=3;
[Pa,Qa]=lp2lp(Pn,Qn,Wc);
w =-5:.1:5;
h = freqs(Pn,Qn,w);


figure(1)
subplot (1,2,1)                    % plot of normalized butt. LPF
plot (w, abs(h));
axis([-5 5 0 1.2]);
title (' Mag.analog norm. LPF')
xlabel ('w in rad/sec')
ylabel ('Magnitude')
subplot (1,2,2)
plot(w,angle(h))
xlabel ('w in rad/sec')
ylabel ('angle')
title('Phase analog norm LPF')


figure(2)
subplot (1,2,1)                    % plot of LPF, wc =3
h2 = freqs(Pa,Qa,w);
plot (w, abs(h2));
title (' Mag. of LPF with wc=3')
xlabel ('w in rad/sec')
ylabel ('magnitude '); axis([-5 5 0.2 1.1]);
subplot (1,2,2)
```

```
phase2=angle(h2)*180/pi;
plot (w, phase2)
title ('Phase of LPF with wc=3')
xlabel ('w in rad/sec')
ylabel ('degrees')
[Pz,Qz]=bilinear (Pa,Qa,1/T);      % discrete transfer function
disp('*******************************************')
disp('**************RESULTS****************** ')
disp('*******************************************')
disp('Digital filter /numerator coefficients of the transfer
function :')
disp(Pz)
disp(' Digital filter / denominator coefficients of the transfer func-
tion :')
disp(Qz)
disp('*******************************************')


figure(3); zplane (Pz,Qz)
title ('zero-pole plot of H(z)')


figure(4);pzmap(Pa,Qa)
title ('zero-pole plot of H(s)')


figure(5)
[magz, phasez] = freqz (Pz, Qz, w*T/pi);
subplot (1,2,1)                              % digital filter
plot (w*T, magz)
xlabel ('W in rad'); title('Mag.of digital filter H(Z)')
ylabel ('magnitude');axis([-2 2 0.4 1.1]);
subplot(1,2,2)
plot(w*T,phasez*180/pi)
xlabel('W in rad'); ylabel('degrees');
title('Phase of digital filter H(z)')
```

Back in the command window, the script file *analog_digital* is executed and the results are shown as follows (Figures 6.53 through 6.57):

```
>> analog _ digital
*****************************************
**************RESULTS*****************
*****************************************
Digital filter /numerator coefficients of the transfer function :
   0.0380   0.1141   0.1141   0.0380
Digital filter / denominator coefficients of the transfer function :
   1.0000 -1.3445   0.8215 -0.1727
*****************************************
```

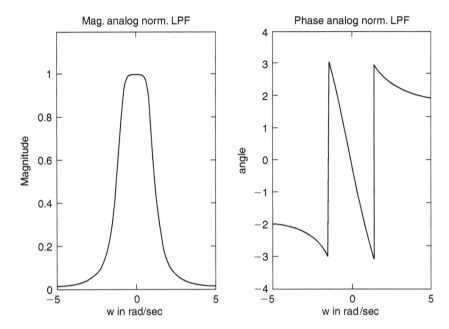

FIGURE 6.53
Magnitude and phase plots of the analog normalized LPF of Example 6.7.

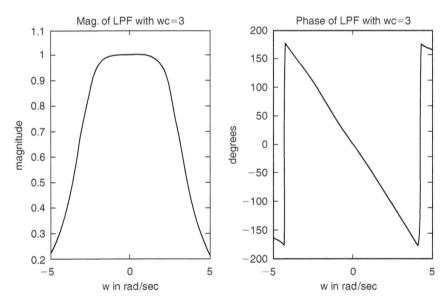

FIGURE 6.54
Magnitude and phase plots of the analog denormalized LPF of part (b) of Example 6.7.

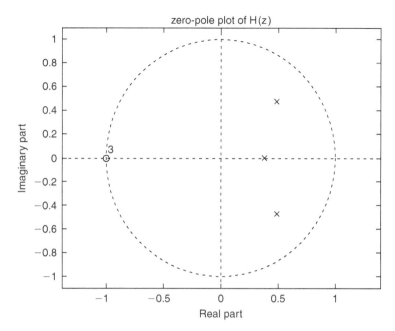

FIGURE 6.55
Zero/pole plot of the discrete IIR transfer function $H(z)$ for $w_c = 3$ rad/s of part (b) of Example 6.7.

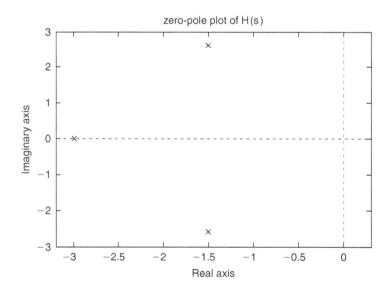

FIGURE 6.56
Zero/pole plot of the analog transfer function $H(s)$ for $w_c = 3$ rad/s of part (b) of Example 6.7.

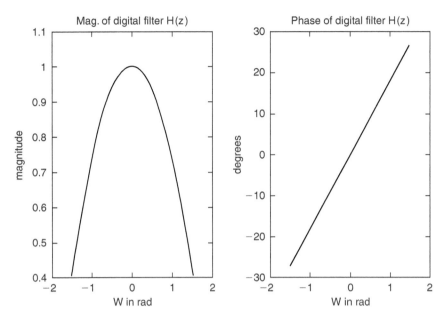

FIGURE 6.57
Magnitude and phase plots of *H(z)* of Example 6.7.

Example 6.8

Create the script file *direc_butt_disig* that solves Example 6.7 by directly obtaining the magnitude and phase plots of the discrete transfer function *H(z)* (LPF third order, with $T = 0.3$ and a cutoff frequency of $w_c = 3$ rad/s).

```
% Script file : direc _ butt _ disig
n =3;
T=.03;wc=3;w =-5:.1:5;
wz = w*T;
[Pz,Qz] = butter(n,wc*T/pi);
[magz,phasez]=freqz(Pz,Qz,wz/pi);
subplot(1,2,1)
plot(wz,magz);
title('Mag. of digital LPF');axis([-5 5 0.4 1.1]);
xlabel('w in rad')
ylabel('magnitude')
subplot(1,2,2)
plot(wz,phasez*180/pi);
title('Phase of digital LPF vs w')
xlabel('w in rad')
ylabel('phase in degrees')
```

Back in the command window, the script file *direc_butt_disig* is executed and the results are shown as follows (Figure 6.58):

```
>> direc _ butt _ disig
```

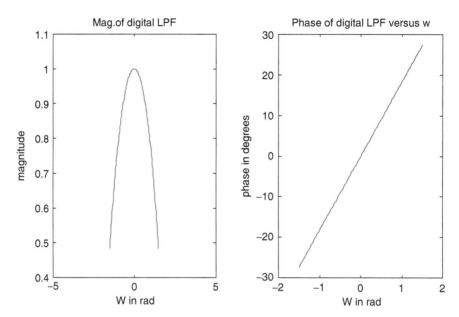

FIGURE 6.58
Magnitude and phase plots of *H(z)* of Example 6.8.

Example 6.9

Create the script file *cheby_HPF* that returns the coefficients of the digital transfer function and the magnitude and phase plots of a third-order Chebyshev type-1 and -2 digital HPFs with the following specs:

1. $w_c = 0.5$ rad/s
2. $T = 0.3$
3. $R_p = 2$ dB
4. $R_s = 3$ dB

MATLAB Solution
```
% Script file: cheby _ HPF
N =3;                            % order
R_p = 2; R_s = 3;                % ripples
T = .3;                          % sampling period
Wc = .5;                         % cut off frequency
Wz = -2*pi:.01*pi:3*pi;
wzc = Wc*T/pi;                   % normalized dig. frequency
[Pz1, Qz1] = cheby1(N, R_p, wzc, 'high');
disp('*********************************************')
disp('*************R E S U L T S ****************')
disp('*********************************************')
disp('For the transfer function H(z) of the Cheby/type1 HPF')
disp('the numerators coefficients are:')
disp(Pz1)
disp('The denominator coefficients are:')
disp(Qz1)
[Pz2, Qz2] = cheby2(N, R_s, wzc, 'high');
disp(' For the transfer function H(z) of the Cheby/type2 HPF')
disp('The numerators coefficients are:')
```

```
disp(Pz2)
disp('The denominator coefficients are:'); disp(Qz2)
disp('*********************************************')
[mag1, phase1] = freqz (Pz1,Qz1,Wz);
[mag2, phase2] = freqz (Pz2,Qz2,Wz);
subplot (2,2,1)
plot (Wz, mag1)
title ('Mag of Cheby. type/1, HPF');
axis([0 5 0 1.1]);
xlabel (' W in rad')
ylabel ('Magnitude')
subplot (2,2,2)
plot (Wz, phase1* 180/pi)
title ('Phase of Cheby. type/1,HPF ')
ylabel ('Angle in degrees')
xlabel (' W in rad')
subplot (2,2,3)
plot (Wz, mag2)
title ('Mag.of Cheby. type/2, HPF'); axis([0 1 0 1.5]);
xlabel ('W in rad'); ylabel ('Magnitude')
subplot (2,2,4)
plot (Wz, phase2 * 180/ pi);
xlabel ('W in rad')
ylabel ('Angle in degrees')
title('Phase. of Cheby. type/2, HPF')
disp('****** FILTERS TRANSFER FUNCTIONS*******')
 disp('*********************************************')
disp('Chebyshev type 1')
tf(Pz2,Qz2,T)
disp('^^^^^^^^^^^^^^^^^^^^^^,')
disp('Chebyshev type 2')
tf(Pz1,Qz1,T)
disp('*****************************************************')
```

The script file *cheby_HPF* is executed and the results are shown as follows (Figure 6.59):

```
>> cheby _ HPF

**********************************************
************* R E S U L T S ***************
**********************************************
For the transfer function H(z) of the Cheby/type1 HPF
the numerators coefficients are:
  0.8006 -2.4019  2.4019 -0.8006

The denominator coefficients are:
  1.0000 -2.5767  2.2067 -0.6217

For the transfer function H(z) of the Cheby/type2 HPF
The numerators coefficients are:
  0.9564 -2.8530  2.8530 -0.9564

The denominator coefficients are:
  1.0000 -2.8943  2.8098 -0.9147
```

```
**********************************************
*****FILTERS TRANSFER FUNCTIONS **********
**********************************************
Chebyshev type 1

Transfer function:
0.8006 z^3 - 2.402 z^2 + 2.402 z - 0.8006
----------------------------------------
  z^3 - 2.577 z^2 + 2.207 z - 0.6217

Sampling time: 0.3
^^^^^^^^^^^^^^^^^^^^
Chebyshev type 2

Transfer function:
0.9564 z^3 - 2.853 z^2 + 2.853 z - 0.9564
----------------------------------------
  z^3 - 2.894 z^2 + 2.81 z - 0.9147

Sampling time: 0.3
*****************************************************
```

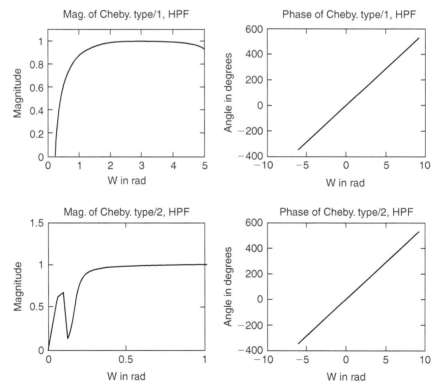

FIGURE 6.59
Magnitude and phase plots of Chebyshev type-1 and -2 digital HPFs of Example 6.9.

Example 6.10

Test the Butterworth LP digital filter of Example 6.7 by applying as an input the signal $x(t) = 2 + \cos(0.2*t) + \cos(4*t)$ sampled every $T = 0.3$ s.
Create the script file *inp_out* that returns the following plots:

a. *x(t)* versus *t*

b. *y(t)* versus *t*

c. *x(nT)* versus *nT*

d. *y(nT)* versus *nT*

where *y* denotes the filter's output.

MATLAB Solution
```
% Script file: inp _ out
Pz = [0.0380   0.1141   0.1141   0.0380];
Qz = [1.0000 -1.3445   0.8215 -0.172];
n = 0:1:100; %Discrete time sequence
T = 0.3;
input = 2 + cos(0.2*T.*n)+cos(4*T.*n);
output= filter(Pz,Qz,input);
t = 0:.05:8;                                    % time range
inputa = 2 + cos(0.2.*t) + cos(4.*t);
subplot(2,2,1)
plot(t,inputa)
xlabel('t in sec')
ylabel('Amplitude')
title('[x(t) = 2 + cos(0.2.*t) + cos (4.*t)] vs. t')
axis([0 5 0 5])
subplot(2,2,2)
plot(n*T,output)
xlabel('t in sec')
ylabel('Amplitude')
title('y(t) vs. t');axis([0 8 0 4])
subplot(2,2,3)
stem(n*T,input)
xlabel('time index nT')
ylabel('Amplitude')
title('x(nT) vs. nT');
axis([0 5 0 5])
subplot(2,2,4)
stem(n*T,output)
xlabel('time index nT')
ylabel('Amplitude')
title('y(nT) vs. nT');
axis([0 8 0 4])
```

The script file *inp_out* is executed and the results are shown in Figure 6.60.

```
>> inp _ out
```

Note that the filter's output presents a high DC component plus the low frequency of the input with some traces of the high frequency.

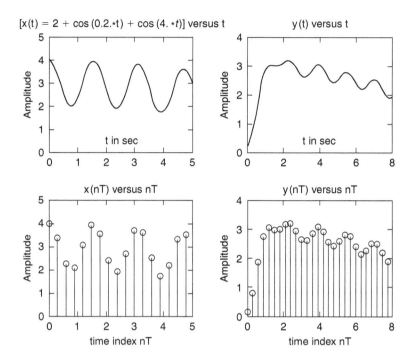

FIGURE 6.60
Analog and discrete filter's input and output of Example 6.10.

Example 6.11

Create the script file *Cheby_1_2* that tests the design of the Chebyshev/types 1 and 2 HO digital filter of Example 6.9 and returns their outputs by applying as input the signal $x(t) = 2 + cos(0.2 * t) + cos(0.5 * t)$ sampled at the rate $T = 0.3$.

a. Obtain plots of each component of $x(t)$.
b. Obtain plots of the discrete input signal and the corresponding filter's outputs.
c. Analyze and discuss each filter's output and state which of the two has a better performance.

MATLAB Solution
```
% Script file: Cheby _ 1 _ 2,
Pz1 = [0.8006 -2.4019  2.4019 -0.8006]; % num. and den. of digital filter
type 1
Qz1 = [1.0000 -2.5767  2.2067 -0.6217];
Pz2 = [0.9564 -2.8530  2.8530 -0.9564]; % num. and den. of digital filter
type 2
Qz2 = [1.0000 -2.8943  2.8098 -0.9147];
n = 0:1:300;                            % discrete time sequence
T = 0.3;
Dc = 2.*ones(1,301);
f1 = 2*cos(.2*T.*n);
f2 = cos(.5*T.*n);
figure(1)
subplot(3,1,1)
plot(n*T,dc);
```

```
ylabel('Input:DC component');
subplot(3,1,2)
plot(n*T,f1);
ylabel('Input:2cos(0.2 t)');
subplot(3,1,3)
plot(n*T,f2);
ylabel('Input:cos(0.5t)');
xlabel('time')
figure(2)
input = 2.*ones(1,301) + 2*cos(.2*T.*n) + cos(.5*T.*n);
output1 = filter(Pz1,Qz1,input);
output2 = filter(Pz2,Qz2,input);
subplot(3,1,1)
plot(n*T,input), hold on; stem(n*T,input)
ylabel('Input sequence')
subplot(3,1,2);
plot(n*T,output1); hold on; stem(n*T,output1);
ylabel('Outputs Chev 1')
subplot(3,1,3);
plot(n*T,output2);hold on;stem(n*T,output2)
ylabel('Outputs Chev 2')
xlabel('nT');
```

The script file *Cheby_1_2* is executed and the results are shown in Figures 6.61 and 6.62.

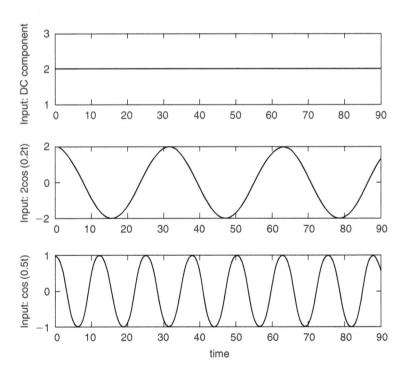

FIGURE 6.61
Component plots of the input of *x(t)* of Example 6.11.

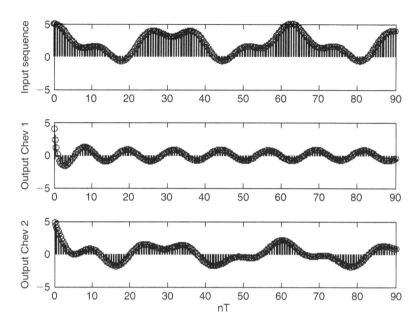

FIGURE 6.62
Plots of the discrete input and the filter's outputs of Example 6.11.

Note that the output of the *Cheby type-1* filter is truly an HPF (only the frequency $w = 0.5$ is present), whereas the output of the *Cheby type-2* filter is similar to the composite input signal in which the two input frequencies are present.

Example 6.12

Example 6.12 involves an FIR filter, type 1 with symmetry, and length $N = odd$. Obtain by hand expressions of the amplitude and phase of $H(e^{jW})$, verifying that the phase is linear if the impulse response sequence is given by

$$h(n) = [.5\ 1\ .5] \quad \text{for } n = 0, 1, \text{ and } 2$$

ANALYTICAL Solution

$$H(e^{jW}) = \sum_{n=0}^{n=2} h(n)z^{-n} = 0.5 + 1e^{-jW} + 0.5e^{-j2W}$$

$$H(e^{jW}) = [0.5e^{+jW} + 0.5e^{-jW}]e^{-jW}$$

$$H(e^{jW}) = [1 + \cos W]e^{-jW}$$

$$|H(e^{jW})| = 1 + \cos W \quad \text{for } -\pi < W < +\pi$$

$$\angle H(e^{jW}) = -\omega \quad \text{for } -\pi < W < +\pi$$

Example 6.13

Given the following impulse response sequence: $hn = [-5\,2\,3\,4\,2\,-1\,3\,-1\,2\,4\,3\,2\,-5]$ that corresponds to a digital filter.

Determine the following:

a. The type of filter
b. By hand calculate its frequency response
c. Create the script file *Fir_typ1_1* that returns the following plots:
    i. $h(n)$ versus n
    ii. The first seven $a(n)'s$ terms of $H(W)$, in a tablelike format, and verify the results obtained in part (b)
    iii. $H(W)$ versus W
    iv. Zero–pole

ANALYTICAL Solution

Clearly, $h(n) = h(N - 1 - n)$ is a symmetric sequence with N = odd (13), then the filter is an FIR type 1. Then from R.6.120

$$H_1(W) = \sum_{n=0}^{6} a_1(n)\cos(nW)$$

where

$$a_1(0) = h(6) = 3$$

$$a_1(1) = 2h(6 - 1) + 2h(5) = 2(-1) = -2$$

$$a_1(2) = 2h(6 - 2) = 2h(4) = 2(2) = 4$$

$$a_1(3) = 2h(6 - 3) = 2h(3) = 2(4) = 8$$

$$a_1(4) = 2h(6 - 4) = 2h(2) = 2(3) = 6$$

$$a_1(5) = 2h(6 - 5) = 2h(1) = 2(2) = 4$$

$$a_1(6) = 2h(6 - 6) = 2h(0) = 2(-5)$$

$$a_1(6) = -10$$

Then

$$H_1(w) = a_1(0) + a_1(1)\,\cos(W) + a_1(2)\,\cos(2W) + a_1(3)\cos(3W)$$
$$+ a_1(4)\cos(4W) + a_1(5)\cos(5W) + a_1(6)\cos(6W)$$
$$H_1(\omega) = 3 - 2\cos(W) + 4\cos(2W) + 8\cos(3W) + 6\cos(4W) + 4\cos(5W) - 10\cos(6W)$$

MATLAB Solution

```
% Script file : Fir _ typ1 _ 1
h = [-5 2 3 4 2 -1 3 -1 2 4 3 2 -5];
N = length(h);
n = 0:N-1;
mid = (N-1)/2;
disp('****coeff. of H1(w)****')
disp('index n coeff. a(n)')
a = [h(mid+1) 2*h(mid:-1:1)];            % generates the a's coeff.
nn = (0:6)';
```

```
aa = a(1:7)';
[nn aa]
disp('********************')
m = 0:1:mid;
w = -pi:.01*pi:pi;
subplot(2,2,1)
stem(n,h)
axis([0 15 -6 5])
xlabel('time index n')
ylabel('Amplitude ')
title('[Impulse sequence h(n)] vs. n')
subplot(2,2,2)
stem(0:mid,a)
xlabel(' index n')
ylabel('Amplitude')
title('[coeff. a1(n)] vs. n')
H1 = a(1)+a(2)*cos(w)+a(3)*cos(2*w)+a(4)*cos(3*w)+a(5)*cos(4*w);
H2 = a(6)*cos(5*w)+a(7)*cos(6*w);
H = H1+H2;
subplot(2,2,3)
plot(w,H)
xlabel('W');ylabel('Amplitude')
title('H1(W) vs. W')
subplot(2,2,4)
zplane(h,1)
title('zero-pole plot')
```

The script file *Fir_typ1_1* is executed and the results are shown as follows:

```
>> Fir _ typ1 _ 1

        ****coeff. of H1(w)****
          index n coeff. a(n)
        ans =
              0     3
              1    -2
              2     4
              3     8
              4     6
              5     4
              6   -10
        **********************
```

Example 6.14

Create the script file *filter_wind* that returns the normalized LP-FIR filters using the command *fir1* in conjunction with the following windows:

   a. *Kaiser (β = 3.5)*
   b. *Hanning*
   c. *Blackman*
   d. *Hamming*
   e. *Triangular*
   f. *Bartlett*

The filter's specs are $N = 64$ and $R_p = 1.4$ dB.

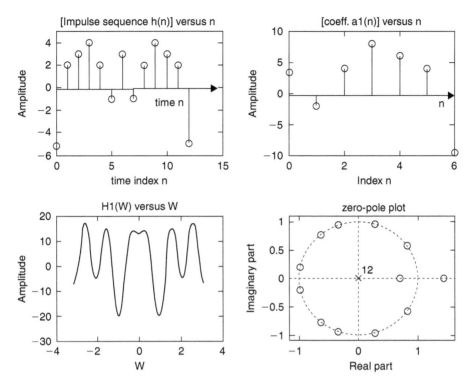

FIGURE 6.63
Filter's plots of Example 6.13.

MATLAB Solution
```
% Script file: filter _ wind
format long
hnk = fir1(64,.4,kaiser(65,3.5));
[h1,w1] = freqz(hnk,1,256);
magk = 20*log10(abs(h1));
subplot(3,2,1)
plot(w1,magk)
title('Filter using kaiser win.')
ylabel('gain in db')
subplot(3,2,2)
hnhan = fir1(64,.4,hanning(65));
[h2,w2] = freqz(hnhan,1,256);
maghan = 20*log10(abs(h2));
plot(w2,maghan)
title('Filter using hanning win.')
ylabel('gain in db')
subplot(3,2,3)
hnbla = fir1(64,.4,blackman(65));
[h3,w3] = freqz(hnbla,1,256);
magblac=20*log10(abs(h3));
plot(w3,magblac)
title('Filter using blackman win.')
ylabel('gain in db')
subplot(3,2,4)
% hamming window is the default
```

```
hnham = fir1(64,.4);
[h4,w4] = freqz(hnham,1,256);
magham = 20*log10(abs(h4));
plot(w4,magham)
title('Filter using hamming win.')
ylabel('gain in db')
subplot(3,2,5)
hntri = fir1(64,.4,triang(65));
[h5,w5] = freqz(hntri,1,256);
magtri = 20*log10(abs(h5));
plot(w5,magtri)
title('Filter using triang. win.')
xlabel('frequency in rad.')
ylabel('gain in db')
axis([0 3 -60 10])
subplot(3,2,6)
hnbar = fir1(64,.4,bartlett(65));
[h6,w6] = freqz(hnbar,1,256);
magbar = 20*log10(abs(h6));
plot(w6,magbar)
title('Filter using bartlett win.')
xlabel('frequency in rad.')
ylabel('gain in db')
axis([0 3 -60 10])
```

The script file *filter_wind* is executed and the results are shown in Figure 6.64.

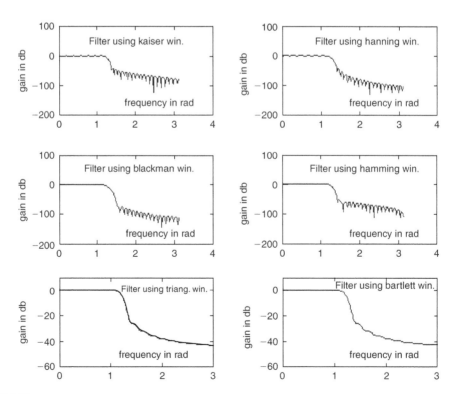

FIGURE 6.64
Normalized LP FIR filters' plots using *fir1* with the window arguments of Example 6.14.

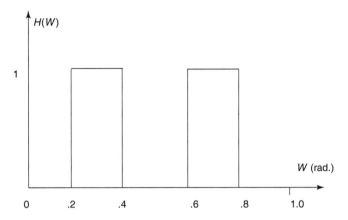

FIGURE 6.65
Filter's specs of Example 6.15.

Example 6.15

Create the script file *multiban* that returns the realization of a digital multiband normalized filter with the specs indicated by the sketch shown in Figure 6.65, using the following functions:

a. *fir2* of order 50, with a Bartlett *window (FIR)*

b. *remez (FIR)* of order 50

c. *yulewalk* digital *IIR* of order 15

MATLAB Solution
```
% Script File: multiban
w = [0 .1 .2 .3 .4 .5 .6 .7 .8 1];
mag = [0 0 1 1 0 0 1 1 0 0];
hnfir = fir2(50,w,mag,bartlett(51));
[h1, w1] = freqz(hnfir,1,256);
subplot (3,1,1)
plot (w1/pi, abs(h1))
ylabel ('gain')
title ('Multiban filter using fir2 (with bartlett window)')
xlabel('frequency W (in rad)')
subplot (3,1,2)
hnrem = remez(50,w,mag);
[h2, w2] = freqz (hnrem,1,256);
plot (w2/pi, abs(h2))
ylabel ('gain')
title ('Multiban filter using remez')
xlabel('frequency W (in rad)')
subplot(3,1,3)
[num,den] = yulewalk(15,w,mag);
[hiir,w] = freqz(num,den,256);
plot(w/(pi),abs(hiir))
ylabel ('gain')
title ('Multiban filter using yulewalk')
xlabel('frequency W (in rad)')
```

The script file *multiban* is executed and the results are as follows (Figure 6.66):

```
>> multiban
```

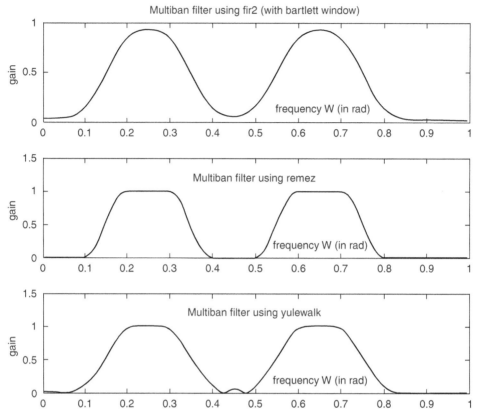

FIGURE 6.66
Multiband filter's plots of Example 6.15.

Example 6.16

Rerun Example 6.15 given by the script file *multiban* for the filter's specs given by

  a. $0.2 \leq W \leq 0.4$, $H(W) = 0.3$
  b. $0.6 \leq W \leq 0.8$, $H(W) = 0.6$
  c. Zero everywhere else, that is,
    $0 \leq W < 0.2$, $H(W) = 0$
    $0.4 < W < 0.6$, $H(W) = 0$
    $0.8 < W \leq 1$, $H(W) = 0$

MATLAB Solution
The new filter's specs translate in modifying the frequency–magnitude vectors of the script file *multiban* as follows:

```
w = [0 .1 .2 .3 .4 .5 .6 .7 .8 1]
mag = [0 0 .3 .3 0 0 .6 .6 0 0]
```

The modified script file *multiban* is executed and the resulting plots are shown in Figure 6.67.

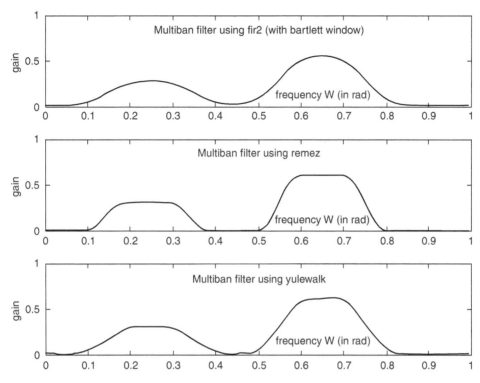

FIGURE 6.67
Multiband filter's plots of Example 6.16.

Example 6.17

Create the script file *butt_implem* that returns a third-order analog Butterworth LPF, at the component level, with $w_c = 10^4$ rad/s and a load impedance of 100 Ω.

    a. Start the design with a third-order normalized prototype LPF and obtain the corresponding magnitude and phase plots.
    b. Obtain the corresponding transfer function coefficients and the system poles (check for stability).
    c. Obtain the normalized filter's transfer function.
    d. Use frequency and magnitude scaling to obtain the desired filter transfer function.

ANALYTICAL Solution

The structure chosen for the desired LP normalized, third-order filter prototype (see R.6.49) is the π section shown in Figure 6.68.
    The transfer function is indicated as follows:

$$H_N(s) = \frac{1}{(L_1 L_2 C)s^3 + (L_1 C)s^2 + (L_1 + L_2)s + 1} \qquad (6.3)$$

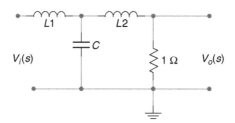

FIGURE 6.68
Filter's structure of Example 6.17.

```
MATLAB Solution
% Script file : butt _ implem
[z,p,k] = buttap(3);
disp(' ******* R E S U L T S ********')
disp('*********************************')
disp('the poles are :');disp(p);
[Pn,Qn] = zp2tf(z,p,k);                    % returns the transf. function
disp('The coefficients of the numerator of Hn(s) are: '); disp(Pn);
disp('The coefficients of the denominator of Hn(s) are: '); disp(Qn);
system = tf(Pn,Qn);
system
disp('*********************************')
w =-3:.1:3;
hn = freqs(Pn,Qn,w);
gain=20*log10(abs(hn));
subplot(1,2,1)
plot(w,gain)
xlabel('w in rad/sec')
ylabel('gain in db')
title('Normalize filter/magnitude')
subplot(1,2,2)
plot(w,angle(hn)*180/pi);
xlabel('w in rad/sec');ylabel('Phase in degrees')
title('Normalize filter/phase')
```

Back in the command window the script file *butt_implem* is executed and the results are shown as follows (Figure 6.69):

```
>> butt _ implem

        ******* R E S U L T S ********
        *********************************
the poles are :
 -0.5000 + 0.8660i
 -0.5000 - 0.8660i
 -1.0000
The coefficients of the numerator of Hn(s) are:
     0    0    0   1.0000
The coefficients of the denominator of Hn(s) are:
   1.0000   2.0000   2.0000   1.0000
  Transfer function:
      1
 --------------------
 s^3 + 2 s^2 + 2 s + 1
 *********************************************
```

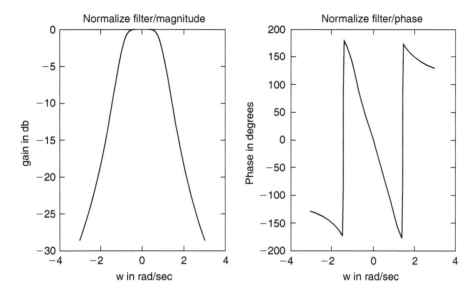

FIGURE 6.69
Normalized filter's plot of Example 6.17.

The transfer function is given by

$$H(s) = \frac{1}{s^3 + 2s^2 + 2s + 1} \tag{6.4}$$

Equating the coefficients of Equation 6.3 with the coefficients of Equation 6.4 the following relations are obtained:

$$L_1 L_2 C = 1$$
$$L_1 C = 2$$
$$L_1 + L_2 = 2$$

Then solving for L_1, L_2, and C the following is obtained:

$$L_1 = 3/2 \text{ H}, L_2 = 1/2 \text{ H}, \text{ and } C = 4/3 \text{ F}$$

Then the normalized LPF at the component level is shown in Figure 6.70.
    Frequency and magnitude scaling results in

$$\text{new-}R_0 = R * R_0 = 1 * 100 = 100 \ \Omega$$

$$L_{01} = \frac{L_1 * R_0}{10^4} = \frac{3 * 100}{2 * 10^4} = \frac{3 * 10^{-2}}{2} = 15 \text{ mH}$$

$$L_{02} = \frac{L_2 * R_0}{10^4} = \frac{100}{2 * 10^4} = 0.5 * 10^{-2} = 5 \text{ mH}$$

$$C_0 = \frac{C}{w_c * R_0} = \frac{4}{3} * \frac{1}{10^4 * 10^2} = \frac{4}{3} * 10^{-6} = 1.33 \ \mu\text{F}$$

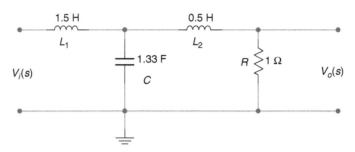

FIGURE 6.70
Elements of normalized LPF of Example 6.17.

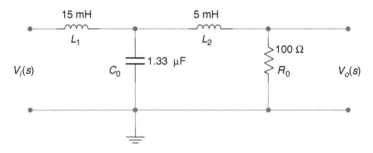

FIGURE 6.71
Scaled elements of the Butterworth LPF of Example 6.17.

The Butterworth LPF satisfying the given specs is shown in Figure 6.71.

```
% frequency plots for the network of Figure 6.71
L01=15e-3;L02=5e-3; C0=1.33e-6;
num=1;
den=[L01*L02 L01*C0 L01+L02 1];w = -155000: 250 : 155000;
hnn = freqs(num,den,w);
gain = 20 * log10 (abs (hnn));
f = w./(2 * pi);
subplot (2,1,1)
plot (f, gain)
ylabel ('Gain in db')
title ('Magnitude of [H(f)] vs. f')
subplot (2,1,2)
plot ( f,angle(hnn).*180/ pi)
xlabel (' f in Hertz ')
ylabel ( ' Angle in degrees')
title (' Phase of [H(f)] vs. f ');
```

Example 6.18

Create the script file *IIR-yul* that returns the magnitude and phase plots of IIR normalized LP digital filter of orders 4, 6, 8, and 10 using the function *yulewalk* with the magnitude–frequency specs given by the following MATLAB vectors:

$$f = [0:0.1:1.0]$$

$$mag = [1\ 1\ 1\ .707\ 0\ 0\ 0\ 0\ 0\ 0\ 0]$$

See Figure 6.72.

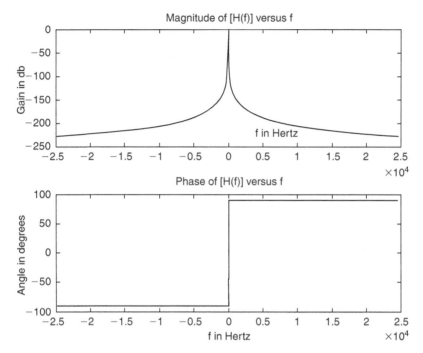

FIGURE 6.72
Gain and phase plots of the transfer function of the filter of Example 6.17.

MATLAB Solution
```
% Script file: IIR-yul
% IIR/YULE-WALKER
% for N = 4, 6, 8, and 10
f = [0:.1:1.0];
omega = 0:1/512:511/512;
mag = [1 1 1 .707 zeros(1,7)];
[num4,den4] = yulewalk(4,f,mag);
[num6,den6] = yulewalk(6,f,mag);
[num8,den8] = yulewalk(8,f,mag);
[num10,den10] = yulewalk(10,f,mag);
h4 = freqz(num4,den4);
h6 = freqz(num6,den6);
h8 = freqz(num8,den8);
h10 = freqz(num10,den10);

figure(1)
subplot(2,2,1)
plot(f,mag,'o',omega,abs(h4))
ylabel('Magnitude(H)')
title('Order=4')
subplot(2,2,2)
plot(f,mag,'o',omega,abs(h6))
ylabel('Magnitude(H)')
title('Order=6')
subplot(2,2,3)
plot(f,mag,'o',omega,abs(h8))
```

```
title('Order=8')
ylabel('Magnitude(H)');
xlabel('W in rad')
subplot(2,2,4)
plot(f,mag,'o',omega,abs(h10))
title('Order=10')
ylabel('Magnitude(H)');
xlabel('W in rad')

figure(2)
subplot(2,2,1)
plot(omega,angle(h4)),
title('Order=4');ylabel('Angle')
subplot(2,2,2)
plot(omega,angle(h6));
title('Order=6');ylabel('Angle')
subplot(2,2,3)
plot(omega,angle(h8))
title('Order=8');
xlabel('W in rad');ylabel('Angle')
subplot(2,2,4)
plot(omega,angle(h10));
title('Order=10')
xlabel('W in rad');ylabel('Angle')
```

Back in the command window, the script file *IIR-yul* is executed and the results are shown as follows (Figures 6.73 and 6.74):

```
>> IIR-yul
```

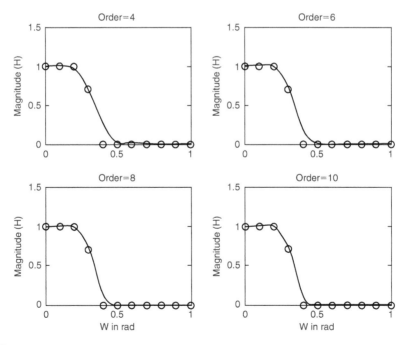

FIGURE 6.73
Magnitude plots of IIR normalized LP digital filter of order: 4, 6, 8, and 10 using the function *yulewalk* of Example 6.18.

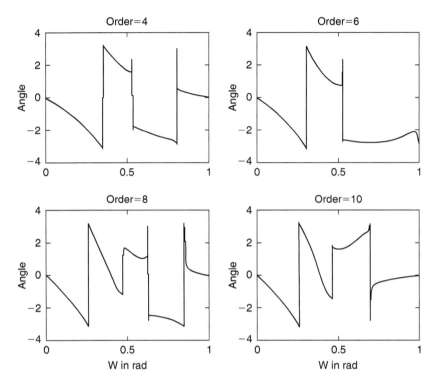

FIGURE 6.74
Phase plots of IIR normalized LP digital filter of order: 4, 6, 8, and 10 using the function *yulewalk* of Example 6.18.

Example 6.19

Create the script file *FIR_IIR* that returns the magnitude plots of a 17th-order normalized FIR filter by truncating a third-order IIR Butterworth digital filter with a cutoff frequency at 0.25 rad.
 Compare the plots of the 17th-order normalized FIR filter with the third-order IIR filter.

MATLAB Solution
```
% Script file: FIR _ IIR
% FIR from a third order normalized
% butterworth IIR FILTER with Wc=.25
[num,den] = butter(3,.25);
numad = [num zeros(1,30)];
divnumbyden = deconv(numad,den);
imp = dimpulse(num,den,30);
num9 = imp(1:19);                        % first 18 coefficients
                                         % for the FIR implementation
h18 = freqz(numad,1);h = freqz(num,den);
W = 0:1/512:511/512;
plot(W,20*log10(abs(h)),W,20*log10(abs(h18)))
ylabel('Magnitude in db')
xlabel('W in rad');
title('FIR IMPLEMENTATION FROM IIR');
axis([0 1 -100 20]);grid on;
gtext('IIR FILTER')
gtext('FIR FILTER')
```

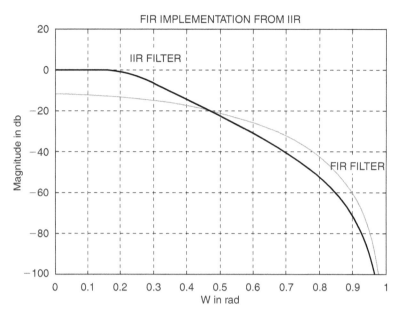

FIGURE 6.75
IIR and FIR filter's plots of Example 6.19.

Back in the command window, the script file *FIR_IIR* is executed and the result is shown in Figure 6.75.

Example 6.20

Design the following digital filters:

a. An 18th-order normalized LPF with a cutoff frequency at 0.3 rad using the function *fir1*, truncated by a *Hamming window*.

b. An 8th-order normalized BPF with a pass band between 0.3 and 0.6 rad using the function *fir1*.

c. Repeat part (b) using the *remez* exchange algorithm that employs the Park–McClellan approach.

For each one of the preceding filters, return the coefficients of the polynomials that make up the transfer function, their transfer functions, and their respective magnitude and phase plots.

The solution is given by the script file *designs* as follows:

MATLAB Solution
```
% Script file : designs
% FIR/LPF design using fir1/part (a)
coeflp = fir1(18,.3);     % 18 th..order FIR filter with cutoff of .3
Hlp = freqz(coeflp,1);    % coef. of FIR filter

figure(1)
w1= 0:1/512:511/512 ;
subplot(1,2,1)
plot(w1,abs(Hlp));grid on;
xlabel(' frequency W in rad.')
```

```
ylabel('Magnitude')
title('Mag. of LPF using fir1/order=18')
subplot(1,2,2)
plot(w1,angle(Hlp)*180/pi);grid on;
xlabel('frequency W in rad.')
ylabel('Angle in degrees')
title('Phase of LPF using fir1/order=18')
% FIR/BPF using fir1/part(b)
f = [.3 .6];                        % passband region
coefbp = fir1(8,f,'noscale') ;      % noscale, BP 8th. order filter
Hbp=freqz(coefbp,1);

figure(2)
subplot(1,2,1)
plot(w1,abs(Hbp));grid on;
xlabel(' frequency W in rad.')
ylabel('Magnitude')
title('Mag. of BPF using fir1/order=8')
subplot(1,2,2)
plot(w1,angle(Hbp)*180/pi);grid on;
xlabel(' frequency W in rad.')
ylabel('Angle in degrees')
title('Phase of BPF using fir1/order=8')
%***********************************
%FIR/BPF using remez
freq = [0 .25 .3 .6 .65 1.0];       % specifies normalized freqs.
mag = [0 0 1 1 0 0];
error = [2.0 1.5 2.0];              % error weights per band
num = remez(8,freq,mag,error) ;     % 8th. order BPF
Hrem=freqz(num,1);

figure(3)
subplot(1,2,1)
plot(w1,abs(Hrem));grid on;
xlabel('frequency W in rad.')
ylabel('Magnitude')
title('Mag. of BPF using remez/order=8')
subplot(1,2,2)
plot(w1,angle(Hrem)*180/pi);grid on;
xlabel('frequency W in rad.')
ylabel('Angle in degrees')
title('Phase of BPF using remez/order=8');
den=1;
disp('***********************************************')
disp('     * * * * R E S U L T S * * * *         ')
disp('***********************************************')
disp('***********************************************')
disp(' The coefficients of the FIR/LPF using fir1 are:')
disp(coeflp)
tf(coeflp,1,-1)
disp('***********************************************')
disp(' The coefficients of the FIR/BPF using fir1 are:')
disp(coefbp)
```

```
tf(coefbp,1,-1)
disp('************************************************')
disp(' The coefficients of the FIR/BPF using remez are:')
disp(num)
tf(num,den,-1)
disp('************************************************')
```

Back in the command window, the script file *designs* is executed, and the results are indicated as follows (Figures 6.76 through 6.78):

```
>> designs

************************************************
            * * * * R E S U L T S * * * *
************************************************
************************************************
 The coefficients of the FIR/LPF using firl are:
 Columns 1 through 8
  0.0023   0.0041   0.0026 -0.0096 -0.0292 -0.0289   0.0252   0.1346
 Columns 9 through 16
  0.2495   0.2990   0.2495   0.1346   0.0252 -0.0289 -0.0292 -0.0096
 Columns 17 through 19
  0.0026   0.0041   0.0023

 Transfer function:
 0.002281 z^18 + 0.004063 z^17 + 0.002627 z^16 - 0.009633 z^15 - 0.02919 z^14

    - 0.02889 z^13 + 0.02516 z^12 + 0.1346 z^11 + 0.2495 z^10

    + 0.299 z^9 + 0.2495 z^8 + 0.1346 z^7 + 0.02516 z^6 - 0.02889 z^5

    - 0.02919 z^4 - 0.009633 z^3 + 0.002627 z^2 + 0.004063 z + 0.002281

Sampling time: unspecified
************************************************
 The coefficients of the FIR/BPF using firl are:
 Columns 1 through 8
  0.0098 -0.0204 -0.1323   0.0391   0.3000   0.0391 -0.1323 -0.0204
 Column 9
  0.0098

Transfer function:
0.009797 z^8 - 0.02043 z^7 - 0.1323 z^6 + 0.03912 z^5 + 0.3 z^4 + 0.03912 z^3

             - 0.1323 z^2 - 0.02043 z + 0.009797

Sampling time: unspecified
************************************************
 The coefficients of the FIR/BPF using remez are:
 Columns 1 through 8
  0.1603 -0.1661 -0.3275   0.0244   0.3346   0.0244 -0.3275 -0.1661
 Column 9
  0.1603

Transfer function:
0.1603 z^8 - 0.1661 z^7 - 0.3275 z^6 + 0.02438 z^5 + 0.3346 z^4 + 0.02438 z^3

                   - 0.3275 z^2 - 0.1661 z + 0.1603

Sampling time: unspecified
************************************************
```

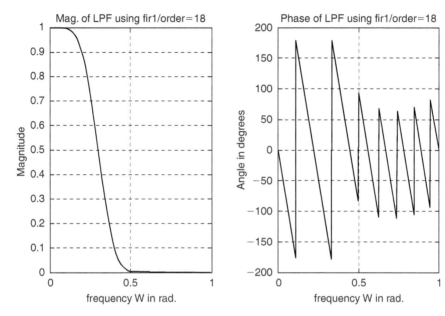

FIGURE 6.76
Filter's plot of part (a) of Example 6.20.

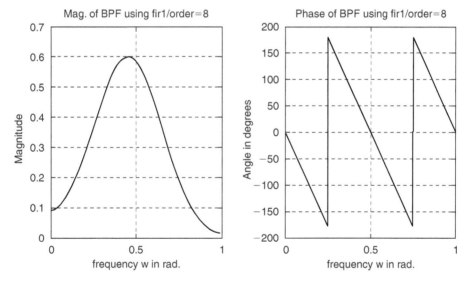

FIGURE 6.77
Filter's plot of part (b) of Example 6.20.

Example 6.21

Design a 25th-order normalized FIR/LPF with a BW of 0.25 rad by using

a. The *remez* algorithm
b. *fir1* limited by a *Hamming* window

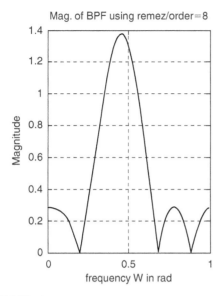
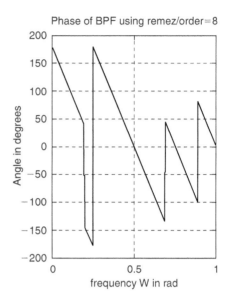

FIGURE 6.78
Filter's plot of part (c) of Example 6.20.

For each of the filters obtain the magnitude and phase plots.
   The solution is given by the script file *review_FIR_LPF* as follows:

MATLAB Solution
```
% Script file : review _ FIR _ LPF
% Design of FIR using remez
f = linspace(0,1,10);
mag = [ones(1,3) .707 zeros(1,6)];
Pz = remez(25,f,mag);
Qz =1;
Hremez = freqz(Pz,Qz);
omega = 0:1/512:511/512;

figure(1)
subplot(2,1,1)
plot(omega,abs(Hremez))
xlabel('frequency in rad')
ylabel('magnitude')
title('Mag./FIR using remez')
grid on;
subplot(2,1,2)
plot(omega,unwrap(angle(Hremez)))
xlabel('frequency in rad')
ylabel('phase')
title('Phase/FIR using remez')
grid on;
% Design of FIR using fir1/Hamming window
Pz1 = fir1(25,0.3);
win = hamming(26);
Pz1= Pz1.*win';
Hfir1= freqz(Pz1,1);
```

```
figure(2)
subplot(2,1,1)
plot(omega,abs(Hfir1))
xlabel('frequency in rad')
ylabel('magnitude')
title('Mag./FIR using fir1')
grid on;
subplot(2,1,2)
plot(omega,unwrap(angle(Hfir1)))
xlabel('frequency in rad')
ylabel('phase');
title('Phase/FIR using fir1');grid on;
```

Back in the command window, the script file *review_FIR_LPF* is executed and the results are as follows (Figures 6.79 and 6.80):

```
>> review _ FIR _ LPF
```

Example 6.22

Create the script file *revisit_analog_filters* that returns the following filter implementations:

- a. A four-pole analog Butterworth and Chebyshev/type-1 normalized filters with a pass-band ripple of $R_p = 2$ dB.
- b. Transform the LPF of part (a) to an LPF with a BW of 2 rad/s using the function *lp2lp*.
- c. Convert the Butterworth LPF of part (a) to an HPF with a cutoff frequency of 2 rad/s.
- d. Convert the Butterworth LPF of part (a) to a BPF with a center frequency at 2 rad/s and a BW of 1 rad/s.
- e. Convert the Butterworth and Chebyshev LPFs of part (a) to BSFs with a center frequency at 2 rad/s and a BW of 1 rad/s (from $w = 1.5$ rad/s to $w = 2.5$ rad/s).

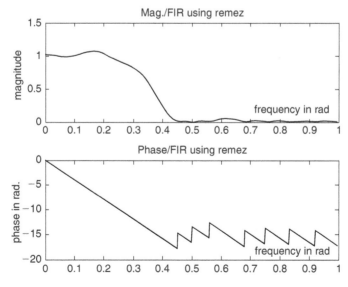

FIGURE 6.79
Twenty-fifth-order normalized FIR/LPF using the *remez* algorithm of Example 6.21.

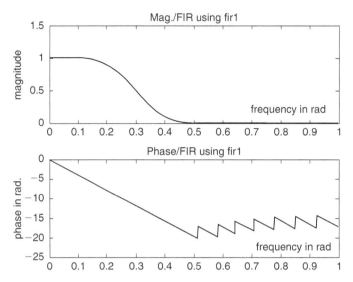

FIGURE 6.80
Twenty-fifth-order normalized FIR/LPF using the MATLAB *fir1* function of Example 6.21.

MATLAB Solution
```
% Script file: revisit _ analog _ filters
% part (a)
% normalize low pass filters: Butt. and Cheby.
[z1,p1,k1] = buttap (4);
[num1,den1] = zp2tf(z1,p1,k1);
w =0:.01:6;

figure(1)
subplot(2,2,1);
H1= freqs(num1,den1,w);
plot(w,abs(H1));
ylabel('magnitude')
title('Butt. norm. LPF')
subplot(2,2,2);
plot(w,angle(H1)*180/pi)
ylabel('phase (deg)')
title('Butt. LPF')
[z2,p2,k2] = cheb1ap(4,2);                    % type 1,with 2 db passband
[num2,den2] = zp2tf(z2,p2,k2);
H2 = freqs(num2,den2,w);
subplot(2,2,3);
plot(w,abs(H2))
ylabel('magnitude')
title('Cheb. norm. LPF')
xlabel('frequency (rad/sec)')
subplot(2,2,4);
plot(w,angle(H2)*180/pi)
ylabel('phase (deg)')
title('Cheb. LPF')
xlabel('frequency (rad/sec)')
```

```
figure(2)
[num3,den3] = lp2lp(num1,den1,2);                    %. transf function /Butt
H3 = freqs(num3,den3,w);
subplot(2,2,1)
plot(w,abs(H3));
ylabel('magnitude')
title('Butt. LPF ,w_p=2')
subplot(2,2,2);
plot(w,angle(H3)*180/pi)
ylabel('phase (deg)')
title('Butt. LPF')
    [num4,den4] = lp2lp(num2,den2,2);                % transf. function/Cheby
H4 = freqs(num4,den4,w);
subplot(2,2,3)
plot(w,abs(H4));
ylabel('magnitude')
xlabel('frequency(rad/sec)')
title('Cheb. LPF ,w_p=2')
subplot(2,2,4);
plot(w,angle(H4)*180/pi)
ylabel('phase (deg)')
xlabel('frequency (rad/sec)')
title('Cheb. LPF')

figure(3)
[num5,den5] = lp2hp(num1,den1,2);                    %. transf. function /Butt
H5 = freqs(num5,den5,w);
subplot(2,2,1)
plot(w,abs(H5));
ylabel('magnitude')
title('Butt. HPF ,w_p=2')
subplot(2,2,2);
plot(w,angle(H5)*180/pi)
ylabel('phase (deg)')
title('Butt. HPF')
[num6,den6] = lp2bp(num1,den1,2,1);
                                              % transf function/Butt. cen-
                                              ter freq of 1.5 and Bw =1
H6 = freqs(num6,den6,w);
subplot(2,2,3)
plot(w,abs(H6));
ylabel('magnitude')
xlabel('frequency(rad/sec)')
title('But.BPF ,cent freq=2,BW=1')
subplot(2,2,4);
plot(w,angle(H6)*180/pi)
ylabel('phase (deg)')
xlabel('frequency (rad/sec)')
title('Butt. BPF')
% part(e)

figure(4)
[num7,den7] = lp2bs(num1,den1,2,1);
                % trasf. function/Butt.center freq at 1.5 and stop Bw =1
H7 = freqs(num7,den7,w);
subplot(2,2,1)
```

```
plot(w,abs(H7));axis([0 4 -.5 1.5]);
ylabel('magnitude')
title('Butt. BS,center freq,=2,stopband=1Butt.')
subplot(2,2,2);
plot(w,angle(H7)*180/pi)
ylabel('phase (deg)')
title('Butt.BSF')
[num8,den8] = lp2bs(num2,den2,2,1);
H8 = freqs(num8,den8,w);
subplot(2,2,3)
plot(w,abs(H8));axis([0 6 -.5 1.5]);
ylabel('magnitude')
xlabel('frequency(rad/sec)')
title('Cheb.BS ,cent freq=2,BW stopband=1')
subplot(2,2,4);
plot(w,angle(H8)*180/pi)
ylabel('phase (deg)')
xlabel('frequency (rad/sec)')
title('Cheb. BSF')
```

Back in the command window, the script file *revisit_analog_filters* is executed and the results are as follows (Figures 6.81 through 6.84):

```
>> revisit _ analog _ filters
```

Example 6.23

Indicate the steps and compute by hand the discrete IIR filter's transfer function from the analog RC-LPF of Example 6.1 ($R = 1\Omega$ and $C = 1$ F) by employing a sampling period of T = 0.01 s.

ANALYTICAL Solution

The solution follows the steps indicated:

1. Determine the analog filter's transfer function $H(s)$.
2. From the transfer function $H(s)$, obtain the system impulse response $h(t) = \pounds^{-1}[H(s)]$.

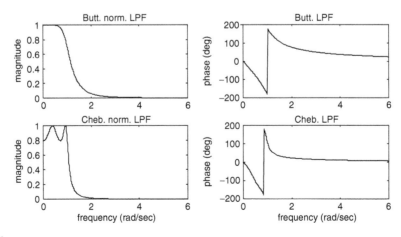

FIGURE 6.81
Four-pole analog Butterworth and Chebyshev/type-1 normalized LPFs of Example 6.22.

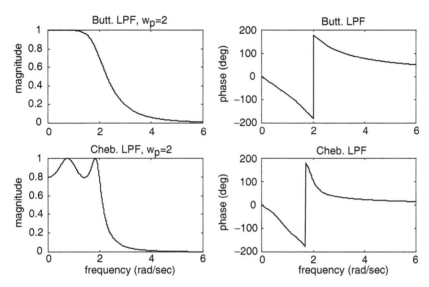

FIGURE 6.82
Four-pole analog Butterworth and Chebyshev/type-1 denormalized filters with $w_c = 2$ rad/s of Example 6.22.

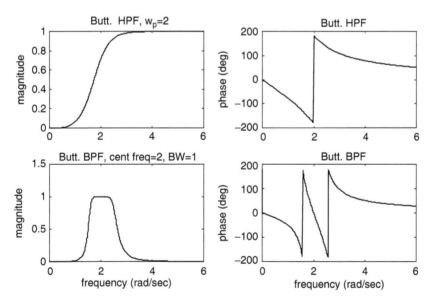

FIGURE 6.83
Plots of Butterworth denormalized HP and BP filters of Example 6.22.

3. Replace t by the discrete time nT in the impulse system response $h(t)$.
4. Take the z transform of $h(nT)$ times T given by $Z[h(n)] * T$.
   a. The analog transfer function of the RC-LPF of Example 6.1 is given by

$$H(s = jw) = \frac{1}{s+1}$$

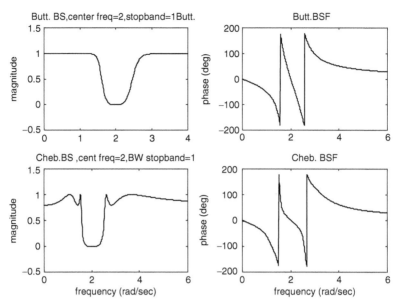

FIGURE 6.84
Plots of Butterworth and Chebyshev BSFs of Example 6.22.

b. Then $h(t) = \pounds^{-1}\left[\dfrac{1}{s+1}\right]$

$$h(t) = e^{-t} \quad \text{for } t > 0$$

and

c. $h(nT) = e^{-nT} \quad \text{for } t > 0$

b. $H(z) = Z\{e^{-nT}\} * T$

$$H(z) = Z\{e^{-n(0.01)}\} * \{0.01\}$$

$$H(z) = Z\{e^{-(0.01)}\}^n * \{0.01\}$$

$$H(z) = \frac{(0.01) * z}{z - e^{-0.01}}$$

6.5   Application Problems

P.6.1   Verify that the transfer function given by

$$H(jw) = \frac{a}{jw + a} \quad \text{for } a > 0$$

is of an LPF, and evaluate by hand the following

a. The gain at $H(w = 0)$ and $H(w = \infty)$

b. The cutoff frequency w_c

c. The gain at $|H(w = w_c)|$ and phase at $H(w = w_c)$

d. The filter's BW

P.6.2 Given the following filter's transfer functions:

a. $H(jw) = \dfrac{jw}{jw + a}$ for $a > 0$

b. $H(jw) = \dfrac{(jw)^2 + w_0^2}{(jw)^2 + b\ jw + w_0^2}$ for $b > 0$ (notch filter)

Evaluate by hand:

a. The filter type

b. The gain at $H(w = 0)$ and $H(w = \infty)$

c. The cutoff frequency w_c

d. The gain at $|H(w = w_c)|$ and phase at $H(w = w_c)$

e. The filter's BW

P.6.3 Obtain the magnitude and phase plots for the transfer functions of P.6.1 for the following cases: $a = 2, 6,$ and 10, and discuss the effect of a.

P.6.4 Obtain the magnitude and phase plots of the transfer functions of P.6.2 for the following cases:

a. $a = 2, b = 2$

b. $a = 2, b = 5,$ and $w_o = 15$

P.6.5 Verify that the circuit shown in Figure 6.85 is of a normalized LPF, where all the corresponding element values are in ohms, henries, and farads.

P.6.6 Transform the LPF of P.6.5 (Figure 6.85) into an LPF with a cutoff frequency at 5 kHz, and a load impedance of $R = 600\ \Omega$ connected at its output terminals.

P.6.7 Transform the LPF of Figure 6.85 into a normalized HPF ($w_c = 1$ and a load impedance of $R = 1\ \Omega$).

P.6.8 Analyze the circuits shown in Figure 6.86 and

a. Verify the given expressions of $H(w) = V_o(w)/V_i(w)$.

b. Obtain analytical expressions of the magnitudes $[H(w)]$ and phases of $H(w)$.

c. Let the values of the electrical elements be $R_1 = 1\ \Omega, R_2 = 2\ \Omega, R_3 = 3\ \Omega, C_1 = 1\ F,$ and $C_2 = 2\ F$.

Then create the script file *plots* that returns the plots of magnitude $[H(w)]$ versus w and phase $[H(w)]$ versus w.

d. For each circuit evaluate or determine the following:

    i. The BW

    ii. The type of filter

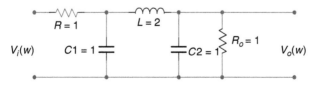

FIGURE 6.85
Circuit diagram of P.6.5.

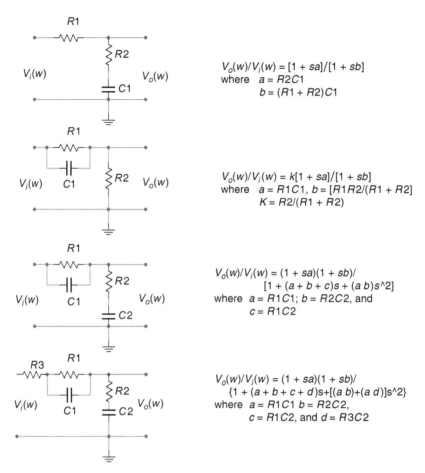

FIGURE 6.86
Circuit diagrams of P.6.8 (where $s = jw$).

    iii. Regions of attenuation

    iv. Regions of distortion

P.6.9   Design, at the circuit level, a Chebyshev type-1 and -2 BPF with $w_p = 2000$ Hz connected to a load of $R = 600\ \Omega$. The schematic diagram is shown in Figure 6.87 and verify that its transfer function is given by

$$H(s) = \frac{\dfrac{1}{LC}}{s^2 + \dfrac{s}{CR} + \dfrac{1}{LC}}$$

P.6.10   Obtain expressions of the 10th-order polynomials of the analog, Butterworth and Chebyshev (types 1 and 2) LPFs with a cutoff frequency at 4 kHz.

P.6.11   Analyze the circuit diagram shown in Figure 6.88 and

    a. Obtain the system transfer function.

    b. Obtain plots of magnitude and phase of its transfer function for $L = 1$, and $C = 1/2$.

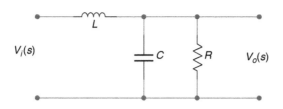

FIGURE 6.87
Circuit diagrams of P.6.9.

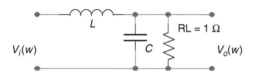

FIGURE 6.88
Circuit diagrams of P.6.11.

    c. Transform the circuit into a PBF, where the pass-band region is between 2000 and 4000 Hz.

    d. Repeat part (c) for the case of a BSF.

P.6.12 The circuit diagram shown in Figure 6.89 is called a notch filter tuned at 2 rad/s.

    a. Verify the preceding statement by obtaining plots of $H(w) = V_o(w)/V_i(w)$.

    b. Modify the circuit to a notch filter tuned at 10 rad/s.

P.6.13 Determine the lowest order of the analog transfer function of an LPF with the following specs: −2 dB at 2000 Hz, a minimum attenuation of 30 dB at 7000 Hz, for the following filters types:

    a. Butterworth

    b. Chebyshev (types 1 and 2)

    c. Elliptic

    d. Bessel

P.6.14 Obtain the transfer function coefficients and the frequency response plots of a 10th-order maximally flat analog LPF with a 3 dB cutoff angular frequency at 1 rad/s.

P.6.15 State and discuss the order of the analog filter obtained as a result of the LP-to-LP filter transformation. Illustrate your conclusion with a 9th-order LP Butterworth normalized filter with a cutoff frequency of $w_c = 0.35$ rad/s that is transformed to an equivalent LPF with $w_c = 3$ rad/s.

P.6.16 State and discuss the order of the analog filter obtained as a result of the LP-to-HP filter transformation. Illustrate your conclusion with a 9th-order LP Butterworth normalized filter with a cutoff frequency $w_c = 0.4$ rad/s that is transformed to an equivalent HPF with $w_c = 4$ rad/s.

P.6.17 State and discuss the order of the analog filter obtained as a result of the LP-to-BP filter transformation. Illustrate your conclusion with a 10th-order low-pass Butterworth normalized filter with a cutoff frequency $w_c = 0.45$ rad/s that is transformed to an equivalent BPF with $w_c = 10$ rad/s and BW of 2 rad/s.

P.6.18 Repeat problem P.6.17 for the case of a BPF with a stop-band range given by 9 rad/s < w < 11 rad/s.

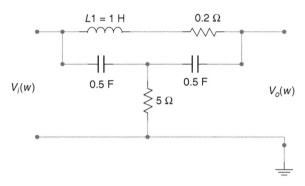

FIGURE 6.89
Circuit diagrams of P.6.12.

P.6.19   Design, using the zero–pole–gain command, an analog normalized LPF of order 5 for the following filter types:

     a. Butterworth

     b. Chebyshev (types 1 and 2)

     c. Elliptic

     d. Bessel

     with the following filter's specs:

     $w_c = 0.3$ rad/s

     $R_p = 0.5$ dB

     R_s = Minimum attenuation of -30 dB

P.6.20   Repeat P.6.19 using the transfer function approach.

P.6.21   Verify the following statement by means of an example plot: an elliptic filter of order 6 is equivalent to a Chebyshev filter (type 1 or 2) of order 10 and a Butterworth filter of order 29.

P.6.22   Design an analog elliptic filter of minimum order that satisfies the following specs:

     • Pass band: 0–1 rad/s and 7–10 rad/s

     • Stop band: 3–6 rad/s with maximum gain of 0.1 dB

     Obtain the filter coefficients as well as the gain and phase plots.

P.6.23   Determine the minimum length of an FIR filter with the following specs:

     $f_c = 200$ Hz (pass-band edge)

     $f_s = 300$ Hz (stop-band edge)

     $R_p = 0.01$ dB

     $R_s = 0.02$ dB

P.6.24   Determine and plot the transfer function (gain and phase) as well as the impulse response of a linear FIR BPF with the following specs:

     1. Order 30

     2. Pass band: from 0.4 to 0.6

     3. Stop band: from 0.2 to 0.3 and from 0.8 to 0.9

4. Desire pass-band magnitude: 1

5. Desire stop-band magnitude: 0

P.6.25 Design an FIR LPF and obtain magnitude and phase plots of the transfer function using the following window types:

- Kaiser ($\beta = 3.3$)
- Hanning
- Blackman
- Hamming
- Triangular
- Bartlett

with the following filter's specs:

1. $w_p = 0.2\pi$

2. $w_s = 0.5\pi$

3. $w_c = 0.35\pi$, [where $(w_p + w_s)/2 = w_c$]

4. $R_s = 60$ dB

P.6.26 Transform the analog filters shown in Figure 6.86 into digital filters using the impulse invariance method with $T_s = 0.5$ s.

Note: A convenient way to perform the transformation is to decompose the gain function into first-order partial fraction expansion blocks.

Obtain plots of the gain response and pole/zero of the resultant digital filter and discuss its stability.

P.6.27 Determine the transfer function and obtain plots of the gain (magnitude and phase) and the pole–zero of an FIR BPF of order 35 with pass band from 0.4 to 0.6 rad and a stop band from 0 to 0.2 and from 0.8 to 1.0 rad. Use the *remez* function (Parks–McClellan Algorithm) and represent the stop- and pass band with 1 and 0, respectively.

Bibliography

Adamson, T., *Structured Basic*, 2nd Edition, Merrill Publishing Co., New York, 1993.

Ana, B., The state of the research isn't all that grand, *The New York Times*, September 3, 2006.

Attia, J.O., *Electronics and Circuits, Analysis Using Matlab*, CRC Press, Boca Raton FL, 1999.

Austin, M. and Chancogne, D., *Engineering Programming C, Matlab, Java*, Wiley, New York, 1999.

Ayres, A., *Theory and Problems of Differential Equations*, Schaum's Outlines Series, McGraw-Hill, New York, 1952.

Balador, A., *Algebra*, Deama Octava Re-impression, Publications Cultural, Mexico, 2000.

Biran, A. and Breiner, M., *Matlab for Engineers*, Cambridge, Great Britain, 1995.

Bogard, T.F., Beasley, J.S., and Rico, G., *Electronic Devices and Circuits*, 5th Edition, Prentice-Hall, New York, 1997.

Borse, G.J., *Numerical Methods with Matlab*, P.W.S. Publishing Co., Boston, 1997.

Brooks, D., The populist myths on income inequality, *The New York Times*, September 7, 2006.

Buck, J., Daniel, M., and Singer, A., *Computer Explorations in Signals and Systems Using Matlab*, 2nd Edition, Prentice-Hall, Upper Saddle River, NJ, 2002.

Caputo, D.A., The world's best education; remade in America, *The New York Times*, A29, December 6, 2006.

Carlson, A.B., *Circuits*, Brooks/Cole, Pacific Grove, CA, 2000.

Cartinhour, J., *Digital Signal Processing*, Prentice-Hall, New York, 2000.

Chapman, S., *Matlab for Engineers*, Brooks/Cole, Pacific Grove, CA, 2000.

Clawson, C.C., *Mathematical Mysteries*, Perseus Books, Cambridge, MA, 1996.

Connor, F.R., *Circuits*, Calabria, Barcelona, Spain, 1976.

Cyganski, D., Orr, J.A., and Vaz, R.F., *Information Technology Inside and Outside*, Prentice-Hall, New York, 2001.

Deziel, P.J., *Applied Digital Signal Processing*, Prentice-Hall, New York, 2001.

Dwyer, D. and Gruenwald, M., *Precalculus*, Thomson-Brookes/Cole, Pacific Grove, CA, 2004.

Edminister, J., *Theory and Problems of Electric Circuits*, Schaum's Outlines Series, McGraw-Hill, New York, 1965.

Etter, D.M., *Introduction to Matlab for Engineers and Scientist*, Prentice-Hall, Upper Saddle River, NJ, 1996.

Etter, D.M. and Kuncicky, D.C., *Introduction to Matlab*, Prentice-Hall, New York, 1999.

Exxon Mobil, Multiplier effects, *The New York Times*, A33, December 19, 2006.

Friedman, T.L., Learning to keep learning, *The New York Times*, A33, December 13, 2006.

Gabel, R. and Roberts, R., *Signals and Linear Systems*, Wiley, New York, 1973.

Gawell, K., report available at http:/www.geo-energy.org/publications/reports.asp. 2007.

Grover, D. and Deller, J.R., *Digital Signal Processing and the Microcontroller*, Prentice-Hall, New York, 1999.

Grunwald, M., *The Clean Energy Scan*, Time Magazine, 2008.

Gustafsson, F. and Bergman, N., *Matlab for Engineers Explained*, The Cromwell Press, Springer-Verlag, London, U.K., 2004.

Hanselman, D. and Littlefield, D., *The Student Edition of Matlab*, Version 5, User's Guide, Prentice-Hall, Upper Saddle River, NJ, 1997.

Hanselman, D. and Littlefield, D., *Mastering Matlab 7*, Prentice-Hall, Upper Saddle River, NJ, 2005.

Harman, T.L., Dabney, J., and Richer, N., *Advanced Engineering Mathematics Using Matlab*, Brooks/Cole, Pacific Grove, CA, 2000.

Hayt, W. and Kemmerly, J.E., *Engineering Circuit Analysis*, McGraw-Hill, New York, 1962.

Hill, D. and David, E.Z., *Linear Algebra Labs with Matlab*, Second Edition, Prentice-Hall, Upper Saddle River, NJ, 1996.

Hodge, N., Solar energy stock, *Wealth Daily*, Anger Publishing LLC, 2007.

Hsu, H.P., *Analysis De Fourier*, Fondo Education Interamericano, S.A., Bogata, Colombia, 1973.

Ingle, V. and Proakis, J., *Digital Signal Processing*, Brooks/Cole, Pacific Grove, CA, 2000a.

Ingle, V. and Proakis, J., *Digital Signal Processing Using Matlab*, Brooks/Cole, Pacific Grove, CA, 2000b.

Jack, K., *Engineering Circuit Analysis*, McGraw-Hill, New York, 1962.

Jairam, A., *Companion in Alternating Current Technology*, Prentice-Hall, Upper Saddle River, NJ, 1999.

Jairam, A., *Companion in Direct Current Technology*, Prentice-Hall, Upper Saddle River, NJ, 2000.

Jensen, G., *Using Matlab in Calculus*, Prentice-Hall, Upper Saddle River, NJ, 2000.

Joseph, J.D. III, Allen, R.S., and Ivan, J.W., *Feedback and Control Systems*, Schaum Publishing Co., New York, 1967.

Judith, M. and Muschle, G.R., *The Math Teacher's Book of Lists*, Prentice-Hall, Upper Saddle River, NJ, 1995.

Kamen, E.W. and Heck, B.S., *Fundamentals and Systems Using the Web and Matlab*, 2nd Edition, Prentice-Hall, Upper Saddle River, NJ, 2000.

Kay, D.A., *Trigonometry*, Cliffs Quick Review, First Edition, Cliffs Notes, Lincoln, NE, 1994.

Keedy, M., Bittinger, M.L., and Rudolph, W.B., *Essential Mathematics for Long Island University*, Brooklyn Campus, Pearson-Addison-Wesley, Custom Publishing, Boston, 1992.

Keedy, M., Griswold, A., Schacht, J., and Mamary, A., *Algebra and Trigonometry*, Holt, Rinehart and Winston Inc., New York, 1967.

Krauss, C., Move over, Oil. There's Money in Texas Wind, *The New York Times* (pp. A1, A15), February 23, 2008.

Kurtz, M., *Engineering Economics for Professional Engineer's Examinations*, Third Edition, McGraw-Hill, New York, 1985.

Lathi, B.P., *Modern Digital and Analog Communication Systems*, 3rd Edition, Oxford University Press, New York, 1998.

Leon, S., Eugene, H., and Richard, F., *ATLAST, Computer Exercises for Linear Algebra*, Prentice-Hall, Upper Saddle River, NJ, 1996.

Linderburg, M., *Engineer in Training Review*, 6th Edition, Belmont, CA, 1982.

Lindfield, G. and Penny, J., *Numerical Methods Using Matlab*, Prentice-Hall, Upper Saddle River, NJ, 2000.

Lipschutz, S., *Theory and Problems of Linear Algebra*, Schaum's Outlines Series, McGraw-Hill, New York, 1968.

Lutovac, M.D., Tosic, D.V., and Evans, B.L., *Filter Design for Signal Processing (Using Matlab and Mathematics)*, Prentice-Hall, Upper Saddle River, NJ, 2001.

Lynch, W.A. and Truxal, J.G., *Signals and Systems in Electrical Engineering*, McGraw-Hill, The Maple Press Company, York, PA, 1962.

Magrad, E.B., Azarm, S., Balachandran, B., Duncan, J.H., Herold, K.E., Walsh, G., *An Engineer's Guide to Matlab*, Prentice-Hall, Upper Saddle River, NJ, 2000.

Maloney, T.J., *Electric Circuits: Principles and Adaption*, Prentice-Hall, Upper Saddle River, NJ, 1984.

Markoff, J., Smaller than a pushpin, more power than a PC, *The New York Times*, C3, February 7, 2005.

Markoff, J., A chip that can move data at the speed of laser light, *The New York Times*, C1, September 18, 2006.

Markoff, J., Intel prototype may herald a new age of processing, *The New York Times*, C9, February 12, 2007.

Matt, R., Start up fever shift to energy in Silicon Valley, *The New York Times*, A1/C4, March 14, 2007.

McClellan, J.H., Schafer, R.W., and Yoder, M.A., *DSP First: A Multimedia Approach*, Prentice-Hall, Upper Saddle River, NJ, 1998.

McMenamin, S.M. and John, F.P., *Essentials Systems Analysis*, Yourdon Press Computing Series, Prentice-Hall, Englewood Cliffs, NJ, 1984.

Meador, D., *Analog Signal Processing with Laplace Transforms and Active Filter Design*, Belmont, CA, 2002.

Miller, M.L., *Introduction to Digital and Data Communications*, West Publishing Company, St. Paul, MN, 1992.

Minister, J.A., *Electric Circuits*, Schaum Publishing Co., New York, 1965.

Miroslav, D., Tosic, D., and Evan, B., *Filtering Design for Signal Processing*, Prentice-Hall, Upper Saddle River, NJ, 2001.

Mitra, S.K., *Digital Signal Processing Laboratory Using Matlab*, McGraw-Hill, New York, 1999.

Mitra, S.K., *Digital Signal Processing*, McGraw-Hill, New York, 2001.

Nashelsky, L. and Boylestad, R.L., *Basic Applied to Circuit Analysis*, Merrill Publishing Co., Columbus, OH, 1984.

Navarro, H., *Instrumentacion Electronica Moderna*, Univsidad Central de Venezuela, 1995.

New York City Board of Education, *Sequential Mathematics*, 1989 (Revision).

Newman, J., *The World of Mathematics*, Volumes 1, 2, 3 and 4, Simon and Schuster, New York, 1956 (commentaries).

Novelli, A., *Lecciones De Analisis I*, Impresiones Avellaneda, Buenos Aires, Argentina, 1998.

Novelli, A., *Algebra Lineal Y Geometria*, 2nd Edition, Impresiones Avellaneda, Buenos Aires, Argentina, 2004a.

Novelli, A., *Lecciones De Analisis II*, Impresiones Avellaneda, Buenos Aires, Argentina, 2004b.

O'Brien, M.J. and Larry, S., *Profit from Experience*, Bard & Stephen, Austin, TX, 1995.

Oppenheim, A.V., Schafer, R.W., and Buck, J.R., *Discrete-Time Signal Processing*, 2nd Edition, Prentice-Hall, Upper Saddle River, NJ, 1999.

Oppenheim, A., Willsky, A., and Young, I., *Signals and Systems*, Prentice-Hall, Upper Saddle River, NJ, 1983.

Palm, W.J., III, *Introduction to Matlab for Engineers*, McGraw-Hill, Natick, MA, 1998.

Parson, J.J. and Oja, D., *Computer Concepts*, Thomson Publishing Company, 1996.

Patrick, D.R. and Fardo, S.W., *Electricity and Electronics: A Survey*, 4th Edition, Prentice-Hall, Upper Saddle River, NJ, 1999.

Petr, B., *A History of TT*, St. Martin's Press (Golem), New York, 1971.

Petruzella, F., *Essentials of Electronics*, 2nd Edition, McGraw-Hill, New York, 2001.

Polking, J. and Arnold, D., *Ordinary Differential Equations Using Matlab*, 3rd Edition, Prentice-Hall/Pearson, Upper Saddle River, NJ, 2004.

Pratap, R., *Getting Started with Matlab*, Saunders College Publishing, Orlando, FL, 1996.

Proakis, J. and Salekis, M., *Contemporary Communication Systems Using Matlab*, PWS Publishing Co., Boston, 1998.

Randall, S., It's not who you know. It's where you are, *The New York Times*, B3, December 22, 2006.

Recktenwald, G., *Numerical Methods with Matlab*, Prentice-Hall, Upper Saddle River, NJ, 2000.

Rich, B., *Elementary Algebra*, Schaum's Outline Series, McGraw-Hill, New York, 1960.

Rich, B., *Modern Elementary Algebra*, Schaum's Outline Series, McGraw-Hill, New York, 1973.

Robert, J.P., *Introduction to Engineering Technology*, Third Edition, Prentice-Hall, Englewood Cliffs, NJ, 1996.

Robinson, D., *Fundamentals of Structured Program Design*, Prentice-Hall, Upper Saddle River, NJ, 2000.

Russel, M.M. and Mark, J.T.S., *Digital Filtering: A Computer Laboratory Textbook*, John Wiley, New York, 1994.

Ruston, H. and Bordogna, J., *Electric Networks: Functions, Filters Analysis*, McGraw-Hill, New York, 1966.

Sarachik, P., *Principles of Linear Systems*, Cambridge University Press, New York, 1997.

Schilling, R. and Harris, S., *Applied Numerical Methods for Engineers, Using Matlab and C*, Brooks/Cole, Pacific Grove, CA, 2000.

Schuller, C.A., *Electronics Principles and Applications*, McGraw-Hill, New York, 1989.

Shenoi, K., *Digital Signal Processing in Telecommunications*, Prentice-Hall, Upper Saddle River, NJ, 1995.

Sherrick, J.D., *Concepts in Systems and Signals*, Prentice-Hall, Upper Saddle River, NJ, 2001.

Silverman, G. and Tukiew, D.B., *Computers and Computer Languages*, McGraw-Hill, New York, 1988.

Smith, D.M., *Engineering Computation with Matlab*, Pearson, Boston, 2008.

Smith, M.J.T. and Mersereau, R.M., *Introduction to Digital Signal Processing*, Wiley, New York, 1992.

Spasov, P., *Programming for Technology Students Using Visual Basic*, 2nd Edition, Prentice-Hall, Upper Saddle River, NJ, 2002.

Spiegel, M.R., *Laplace Transforms*, Schaum's Outline Series, McGraw-Hill, New York, 1965.

Sprankle, M., *Problem Solving and Programming Concepts*, 5th Edition, Prentice-Hall, Upper Saddle River, NJ, 2001.

Sprankle, M., *Problem Solving for Information Processing*, Prentice-Hall, Upper Saddle River, NJ, 2002.

Stanley, W.D., *Network Analysis with Applications*, 4th Edition, Prentice-Hall, Upper Saddle River, NJ, 2003.

Stanley, W.D., *Technical Analysis and Applications With Matlab*, Thomson V Delmar Learning, New York, 2005.

Stearns, S. and David, R., *Signal Processing Algorithms in Matlab*, Prentice-Hall, Upper Saddle River, NJ, 1996.

Stein, E.I., *Fundamentals of Mathematics*, Modern Edition, Allyn and Bacon Inc., Boston, 1964.

Steve, L., Parsing the truths about visas for tech workers, *The New York Times*, April 15, 2007.

Sticklen, J. and Taner, E.M., *An Introduction to Technical Problem Solving with Matlab*, Volume 7, Second Edition, Great Lakes Press Inc., Wildwood, MO, 2006.

Strum, R. and Kirk, D., *Contemporary Linear Systems Using Matlab*, Brooks/Cole, Pacific Grove, CA, 2000.

Tagliabue, J., The eastern bloc of outsourcing, *The New York Times*, C1/5, April 19, 2007.

Tahan, M., *EL Hombre the Calculaba Segunola*, Edition Ampliaoa, Buenos Aires, Argentina (Traduudo por Mario Cappetti), 1938.

The Math Work Inc, *The Student Edition of Matlab*, Version 4, User's Guide, Prentice-Hall, Englewood Cliffs, NJ, 1995.

Theodore, F.B., *Basic Programs for Electrical Circuit Analysis*, Reston Publishing Company, Inc., Reston, VA, 1985.

Van de Vegte, J., *Fundamentals of Digital Signal Processing*, Prentice-Hall, Upper Saddle River, NJ, 2002.

Van Valkenburg, M.E., *Network Analysis*, Third Edition, Prentice-Hall, Englewood Cliffs, NJ, 1974.

Von Seggern, D., *Standard Curves and Surfaces*, CRC Press, Boca Raton, FL, 1993.

White, S., *Digital Signal Processing*, Thomson Learning, Albany, NY, 2000.

Young, P.H., *Electronic Communication Techniques*, 4th Edition, Prentice-Hall, Upper Saddle River, NJ, 1999.

Zabinski, M.P., *Introduction to TRS-80 Level II Basic*, Prentice-Hall, Upper Saddle River, NJ, 1980.

Index

A

AC (alternating current). *See* Alternating current (AC) analysis
Acceleration, 105
Active filter
  features of, 563, 569
  transfer function, 573, 620–622
Active power, 242–244
Admittance *G*, 235
Algorithms
  Cooley-Tukey, 498
  fft, 498
  Goertzel, 499
  Parks-McClellan, 595
  prime factor, 498
  remez exchange, 595
  Yule-Walker, 584
Aliasing, 586
All-pole-prototype LPF, 571
Alternating current (AC) analysis
  application problems, 310–317
  balance system, 263–264
  circuits, types of, 234–246, 252–254, 256
  connections, 260
  defined, 102
  DC circuits and, 224
  examples, 267–309
  frequency, 245–246
  generators, 261–267
  loop equations, 248–250
  measurements, 230
  mesh equation, 248–250
  node equation, 251–252
  power, 231–232, 242–244, 259–260
  resistors, 230–231
  sinusoid function, 224, 230
  source transformation, 253–255
  superposition principle, 257–259
  terminology, 223–224
  types of signals, 226–228
  wave model, 224
Alternating waveforms, 223
AM signals
  bandwidth, 22
  features of, 21
  time domain representations, 22
AM spectrum, 443–447

Ammeter, 106
Ampere, Andres Maria, 101
Ampere units, as unit measure, 101, 106
Amplification, 569
Amplitude
  even function, 75–76
  filter transfer function, 590
  Fourier transform, 321
  implications of, 5, 563, 583
  impulse function, 71
  Laplace transform, 338
  odd function, 75–76
  scaled step sequence, 11
  shifting, continuous time signal, 25
  spectrum, 327
  step function, 71
  transformation, 23
Amplitude shift keying (ASK)
    signals, 21, 87–88
Amps, as unit measure, 102
Analog filters
  digital computers distinguished from, 570
  features of, 569–570, 586
  frequencies, 583–585
  low-pass, 622–624
  LP normalized prototype, 584
  RC-LPF, 658–660
  stable, 584
  synthesis process, 599
  transfer function, 586, 596
  transformation, complex, 585
  see also Active filters; Passive filters
Analog FT, 457
Analog function
  defined, 16
  impulse, 68
  pulse, 12
Analog signals
  conversion to discrete, 3
  defined, 2–3
  examples of, 9
  exponential, 14, 64
  reconstruction from discrete sequence, 44
Analog step function, 12, 68
Analog-to-digital converters/encoders, 3, 570
Analog unit
  parabolic function, 14
  ramp function, 12–13

Ancient history of electricity
    China, 101
    Greece, 101, 104
Angle modulation, 22, 443–447
Angular frequency, 16–17
Animation, 56
Apparent power, 225, 232, 242–244
Approximations, 6, 170–171, 223, 326, 338, 588
ASCII, 9
Asymmetric function, 22
Atoms, 104
Attenuation, 21, 23, 561, 563, 565, 568, 581
Attraction force, 105
Autocorrelations, 336, 502–504
Average power, 225, 231, 245, 324–325, 491–492

B

Balanced load, 225
Balance system
    polyphase network, 265
    three-phase generator system, 263
Band-pass filter (BPF), 565–566, 572, 590
Band-reject filter (BRF), 565–566, 573
Band-stop filter (BSF), 565
Bandwidth (BW)
    AM signals, 22
    angle modulation, 22
    implications of, 225, 329, 562, 565, 567, 583
    low-pass filters and, 583
    parallel RLC, 246–247
    series RLC, 245–246
Bank reject filter, 562
Bartlett(N) function, 48
Bartlett window, 48, 52, 93
Base band signal, 21
Batteries, 102
Bell, Alexander Graham, 102
Bessel filters, 562, 573, 578–579, 582
Bessel functions
    besselap, 582
    besself, 582
    impact of, 47
bilinear function, 585–586
Bilinear transformation, 584–586, 588
Binary codes, 9
Binary signal, 9
Blackman window, 47–50
Block box system
    examples, 412–414
    representations, 338
    transformation, 337
Bode, Hendrik Wade, 339n

Bode plots, 339–340, 603, 608, 620, 622
Bounded input-bounded output (BIBO)
        system, 458, 460
boxcar(N) function, 48
Boxcar window, 48, 93
Branch currents, 111, 116–117
Brickwall conditions, 593
buttap functions, 580
butter functions, 580, 588
Butterworth filters
    IIR, 596, 649–650
    analog, 658–660
    features of, 562, 573–575, 577, 580, 596
    low-pass filter (LPF), 580, 646
    LP digital filter, 633–634
    LPF transfer function, 580
    second-order low pass-band filter, 587
buttord functions, 580

C

Capacitance, equivalent, 114
Capacitative reactive power, 232, 234
Capacitors
    features of, 103–104, 108, 110, 114, 130, 136,
        569
    ideal, 115
    impedance, 235
    LP normalized filter, 568
    parallel RLC, 247
    pure, 116
    storing energy elements, 127
Capacitor C, 355
Carbon, 109
Carrier
 analog high-frequency, 22
    defined, 21
    high-frequency, 21
    signal, 334
Cascade, 110
Cauchy's residue theorem, 483
Cauchy window, 48
Cauer filter, 562, 573, 577–578
Causal sequence, 22
Causal systems
    characteristics of, 346, 349–350
    discrete, 458
    discrete-time invariance linear system
        (LTI), 462–463
Change, creation of, 105
Charges, 110
cheb1 function, 581
cheblap function, 581

cheb1ord, 581
cheb2ap function, 581
cheb2ord function, 581
cheby1 function, 581, 588
cheby2 function, 581, 588
Chebyshev filters
    analog, type 1/type 2, 575–577, 581
    features of, 562, 573, 596
    function files, 582
    type-1, 45, 581, 634–636, 658–660
    type-2 filter, 634–636
Chebyshev FPF transfer function, 581
Circuit analysis, 17, 320
Circuit diagrams
    characteristics of, 122–123
    LC parallel, 239
    LC series, 238–239
    loop equation, 248
    power triangle, 244
    RC connection, 237–238
    RC parallel, 237
    RL, 234, 236
    RL parallel, 237
    source transformation, 122–123, 253–255
    Thevenin's equivalent circuit, 254
    Thevenin's model, 253
    RLC parallel, 242, 246
    RLC series, 245
Circuit laws, 102
Circuit techniques, standard, 604
Circuit theorems, DC, 225
Circular convolution, 493–495, 497–498
Circular shift, 493
Close loop system, 108
Closed system, 108
Coherent relations, 374
Colormap, 55
Commands
    Blackman(N), 47
    del1(matrix_A), 56
    fft, 459, 490–491, 495–496, 499–500, 502, 555–556
    fftn(f), 500
    fftn(F), 500
    fftshift(F), 500
    filter, 489
    fir1, 594, 653–655
    fir2, 588, 594–595
    Frms, 326
    fourier(f,x), 342
    gensig, 38–39
    getframe, 56–57
    gradianr(matrix_A), 56
    Hanning(N), 47
    ifftshift(F), 500

ilaplace, 320
image(matrix_A), 55
impulse(P, Q, t), 40
impz, 489
iztrans(Fz), 488
k = nextpower2(N), 499
Kaiser(N), 47
laplace, 320
movie(M), 57
nyquist, 339
plot, 27
sawtooth(t, b), 33–35
square(t), 32
step(P, Q, t), 40
wavplay(y, Fs), 54
wavread, 55
wavrecord(N, Fs), 54
wavwrite, 55
zplane, 489
zp2tf(zeros, poles k), 489
ztrans(f), 484–485
Communication system
    analog, 86–88
    analysis, 335
    broadcasting, 22
Communication theory, 334
Complex circuits, 127
Complex conjugate, 242
Complex frequencies
    implications of, 133, 197, 201
    natural, 131
    network, 134
Complex power, 232
Complex sequences, continuous time signals,
    25–26
Complex wave, 17
Composite function, 14
Compression, 23–25, 569
Computer chips, 570
Computer software programs, digital filters, 570
Conductance, 109
Conductors, 105–106, 109
Conjugates
    complex roots, 131
    ZT, 475–476
Conservation of energy, 107–108
Continuous and discrete signals, time domain
    representation
    application problems, 93–100
    examples, 58–93
    expression evaluation, 29–31
    types of signals and sequences,
        overview, 1–27
    windows, 47–52

Continuous time
  function (*f(t)*), 4, 322, 499, 538–541
  time signals, 23–24
  time systems, 3
Convergence, 477–479
Conversion tables, transform, 320, 347, 483
Convolution integral, 335–336
Convolution process
  characteristics of, 337, 493, 502
  discrete systems, 465–468
  in frequency, 471
  time functions, 335, 348, 471
  ZT, 476
Cooley-Tukey algorithms, 498
Copper, 105, 109, 261
Correlation function, 502
Cosine functions, 47, 226, 252, 321, 491
Cosine-sine series, 322
Coulomb, Charles, 101
Coulomb units, 101, 106
Coulomb's law, 103, 105
cps, 226
Crest factor, 230
Critical damped system
  characteristics of, 131
  examples of, 199–200, 203–204
  parallel, 133
  series, 135
Cross-correlations, 335–336, 503–504
Current
  circuit elements, 103
  defined, 102
  divider network, 112
  divider rule, 111, 136–138, 225
  flow, 106
  Norton short-circuit, 119–120
  sources, 119, 122–123
Cutoff frequency
  Butterworth filters, 580
  filters and, 565
  implications of, 4, 44–45, 582
  parallel RLC, 246
  series RLC, 245
  stop-band, 577
Cycles per second, 16

D

Damping coefficient, 14, 131
Damping constant, 14
Daulity, 334
DC (direct current)
  component, 323

current source, 102
  defined, 102
  electrical systems, components of, 102
  linear network, 118–119
  Fourier transform, 393
  production of, 102
  shifting, 23
  theorem, 225
  transient analysis, *see* Direct current (DC)
    transient analysis
  transient response, 103
  voltage sources, 102
Decaying exponentials, 131–132
Decay time exponentials, 127
Decibels, 107, 565
decimate function, 44
Decimation process, 44–45, 83–86
Decomposition, 329
De Forest, Lee, 102
Delay
  distortion, 321
  impact of, 337
  propagation, 564
  time, 338
  unit, 458
Delta function, 5
Democritus, 104
denz function, 585
Derivatives
  higher-order, 348
  sinusoid, 225
Deterministic signals, 2, 45
Dielectric constant, 110
Difference equations, 458, 461, 463–464, 552–555
Differential equations (DEs), 3, 15, 320, 361, 589
Differentiation, 348
Digital filters
  characteristics of, 569–570
  design examples, 650–653
  frequencies, 583
  multiband normalized, 641–642
  stable, 584
  synthesis process, 599
  see also Finite impulse response (FIR) filters;
    Infinite impulse response (IIR) filters
Digital-filter theory, 457
Digital frequency, 583–585
Digital information transmission, 21
Digital signals
  features of, 9
  processing (DSP) special chip, 570
Digital-to-analog (D/A) converter, 9
dimpulse function, 40, 70–71
Dirac function, 5–6

Dirac(t) function, 30
Direct current (DC) transient analysis
    application problems, 208–222
    examples of, 138–208
    overview of circuits, 104–138
    response to, 127–138
Direct equation, 321, 330
Dirichlet's conditions, 323, 331
Discontinuity, 14, 326
Discrete equivalent transfer function, 586
Discrete filter transfer function, 584
Discrete Fourier transform (DFT)
    average power, 491–492
    convolution, 493–499, 502–503, 534–538
    correlation functions, 502–503
    discrete sequences, 499–500
    evaluation of *f(n)*, 495–496
    examples, 515–556
    features of, 457, 459
    frequencies, 489, 499, 530–534
    inverse (IDFT), 468–469, 490
    length sequences, 489
    periodic functions, 500
    power spectrum, 492–493
    properties of, 493
    relationship with DTFT and ZT, 490
    sequences, 499–502
Discrete mapping 584
Discrete sequence
    components of, 43, 76
    exponential, 16
    pulse, 12
    ramp, 13
    reverse time, 24
    sinusoidal, 16
    step, 12
    time, 17, 457
Discrete shifted impulse function, 9–10
Discrete signals
    defined, 2–3
    time, 9, 44
Discrete spectrum, 327
Discrete-step function, 480–481
Discrete systems
    analysis synthesis, 458
    application problems, 556–560
    block diagram, 463–464
    bounded input-bounded output (BIBO)
        system, 457, 460
    causality, 460, 462–463
    complex, 462
    components of, 458–459
    continuous, 475
    convolution, 465–468

difference equations, 552–553
    input sequence, 459, 543–544
    invariance, time- or shift-, 460
    linearity, 459
    linear and time invariant (LTI), 461–463,
        467–468, 541–545, 556
    memoryless, 464
    passive, 460
    output sequence, 459, 489
    signals, 459
    single-input, single-output (SISO) systems,
        457–458, 460
    stability, 460
    transfer function, 458, 475, 548–555
Discrete-time Fourier transform (DTFT)
    characteristics of, 457, 459, 468–469
    comparison with Fourier transform (FT), 472
    continuous functions, 472
    convergence, 470, 473
    examples, 505–556
    frequency domains, 470
    length sequences, 489–491
    magnitude, 469
    periodic function, 470, 472
    transfer function, 472–473
    transform pairs, 470
Discrete time
    signals, 9, 44
    transfer function, 472
Discrete transfer function, 588
Discrete unit
    parabolic function, 14
    step function, 11
Distortion
    filters and, 563
    impact of, 563
    process, 338
    sources of, 330
    types of, 321
Distortionless
    time-frequency relation, 337
    transmission, 337
Distributional calculus, 10
Distribution functions, 5
Don't care condition, 595
Down-sampling, 44–45
dstep function, 42

E

Edison, Thomas, 102
Effective power, 225
Effective value, 225

Efficient system, components of, 116
Electrical energy, transformation of, 107
Electrical networks, 17, 116, 118
Electrical symbols, 108
Electrical waveforms, 323
Electric circuit
    components of, 102–103
    conservation of energy, 108
    diagram, *see* Circuit diagram
    passive, 103
    theory, 14
Electric field
    characteristics of 110, 115
    radial, 105
Electricity
    electric units, 101–102
    historical perspectives, 101
Electric power, 107
Electrodes, 110
Electromagnetic theory, 102
Electromagnetic waves, 102
Electrons
    characteristics of, 102, 106, 138
    discovery of, 104
    electric charge and, 106
    flow of, 105–106
    free, 105
Element transformation, 567–568
ellip function, 588
Elliptic filter
    characteristics of, 562, 573, 577–578, 596
    implementation of, 583
    transfer function, 582
Empedocles, 104
Energy
    circuit elements, 103
    signal, 2, 26–27
Enhancement devices, 569
Equalizer
    circuit, 321
    components of, 338, 569
Equivalent circuit
    characteristics of, 239–240
    DC, 145
    model, 355–356
    Norton's, 104, 120–121, 125–126, 252, 256
    Thevenin's, 104, 120, 125–126, 169, 252, 254,
        256–257
Equivalent digital FIR, 586
Equivalent resistance, 113, 126
Euler's equalities, 15
Euler's identities, 16, 73, 226–227, 323
Even function
    algebraic rules, 23

analog and discrete cases, 22
    energy analysis, 75–76
    Fourier transform, 329, 336, 470
    symmetric, 23
Even signals, 2
Expansion
    continuous time signal, 25
    devices, 569
    Fourier transform, 323–324, 326, 329–330, 332
Expected value, 45
Exponential
    coefficient, 127
    complex, 73
    functions, 15–16
    sequence, 15
    signals, 14
ezplot, 31–32

F

Faraday, Michael, 102
Faradays, 108
Fast Fourier transform (FFT), 457–459, 499
Feedback, 569
Feedback loops, 595–596
fft algorithms, 498
Fidelity, angle modulation, 22
filter function, 43
Filtering process, 56
Filters
    application problems, 660–665
    characteristics of, 40
    classification of, 561
    complexity, 583
    design specs, 568–569
    examples, 599–660
    frequency-magnitude vectors, 642–643,
        646–649
    ideal, 561–563
    implementation examples, 655–658
    impulse response, 594
    in frequency domain, 563
    order, 568–569
    performance test example, 617–620
    prototypes, 562–563, 566–567, 569, 588
    theory, 561, 578
Final value theorem, 321, 349
Finite discrete sequence, 26
Finite energy, 26, 331
Finite impulse response (FIR) filters
    characteristics 45, of, 595–596
    multiband, 595
    symmetry conditions, 592–593, 636

transfer functions, 563, 569, 583, 585, 589
    Type-1, 590–592, 636–638
    Type-2, 590–592
    Type-3, 590, 592–593
Finite length sequence, 43
First-order circuit, 116n
First-order equations
    linear, 129
    ordinary, 127
First-order filters
    characteristics of, 572
    defined, 569
    examples, 599–609
    types of, 573
First-to-order analog (RC) LPF, 585
FM signals
    bandwidth, 22
    features of, 21–22
    time domain representations, 22
FM spectrum, 443–447
f(n) function, 495–497
Force field, 105
Forcing function, 248
Form factor, 230
Fourier, Jean Baptiste Joseph, 319
Fourier analysis, 326. *See also* Fourier series
Fourier coefficients, 322, 324
Fourier series (FS)
    application problems, 447–451
    approximation, 170–171, 223, 326
    characteristics of, 468
    examples, 376–397
    expansion, 323–324, 326, 329–330, 332, 593
    trigonometric, 322–324, 396–400
Fourier transform (FT)
    analog signal, 469
    characteristics of, 468
    direct, 330–331
    examples, 400–413, 443–447
    inverse, 344–345
    inverse equation (IFT), 321–322, 330
    modulation, 330
    notation, 331
    pair time/frequency, 333
    series, 319–320
    time functions, 342–344
    value theorem, 321
Franklin, Benjamin, 101
Free space, 105
Frequency
    differentiation, 334–335, 470, 476
    domain, *see* Frequency domain
    equivalent circuits, 252
    FM signal, 21

    generator, 261
    harmonic-related, 319
    integration, 335
    linear phase shift, 334
    prewarping, 586
    response, 583
    scaling transformation, 567–568
    shifting, 470, 476
    shift keying (FSK) signal, 21, 87–88
    sinusoid derivatives, 225
    value, initial, 471
    wrapping, 584
Frequency domain
    discrete systems, 468, 545
    DTFT, 472
    domain, in discrete system, 458
    equation, 320
    filter transfer function, 586
    implications of, 324–325, 354
    nonperiodic signal, 321
    periodic signal, 321
    time-domain compared with, 321
freqz function, 472–473
Friedrichs, K. O., 5
F(s) function, 346
f(t). *See* Fourier transform (FT); Laplace
    transform (LT)
Function files, *Impfun,* 27
Fundamental frequency, 17, 323

G

Gain, filter transfer function, 581–582, 586
Gain function, 337–339, 561, 564–565
Gallium arsenate (GaAs), 106
Gauss, Carl Friedrich, 458
Gaussian noise, 52
Gaussian signal, 17
Gaussian window, 48
General analog ramp function, 13
Generalized calculus, 10
Generalized functions, 5
Generators
    functions of, 102
    monophase, 261
    polyphase system, 263–264
    single-phase, 261–262
    three-phase system, 261–266
    two-phase, 261–262
gensig function, 38
Geometric mean, 246
Germanium (Ge), 106
Gibbs, Josiah Willard, 326, 381

Gibb's effect, 46, 593
Gibb's phenomenon, 46, 326, 381
Gilbert, William, 101
Goertzel algorithm, 499
Gravity, 105
Growing exponentials, 131–132
Growth time exponentials, 127

H

Half wave symmetry, 329
Hamming(N) function, 47
Hamming window, 47–50, 397, 594, 653
Hanning window, 47–50
Harmonic distortion, 321, 326
Harmonic frequencies, 17, 323, 326, 381–386,
     388–394
Heaviside, Oliver, 320
Heaviside function, 29–30
Henries, as unit measure, 108
Hermitian orthonormal signal family, 17–18
Hermitian polynomials, 18
Hertz, Heinrich Rudolph, 102
Hertz, as unit measure, 4, 16, 44, 102, 226, 586
High-pass filter (HPF)
     features of, 565–566, 590
     prototypes, 571–572
Higher-order filters, 569
h(n), 497–498, 547
Hydraulic energy, 107

I

ifourier(F) function, 344
ilaplace(F) function, 364
Image processing techniques, 56
Impedance
     construction of, 248
     equivalent, 240
     implications of, 224, 233, 235, 240
     in complex networks, 372–373, 375
     parallel, 254
     polyphase generator system, 265
     scaling transformation, 567
     series RLC circuit, 245
     Thevenin's, 252–253, 259
     three-phase generator system, 266
Impedance Z, 354, 598–599
impinvar function, 588
Impulse
     family, 7
     function, *see* Impulse function

invariant method, 584, 587
invariance response transforms, 586
response, *see* Impulse response, 461
impulse function, 367
Impulse function
     characteristics of, 5, 7, 10, 30, 68
     discrete, 70–71
Impulse response
     digital filter, 586
     discrete systems, 463–464
     filter transfer function, 593
     impact of, 336–337, 489, 583–584
     symmetric, 589–590
Independent variable, continuous time signal,
     23–25
Inductive reactive power, 232–233, 245
Inductor
     effects of, 569
     equivalent, 114–115
     features of, 104, 108, 130, 136
     functions of, 103, 128
     ideal, 115
     impedance, 235
     LP normalized filter, 568
     parallel RLC, 247
     pure, 116
     storing energy elements, 127
Infinite impulse response (IIR) filters
     bilinear, 588
     Butterworth, 596, 649–650
     characteristics of, 596
     Chebyshev, 596
     digital, 563, 569, 583, 585, 646–649
     discrete, examples of, 658–660
     Elliptic, 596
     ideal frequency, 593
     time domain, 589
Infinite length sequence, 43
Information processing signals, 21–22
Initial value theorem, 321, 349
In phase, 233
Input
     discrete systems, 459, 543–544
     power, 107
     signal, 40
     sinusoidal, 563–564
Instantaneous power, 231
Instrumentation
     ammeter, 106
     voltmeter, 106, 227
Insulators, 105–106, 109
Integrals
     convolution, 335–336
     differential, 321

Integrodifferential equations, 359
Interest, time interval, 40
interp function, 44
Inverse discrete–time Fourier transform
(IDFT), 468–469, 490, 500, 515, 547–548
Inverse *fft* function, 588
Inverse Laplace transform (ILT). *See* Laplace
transform, inverse
Inversion
continuous time signal, 25
inverse transformations, 320, 346, 484–485
returns, 23

J

j operator, 225
Joule, as unit of measure, 26, 106, 107n
Jump conditions, 594
Jump value, 46

K

Kaiser window, 45, 47, 50–51, 93
Kirchhoff, Gustav Robert, 102
Kirchhoff's current law (KCL), 104, 110, 117,
132, 267
Kirchhoff's voltage law (KVL), 104, 110–111,
116, 127–128, 135, 149, 204, 358
Kronecker, Leopold, 9
Kronecker delta sequence, 9

L

Lagging, 225
Laguerre orthonormal signal family, 19
Laplace, Pierre Simon de, 319
laplace(f) function, 363–364
Laplace technique, 354
Laplace transform (LT)
advantages of, 376
application problems, 452–456
causal system, 349–350
differential equations (DEs), 361–363
direct, 352–354
examples, 420–423
filter transfer function, 586
impedance, 354–355, 372–373
implications of, 319–320
integrodifferential system, 359–360
inverse (ILT), 346, 351–354, 359, 365, 367
noncausal system, 349–350

partial fraction expansion (PFE), 351–352,
360, 363
periodic/nonperiodic functions, 367–368
power spectrum density, 369, 371–372
real signal/function of, 345, 351–352
switch, examples of, 423–426, 428–443
time-dependent functions, 364
time-frequency domain, 346, 354–358, 367
unilateral, 346–347
value theorems, 349, 436–439
Laplace variable, 40
Laurent series, 473
LC parallel, 239
Leading, 225
Left-sided sequence, 22
Lerch's theorem, 346
Leyden jar, 101
Limiting process, 46
Linear controls, components of, 335
Linear convolution, 493–499
Linear DC network, 118
Linear discrete system, 458
Linearity
DFT, 524–530
significance of, 347, 470
ZT, 474–475
Linear modulation. *See* AM signals;
Modulation
Linear phase
features of, 337–338
response, 582
shift, 337–338
Linear piecewise magnitude approximation,
588
Linear spectral distribution, 321
Linear system
defined, 17
time invariant, 336
Linear time invariant (LTI) system, 351,
461–463, 467–468, 541–546
Line spectrum, 327–328
Loop
analysis, 116
closed electric, 110
current path, 106
equations, *see* Loop equations
feedback, 595–596
Loop equations
applications, 117, 121, 135–136, 320, 359
construction of, 248–249
parallel RLC, 247
polyphase generator system, 263–264
superposition principal, 259
Loss function, 565

Low-pass filters (LPF)
  Butterworth, 574, 616
  features of, 4, 44–45, 330, 562, 564–566, 570
  FIR, implementation of, 591
  ideal (brick wall), 575
  normalized, third-order filter prototype,
      643–646
  prototypes, 583, 585
LP-to-HP transformation, 568
LP-to-LP transformation, 567

M

Magnetic field, 108, 233
Magnetism, electricity and, 101–102
Magnitude spectrum, 327–328, 469, 588, 592–593
Mag vector, 595
Mapping, 584, 588
Mass, as unit of measure, 105
Mathematical models, 330
MATLAB statements, 27, 52
Matrices, types of, 56
maxflat functions, 580–581
Maxima, 26, 323, 331
Maximum power
  theorems, 126, 225
  transfer, 259–260
Maxwell, James Clark, 102, 116
Mean, 45
Mean square error (MSE), 330
Mechanical energy, 107
Mechanical waveforms, 323
Memoryless systems, 464
Mesh analysis
  defined, 118
  example of, 160–161
  parallel RLC, 247
Mesh equations, 248
Metals, 105. *See also* Copper; Silver
Meters, 109
Minima, 26, 323, 331
Modems, 21
Modern filter theory, 561
Modified Yule-Walker equations, 588
Modulating signal, 21, 334. *See also* Modulation
Modulation
  analog signals, 33
  angle, 22, 443–447
  implications of, 330, 338
  theorem, 334, 470
Modulator-demodulator. *See* Modems
Movie frames, 57
Moving averages, 76, 79–80

Multidimensional signals, 2
Multiple-band regions, 565
Multiplexing, 21, 45, 88–90
Multiplier, 458
Musschenbrock, Pieter van, 101

N

Neper frequency, 132, 134, 197, 201
Neperian constant, 14
Network
  current laws, 102
  power triangle, 225, 242
  time constant, 127
Neutral displacement voltage, 263
Neutron, discovery of, 104
Newtons, 105
Node analysis, 117–118
Node equation
  applications, 135, 251–252, 320, 359
  polyphase generator system, 263–264
  solution, 122
  technique, 250–252
Noise-reducing devices, 569
Noise signal, 76–80
Noncausal sequence, 22
Noncausal system, 350
Nondecaying oscillations, 349
Nonorthogonal signals, 2
Nonperiodic function
  implications of, 330, 339
  frequency domain, 336
  Laplace transform (LT), 367–368
Nonperiodic signals, time, 2, 26, 321
Nonplanar networks, 118
Nonrandom signals, 45
Normalized analog LPFs
  analog Chebyshev, 575–576
  analog elliptic, 578
  Bessel, 579
Normalized frequencies, discrete system, 459
Normalized LP filters
  Butterworth filter, 616
  characteristics of, 566–567
  LP-FIR filters, examples, 638–641
  transformation, 568
Normalized multilevel filter, 595
Normalized passive first-order filters
  HPF, 571
  LPF, 570–571
Norton's theorem, 119–120, 225, 252, 254–255
Nuclear energy, 107
numz function, 585

Nyquist, Harry, 339n
Nyquist plots, 339–340
Nyquist rate, 499
Nyquist/Shannon
    sampling rate, 44
    theorem, 4
Nyquist theorem, 587

O

Odd function
    algebraic rules, 23
    analog and discrete cases, 22
    energy analysis, 75–76
    implications of, 328, 336, 470
    symmetric, 23
Odd signals, 2
Ohm, as unit of measure, 102, 108
Ohm, George, 102
Ohm's law, 104, 233, 265, 354
One-dimensional (1-D) signals, 2
1p2bp function files, 583
1p2hp function files, 583–584
1p2p function files, 583–584
On-off keying (OOK), 21
Open circuit
    defined, 110
    voltage, 119
Operational amplifiers (OA), 569–570, 572
Operational calculus, 320
Optimization process, 595
Orthogonal families, 17, 72–73
Orthogonal signals, 2
Orthonormal families, 17
Oscillations, 15, 46, 131, 330, 349, 573
Output
    defined, 2
    power, 107
    signal, 40
Overdamped system
    examples of, 199–200, 203–204
    features of, 131
    parallel, 133
    series, 135
Overlap-add method, 496–497
Overlap-save method, 496–497

P

Parabolic function, 14
Parabolic window, 48
Parallel connection, 111, 113–115, 140

Parallel RLC (L circuit), 133
Parks-McClellan algorithm, 595
Parseval's theorem, 321, 324–325, 339, 471
Partial-fraction expansion, 322, 351–352, 483, 488
Parzen window, 48
Pass band
    characteristics of, 563–565
    cutoff frequency, 577
    filters, 561–563
    ripples, 565, 588
Passive analog electrical filters, 569
Passive discrete system, 458
Passive filters, 563
Passive systems, 338
Percentage of total harmonic distortion
        (PTHD), 326
Periodic function
    defined, 322
    DTFT, 470
    implications of, 323–324, 332, 348, 500
    Laplace transform (LT), 367–368
    time domain, 336
Periodic sequences, 69–70
Periodic signals, continuous time, 2, 26–27,
        61–62, 321
Periodic wave, 325
Personal computer applications, 570
Phase angle, 226, 245
Phase distortion, 564
phase function, 584–585
Phase response, type-2 FIR filters, 592
Phase shift, 233, 338, 564
Phase shift keying (PSK) signals, 22, 87–88
Phase spectrum, 321, 327–328, 470
Phasor currents, 230, 248–249
Phasor diagram
    LC parallel, 239–240
    power analysis, 244–245
    RLC, 235, 240–242
    RL connection, 236
    two-phase generator system, 262
Phasor network, 225, 249
Phasor representation, 224–225, 251
Physics, principles of, 105
Planar networks, 118
PM signals
    bandwidth, 22
    features of, 21–22
    time domain representations, 22
PM spectrum, 443–447
Point-to-point communication systems, 22
Polarity, 106, 109, 118
Poles, filter transfer function, 581–582, 584, 586–587
Poles, Laplace transform, 349–350

Pole-zero cancellations, 338
Polynomials
    Bessel, 578
    Butterworth, 574
    Chebyshev, 576–577
    function files, 583
    Hermitian, 18
    implications of, 339, 351, 488–489, 563
    Laguerre's, 19
    transfer functions, 585
    vector, 40
Polyphase systems, 261
Positive charge, 106
Pound-force, 108
Power
    analysis, 242–243
    circuit elements, 103
    distribution, 261
    factor, 225
    real, 231
    relations, *see* Active power; Apparent
        power; Reactive power
    signals, 2, 27, 331
    spectral distribution, 321
    spectrum, 327–328, 492, 588
    supplies, 102
    triangle, 243–244
Power spectrum density (PSD)
    characteristics of, 492–493
    examples of, 427–428
    Laplace transform (LT), 369, 371–372
Prime factor algorithms, 498
Probabilistic signals, 2, 45
Programmable filters. *See* Digital filters
Propagation delay, 564
Protons, 104
Prototypes, types of, 562–563, 565–567, 569,
        571–572, 583–585, 588, 643–646
psd function, 369, 371–373
pulstran function, 38
pzmap function, 350

Q

Quality factor
    parallel RLC, 246–247
    series RLC, 245–246

R

Radian, as unit measure, 16, 469
Radio broadcasting, 22

Random signals, 45, 69–70
Rational numbers, 17
RC circuit
    analysis, *see* RC circuit analysis
    features of, 108, 129
    series, 128–130, 238–239
RC circuit analysis
    components of, 356–358
    examples, 415–416
RCL circuit
    parallel circuit, 196–200
    series, source-free, 200–204
    source connected to, 194–196
    source-free parallel, 132
    source-free series, 134
Reactive power, types of, 225, 232–233,
        242–245
Real function/sequence, expression of, 23
Real power, 231
Rectangular filters, 44
Rectangular window, 46, 48, 52
rectpuls function, 36
Recursive optimization techniques, 595
Reference node, 117, 122
Reflection returns, 23
Region of convergence (ROC), 346, 349–350,
        478–484, 548
Remez exchange algorithm, 595
remez functions, 595, 653–655
Repulsion force, 105
resample function, 45
Resampling rate, 44–45
Resistance
    equivalent, 113, 126
    impact of, 109
    internal, 247
    Thevenin, 119
    values, 139–140
Resistive circuit, 146–147, 233
Resistive DC, 247–248, 250
Resistive/reactive (RLC) system
    AC, 225
    power analysis, 245
Resistivity factor, 109
Resistor, functions of, 103, 108
Resistor capacitor (RC), 10, 131. *See also* RC
        circuit; RC circuit analysis
Resistor capacitor network (RLC)
    characteristics of, 104, 563
    circuit analysis, *see* RLC circuit analysis
    examples, 416–420, 423–426
    parallel, 242, 246–247
    series, 240–241, 245, 423–426
Resistor color bands, 139–141, 144

Resistor inductor network (RL), 10, 131. *See also* RL circuit
Resonant circuits
  construction of, 247
  quality factor, 225
Resonant frequency, 132, 134, 197, 201, 245–247
Reversal returns, 23
Reverse-wrap-around, 491
Right-sided sequence, 22
Ripple, filter transfer function, 562, 565, 573, 575, 581, 583, 588–589, 593
RL circuit
  analysis, 358–359
  features of, 108, 178–180
  parallel, 236–237
  series, 128, 236
Root-mean-square (RMS), 225, 228, 230, 326
Rotational symmetry, 329

S

Sampling
  frequency, 4, 55, 459, 588
  interval, 3–4
  period, 3–4, 583–585
  process, 8, 338, 587
  rate, 3, 44, 499, 585
  rules, 458
  theorem, 4, 44
Sawtooth waves, 223, 321, 381–388
Scaled frequency, 583
Scanning parameter, 502
Script files
  audio, 54–55
  compare_win, 49–50
  disc_imp, 41–42
  disc_step, 42–43
  exponentials, 15–16
  F_f, 367–369
  Hermite, 18–19
  Kaisers, 50–51
  Laguerre, 19–20
  map_pz, 350–351
  nyquist_bode, 339, 341–342
  plot_ramp, 31–32
  rect_pulses, 37–38
  sample_cos, 492
  sequence_impulse, 27
  signal_noise, 53
  sinc_n, 20
  sincs, 34–36
  solve_DE, 365–367
  specgrph, 372

specpsd, 369–370
specpsol, 370–371
squares, 32–33
step_imp, 554
sym_z_plots, 486–487
triang, 36–37
triangles, 34
triang-pulses, 39–40
window, 48
win_tri_rect_bar, 51–52
s-domain
  filter transfer function, 586
  time signals model, 320–321, 436–437
Searching parameter, 502
Second-order filters
  defined, 569
  examples, 609–617
  types of, 572–573
Second-order
  linear, homogeneous differential equation, 134
  loop differential equation, 194–195
  system, 131–132
Semiconductors, 105–106
Semilog phase, 339
Series approximation, 326
Series connection, 110–112, 114, 139
Series equivalent resistor, 113
Shifted impulse, 6, 476
Shifted sequences, 471
Shifted step sequence, 11
Shifting operations, 493
Short circuit, 125
Siemen, 235
sign(t) function, 29–30
Signals, classification of, 2
signum(t), 29
Silicon (Si), 106
Silver, 105
Simple-system structures, 478
Sinc family, 20
sinc function, 20, 34
Sine function, 226, 252, 321, 491
Single input-single output (SISO) systems, 457–458, 460
Single signals, 2
Single source networks, 119, 124
Singular functions, 10
Sinusoid AC voltage, 226
Sinusoid sequence, exponential discrete modulated, 21
Sinusoidal frequency, 323
Sinusoidal function, 16–17, 73, 131–132, 223, 248, 320, 370

Sinusoidal input, filters and, 563–564
Sinusoidal signals, 14, 21, 492
Sinusoidal waves, 17, 223, 225–227, 233
Slope, 33–34
Sound files, types of, 54
Sound waves, 54
Source network
    Norton's, 125
    Thevenin's, 124
Source transformation
    AC circuit, 253
    characteristics of, 118–119, 254, 355
    theorem, 225
    Thevenin's, 121
    voltage, 122–123
specplot function, 372
specpsd function, 369, 371
Spectra analysis, 321
Spectral distribution, 321
Spectrogram, 369
Spectrum
    AM, 443–447
    amplitude, 327
    analyzer, 320
    phase, 321, 327–328, 470
    PM, 443–447
    power, 327–328, 492, 588
    two-sided, 327
spectrum function, 371
Speech file, types of, 54–55, 369–370
s-plane, 584
Square waves, 223, 321
s-Shifting, 347
Stable discrete system, 458
Standard deviation, 45
Star, defined, 261
Steady-state conditions
    examples of, 127, 200–204
    impact of, 187, 190
Steady-state response, 102–103, 116, 127, 129
Steady-steady
    current, 116
    voltage, 115
stem function, 27
stepfun(n, no) function, 28
Step functions
    characteristics of, 12, 14, 472
    discrete, 70–71
    in time domain, 357
Step invariant technique, 586
Step response, digital filter, 586
Stochastic signals, 45
Stop band
    characteristics of, 563–565

    filters, 561–562, 577
    frequency, 565
    ripple, 565, 583, 588
Superconductors, 105–106
Superposition principles, 257–258, 330
Superposition solution, 119–120
Superposition theorem, 118–119, 225
Switching, 127
Switch in Laplace transform system, examples
        of, 423–426, 428–443
Symbolic solver, 204–208
Symmetric functions, 22–23
Symmetric signals, 2
Symmetry, 327, 329, 334, 589
Synthesis, defined, 596
System analysis, 17
System function, 564
System transfer function, 337, 561

T

Taylor's expansion series, 226
Telecommunication signals
    analog, 88
    characterized, 21
    discrete, 88–90
Thales, 101
Theorems
    Cauchy's residue, 483
    circuit, 225
    direct current (DC), 225
    Lerch's, 346
    maximum power, 126, 225
    modulation, 334, 470
    Norton's, 119–120, 225, 252, 254–255
    Nyquist, 587
    Nyquist/Shannon, 4
    Parseval's, 321, 324–325, 339, 471
    sampling, 4, 44
    source transformation, 225
    superposition, 118–119, 225
    Thevenin's, 119–120, 124, 127, 169, 225,
        252–254
    value, 321, 349, 436–439
Thermal energy, 107
Thevenin's theorem, 119–120, 124, 127, 169, 225,
        252–254
Thevenin's voltage V_{TH}, 252
Third-order filters
    Butterworth IIR filters, 596, 649–650
    Butterworth LPF, 643–646
    examples of, 625–632
Thomason, J. J., 104

3-D matrix, 55
Three-dimensional (3-D) signals, 2
Three-phase system, 223, 225
Time complex frequency, 376
Time constant, 127, 129
Time continuous signal, 44
Time delays, 338
Time differentiation, 334
Time domain
  block box system diagram, 336–337
  in discrete system, 458
  DTFT, 472
  Fourier transform, 324–325
  frequency-domain compared with, 321
  ZT, 476
Time frequency
  domain, 354–356
  Laplace transform (LT), 346
  transformations, 333
Time function, obtaining, 346
Time integration, 335
Time invariance, 458
Time invariant linear systems, 461
Time-limited signal, 26
Time reversal, 25, 336, 470, 475
Time scaling, 23, 333–334, 347
Time shifting, 23, 25, 334, 347, 470, 474–475, 493
Time transformation, 23
Time value, initial, 471
Total distortion, 321
Total energy, 108
Transfer function
  analog, 68
  application problems, 451–452
  Bessel filter, 578
  Butterworth filters, 580
  characteristics of, 40, 338–339, 349, 564, 583
  discrete, 70–71
  linear time invariant (LTI) system, 351, 461–463, 467–468, 541–546
  normalized, third-order filter prototype, 643–646
Transform equation pair, 336
Transform pairs, 333–334, 347, 470–471
Transform theory, 56
Transient analysis, 127, 135
Transient electrical network, 135
Transient response, types of
  RC, 131
  RL, 131
  RLC, 131
Transient solution
  examples, 193–194
  types of, 127

Transition bands, 562, 565
Transition response, 130
Transmission lines, 261
Transmission signals, 21
Triangle, symmetric, 36
triang(N) function, 47–48
Triangular waves, 33, 223
Triangular window, 47, 52, 93
Trigonometric identity, 231
tripuls function, 36
Truncation process, 46, 330, 499, 593
TV broadcasting, 22
2-D matrices, 54–56
Two-dimensional (2-D) signals, 2
Two-sided sequence, 22
Type-1 filters
  Chebyshev transfer function, 575
  FIR filters, 590–592
  overview, 577, 590
Type-2 filters
  Chebyshev transfer function, 575
  FIR filters, 590–592
  overview, 577, 590
Type-3 FIR filters, 590, 592–593
Type-4 FIR filter
  characteristics of, 591
  frequency response, 593

U

Unbalanced load, 225
Underdamped system
  characteristics of, 131
  examples of, 199–200, 203–204
  parallel, 133
  series, 135
un = dstep function, 42
Unilateral LT, 346–347
Unit conversions, 107–108
Unit delay, 458
Unit discrete step sequence, 11
Unit doublet, 6–7
Unit impulse, 5, 7, 27
Unit measures, 4, 16, 44, 101–102, 235
Unit ramp function, 12–14
Unit shift, 458
Unit step function
  analog, 10
  right–shifted, 11
Unit step sequence, 28
unwrap(phase) function, 585
Up-sampling, 44
Utility companies, 107, 223

V

Vacuum, 110
Value theorems, 321, 349
Van Hann window, 47
Vector
  frequency, 339
  frequency magnitude, 642–643, 646–649
  mag, 595
  1-D, 54
  polynomials, 40
Video signals, 2, 56
Volta, Count Alexander, 102
Voltage
  circuit elements, 103
  constant, 115
  defined, 102
  divider rule, 112, 136–137, 225
  impact of, 106
  laws, 102
  multiple-phase generator system, 261
  nodal, 117
  open circuit, 252
  parallel RLC, 248
  polyphase generator system, 263–264, 267
  sinusoidal, 261
  source, 119, 355, 357
  Thevenin, 119
  three-phase generator system, 262–263
Voltaic cell, 102
Voltmeter, 106, 227
Volts, 102, 106. *See also* Voltage

W

Watt, as unit of measure, 107
Waveforms
  distortion, 338
  features of, 323
  symmetry, 321

Weighted time-shifted impulse
    samples, 43
Window function, 47, 91–93, 321,
    330, 563
Window models, 46, 48, 593–595
Windowing process, 397–398, 499
Windowing technique, 595

Y

Y-Δ structure, 260–261
ynoise function, 369, 372–374
Yule-Walker algorithm, 584
yulewalk function, 588, 646–649

Z

Zero-padding, 491
Zeros
  filter transfer function, 581–582,
      584, 586, 593
  Laplace transform, 349–350
z-plane, 584, 593
Z-transform (ZT)
  causal exponential function, 479–480
  convergence, 477–484
  convolution, 474, 476
  defined, 457–458, 461
  discrete-time system analysis, 477
  evaluation of, 485–486
  inverse, 483–484
  length sequences, 489
  power of, 473–477
  properties of, 474–476
  system transfer function, 482
  time domain, 478–479, 482
  transformation from time to frequency
      domain, 477–478
  transformation pairs, 483

Milton Keynes UK
Ingram Content Group UK Ltd.
UKHW051538141024
449569UK00028B/1520